CHAPTER 8

PERT Equations

$$\text{Expected duration} \quad t_e = \frac{(t_a + 4t_m + t_b)}{6}$$

where t_a is the most optimistic duration estimate, t_m is the most likely duration estimate, and t_b is the most pessimistic duration estimate.

$$\text{Variance for each PERT activity} \quad \sigma^2 = \left[\frac{(t_b - t_a)}{6}\right]^2$$

$$Z = \frac{\text{Mean} - x}{\sqrt{\text{Variance}}} \quad \text{or} \quad Z = \frac{(\bar{X} - x)}{\sigma}$$

where σ is the standard deviation of the cumulative normal distribution.

CHAPTER 9

$$\text{Pay} = 1.25(\text{indirect expense} + \text{direct expense})$$
$$- 0.10[1.25(\text{indirect expense} + \text{direct expense})]$$

Rate of Return

$$\sum^{all\ I} PW[REV(I)] - \sum^{all\ I} PW[EXP(I)] = 0$$

where $REV(I)$ = revenue for period I

$EXP(I)$ = expenditure for period I

PW = present worth of these values

CHAPTER 11

$$\text{Depreciation Cost per Hour} = \frac{\text{Purchase Price} - \text{Tire Value}}{\text{Estimated Service Life in Hours}}$$

$$\text{Average Annual Value (AAV)} = \frac{C(n+1)}{2n},$$

where AAV is the average annual value, C is the initial new value of the asset, and n is the number of service life years.

$$\text{Average Annual Value (AAV), including the salvage value} = \frac{C(n+1) + S(n-1)}{2n}$$

The hourly charge for IIT is calculated as:

$$\text{IIT/hour} = \frac{\text{factor} \times \text{delivery price}}{1000}$$

CHAPTER 12

$$\text{Power Required} = \text{RR} \pm \text{GR}$$

$$\text{Percent swell} = \left[\left(\frac{1}{\text{load factor}}\right) - 1\right] \times 100$$

$$\text{where Load factor} = \frac{\text{pounds per cubic yard-loose}}{\text{pounds per cubic yard-bank}}$$

$$\text{Grade Resistance (GR)} = \text{percent grade} \times 20 \text{ lb/ton/\% grade} \times \text{weight on wheels (tons)}$$

$$\text{Equivalent percent grade} = \frac{\text{RR}}{20 \text{ lb/ton/\% grade}}$$

$$\text{Usable pounds pull} = (\text{coefficient of traction}) \times (\text{weight on drivers})$$

CHAPTER 13

$$\frac{\text{Resource-hours per hour}}{\text{Units per hour}} = RH/\text{unit, where } RH = \text{resource hour}$$

The basic equation for unit pricing:

$$\frac{\text{Resource cost per unit time}}{\text{Production rate}} = \frac{\$/\text{hr}}{\text{unit/hr}} = \$/\text{unit}$$

CHAPTER 15

The five parameters which form the foundation of the "earned" value concept are:
- Budgeted Cost of Work Schedule (BCWS) = Value of the baseline at a given time
- Actual Cost of Worked Performed (ACWP) − Measured in the field
- Budgeted Cost of Worked Performed (BCWP) = [% Complete] × BCAC
- Budgeted Cost At Completion (BCAC) = Estimated Total Cost for the work Package
- Actual Quantity of Worked Performed (AQWP) − Measured in the field
  CV, Cost Variance = BCWP − ACWP
  SV, Schedule Variance = BCWP − BCWS
  CPI, Cost Performance Index = BCWP/ACWP
    - CPI <1.0 indicates cost overrun of budget
    - CPI >1.0 indicates actual cost less than budgeted cost
  SPI, Schedule Performance Index = BCWP/BCWS

$$P_c = \frac{P_s}{(100 - P_s)}$$

where P_c = percentage applied to the project's total direct cost for the coming year

P_s = percentage of total volume in the reference year incurred as fixed or G&A expense

Construction Management

Third Edition

Daniel W. Halpin

Purdue University

John Wiley & Sons, Inc.

ACQUISITIONS EDITOR	Jennifer Welter
MARKETING MANAGER	Frank Lyman
SENIOR PRODUCTION EDITOR	Lisa Wasserman
COVER DESIGNER	Hope Miller
COVER PHOTO	PhotoDisc Inc./Getty Images
MEDIA EDITOR	Tom Kulesa
ILLUSTRATION COORDINATOR	Mary Alma

This book was set in TimesTen by TechBooks and printed and bound by R.R. Donnelley. The cover was printed by Phoenix Color Corp.

This book is printed on acid free paper. ∞

To order books or for customer service please, call 1-800-CALL WILEY (225-5945).

Library of Congress Cataloging in Publication Data:
Halpin, Daniel W.
Construction management/Daniel W. Halpin.– 3rd ed.
    p.   cm.
Includes bibliographical references and index.
ISBN-13   978-0-471-66173-3 (cloth)
ISBN-10   0-471-66173-2 (cloth)

1. Construction industry—United States—Management.
2. Construction industry—Law and legislation—United States.
I. Title.
HD9715.U52H324 2005
624.068—dc22

                                                                        2005026203

Printed in the United States of America

10 9 8 7 6 5 4 3

Preface (3rd Edition)

INTRODUCTION

The first and second editions of this text book have enjoyed a good deal of success and been translated into a number of foreign languages. The third edition builds upon the core material that was well received in the first two editions.

I hope that this text will give those interested in construction engineering a broad coverage of some of the many topics which construction engineers deal with on a daily basis. As I stated in previous editions, a construction manager is like an Olympic decathlon athlete who must show great competence in a multitude of areas ranging from design of construction operations to labor relations. The successful construction manager must be "a jack of many skills and master of all."

Construction, as an engineering discipline, is well characterized by a former President of the United States who was a respected engineer in his own right. Herbert Hoover stated:

"It is a great profession. There is the fascination of watching a figment of the imagination emerge through the aid of science to a plan on paper. Then it moves to realization in stone or metal or energy. Then it brings jobs and homes to men. Then it elevates the standards of living and adds to the comforts of life. That is the engineer's high privilege."

As noted in Chapter 6, another U. S. President, Dwight D. Eisenhower, renowned for his ability to manage, stated a concept that is a good guide to all:

"Plans are nothing. Planning is everything!"

NEW MATERIAL IN THIS EDITION INCLUDES:

- Chapter openers added at the beginning of each chapter, describing new technologies or supplemental materials that are relevant to the topic of the chapter.

- Section 1.5, *Construction versus Manufacturing Processes*, which discusses the difference between purchasing construction and purchasing manufactured speculative products.

- Expanded coverage of Value Engineering (Chapter 3), including a demonstration example.

- Additional explanations and a discussion of Product Delivery Systems (Chapter 4), illustrating the different systems available. This material helps students see the benefits/weaknesses of various contract formats and when it is best to use one over another.

- The concept of "scope of work" defining the project and break-down of the project into "work packages" in the context of a "Work Breakdown Structure" (Chapter 6). This material is designed to help students break a project into its component parts and analyze project progress in terms of earned value (Chapter 15). This prepares the student for topics in the Scheduling Chapter (Chapter 7). One of the problems students have with scheduling is how to break a project into pieces/activities. This material helps answer the question, "Where do I start in developing a schedule?"

- In the "Project Scheduling" Chapter 7, the work packaging material is used to develop the activities in the network schedule. This chapter opens with a discussion of the precedence notation (Activity on Node or AON) since this is the basis of present practice.

- Earned value analysis is included in Chapter 15.

- Chapter 18, covering Construction Operations, is on the book website at www.wiley.com/college/halpin. This chapter involves the actual placement of construction in the field, using resources such as equipment and labor.

- Dated material is cut or moved to an appendix, in order to reflect current practice. For example, arrow notation scheduling has been moved to an appendix.

- Web-based resources including Web CYCLONE (permits simulation of construction operations), and examples showing how to build models, from masonry to bridges, are included on the book web site.

STUDENT RESOURCES

The following resources are available from the book website at www.wiley.com/college/halpin. Visit the Student section of the website.

- Information about and support materials for the two leading scheduling software programs, Primavera, and Microsoft Project.

- The author's website "Emerging Construction Technologies"

- Chapter 18, covering Construction Operations. This chapter involves the actual placement of construction in the field, using resources such as equipment and labor. This material covers the business of placing the construction physically in the field, what is done to actually build the project (e.g. putting concrete in place with a crane and bucket and pump) and the sequence of when tasks need to be done. The chapter discusses the concept of sequence and technical logic. It addresses the questions "How and in what sequence am I going to build?" and "What resources will I use"

- The book web site includes a comprehensive Simulation Homepage. This homepage includes material regarding construction process simulation and describing glossary/definition terms specific to CYCLONE modeling format.

- A web-based program, Web CYCLONE permits simulation of construction operations. The web site also includes extensive material regarding the CYCLONE construction modeling system.

- Examples showing how to build models (e.g. a masonry model, an asphalt paving model and a concrete supply model.)

INSTRUCTOR RESOURCES

All instructor resources are available from the book website at www.wiley.com/college/halpin, available only to instructors who adopt the text:

- Solutions Manual
- Image Gallery of Text Figures
- Text Figures in PowerPoint format
- All resources from Student section of the website.

These resources are password-protected. Visit the Instructor section of the book website to register for a password to access these materials.

ACKNOWLEDGEMENTS

I would like to acknowledge my co-author in previous editions, Ron Woodhead, Emeritus Professor at the U. of New South Wales, Australia. Without his vision, this book would not have become a reality.

I would also like to thank the many colleagues and numerous students who have provided valuable feedback regarding various aspects of this text. It is safe to say that well over 100,000 students have used this text in some fashion since it was originally published. I would like to say "many thanks" to all who have provided comments and constructive feedback over the years, including the following instructors who reviewed or provided feedback for this edition:

Irtishad Ahmad, Florida International University

Lansford C. Bell, Clemson University

Paul G. Carr, Cornell University

Kris G. Mattila, Michigan Technological University

Ali Touran, Northeastern University

Marlee Walton, Iowa State University

Trefor P. Williams, Rutgers University

Wesley C. Zech, Auburn University

In particular, I would like to thank the following colleagues for their insights and contributions regarding the material provided in this book:

Dulcy Abraham, Bob Bowen, Peter Dozzi, Jimmie Hinze, Mike Kenig, Jerry Kerr, Bolivar Senior and Joe Sinfield.

Finally and most importantly, I would like to acknowledge my wife, Maria, for her incomparable support. Our many lively discussions, full of questions and insight, have shaped my thinking over the past 40 years. Her interest and support has been the most significant force in realizing this and a lifetime of exciting projects.

Daniel W. Halpin
Purdue University
W. Lafayette, IN

Contents

Chapter 18
Construction Operations
(Available on web site at www.wiley.com/college/halpin)

Chapter 1

History and Basic Concepts

Bridges and History

Water crossings have always been seen as great engineering challenges. Since Roman times, bridges and various river crossings have been linked with great engineering achievements. Apollodorus was chief engineer for the Emperor Trajan and built a bridge across the Danube River in the second century A.D. This bridge allowed Trajan to invade Dacia and annex the territory of modern day Romania.

The length of clear span bridging was greatly increased by the development of the cable supported suspension bridge. The oldest vehicular steel cable suspension bridge in the world in continuous use was built by John A. Roebling in Cincinnati, Ohio during the Civil War. It is still one of the major arteries connecting Cincinnati with Covington, Kentucky.

When construction started in 1856, the charter authorizing the construction required a clear span of 1,000 feet between two towers, with the deck located a minimum of 100 feet above the water's surface. The bridge was completed in December 1866. The 1,057-foot main span was, at the time, the longest in the world. It was one of the first suspension bridges to use both vertical suspenders and diagonal cable stays which radiated from the top of each tower. This innovative use of cable stays gave the bridge great rigidity and resistance to movement during high winds. Roebling used this same concept later when building the Brooklyn Bridge.

The bridge was upgraded to its present configuration in 1894. A second set of 10.5-inch cables were added to carry heavier decks. This reconstruction increased the carrying capacity of the bridge to a 30-ton limit. As a native of Covington, the author rode both trolley (street) cars and electrically powered buses hundreds of times to the transit terminal in Cincinnati located at the north end of the bridge. In 1984 the bridge was renamed the John A. Roebling Bridge.

Bridges Today

John A. Roebling Bridge
Covington, Ky. Side

World famous bridges have become a symbol of Civil Engineering. The Golden Gate Bridge in San Francisco has not only been hailed a tremendous engineering achievement, but also a beautifully balanced aesthetic achievement. Plans are now underway to bridge the famous Straits of Messina between the toe of Italy and the Island of Sicily. This bridge will have a clear span of almost 2 miles, approximately 10 times the span of the Roebling Bridge in Cincinnati. It will also be designed to resist hurricane-force winds. Construction of this bridge will rival the construction of the Channel Tunnel connecting England and France.

Figure 1.1 The Parthenon in Athens.

1.1 HISTORICAL PERSPECTIVE

Construction and the ability to build things is one of the most ancient of human skills. In prehistorical times, it was one of the talents that set *Homo sapiens* apart from other species. Humans struggled to survive and sought shelter from the elements and the hostile environment that surrounded them by building protective structures. Using natural materials such as earth, stone, wood, and animal skins, humans were able to fabricate housing that provided both shelter and a degree of protection.

As society became more organized, the ability to build things became a hallmark of the sophistication of ancient civilizations. The wonders of the ancient world reflect an astounding ability to build not only structures for shelter but monuments of gigantic scale. The pyramids and Greek temples such as the Parthenon (Fig. 1.1) are an impressive testimony to the building skills of the civilizations of antiquity. Great structures punctuate the march of time and many of the structures of ancient times are impressive even by modern standards. The great Church of Hagia Sophia in Constantinople, constructed during the sixth century, was the greatest domed structure in the world for nine centuries. It is an impressive example of the ingenuity of the builders of that time and their mastery of how forces can be carried to the ground using arches in one dimension and in three dimensions as domes.

In modern times, the Brooklyn Bridge and the Panama Canal stand as legendary feats of engineering achievement. They are also testimonies to the fact that realizing a construction project involves solving a multitude of problems, many of which are not technical. In both the Brooklyn Bridge and Panama Canal projects, people problems requiring great innovation and leadership were just as formidable as the technical problems encountered. To solve them, the engineers involved accomplished "heroic" feats.

1.2 GREAT CAPTAINS OF CONSTRUCTION

The Roebling family as a group can be credited with building the Brooklyn Bridge between 1869 and 1883. It was the greatest project of its time and required the use of technology

Figure 1.2 John A. Roebling, designer of the Brooklyn Bridge. (American Society of Civil Engineers)

at a scale never before tried. The concept of a cable-supported suspension bridge was literally invented by John A. Roebling (see Fig. 1.2). Roebling was born in Germany and was the favorite student of the famous philosopher Hegel. Roebling was a man of tremendous energy and powerful intellect. He built a number of suspension bridges, notably the John A. Roebling Bridge in Cincinnati (which is still in daily use), that demonstrated the cable-supported concept prior to designing the Brooklyn Bridge. Upon his death (precipitated by an accident that occurred during the initial survey of the centerline of the bridge) his son Washington took charge.

Washington Roebling (see Fig. 1.3) was a decorated hero of the Civil War who had received his training in civil engineering at Rensselaer Polytechnic Institute. Like his father he was a man of great vision and courage. He refined the concepts of caisson construction and solved numerous problems as the great towers of the bridge rose above New York City (see Fig. 1.4). Since he would not require anyone to work under unsafe conditions, he entered the caissons and supervised the work personally. He ultimately suffered from a mysterious illness related to the fact that the work was carried out under elevated air pressure in the caissons. We now know that this illness, called "the bends," was caused by the absorption and rapid exit of nitrogen from the bloodstream when workers entered and exited the pressurized caissons.

Although incapacitated, Washington continued to supervise the work from an apartment that overlooked the site. At this point, Emily, Washington's wife and the sister of an army general, entered the picture (see Fig. 1.5). Emily carried information to Roebling's supervising engineers on the site. She became the surrogate chief engineer and gave directives in the name of her husband. She was able to gain the confidence and respect of the site engineers and was instrumental in carrying the project through to successful accomplishment. The tale of the building of the great bridge (see *The Great Bridge* by McCullough, 1972) is one of the most extraordinary stories of technical innovation and personal achievement in the annals of American history.

Figure 1.3 Washington A. Roebling, chief engineer of the Brooklyn Bridge. (Special Collections and University Archives, Rutgers University Libraries)

1.3 PANAMA CANAL

The end of the nineteenth century was a time of visionaries who conceived of projects that would change the history of humankind. Since the time Balboa crossed Panama and discovered a great ocean, planners had conceived of the idea of a water link between the Atlantic and the Pacific Oceans. Having successfully connected the Mediterranean with the Red Sea at Suez, in 1882 the French began work on a canal across the narrow isthmus of Panama, which at that time was part of Colombia. After struggling for 9 years, the French were ultimately defeated by the formidable technical difficulties as well as the hostile climate and the scourge of yellow fever.

Theodore Roosevelt became president during this period and his administration decided to take up the canal project and carry it to completion. Using what would be referred to as "gun-boat" diplomacy, Roosevelt precipitated a revolution that led to the formation of the Republic of Panama. Having clarified the political situation with this stratagem, the famous "Teddy" then looked for the right man to actually construct the canal. That right man turned out to be John F. Stevens, a railroad engineer who had made his reputation building the Great Northern Railroad (see Fig. 1.6). Stevens proved to be the right man at the right time.

Stevens understood the organizational aspects of large projects. He immediately realized that the working conditions of the laborers had to be improved. He also understood that measures had to be taken to eradicate the fear of yellow fever. To address the first problem, he constructed large and functional camps for the workers in which good food was available. To deal with the problem of yellow fever, he enlisted the help of an army doctor named William C. Gorgas. Prior to being assigned to Panama, Dr. Gorgas had worked with Dr. Walter Reed in wiping out yellow fever in Havana, Cuba. He had come to understand that the key to controlling and eliminating this disease was, as Dr. Reed had shown, the control of the mosquitoes that carried the dreaded infection and the elimination of their breeding places (see *The Microbe Hunters* by Paul DeKruif). Gorgas was successful in effectively controlling the threat of yellow fever, but his success would not have been possible without the total commitment and support of John Stevens.

Figure 1.4 Brooklyn Bridge under construction 1881. (Museum of the City of New York, Print Archives, Gift of the Essex Institute.)

Having established an organizational framework for the project and provided a safe and reasonably comfortable environment for the workers, Stevens addressed the technical problems presented by the project. The French had initially conceived of a canal built at sea level and similar to the Suez Canal. That is, the initial technical concept was to build a canal at one elevation. Due to the high ground and low mountains of the interior portion of the isthmus, it became apparent that this approach would not work. To solve the problem

Figure 1.5 Emily Warren Roebling, wife of Washington Roebling. (Special Collections and University Archives, Rutgers University Libraries)

of moving ships over the "hump" of the interior, it was decided that a set of water steps, or locks, would be needed to lift the ships transiting the canal up and over the high ground of Central Panama and down to the elevation of the opposite side. The construction of this system of locks presented a formidable challenge. Particularly on the Atlantic side of the canal, the situation was complicated by the presence of the wild Chagres River, which flowed in torrents during the rainy season and dropped to a much lower elevation during the dry season.

Figure 1.6 John F. Stevens, chief engineer of the Panama Canal. (National Archives, Washington, D.C.)

The decision was made to control the Chagres by constructing a great dam that would impound its water and allow for control of its flow. The dam would create a large lake that would become one of the levels in the set of steps used to move ships through the canal. The damming of the Chagres and the creation of Lake Gatun itself was a project of immense proportions requiring concrete and earthwork structures of unprecedented size (see Fig. 1.7).

The other major problem had to do with the excavation of a great cut through the highest area of the canal. The Culebra cut, as this part of the canal was called, required the excavation of earthwork quantities that even by today's standards stretch the imagination. Stevens viewed this part of the project as the construction of a gigantic railroad system that would operate continuously (24 hours a day) moving earth from the area of the cut to the

Figure 1.7 Work in progress on the Great Gatun lock gates. (The Bettmann Archive)

Chagres dam construction site. The material removed from the cut would provide the fill for the dam. It was an ingenious idea.

To realize this system, Stevens built one of the great rail systems of the world at that time. Steam-driven excavators (shovel fronts) worked continuously loading railcars. The excavators worked on flexible rail spurs that could be repositioned by labor crews to maintain contact with the work face. In effect, the shovels worked on sidings that could be moved many times each day to facilitate access to the work face. The railcars passed continuously under these shovels on parallel rail lines.

Stevens's qualities as a great engineer and leader were on a level with those of the Roeblings'. As an engineer, he understood that planning must be done to provide a climate and environment for success. Based on his railroading experience, he knew that a project of this magnitude could not be accomplished by committing resources in a piecemeal fashion. He took the required time to organize and mass his forces. He also intuitively understood that the problem of disease had to be confronted and conquered. Some credit for Stevens's success must go to Theodore Roosevelt and his secretary of war, William Howard Taft. Taft gave Stevens a free hand to make decisions on the spot and, in effect, gave him total control of the project. Stevens was able to be decisive and was not held in check by a committee of bureaucrats located in Washington (i.e., the situation present prior to his taking charge of the job).

Having set the course that would ultimately lead to successful completion of the canal. Stevens abruptly resigned. It is not clear why he decided not to carry the project through to completion. President Roosevelt reacted to his resignation by appointing a man who, as Roosevelt would say, "could not resign." Roosevelt selected an army colonel and West Point graduate named George Washington Goethals to succeed Stevens. Goethals had the managerial and organizational skills needed to push the job to successful completion. Rightfully so, General Goethals received a great deal of credit for the construction of the Panama Canal. However, primary credit for pulling the job "out of the mud," getting it on track, and developing the technical concept of the canal that ultimately led to success must be given to Stevens—a great engineer and a great construction manager.

1.4 OTHER HISTORIC PROJECTS

Much can he learned from reading about and understanding projects like the Brooklyn Bridge and the Panama Canal. David McCullough's books *The Great Bridge* and *The Path Between the Seas* are as exciting and gripping as any spy novel. They also reflect the many dimensions of great and small construction projects. Other projects such as the building of the Hoover Dam on the Colorado River have the same sweep of adventure and challenge as the construction of the Panama Canal. The construction of the Golden Gate Bridge in San Francisco was just as challenging a project as the construction of the Brooklyn Bridge in its time.

The construction of the Empire State Building in only 18 months is another example of a heroic engineering accomplishment. Realization of great skyscrapers such as the Empire State Building and the Chrysler Building in New York was made possible by the development of technologies and techniques in the construction of earlier projects. The construction of the Eiffel Tower in Paris and the towers of the "miracle mile" in Chicago in the early 1900s demonstrated the feasibility of building tall steel-frame-supported structures. Until the advent of the steel frame with its enclosing "curtain" walls, the height of buildings had been limited based on the strength of materials used in the bearing walls, which carried loads to the ground.

The perfection of the concept of steel-frame-supported structures and the development of the elevator as a means of moving people vertically in tall buildings provided the necessary technologies for the construction of the tall buildings that we take for granted today. Modern-day city skylines would not have been possible without these engineering innovations.

More recently, a project of historical proportions was realized with the completion of the Eurotunnel connecting the British Isles and France. This project has been dreamed of for many centuries. Through the skill and leadership of a large team of engineers and managers, it has now become a reality. Great projects are still being proposed and constructed. For the interested reader, brief coverage of many historical projects is given in *The Builders— Marvels of Engineering* published by the National Geographic Society (editor: Elizabeth L. Newhouse, 1992).

1.5 CONSTRUCTION VERSUS MANUFACTURING PROCESSES

Construction is the largest product-based (as opposed to service-oriented) industry in the U.S. The dollar volume of the industry is on the order of one trillion (1,000 billion) dollars annually. The process of realizing a constructed facility such as a road, bridge, or building, however, is quite different from that involved in manufacturing an automobile or a television set.

Manufactured products are typically designed and produced without a designated purchaser. In other words, products (e.g., automobiles or TV sets) are produced and then presented for sale to any potential purchaser. The product is produced on the speculation that a purchaser will be found for the item produced. A manufacturer of bicycles, for instance, must determine the size of the market, design a bicycle which appeals to the potential purchaser, and then manufacture the number of units which market studies indicate can be sold. Design and production are done prior to sale. In order to attract possible buyers, advertising is required and is an important cost center.

Many variables exist in this undertaking, and the manufacturer is "at risk" of failing to recover the money invested once a decision is made to proceed with design and production of the end item. The market may not respond to the product at the price offered. Units may remain unsold or sell at or below the cost of production (i.e., yielding no profit). If the product cannot be sold so as to recover the cost of manufacture, a loss is incurred and the enterprise is unprofitable. When pricing a given product, the manufacturer must not only recover the direct (labor, materials, etc.) cost of manufacturing, but also the so-called indirect and General and Administrative (G&A) costs such as the cost of management and the implementation of the production process (e.g., legal costs, marketing costs, supervisory costs, etc.). Finally, unless the enterprise is a "non-profit," the desire of the manufacturer is to increase the value of the firm. Therefore, profit must be added to the direct, indirect, and G&A costs of manufacturing.

Manufacturers offer their products for sale either directly to individuals (e.g., by mail order or directly over the web), to wholesalers who purchase in quantity and provide units to specific sales outlets, or to retailers who sell directly to the public. This sales network approach has developed as the framework for moving products to the eventual purchaser. (See if you can think of some manufacturers who sell products directly to the end user, sell to wholesalers, and/or sell to retail stores.)

In construction, projects are sold to the client in a different way. The process of purchase begins with a client who has need for a facility. The purchaser typically approaches a design professional to more specifically define the nature of the project. This leads to a conceptual definition of the scope of work required to build the desired facility. Prior to the age of mass production, purchasers presented plans of the end object (e.g., a piece of furniture) to a craftsman for manufacture. The craftsman then proceeded to produce the desired object. For example, if King Louis XIV desired a desk at which he could work, an artisan would design the object, and a craftsman would be selected to complete the construction of the desk. In this situation, the purchaser (King Louis XIV) contracts with a specialist to construct a unique object. The end item is not available for inspection until it is fabricated. That is, since the object is unique, it is not sitting on the show room floor and must be specially fabricated.

Due to the "one of a kind" unique nature of constructed facilities, this is still the method used for building construction projects. The purchaser approaches a set of potential contractors. Once an agreement is reached among the parties (e.g., clients, designers, etc.) as to the scope of work to be performed, the details of the project or end item are designed and constructed. Purchase is made based on a graphical and verbal description of the end item, rather than the completed item itself. This is the opposite of the speculative process where design and manufacture of the product are done prior to identifying specific purchasers. A constructed facility is not commenced until the purchaser has been identified. For instance, it would be hard to imagine building a bridge without having identified the potential buyer. (Can you think of a construction situation where the construction is completed prior to identifying the buyer?)

The nature of risk is influenced by this process of purchasing construction. For the manufacturer of a refrigerator, risk relates primarily to being able to produce units at a competitive price. For the purchaser of the refrigerator, the risk involves mainly whether the appliance operates as advertised.

In construction, since the item purchased is to be produced (rather than being in a finished state), there are many complex issues which can lead to failure to complete the project in a functional and/or timely manner. The number of stake holders and issues that must be dealt with prior to project completion lead to a complex level of risk for all parties involved (e.g., designers, constructors, government authorities, real estate brokers, etc.). A manufactured product is, so to say, "a bird in the hand." A construction project is "a bird in the bush."

The risks of the manufacturing process to the consumer are somewhat like those incurred when a person goes to the store and buys a music CD. If the recording is good and the disk is serviceable, the risk is reduced to whether the customer is satisfied with the musical group's performance. The client in a construction project is more like a musical director who must assemble an orchestra and do a live performance hoping that the recording will be acceptable. The risks of a failure in this case are infinitely greater. A chronological diagram of the events involved in the manufacturing process versus those in the construction process are shown schematically in Figure 1.8.

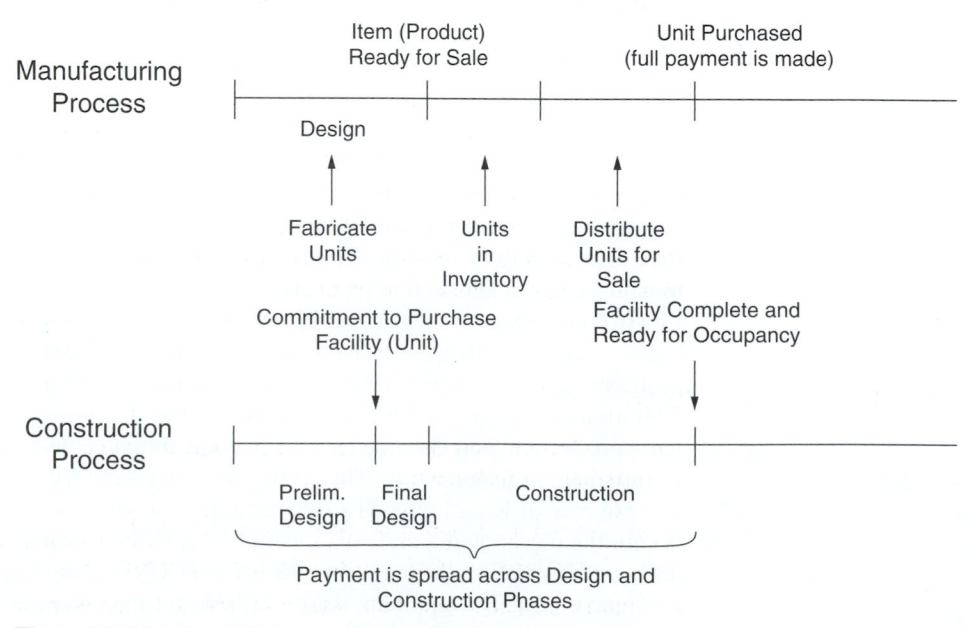

Figure 1.8 Manufacturing versus Construction Process.

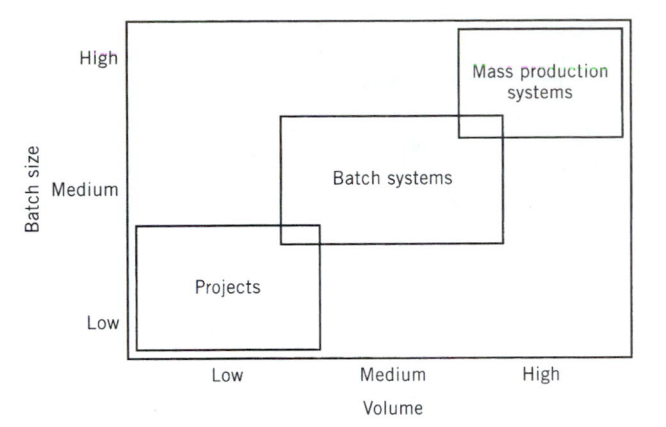

Figure 1.9 Comparison of production systems.

1.6   PROJECT FORMAT

In contrast to other manufacturing industries that fabricate large numbers of units such as automobiles or television sets, the construction industry is generally focused on the production of a single and unique end product. That is, the product of the construction industry is a facility that is usually unique in design and method of fabrication. It is a single "one-off" item that is stylized in terms of its function, appearance, and location. In certain cases, basically similar units are constructed as in the case of town houses or fast-food restaurants. But even in this case, the units must be site adapted and stylized to some degree.

Mass production is typical of most manufacturing activities. Some manufacturing sectors make large numbers of similar units or batches of units that are exactly the same. A single item is designed to be fabricated many times. Firms manufacture many repetitions of the same item (e.g., telephone instruments, thermos bottles, etc.) and sell large numbers to achieve a profit. In certain cases, a limited number or batch of units of a product is required. For instance, a specially designed transformer or hydropower turbine may be fabricated in limited numbers (e.g., 2, 3, or 10) to meet the special requirements of a specific client. This production of a limited number of similar units is referred to as batch production.

Mass production and batch production are not typical of the construction industry (see Fig. 1.9). Since the industry is oriented to the production of single unique units, the format in which these one-off units are achieved is called the project format. Both the design and production of constructed facilities are realized in the framework of a project. That is, one speaks of a project that addresses the realization of a single constructed facility.

The focus of construction management is the planning and control of resources within the framework of a project. This is in contrast to other manufacturing sectors that are interested in the application of resources over the life of an extended production run of many units.

1.7   PROJECT DEVELOPMENT

Construction projects develop in a clearly sequential or linear fashion. The general steps involved are as follows:

- A need for a facility is identified by the owner.
- Initial feasibility and cost projections are developed.
- The decision to proceed with conceptual design is made, and a design professional is retained.

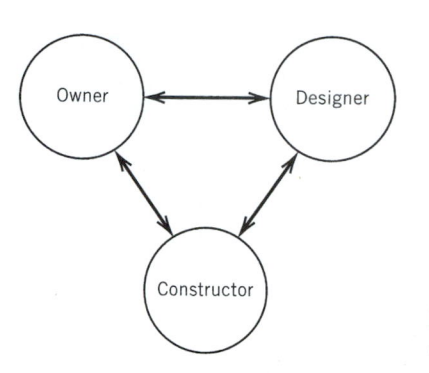

Figure 1.10 Relationship between owner, designer, and constructor.

- The conceptual design and scope of work are developed to include an approximate estimate of cost.
- The decision is made to proceed with the development of final design documents, which fully define the project for purposes of construction.
- Based on the final design documents, the project is advertised and proposals to include quotations for construction of the work are solicited.
- Based on proposals received, a constructor is selected and a notice to the constructor to proceed with the work is given. The proposal and the acceptance of the proposal on the part of the owner constitute the formation of a contract for the work.
- The process of constructing the facility is initiated. Work is completed, and the facility is available for acceptance and occupancy/utilization.
- In complex projects, a period of testing decides if the facility operates as designed and planned. This period is typical of industrial projects and is referred to as project start-up.
- The facility operates and is maintained during a specified service life.
- The facility is disposed of if appropriate or maintained in perpetuity.

These steps must be modified on a case-by-case basis to address the special aspects of a given project. Topics relating to items 1 through 8 will be discussed in detail in Chapters 2 and 3.

The key players in this developmental sequence are:

1. The owner
2. The designer or design professional
3. The constructor

The interaction of these three major entities is shown in Figure 1.10.

Although other entities such as regulators, subcontractors, materials vendors, and so forth are important supporting players in this sequence, the major development of the project revolves about these three major entities. The legal definition of this interaction is established in the general conditions of the contract. This interaction will be described in detail in the following chapters.

1.8 CONSTRUCTION TECHNOLOGY AND CONSTRUCTION MANAGEMENT

The study of construction as a discipline can be broadly structured into two general themes:

1. Construction technology
2. Construction management

As the name implies, construction technology relates to the methods or techniques used to place the physical materials and elements of construction at the job site. The word *technology* can be broken into two subwords—*technical* from "techno" and *logic*. *Logic* addresses the concept of sequence or procedure. That is, logic addresses the order of things—something is done first, another thing second, and so on until a result is achieved. Adding *technical* to this leads to the idea that technology has to do with the technical sequence in which something is done to produce an end result. It is possible to talk about a technology that applies to placing concrete, cladding a building, excavating a tunnel, and so on.

Once a project has been defined, one of the most critical questions facing the construction manager is "What construction technique or method should be selected?" The types of methods for placing construction are diverse. New methods are continuously being perfected, and a construction manager must weigh the advantages and disadvantages of a given method or technique.

In contrast to construction technology, construction management addresses how the resources available to the manager can be best applied. Typically, when speaking of resources for construction, we think of the four Ms of construction: **m**anpower, **m**achines, **m**aterials, and **m**oney. Management involves the timely and efficient application of the four Ms to construct a project. Many issues must be considered when managing a project and successfully applying the four Ms. Some are technical (e.g., design of formwork, capacities of excavators, weather tightness of exterior finishes, etc.). Many issues, however, are more qualitative in nature and deal with the motivation of workers, labor relations, the form of contracts, legal liability, and safety on the job site. As noted, in discussing the Panama Canal, organizational issues can be very critical to the success of any project. This book will focus mainly on the topic of construction management. Therefore, we will be talking about the four Ms and subjects that relate to management and the timely and cost-effective realization of a project.

1.9   CONSTRUCTION MANAGEMENT IS RESOURCE DRIVEN

The job of a construction manager is to efficiently and economically apply the required resources to realize a constructed facility of acceptable quality within the time frame and budgeted cost specified. Among the many watch words within the construction industry is the expression "on time and within budget." More recently, the concept of quality as a requirement has become an increasingly important aspect of the construction process. So this old adage can be expanded to say "a quality facility on time and within budget."

The construction manager is provided with resources such as labor, equipment, and materials and is expected to build a facility that meets the specifications and is consistent with the drawings provided for the project. The mission of construction is constrained in terms of the available time and amount of money available. The challenge faced by the construction manager is to apply the resources of workers, machines, and materials within the limited funding (money) and time available. This is the essence of construction.

The manager must be clever and innovative in the utilization of resources available. Somewhat like a general in battle, the manager must develop a plan of action and then direct and control forces (resources) in a coordinated and timely fashion so that the objective is achieved.

This requires a variety of skills. A high level of competency is needed in a broad range of qualitative and quantitative subjects. A manager must be like a decathlon athlete. A strong ability in many areas is a necessity. Being outstanding in one area (e.g., engineering) but weak in a number of others (e.g., interpersonal relationships, contract law, labor relations, etc.) is not enough to be a successful construction manager. A strong performance across the board is required.

1.10   CONSTRUCTION INDUSTRY

The construction industry has been referred to as the engine that drives the overall economy. It represents one of the largest economic sectors in the United States. Until the early 1980s the construction industry accounted for the largest percent of the gross domestic product (GDP) and had the highest dollar turnover of any U.S. industry. Presently, construction is still the largest manufacturing industry in the United States. New construction accounts for approximately 8% of the GDP and retrofit projects contribute an additional 5%. As noted above, the total annual volume of activity in the construction sector is estimated to be well in excess of $800 billion. More than a million firms operate in the construction sector, and the number of people employed in construction is estimated to be 10 million.

The industry consists of very large and very small firms. The largest firms sign contracts in excess of $20 billion annually and consist of thousands of employees. Many of the largest firms work both domestically and in the international market. In contrast to the large companies, statistics indicate that over two-thirds of the firms have less than five employees. The spectrum of work ranges from the construction of large power plants and interstate highways costing billions of dollars to the construction of single-family houses and the paving of driveways and sidewalks. The high quality of life available in the United States is possible in large part because of the highly developed infrastructure. The American infrastructure, which consists of the roads, tunnels, bridges, communications systems, power plants and distribution networks, water treatment systems, and all of the structures and facilities that support daily life, is without peer. The infrastructure is constructed and maintained by the construction industry. Without it, the country would not be able to function.

1.11   STRUCTURE OF THE CONSTRUCTION INDUSTRY

Since the construction sector is so diverse, it is helpful to look at the major types of projects typical of construction in order to understand the structure of the industry. Construction projects can be broadly classified as (1) building construction, (2) engineered construction, and (3) industrial construction, depending on whether they are associated with housing, public works, or manufacturing processes.

The building construction category includes facilities commonly built for habitational, institutional, educational, light industrial (e.g., warehousing, etc.), commercial, social, and recreational purposes. Typical building construction projects include office buildings, shopping centers, sports complexes, banks, and automobile dealerships. Building construction projects are usually designed by architects or architect/engineers (A/Es). The materials required for the construction emphasize the architectural aspects of the construction (e.g., interior and exterior finishes).

Engineered construction usually involves structures that are planned and designed primarily by trained professional engineers (in contrast to architects). Normally, engineered construction projects provide facilities that have a public function relating to the infrastructure and, therefore, public or semipublic (e.g., utilities) owners generate the requirements for such projects. This category of construction is commonly subdivided into two major subcategories; thus, engineered construction is also referred to as (1) highway construction and (2) heavy construction.

Highway projects are generally designed by state or local highway departments. These projects commonly require excavation, fill, paving, and the construction of bridges and drainage structures. Consequently, highway construction differs from building construction in terms of the division of activity between owner, designer, and constructor. In highway construction, owners may use in-house designers and design teams to perform the design so that both owner and designer are public entities.

Heavy construction projects are also typically funded by public or quasi-public agencies and include sewage plants, flood protection projects, dams, transportation projects (other than highways), pipelines, and waterways. The owner and design firm can be either public or private depending on the situation. In the United States, for instance, the U.S. Army Corps of Engineers (a public agency) has, in the past, used its in-house design force to engineer public flood protection structures (i.e., dams, dikes) and waterway navigational structures (e.g., river dams, locks, etc.). Due to the trend toward downsizing government agencies, more design work is now being subcontracted to private design engineering firms. Public electrical power companies use private engineering firms to design their power plants. Public mass-transit authorities also call on private design firms (design professionals) for assistance in the engineering of rapid-transit projects.

Industrial construction usually involves highly technical projects in manufacturing and processing of products. Private clients retain engineering firms to design such facilities. In some cases, specialty firms perform both design and construction under a single contract for the owner/client.

1.12   DIFFERING APPROACHES TO INDUSTRY BREAKDOWN

Figure 1.11 represents one of many ways in which the industry can be divided into a number of sectors. This breakdown includes single-family houses within the residential construction sector. In some breakdowns, one- and two-family houses are considered to be a separate industry, and this residential activity is not reported as part of the construction industry. As can be seen from the pie chart, residential and building construction account for between 65 and 75% of the industry. Industrial construction and heavy engineering construction (which are more closely related to the infrastructure) account for 25 to 35% of industry activity.

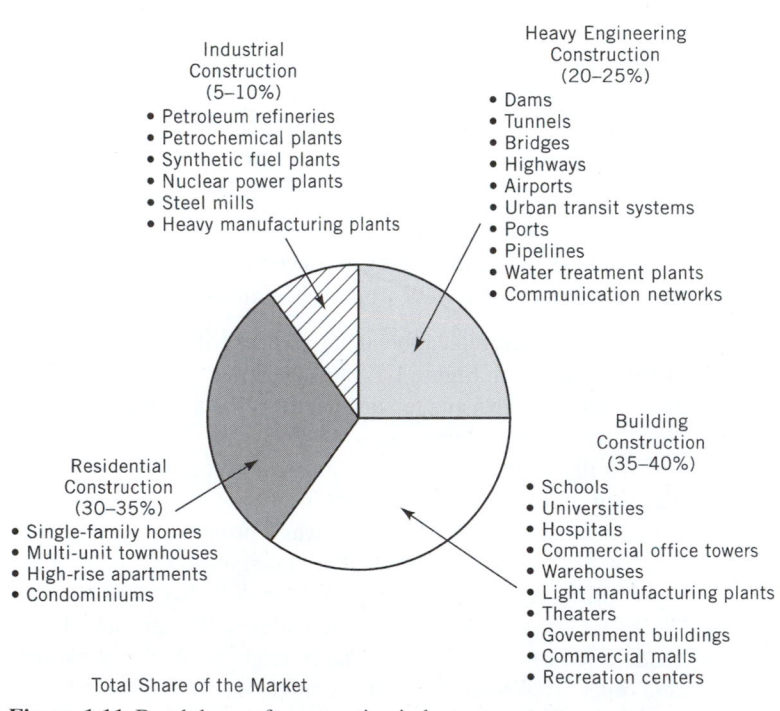

Figure 1.11 Breakdown of construction industry segments.

A slightly different approach to project classification is used by the *Engineering News Record* (ENR) magazine, which reflects the weekly dynamics of the construction industry in the U.S. This breakdown identifies three major construction categories as follows:

1. Heavy and highway
2. Nonresidential building
3. Multiunit housing

The nonresidential building category includes building and industrial construction as defined above. These overall categories are further dissected to reflect the major areas of specialization within the construction industry. The ENR publishes a web-based update of information based on this set of construction categories each week.

1.13   MANAGEMENT LEVELS OF CONSTRUCTION

Organizational considerations lead to a number of hierarchical levels that can be identified in construction. This derives from the project format. Decision making at levels above the project relate to company management considerations. Decisions within the project relate to operational considerations (e.g., selection of production methods) as well as the application of resources to the various construction production processes and work tasks selected to realize the constructed facility. Specifically, four levels of hierarchy can be identified as follows:

1. *Organizational*   The organizational level is concerned with the legal and business structure of a firm, the various functional areas of management, and the interaction between head office and field managers performing these management functions.
2. *Project Project-level*   Vocabulary is dominated by terms relating to the breakdown of the project for the purpose of time and cost control (e.g., the project activity and the project cost account). Also, the concept of resources is defined and related to the activity as either an added descriptive attribute of the activity or for resource scheduling purposes.
3. *Operation (and Process)*   The construction operation and process level is concerned with the technology and details of how construction is performed. It focuses on work at the field level. Usually a construction operation is so complex that it encompasses several distinct processes, each having its own technology and work task sequences. However, for simple situations involving a single process, the terms are synonymous.
4. *Task*   The task level is concerned with the identification and assignment of elemental portions of work to field units and work crews.

The relative hierarchical breakout and description of these levels in construction management are shown in Figure 1.12. It is clear that the organizational, project, and activity levels have a basic project and top management focus, while the operation, process, and work task levels have a basic work focus.

To illustrate the definitions given above, consider a glazing subcontract for the installation of glass and exterior opaque panels on the four concourses of Hartsfield International Airport in Atlanta, Georgia. This was a project requiring the installation of five panels per bay on 72 bays of each of the four concourses. Figure 1.13 shows a schematic diagram of the project. A breakout of typical items of activity at each level of hierarchy is given in Table 1.1. At the project level, activities within the schedule relate to the glass and panel installation in certain areas of the concourses. At the work task level, unloading, stripping, and other crew-related activity is required.

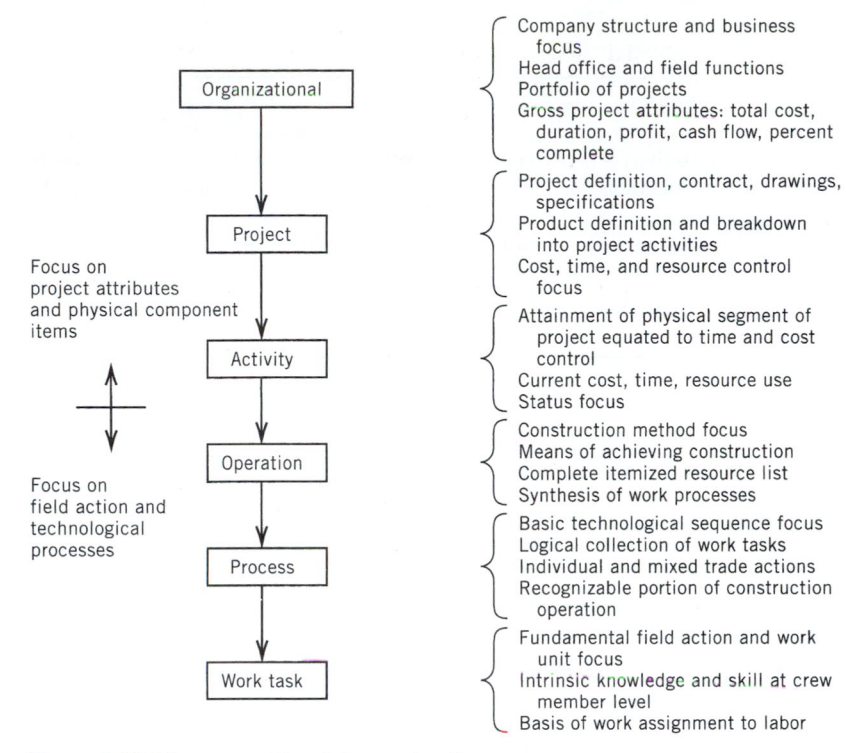

Hierarchical Level	Description and Basic Focus

Organizational
- Company structure and business focus
- Head office and field functions
- Portfolio of projects
- Gross project attributes: total cost, duration, profit, cash flow, percent complete

Project
- Project definition, contract, drawings, specifications
- Product definition and breakdown into project activities
- Cost, time, and resource control focus

Focus on project attributes and physical component items

Activity
- Attainment of physical segment of project equated to time and cost control
- Current cost, time, resource use
- Status focus

Operation
- Construction method focus
- Means of achieving construction
- Complete itemized resource list
- Synthesis of work processes

Focus on field action and technological processes

Process
- Basic technological sequence focus
- Logical collection of work tasks
- Individual and mixed trade actions
- Recognizable portion of construction operation

Work task
- Fundamental field action and work unit focus
- Intrinsic knowledge and skill at crew member level
- Basis of work assignment to labor

Figure 1.12 Management levels in construction.

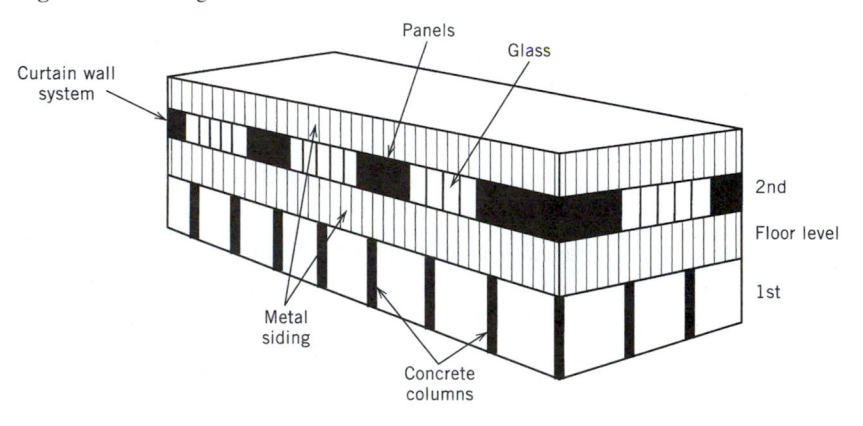

Figure 1.13 Schematic of concourse building.

Table 1.1 Example of Hierarchical Terms

Project	Installation of all exterior glass and panel wall construction on Concourses A–D of the Hartsfield International Airport, Atlanta, GA
Activity	Glass and panel installation on Concourse A, Bays 65–72
Operation	Frame installation to include preparation and installation of five panel frames in each concourse bay; column cover plate installation
Process	Sill clip placement; mullion strips installation
	Glass placement in frame; move and adjust hanging scaffold
Work task	Locate and drill clip fastener; unload and position mullion strips; strip protective cover from glass panel; secure scaffold in travel position

Chapter 2

Preparing the Bid Package

Online Plan Room

The Need

Plan rooms are typically conveniently located and can be visited by contractors in large cities. These facilities provide access to bidding documents in order to facilitate the bidding process. Sets of project plans, specifications, bidding information, and general contractor lists for jobs being let for bid are available at plan rooms. However, utilizing these plan room services can be time consuming and costly due to the need to travel to the physical location of the plan room. Even though many plan rooms mail their members a weekly newsletter for project update, a visit to the facility is required to check out the project information in detail.

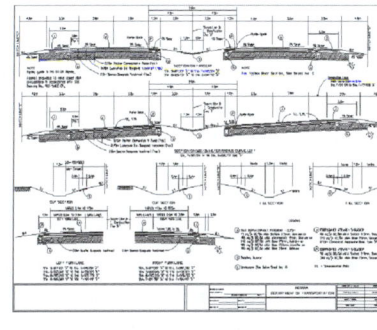
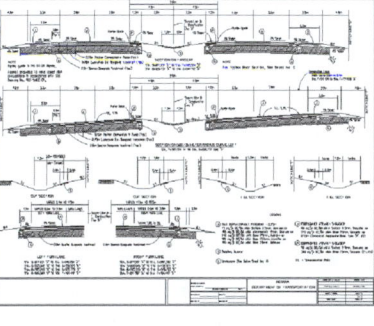

The Technology

Online plan rooms enable their members to conveniently view plan room projects 24 hours a day, seven days a week from the member's home or office location via the Internet. Members can access, look, and download useful information for various construction projects, and even print out actual plans using plotters or printers. Contractors and estimators can get information and prepare for a bid in a timely and efficient manner. In addition, establishing an online plan room enables owners, architects, and contractors to control and manage their bid process more effectively.

Additional features augmenting the online plan room services include Internet access to the plan room via secure login, computerized estimating tools, worksheet build-up function, real-time bidder notification system (email/fax), project search with specific criteria, and blueprint measurement on-screen, etc.

Utilization of these online plan rooms provides prospective bidders with 24 hour accessibility. They also make the bidding process efficient and well-organized, which leads to reduced costs.

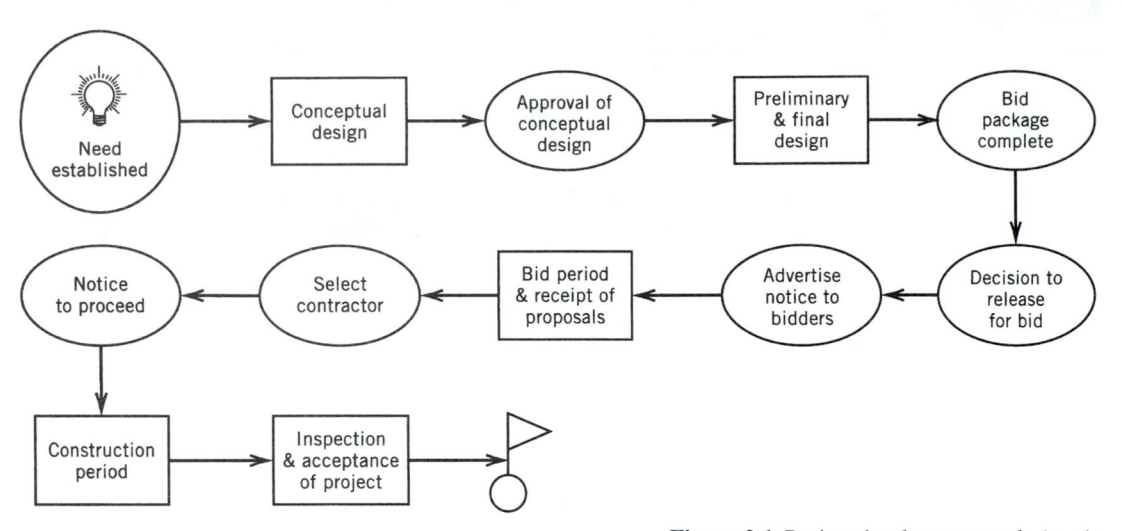

Figure 2.1 Project development cycle (new).

2.1   PROJECT CONCEPT AND NEED

Since the constructed environment in which we live is realized in a project format, the construction process can be best understood by examining the steps required to realize a complete project. In Chapters 2 and 3, we will examine the step-by-step development of a project. A schematic flow diagram of the sequential actions required to realize a project is shown in Figure 2.1. The framework for this discussion will be the development of a project for competitive bid. As we will see in Chapter 4, this is the delivery system characteristic of publicly contracted work. This approach requires that a full set of project documents be developed before the project is offered for bid and construction.

Each project has a life cycle triggered by the recognition of a need that can best be addressed with the construction of a facility. In a complex society, the number of entities generating needs that will shape the built environment is very diverse. Private individuals seek to construct housing that is functional and comfortable (e.g., home or residential construction). Public entities such as city, state, and federal governments construct buildings and required public structures to enhance the quality of life. Many public projects relate to the development of the infrastructure. Bridges, tunnels, transportation facilities, dikes, and dams are typical of public projects designed to meet the needs of a community and the society in general.

Private entities such as commercial firms build facilities that provide goods and services to the economy. These entities are typically driven by the objective of realizing a profit. Facilities constructed by private owners include manufacturing plants, hospitals, research laboratories, hotels and commercial buildings, communications networks, and a host of other project types.

2.2   ESTABLISHING NEED

The first step in any project is the establishment of a need and a conceptual definition and refinement of the facility that will meet that need. If the need has a commercial basis, it is normally defined in terms of a market analysis that establishes the profitability of the proposed project. For instance, if the need relates to the construction of a chemical plant in Spain, the firm constructing the plant will want to establish that a market exists that can be profitably accessed once the plant is in operation.

The economic basis for the plant must be established based on market studies projecting the demand for the plant's product mix across the planning horizon under consideration. In many cases, these studies recommend optimal time frames for the plant construction to meet the market in advance of competition. Plant size, site location, availability of labor and supporting resources such as energy, water, and shipping connections are considered. This study is sometimes referred to as a feasibility study.

This type of information must be developed so that planning decisions by senior management within the company can be made. Typically, feasibility information and supporting cost analyses are submitted to the board of directors. The board then must decide whether the investment required to build a plant is justified.

Similar analysis is necessary for any project. If a group of entrepreneurs decides to build a hotel in Phoenix, Arizona, the basic economic considerations to determine the potential profitability of this venture must be examined. If the economic study supports the idea of a hotel, a need is established. In this case, the financial institutions that lend the money for the development of the hotel typically require certain justification before providing the financing. Therefore, the structure of the feasibility study is dictated, in large part, by the requirements of the lending institution. The types of information required for developing a commercial building project will be addressed in Chapter 10.

Public and community-service-related projects do not typically involve profit and, therefore, are triggered by other considerations. If the church board of the Smallville Methodist Church decides to add a wing to provide a larger area for the Sunday school, this decision is based on improving the quality of services provided by the church. Since funds must be developed for such an addition, the church board will seek the assistance of a consultant (e.g., architect or architect/engineer) to better define the scope of the new addition. Design and cost information are required to approach a bank or lending agency regarding financing. Based on information from design and cost consultants, the church board must decide whether to proceed to development of final design documents or place the project on hold.

Public entities such as city, state, and federal governments are continuously reviewing societal needs. The annual cycle of activity for public agencies looks at the changing demands of constituents with the objective of developing a plan (e.g., a set of projects) that will improve services. State highway departments, for instance, have annual budgets based on existing strategic plans. These master plans envision the construction of new roads and bridges and the maintenance of existing infrastructure. Such plans are reviewed annually and projects to repair and enhance the transportation network of each state are budgeted. In this situation, the needs of the state are under continuous review. A balance between funds available and transportation needs must be maintained.

2.3   FORMAL NEED EVALUATION

In deciding whether or not to proceed with preliminary and final design of a given project, three items should be developed during the conceptual portion of the project cycle. The following elements provide input to the decision process:

1. Cost/benefit analysis
2. Graphical representation of the project (e.g., sketch or artist's rendering) and a layout diagram of the facility
3. Cost estimate based on the conceptual-level information available

These documents assist the key decision maker(s) in deciding whether to proceed with a proposed project.

The cost/benefit analysis in the case of commercial or profit-based projects is simply a comparison of the estimated cost of the project against the revenues that can be reasonably expected to be generated. In public and other non-profit-based projects (e.g., monuments, churches, museums, etc.), development of the benefit to be achieved is more difficult to pin down.

For instance, if a dam is to be constructed on the Colorado River, part of the benefit will be tangible (i.e., developed in dollars) and part will be intangible (i.e., related to the quality of life). If power is to be generated by the dam, the sale of the electricity and the revenues generated therefrom are tangible and definable in dollars amounts. Much of the benefit may, however, derive from control of the river and the changing of the environment. This dam will prevent flooding of downstream communities and form a lake that can be used as a recreational resource.

The recreational aspects of the project and the protection of communities from flooding are difficult to characterize in dollars and cents. They can be viewed as intangible benefits related to improvement of the quality of life. Protocols for converting intangible aspects of a dam project into benefits have been developed by the Bureau of Reclamation and the Army Corps of Engineers (both government agencies involved in water resource development). At best, however, evaluating intangibles is a judgment call and subject to review and criticism.

2.4   CONCEPTUAL DRAWINGS AND ESTIMATES

In seeking funding for entrepreneurial projects such as hotels, apartment buildings and complexes, shopping malls, and office structures, it is common practice to present conceptual documentation to potential funding sources (e.g., banks and investors). In addition to a cost/benefit analysis, graphical information to include architect's renderings or sketches as well as layout drawings and 3D computer models assist the potential investor in better understanding the project. For this reason, such concept drawings and models are typically part of the conceptual design package. A cost estimate based on the conceptual drawings and other design information (e.g., square footage of roof area, floor space, size of heating and air conditioning units, etc.) is prepared.

Government projects at the federal level require similar supporting analysis and are submitted with budget requests each year for congressional action. Supporting documentation includes layout sketches and outline specifications such as those shown in Figure 2.2. The supporting budget for this project is shown in Figure 2.3. These projects are included as line items in the budget of the government agency requesting funding. In this case, the requestor would be the post engineer, Fort Campbell, Kentucky. This request would be consolidated with requests at the Army and Department of Defense level and forwarded to the Bureau of the Budget to be included in the budget submitted to Congress.

It is of interest to note that since the post office project will not be built for at least a year (assuming it is approved), a projection of cost to the future date on which construction will begin is required. The projection is made using the *Engineering News Record* (ENR) indexes of basic construction cost. Construction cost indexes such as the ENR index allow estimators to project costs into the future. The building and construction cost indexes through March 2005 are shown in Figure 2.4.

The summary on the last page of the estimate indicates that the baseline cost of the project will be $909,050. The reserve for contingency is 10% of the base cost, or $90,900. The amount budgeted for supervision of the work by the Corps of Engineers is $54,800. The design cost is projected to be $70,000.

The amount of conceptual design documentation varies based on the complexity of the project. Fairly simple building projects such as the Sunday school addition or the military post office can be conceptually defined in terms of drawings, guide specifications, and a cost estimate such as those shown in Figures 2.2 and 2.3. Large and complex projects such as

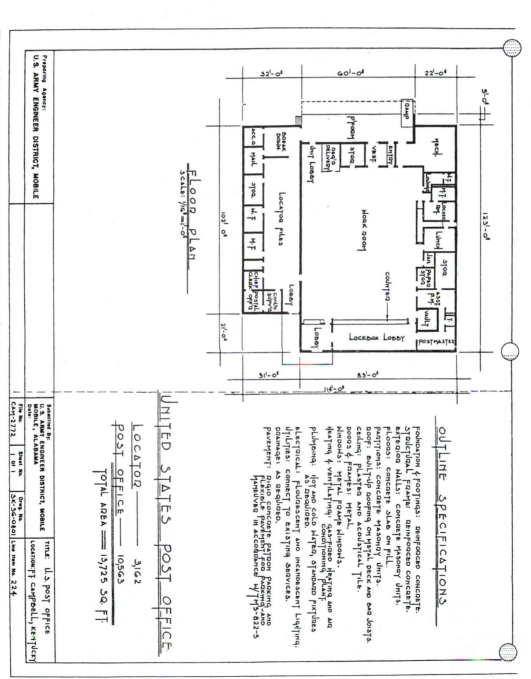

Figure 2.2 Project proposal: layout sketch and outline specifications.

TO: Chief of Engineers
      Department of the Army
      Washington, D.C.

FROM: Louisville District
        Corps of Engineers
        Louisville, KY

Fiscal Year
2XXX

Date Prepared:
14 Oct 2XXX

Name and Address of A.E.
N.A.

Basis of Estimate
Budget Sketch & 1391

A.E. Fee
N.A.

Name and Location of Installation
Ft. Campbell, Kentucky

Type of Construction
Permanent

Status of Design
Preliminary 0% complete

Final 0% complete

Line Item Number
224

Description of Facility
Post Office

Final Design Completion Date
Not Authorized

Description	Quantity	Unit	Unit Price	Totals ($000)
1. *Building*				
General construction	13,725	Sq ft	$42.24	$579.7
Plumbing	13,725	Sq ft	2.42	33.2
Heating and ventilating	13,725	Sq ft	2.68	36.8
Air conditioning (50-ton)	13,725	Sq ft	7.62	104.6
Electrical	13,725	Sq ft	5.76	79.0
Subtotal	13,725	Sq ft	60.72	833.3
2. *Utilities*				
a. *Electrical*				
Transformers	112.5	kVA	50	5.6
Poles with X-arms, pins, insulation, etc.	4	Each	720	2.9
Dead ends	6	Each	80	.5
Down guys and anchors	4	Each	180	.7
Fused cutouts and L.A.	6	Each	60	.4
#6 Bar Cu. conductor	2,400	lin ft	.30	0.7
#3/0 Neoprene covered service	160	lin ft	1.48	0.2
Parking area lights on aluminum pole 3C #8 DB 600-V	7	Each	933.41	6.5

TO: Chief of Engineers
      Department of the Army
      Washington, D.C.

FROM: Louisville District
        Corps of Engineers
        Louisville, KY

Figure 2.3 Current working estimate for budget purposes.

Description	Quantity	Unit	Unit Price	Totals ($000)
3-in. duct conc. encased U.G.	100	lin ft	4.75	0.5
Subtotal				19.3
b. *Water*				
3-in. Water line	365	lin ft	8.60	3.1
3-in. Gate valve and box	1	Each	200.00	.2
Fire hydrants	2	Each	1000.00	2.0
Connections to existing lines	3	Each	500.00	1.5
Subtotal				6.8
c. *Sewer*				
6-in. sanitary sewer	215	lin ft	12.00	2.6
8-in. sanitary sewer	375	lin ft	14.00	5.3
Manhole	2	Each	1000.00	2.0
Connection to exist. manhole	1	Each	250.00	.25
Subtotal				10.15
d. *Gas*				
$1\frac{1}{4}$ in. gas line	1,000	lin ft	6.00	6.0
$1\frac{1}{4}$ in. plug valve and box	1	Each	200.00	0.2
Connect to existing	1	Each	237.60	0.2
street and parking area crossing	280	lin ft	2.0	0.6
Subtotal				7.0
3. *Site Work*				
Clearing and grubbing	2.4	Acre	500	1.2
Borrow excavation	10,000	cu yd	8.00	80.0
Remove B. T. paving	1,070	sq yd	4.00	4.3
Subtotal				85.3
4. *Paving*				
Paving-$1\frac{1}{2}$ A.C. and 8-in. stab.	3,950	sq yd	10.00	39.5
aggr. base	2,250	lin ft	8.50	19.2
6-in. P.C. concrete paving	380	sq yd	20.00	7.6
3-in. painted parking lines	1,680	lin ft	8.50	0.9
Concrete sidewalk	440	sq yd	18.00	7.9
Subtotal				75.1
5. *Storm Damage*				
15-in. concrete Cl. II pipe	40	lin ft	18.00	0.7
15-in. concrete Cl. III pipe	20	lin ft	20.00	0.4
Reinf. drainage structure concrete	8	cu yd	300.00	2.4
C.I. grates and frames	1,900	lb	1.00	1.9
Subtotal				5.4
6. *Landscaping*				

Figure 2.3

Description	Quantity	Unit	Unit Price	Totals ($000)
Sprigging and seeding	1.6	Acre	$1500.00	2.4
Landscaping		Job		3.8
Subtotal				6.2
7. *Communications*				
a. Telephone		LS	$1400.00	1.4
b. Support (within building)				
100 Pr. DB Pic Cable	600	LF	1.26	0.8
51 Pr. DB Pic Cable	550	LF	0.72	0.4
Splicing sleeves and material		LS	$900.00	0.9
Labor		LS		2.5
Subtotal				6.0
Total estimated cost (excluding design, but including reserve for contingencies and supervision and administration (S&A)				1054.75
1. Estimated contract cost				909.05
2. Reserve for contingencies	10	percent		90.90
3. Supervision and administration (S&A); total estimated cost (excluding design, but including reserve for contingencies and supervision and administration)				54.8
				1054.75
4. *Design*				
District expenses (preliminary and final)				70.0
Subtotal				70.0

Figure 2.3

petrochemical plants and power generation facilities require expanded documentation (e.g., hundreds of pages) to define the scope of work. For this reason, the number of engineering man-hours required to develop conceptual design documentation for the chemical plant in Spain would be significantly greater than that required for a small commercial building.

2.5   PRELIMINARY AND DETAIL DESIGN

Once the concept of the project has been approved, the *owner* desiring the construction retains an engineer, an architect, or a combination of the two, called an architect/engineer (A/E).[1] The end product of the design phase of project development is a set of plans and specifications that define the project to be constructed. The drawings are a graphical

---

[1] The A/E is typically a firm and is commonly referred to as "the design professional."

Building Cost Index History (1923-2005)

▶ HOW ENR BUILDS THE INDEX: **68.38 hours of skilled labor** at the 20-city average of bricklayers, carpenters and structural ironworkers rates, plus **25 cwt of standard structural steel shapes** at the mill price prior to 1996 and the fabricated 20-city price from 1996, plus **1.128 tons of portland cement** at the 20-city price, plus **1,088 board ft of 2x4 lumber** at the 20-city price.

ANNUAL AVERAGE

1923	186	1946	262	1969	790
1924	186	1947	313	1970	836
1925	183	1948	341	1971	948
1926	185	1949	352	1972	1048
1927	186	1950	375	1973	1138
1928	188	1951	401	1974	1205
1929	191	1952	416	1975	1306
1930	185	1953	431	1976	1425
1931	168	1954	446	1977	1545
1932	131	1955	469	1978	1674
1933	148	1956	491	1979	1819
1934	167	1957	509	1980	1941
1935	166	1958	525	1981	2097
1936	172	1959	548	1982	2234
1937	196	1960	559	1983	2284
1938	197	1961	568	1984	2417
1939	197	1962	580	1985	2428
1940	203	1963	594	1986	2483
1941	211	1964	612	1987	2541
1942	222	1965	627	1988	2598
1943	229	1966	650		
1944	235	1967	676		
1945	239	1968	721		

	JAN.	FEB.	MARCH	APRIL	MAY	JUNE	JULY	AUG.	SEPT.	OCT.	NOV.	DEC.	ANNUAL AVG.
1989	2615	2608	2612	2615	2616	2623	2627	2637	2660	2662	2665	2669	2634
1990	2664	2668	2673	2676	2691	2715	2716	2716	2730	2728	2730	2720	2702
1991	2720	2716	2715	2709	2723	2733	2757	2792	2785	2786	2791	2784	2751
1992	2784	2775	2799	2809	2828	2838	2845	2854	2857	2867	2873	2875	2834
1993	2886	2886	2915	2976	3071	3066	3038	3014	3009	3016	3029	3046	2996
1994	3071	3106	3116	3127	3125	3115	3107	3109	3116	3116	3109	3110	3111
1995	3112	3111	3103	3100	3096	3095	3114	3121	3109	3117	3131	3128	3111
1996	3127	3131	3135	3148	3161	3178	3190	3223	3246	3284	3304	3311	3203
1997	3332	3333	3323	3364	3377	3396	3392	3385	3378	3372	3350	3370	3364
1998	3363	3372	3368	3375	3374	3379	3382	3391	3414	3423	3424	3419	3391
1999	3425	3417	3411	3421	3422	3433	3460	3474	3504	3505	3498	3497	3456
2000	3503	3523	3536	3534	3536	3553	3545	3546	3539	3547	3541	3548	3539
2001	3545	3536	3541	3541	3547	3572	3625	3605	3597	3602	3596	3577	3574
2002	3581	3581	3597	3583	3612	3624	3652	3648	3655	3651	3654	3640	3623
2003	3648	3655	3649	3652	3660	3677	3684	3712	3717	3745	3766	3758	3694
2004	3767	3802	3859	3908	3955	3996	4013	4027	4103	4129	4128	4123	3984
2005	4112	4116	4127										

BASE: 1913=100.

(a)

Construction Cost Index History (1918-2005)

▶ HOW ENR BUILDS THE INDEX: **200 hours of common labor** at the 20-city average of common labor rates, plus **25 cwt of standard structural steel shapes** at the mill price prior to 1996 and the fabricated 20-city price from 1996, plus **1.128 tons of portland cement** at the 20-city price, plus **1,088 board ft of 2x4 lumber** at the 20-city price.

ANNUAL AVERAGE

1918	189	1942	276	1966	1019
1919	198	1943	290	1967	1074
1920	251	1944	299	1968	1155
1921	202	1945	308	1969	1269
1922	174	1946	346	1970	1381
1923	214	1947	413	1971	1581
1924	215	1948	461	1972	1753
1925	207	1949	477	1973	1895
1926	208	1950	510	1974	2020
1927	206	1951	543	1975	2212
1928	207	1952	569	1976	2401
1929	207	1953	600	1977	2576
1930	203	1954	628	1978	2776
1931	181	1955	660	1979	3003
1932	157	1956	692	1980	3237
1933	170	1957	724	1981	3535
1934	198	1958	759	1982	3825
1935	196	1959	797	1983	4066
1936	206	1960	824	1984	4066
1937	235	1961	847	1985	4066
1938	236	1962	872	1986	4295
1939	236	1963	901	1987	4406
1940	242	1964	936	1988	4519
1941	258	1965	971		

	JAN.	FEB.	MARCH	APRIL	MAY	JUNE	JULY	AUG.	SEPT.	OCT.	NOV.	DEC.	ANNUAL AVG.
1989	4580	4573	4574	4577	4578	4599	4608	4618	4658	4658	4668	4685	4615
1990	4680	4685	4691	4693	4707	4732	4734	4752	4774	4771	4787	4777	4732
1991	4777	4773	4772	4766	4801	4818	4854	4892	4891	4892	4896	4889	4835
1992	4888	4884	4927	4946	4965	4973	4992	5032	5042	5052	5058	5059	4985
1993	5071	5070	5106	5167	5262	5260	5252	5230	5255	5264	5278	5310	5210
1994	5336	5371	5381	5405	5405	5408	5409	5424	5437	5437	5439	5439	5408
1995	5443	5444	5435	5432	5433	5432	5484	5506	5491	5511	5519	5524	5471
1996	5523	5532	5537	5550	5572	5597	5617	5652	5683	5719	5740	5744	5620
1997	5765	5769	5759	5799	5837	5860	5863	5854	5851	5848	5838	5858	5826
1998	5852	5874	5875	5883	5881	5895	5921	5929	5963	5986	5995	5991	5920
1999	6000	5992	5986	6008	6006	6039	6076	6091	6128	6134	6127	6127	6059
2000	6130	6160	6202	6201	6233	6238	6225	6233	6224	6259	6266	6283	6221
2001	6281	6272	6279	6286	6288	6318	6404	6389	6391	6397	6410	6390	6334
2002	6462	6462	6502	6480	6512	6532	6605	6592	6589	6579	6578	6563	6538
2003	6581	6640	6627	6635	6642	6694	6696	6733	6741	6771	6794	6782	6695
2004	6825	6861	6957	7017	7064	7109	7126	7188	7298	7314	7312	7308	7115
2005	7297	7298	7309										

BASE: 1913=100.

(b)

Figure 2.4 *Engineering News Record* Construction Cost Indexes (a) Building Cost Index (b) Construction Cost Index (Reprinted from Engineering News-Record, copyright McGraw-Hill Companies Inc., 21 March 2005, All rights reserved.)

or schematic indication of the work to be accomplished. The specifications are a verbal or word description of what is to he constructed and to what levels of quality. When completed, they are included as legally binding elements of the contract. The production of the plans and specifications usually proceeds in two steps. The first step is called *preliminary design* and offers the owner a pause in which to review the plan before detail design commences. A common time for this review to take place is at 40% completion of the total design. The preliminary design extends the concept documentation. In most projects, a design team leader concept is utilized. The design team leader coordinates the efforts of architects and engineers from differing disciplines. The disciplines normally identified are architectural, civil and structural, mechanical, and electrical. The architect or architectural engineer, for instance, is responsible for the development of floor plans and general layout drawings as well as considerations such as building cladding, exterior effects, and interior finish. The mechanical engineer is concerned with the heating, ventilating, and air conditioning (HVAC), as well as service water systems. At preliminary design, decisions regarding size and location of air conditioning and heating units as well as primary water distribution components (e.g., pumps) are made. Similar decisions regarding the electrical system are made at this point by the electrical engineers. The structural and civil engineers develop the preliminary design of the structural frame and the subsurface foundation support. All of these designs are interlinked. The architectural layout impacts the weight support characteristics of the floor structure and, hence, the selection of structural system. The structural superstructure influences the way in which the foundation of the structure can be handled. The floor plan also determines the positioning of pipes and ducts and the space available for service mains.

Once the preliminary design has been approved by the owner, final or *detail design* is accomplished. This is the second step in the production of the plans and specifications. For the architectural engineer this focuses on the interior finishes, which include walls, floors, ceilings, and glazing. Details required to install special finish items are designed. Precise locations and layout of electrical and mechanical systems as well as the detail design of structural members and connections are accomplished by the appropriate engineers. As noted, the detail design phase culminates in the plans and specifications that are given to the constructor for bidding purposes. In addition to these detailed design documents, the architect/engineer produces a final "owner's" estimate indicating the total job cost minus markup. This estimate should achieve approximately $\pm 3\%$ accuracy, since the total design is now available. The owner's estimate is used (1) to ensure the design produced is within the owner's financial resources to construct (i.e., the architect/engineer has not designed a gold-plated project), and (2) to establish a reference point in evaluating the bids submitted by the competing contractors. In some cases, when all contractor bids greatly exceed the owner's estimate, all bids are rejected and the project is withdrawn for redesign or reconsideration. Once detailed design is completed, the owner again approves the design prior to advertising the project to prospective bidders.

2.6   NOTICE TO BIDDERS

The document announcing to prospective bidders that design documents are available for consideration and that the owner is ready to receive bids is called the *notice to bidders*. Because of his commitment to the owner to design a facility that can be constructed within a given budget and at an acceptable level of quality, the architect/engineer wants to be sure that the lowest bid price is achieved. To ensure this, the job is advertised to those contractors who are capable of completing the work at a reasonable price. All A/E firms maintain mailing lists that contain qualified bidders. When design is complete, a notice to bidders, such as the one shown in Figure 2.5, is sent to all prospective bidders. The notice to bidders contains information regarding the general type and size of the project, the availability of plans and specifications for review, and the time, place, and date of the bid opening. Normally, sets of plans and specifications are available for perusal at the A/E office as well as at *plan rooms*, which are conveniently located and can be visited by contractors in large cities. These facilities have copies of plans and specifications for a large number of jobs being let for bid. They afford contractors the opportunity to go to a central location and look at several jobs without having to drive to the office locations of each architect/engineer. The expenditure on the part of the contractors in going to the A/E office or plans rooms to look at the contract documents amounts only to the price of gas and a small amount of time. If they should decide to bid on a particular job, their commitment increases sharply in terms of money and time invested.

In addition to the mailings made available by the A/E firm as the owner's representative, contractors have other methods of learning about jobs that are available for bid. In some large cities, a builder's exchange may operate to serve the contracting community and keep it appraised of the status of design and bid activity within a given area. In addition to operating plans rooms, these exchanges often publish newsletters such as the one shown in Figure 2.6. These reports indicate what jobs are available for bidding, and architect/engineers make use of such facilities to gain maximum coverage in advertising their jobs. In addition to the basic information describing the job and the time and place of bid opening, these reports include, following each announcement, a statement that the plans are on file, along with the bin location.

Nationwide services such as the Dodge Reporting System are web-based and provide information on projects being let for bid. For a subscription fee such services provide information regarding jobs sorted by type of construction, geographical location, job size, and other parameters directly to the contractor. The information announcements indicate whether the job is under design, ready for bid, or awarded. In the cases of jobs that have been awarded, the low bid and other bid prices submitted are furnished so that the contractor can detect bidding trends in the market. A typical Dodge Reporting System web-based announcement is shown in Figure 2.7.

NOTICE TO BIDDERS
FOR
CONSTRUCTING SEWERAGE SYSTEM IMPROVEMENTS
CONTRACT "B"
CENTRAL STATE HOSPITAL
FOR THE
GEORGIA BUILDING AUTHORITY (HOSPITAL)
STATE CAPITOL—ATLANTA, GEORGIA

Sealed proposals will be received for Constructing Sewerage System Improvements, Contract "B," for the Georgia Building Authority (Hospital), State Capitol, Atlanta, Georgia, at Room 315, State Health Building, 47 Trinity Avenue, S.W., Atlanta, Georgia, until 2:00 P.M., E.S.T., February, 18 __, at which time and place they will be publicly opened and read. Bidding information on equipment in Section No. 10 shall be submitted on or before February 4 __.

Work to Be Done: The work to be done consists of furnishing all materials, equipment, and labor and constructing:

Division One. Approximately 12,400 L.F. 36″ Sewer Pipe, 5,650 L.F. 30″ Sewer Pipe, 7,300 L.F. 24″ Sewer Pipe, 1,160 L.F. 15″ Sewer Pipe, 3,170 L.F. 12″ Sewer Pipe, 300 L.F. 8″ Sewer Pipe, 418 L.F. 36″ C.I. Pipe Sewer, 324 L.F. 30″ C.I. Pipe Sewer, 1,150 L.F. 30″ C.I. Force Main, 333 L.F. 24″ C.I. Force Main, 686 L.F. 24″ C.I. Pipe Sewer, and all other appurtenances for sewers.
Division Two. One Sewage Pumping Station—"Main Pump Station."
Division Three. One Sewage Pumping Station—"Fishing Creek Pump Station."
Division Four. One Sewage Pumping Station—"Camp Creek Pump Station."

Bids may be made on any or all Divisions, any of which may be awarded individually or in any combination.

Figure 2.5 Notice to bidders (courtesy of Georgia Building Authority).

Proposals. Proposals shall contain prices, in words and figures, for the work bid on. All Proposals must be accompanied by a certified check, or a bid bond of a reputable bonding company authorized to do business in the State of Georgia, in an amount equal to at least five (5%) percent of the total amount of the bid.

Upon the proper execution of the contract and required bonds, the checks or bid bonds of all bidders will be returned to them.

If Proposals are submitted via mail rather than delivery they should be addressed to Mr. Smith, Director, Department of Administration and Finance, Georgia Department of Public Health, Room 519, State Health Building, 47 Trinity Avenue, S.W., Atlanta, Georgia 30334.

Performance and Payment Bonds: A contract performance bond and payment bond, each in an amount equal to one hundred (100%) percent of the contract amount, will be required of the successful bidder.

Withdrawal of Bids: No submitted bid may be withdrawn for a period of sixty (60) days after the scheduled closing time for the receipt of bids.

Plans, Specifications, and Contract Documents: Plans, Specifications and Contract Documents are open to inspection at the Office of the Georgia Building Authority (Hospital), State Capitol, Atlanta, Georgia, or may be obtained from Wiedeman and Singleton, Engineers, P.O. Box 1878, Atlanta, Georgia 30301, upon deposit of the following amounts.

Division One: $45.00 for Plans and Specifications.
Divisions Two, Three and Four (*Combined*). $50.00 for Plans and Specifications.
Divisions One to Four, Inclusive. $75.00 for Plans and Specifications.
All Divisions. $20.00 for Specifications only.

Upon the return of all documents in undamaged condition within thirty (30) days after the date of opening of bids, one-half of the deposit will be refunded. No refunds will be made for plans and documents after thirty (30) days.

Wage Schedule: The schedule of minimum hourly rates of wages required to be paid to the various laborers and mechanics employed directly upon the site of the work embraced by the Plans and Specifications as determined by the Secretary of the U.S. Department of Labor, Decision No. AI-971, is included in the General Conditions of the Specifications. This decision, expiring prior to the receipt of bids, will be superceded by a new decision to be incorporated in the Contract before the award is made to the successful bidder.

Acceptance or Rejection of Bids: The right is reserved to accept or reject any or all bids and to waive informalities.

THIS PROJECT WILL BE FINANCED IN PART BY A GRANT FROM THE FEDERAL WATER POLLUTION CONTROL ADMINISTRATION AND WILL BE REFERRED TO AS PROJECT WPC-GA-157.

BIDDERS ON THIS WORK WILL BE REQUIRED TO COMPLY WITH THE PRESIDENT'S EXECUTIVE ORDERS NO. 11246 and NO. 11375. THE REQUIREMENTS FOR BIDDERS AND CONTRACTORS UNDER THESE ORDERS ARE EXPLAINED IN THE SPECIFICATIONS.

GEORGIA BUILDING AUTHORITY (HOSPITAL)

By: _____
Secretary–Treasurer

Figure 2.5

bids

Replace Roof, Building 3245 Ft Benning, GA

2 PM August 20

REPLACE ROOF, BLDG 3245 (B-0037), FT BENNING, GA (CHATTAHOOCHEE CO)

Submit Bids To	Directorate of Contracting, Bldg 6 (Meloy Hall), Mailroom 207, ATZB-KTD, Ft Benning GA 31905-5000 (Isaac D Larry, contact) 706/545-2193 2 PM August 20
Plans Available From	Owner
Bond Info	20% bid bond. 50% pymt bond. 100% perf bond.
Div 2	demol
Div 5	misc mtl
Div 6	rough carp
Div 7	roof insul, b/u rfg, sht mtl wk, caulking & sealants
Div 9	ptg
PLANS ON FILE	ATL Bin 56 - #107293

bids

Family Life Center for First Assembly of God Griffin, GA

2 PM August 22

FAMILY LIFE CENTER FOR FIRST ASSEMBLY OF GOD (2000 W McIntosh Rd) GRIFFIN, GA (SPALDING CO)

Submit Bids To	First Assembly of God, Griffin, GA 2 PM August 22
Archt	Rardin & Carroll Architects, 6105 Preservation Drive, Ste A, Chattanooga, TN 37416 423/(894-2839) 894-3242
Civil Engr	Breelove Land Planners, Atlanta, GA
Struct Engr	Robinson & Associates, Atlanta, GA
Mech/Elect/Plbg Engr	Brewer & Skala Engineers, Atlanta, GA
Plans Available From	Archt $100 plan dep to GC's, one set ref to bona fide GC's within 10 days, all others non-ref cost of reprod.
Bond Info	5% bid bond. 100% pymt/perf bond.
Est Cost	$1,400,000
Scope	Approx 26,000 s.f. 1 story classrooms, offices & gym
Div 2	sel demol, site clrg & prep, earthwk, termite control
Div 3	cast-in-place conc, cementitious wood fiber decking
Div 4	unit masonry, precast conc window stool
Div 5	struct stl, open web joists, composite roof deck assemb, light gauge stl frmg, mtl handrails
Div 6	rough carp, struct glued laminated timber, finish carp & millwk
Div 7	waterprfg, dampprfg, bldg insul, masonry insul, EIFS, roof insul over wood deck, firestpg, fiberglass shingles, vinyl siding, rubber membrane rfg, flash & s/m, int caulking, ext sealant
Div 8	hol mtl wk, alum doors & frames, wood doors, counter doors, alum windows, glass & glzg
Div 9	gyp wallbd, acoust ceilings, poured in place synth sports flrg, resil flrg, cpt (owner provided), ptg, vinyl-coated fabric
Div 10	laminated plastic toilet partns, solid phenolic core partns, toilet & bath acces, fire extin & cabinets
Div 11	athletic equip
Div 15	hvac, ductwk & acces, computerized damper (VVT) sys, louvers-grilles-registers & diffusers, unitary exhaust & supply fans & vents, split sys a/c, refrigerant piping, pkg rooftop htg & vent units, auto controls, high efficiecy gas-fired duct heater, kitchen vent equip, test & balance, plbg, natl gas piping sys, fire prot, roof curbs
Div 16	elect, lighting, fire alarm sys
PLANS ON FILE	ATL Bin 38 - #107182

bids

Misc Elect, Bldg 2752 Ft Benning, GA

2 PM August 22

MISCELLANEOUS ELECTRICAL REPAIRS, BLDG 2752 (B-0051), FT BENNING, GA (CHATTAHOOCHEE CO)

Submit Bids To	Directorate of Contracting, Bldg 6 (Meloy Hall), Mailroom 207, ATZB-KTD, Ft Benning GA 31905-5000 (Sabra A Boynton, contact) 706/545-2221 2 PM August 22
Plans Available From	Owner
Bond Info	20% bid bond. 50% pymt bond. 100% perf bond.
Div 2	demol
Div 16	int elect wk
PLANS ON FILE	ATL Bin 55 - #107292

Figure 2.6 Typical Daily Building Report.

2.7 BID PACKAGE

The documents that are available to the contractor and on which he must make a decision to bid or not to bid are those in the *bid package*. In addition to the plans and technical specifications, the bid package prepared by the A/E consists of a *proposal* form, *general conditions* that cover procedures common to all construction contracts, and *special conditions*, which pertain to procedures to be used that are unique to this particular project. All supporting

Welcome!
Please select the reports that you would like to view or purchase, or you may redefine your
search criteria by using the advanced search option to the left!
(**Please remember to print your viewed reports.**)

Indiana Projects
(In the bidding stages only)
    308 #General Building
    262 #Engineering
     81 #Utility
     13 #Projects Residential
     78 #Bidding in 7 Days
    983 $(mil)Bidding in 7 Days

- **Bidders List Now Available.**
- Unlimited searching.
- Find the jobs you want to bid on.
- Pay only for what you want. Not what you don't.

195 projects were retrieved

D.R. #	Last Update	Bid Date	Project Type	Valuation	State	County	Sub Projects
2004-0083-2502	03/22/05 11:03 AM	03/31/05	Primary School	500,000	IN	Boone	0
2005-0066-9864	03/22/05 11:03 AM	04/07/05	Warehouse (Refrigerated)	1,500,000	IN	Spencer	0
2005-0065-5840	03/22/05 09:03 AM	03/24/05	Food-Beverage Service	90,000	IN	Lake	0
2004-0078-6586	03/22/05 08:03 AM	03/30/05	Supermarket-Convenience Store	700,000	IN	La Porte	10
2005-0066-5962	03/22/05 08:03 AM	04/07/05	College-University	95,000	IN	Lake	0
2005-0063-6114	03/22/05 08:03 AM	03/29/05	Office	1,500,000	IN	Tippecanoe	0
2004-0088-5958	03/22/05 04:03 PM	03/18/05	Retail (Other)	5,000,000	IN	Porter	0
2005-0066-8774	03/22/05 04:03 PM	04/05/05	Primary School	300,000	IN	Porter	0
2005-0060-8359	03/22/05 04:03 PM	03/30/05	Testing-Research-Development Lab	4,000,000	IN	St Joseph	0
2001-0077-8164	03/22/05 03:03 PM	03/24/05	Museum	1,000,000	IN	Jefferson	0

Home   View Next 10

Figure 2.7 Typical Web-Based Dodge Reporting System announcement (copyright © 2005
McGraw-Hill Companies, Inc.).

documents are included by reference in the proposal form. The bid package layout is shown
schematically in Figure 2.8.

The proposal form as designed and laid out by the A/E is the document that, when
completed and submitted by the contractor, indicates the contractor's desire to perform the
work and the price at which he will construct the project. A typical example of a proposal
is shown in Figure 2.9.

The proposal form establishes intent on the part of the contractor to enter into a contract
to complete the work specified at the cost indicated in the proposal. It is an offer and by itself
is not a formal contract. If, however, the owner responds by awarding the contract based on
the proposal, an acceptance of the offer results and a contractual relationship is established.
The prices at which the work will be constructed can be stated either as lump-sum or as

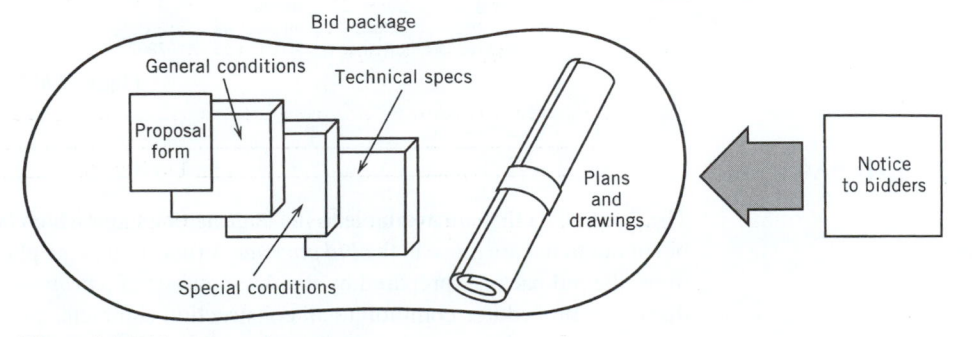

Figure 2.8 Bid package documents.

PROPOSAL
TO THE GEORGIA BUILDING AUTHORITY (HOSPITAL)
STATE CAPITOL
ATLANTA, GEORGIA

Submitted: _____(date)_____, 2XXX

The undersigned, as Bidder, hereby declares that the only person or persons interested in the Proposal as principal or principals is or are named herein and that no other person than herein mentioned has any interest in this Proposal or in the Contract to be entered into; that this Proposal is made without connection with any other person, company, or parties making a bid or Proposal; and that it is in all respects fair and in good faith without collusion or fraud.

The Bidder further declares that he has examined the site of the work and informed himself fully in regard to all conditions pertaining to the place where the work is to be done; that he has examined the plans and specifications for the work and contractual documents relative thereto, and has read all Special Provisions and General Conditions furnished prior to the opening of bids; and that he has satisfied himself relative to the work to be performed.

The Bidder proposes and agrees, if this Proposal is accepted, to contract with the Georgia Building Authority (Hospital), Atlanta, Georgia, in the form of contract specified, to furnish all necessary material, equipment, machinery, tools, apparatus, means of transportation, and labor, and to finish the construction of the work in complete accordance with the shown, noted, described, and reasonable intended requirements of the plans and specifications and contract documents to the full and entire satisfaction of the Authority with a definite understanding that no money will be allowed for extra work except as set forth in the attached General Conditions and Contract Documents, for the following prices:

CAMP CREEK PUMP STATION

Section 1: Unit Price Work

(For part payment—except rock excavation—by unit prices, to establish price for variation in quantities. Include balance of quantities for these items—except rock excavation—in lump sum bid for Section 2.)

Item Number	Quantity	Unit	Description	Unit Price	Total Amount
1.	550	cubic yard (cu yd)	Rock excavation (for structures and pipes only)	$_____	$_____
2.	50	linear foot (lin ft)	8″ C.I. force main	$_____	$_____
3.	20	cubic yard (cu yd)	Trench excavation for pipes	$_____	$_____
4.	200	square yard (sq yd)	Paving	$_____	$_____
Subtotal, Section 1, Item Nos. 1 to 4, Inclusive _____					

_____ Dollars ($_____)

Figure 2.9 A typical proposal (courtesy of Georgia Building Authority).

Section 2: Lump Sum Work

Item No.	Description		Total Amount
5.	Excavation and Fill		
	(a) Access Roadway	$	
	(b) Structure Excavation and Backfill	$	
	(c) Finish Grading	$	
	Total for Item No. 5		$
6.	Paving		
	(a) Access Roadway	$	
	(b) Station Area	$	
	Total for Item No. 6		$
7.	Concrete Work		$

The Bidder further proposes and agrees hereby to commence work under his contract, with adequate force and equipment, on a date to be specified in a written order of the Engineer, and shall fully complete all work thereunder within the time stipulated, from and including said date, in 300 consecutive calendar days.

The Bidder further declares that he understands that the quantities shown in the Proposal are subject to adjustment by either increase or decrease, and that should the quantities of any of the items of work be increased, the Bidder proposes to do the additional work at the unit prices stated herein; and should the quantities be decreased, the Bidder also understands that payment will be made on the basis of actual quantities at the unit price bid and will make no claim for anticipated profits for any decrease in quantities, and that actual quantities will be determined upon completion of the work, at which time adjustment will be made to the Contract amount by direct increase or decrease.

The Bidder further agrees that, in case of failure on his part to execute the Construction Agreement and the Bonds within ten (10) consecutive calendar days after written notice being given of the award of the Contract, the check or bid bond accompanying this bid, and the monies payable thereon, shall be paid into the funds of the Georgia Building Authority (Hospital), Atlanta, Georgia, as liquidated damages for such failure, otherwise the check or bid bond accompanying his Proposal shall be returned to the undersigned.

Attached hereto is a bid bond by the _____

_____ in the amount of

_____

_____ Dollars ($_____)

made payable to the Georgia Building Authority (Hospital), Atlanta, Georgia, in accordance with the conditions of the Advertisement for Bids and the provisions herein.

Submitted: _____ L.S.

By: _____ L.S.

Title: _____

Note: If the Bidder is a corporation, the Proposal shall be signed by an officer of the corporation; if a partnership, it shall be signed by a partner. If signed by others, authority for signature shall be attached.

ADDRESS: _____

_____

Figure 2.9

unit-price figures. Only a portion of the price schedule (see Fig. 2.9) is shown in this example proposal. As shown in the figure, two methods (lump sum and unit price) of quoting price are illustrated. Items 1 to 4 require the bidder to specify unit price (i.e., dollar per unit) for the guide quantities specified. Therefore, if the contractor will do the rock excavation for $80.00 per cubic yard, this price is entered along with the total price ($550 \times \$80.00 = \$44,000$). Items 5, 6, and 7 require lump- or stipulated-sum quotations. Therefore, the contractor states a single price for the access road, finish grading, and so on.

In the proposal form shown in Figure 2.9, the contract duration is also specified, although this is not always the case. In many instances, the project duration in working or calendar days is specified in the *special conditions* portion of the bid package. The proposal form indicates also that the contractor is to begin work within 10 calendar days after receipt of written notice of award of contract. Award of contract is usually communicated to the contractor in the form of a *notice to proceed*. Response by the owner to the contractor's *proposal* with a notice to proceed establishes a legally binding contractual relationship. Legal signatures (L.S.) by individuals empowered to represent (i.e., commit contractually) the firm making the proposal must be affixed to the proposal.

2.8   GENERAL CONDITIONS

Certain stipulations regarding how a contract is to be administered and the relationships between the parties involved are often the same for all contracts. An organization that enters into a large number of contracts each year normally evolves a standard set of stipulations that establishes these procedures and applies them to all construction contracts. This set of provisions is normally referred to as the *general conditions*. Large government contracting organizations such as the U.S. Army Corps of Engineers, the Bureau of Reclamation, and the General Services Administration (Public Building Service) have a standard set of *general provisions*. For those organizations that enter into construction contracts on a less frequent basis, professional and trade organizations publish standards that are commonly used in the industry. A committee for engineer documents has been formed jointly by the American Consulting Engineers Council, the National Society of Professional Engineers, and the American Society of Civil Engineers to prepare standard contract documents. The committee is called the Engineers Joint Contract Documents Committee (EJCDC). These documents have been endorsed by the Associated General Contractors (AGC) of America and the Construction Specifications Institute (CSI). Some topics that are typically considered in the general conditions are shown in Table 2.1. The rights, privileges, and responsibilities

Table 2.1   Topics Typically Addressed in General Conditions

1. Definitions
2. Preliminary matters
3. Contract documents
4. Bonds and insurance
5. Contractor's responsibilities
6. Owner's responsibilities
7. Engineer's responsibilities
8. Changes in the work
9. Change of contract price
10. Change of contract times
11. Tests and inspections
12. Payments to contractor and completion
13. Suspension of work and termination
14. Dispute resolution

that accrue to the primary contractual parties in any construction contract are also defined in the general conditions. Therefore, sections pertaining to the (1) owner, (2) architect (or architect/ engineer), (3) contractor, and (4) subcontractors are typically found in the general conditions. Most contractors become thoroughly familiar with the standard forms of general conditions (i.e., Corps of Engineers, etc.) and can immediately pick up any additions or changes. Each of the provisions of a standard set of general conditions has legal implications, and the wording cannot be changed without careful consideration. The contract language embodied in the general conditions has been hammered out over the years from countless test cases and precedents in both claims and civil courts. The wording has evolved to establish a fair and equitable balance of protection for all parties concerned. In cases where a contractor finds considerable deviation from the standard language, he may decline the opportunity to bid fearing costs of litigation in clarifying contractual problems. In areas where small deviation is possible, the language of a given standard form may tend to be protective of one party (e.g., the architect) and to hold others responsible when gray areas arise. Predictably, the AGC standard subcontract protects the contractor in areas in which responsibility is unclear or subject to interpretation.

2.9   SUPPLEMENTARY CONDITIONS

Those aspects of the contractual relationship that are peculiar or unique to a given project are given in the *supplementary conditions*. Items such as the duration of the project, additional instructions regarding commencement of work, owner-procured materials, mandatory wage rates characteristic of the local area, format required for project progress reporting (e.g., a network schedule), and amount of liquidated damages are typical of the provisions included in the supplementary conditions. Items contained in supplementary conditions are of two types:

1. Modifications to the basic articles of the general condition in the form of additions, deletions, or substitutions
2. Additional articles of a contractual-legal nature that may be desirable or necessary for a particular project

Since some of the provisions are extensions or interpretations of the general conditions, some of the major paragraph titles are similar to those used in the general conditions. The contents of a typical set of supplementary or special conditions for a Corps of Engineers channel improvement project are shown in Figure 2.10.

2.10   TECHNICAL SPECIFICATIONS

The contract documents must convey the requirements of the project to potential bidders and establish a legally precise picture of the technical aspects of the work to be performed. This is accomplished visually through the use of drawings. A verbal description of the technical requirements is established in the *technical specifications*. These provisions pertain in large part to the establishment of quality levels. Standards of workmanship and material standards are defined in the specifications. For materials and equipment, this is often done by citing a specific brand name and model number as the desired item for installation. In government procurement, where competitive procurement must take place, a similar approach is utilized. Government specifications usually cite a specific brand or model and then establish the requirement that this or an equal item be used. The fact that equality exists must be established by the bidder.

Often the quality required will be established by reference to an accepted practice or quality specification. The American Concrete Institute (ACI), the American Welding

PART II
SPECIAL CONDITIONS
INDEX

Paragraph No.	Title	Page No.
SC-1.	Commencement, Prosection, and Completion of Work	SC-1
SC-2.	Liquidated Damages	SC-1
SC-3.	Contract Drawings, Maps, and Specifications	SC-1
SC-4.	Construction Drawings	SC-4
SC-5.	Physical Data	SC-5
SC-6.	Rates of Wages	SC-6
SC-7.	Variations in Estimated Quantities	SC-6
SC-8.	Government-Furnished Property	SC-7
SC-9.	Water	SC-7
SC-10.	Electricity	SC-7
SC-11.	Layout of Work and Surveys	SC-7
SC-12.	Payments for Mobilization and Preparatory Work	SC-8
SC-13.	Damage to Work	SC-8
SC-14.	Funds Available for Payments	SC-9
SC-15.	Additional Supervision of Subcontracted Work	SC-11
SC-16.	Scheduling and Determination of Progress	SC-12
SC-17.	Performance of Work by Contractor	SC-12
SC-18.	Certificates of Compliance	SC-12
SC-19.	Plant Lay-out Drawings	SC-12
SC-20.	Approved Aggregate Sources	SC-13
SC-21.	Testing	SC-14
SC-22.	Work Areas	SC-14
SC-23.	Work under Other Contracts	SC-14
SC-24.	Permits	SC-14
SC-25.	Products and Parts of Standard Manufacture	SC-14
SC-26.	Protective Headgear	SC-15
SC-27.	Inspection and Testing of Construction Equipment	SC-15
SC-28.	Work to Be Done by Other Agencies	SC-15
SC-29.	Protection of Constructed Facilities	SC-16
SC-30.	Protection of Utilities	SC-16
SC-31.	Use of Local Roads and Streets	SC-16
SC-32.	Maintenance of Street Traffic	SC-16
SC-33.	Requirements of Pennsylvania Railroad Company and	SC-16
	Westinghouse Electric Corp. Pertaining to Construction Work Within the Limits of Railroad Right-of-Way and Westinghouse Electric Corp. Property	SC-17
SC-34.	Cofferdams and Flood Stages	SC-19
SC-35.	Watchmen and Danger Signs	SC-19
SC-36.	Sequence of Operations	SC-20
SC-37.	Acceptance of Work	SC-21
SC-38.	Insurance Policies to be Furnished to the Government	SC-22
SC-39.	Payment	SC-22

Figure 2.10 Special conditions: typical index of special conditions (courtesy of the Army Corps of Engineers).

Society (AWS), the American Association of State Highway and Transportation Officials (AASHTO), the American Society for Testing and Materials (ASTM), as well as federal procurement agencies publish recognized specifications and guides. A list of some typical references is given in Figure 2.11. The organization of the technical specifications section usually follows the sequence of construction. Therefore, specifications regarding concrete placement precede those pertaining to mechanical installation. A typical index of specifications for a heavy construction project might appear as follows:

Section

1	Clearing and grubbing
2	Removal of existing structures
3	Excavation and fill
4	Sheet steel piling
5	Stone protection
6	Concrete
7	Miscellaneous items of work
8	Metal work fabrication
9	Water supply facilities
10	Painting
11	Seeding

As with the general conditions, most contractors are familiar with the appearance and provisions of typical technical specifications. A contractor can quickly review the specifications to determine whether there appears to be any extraordinary or nonstandard aspects that will have an impact on cost. These clauses or nonstandard provisions are underlined or highlighted to be studied carefully.

2.11 ADDENDA

The bid package documents represent a description of the project to be constructed. They also spell out the responsibilities of the various parties to the contract and the manner in which the contract will be administered. These documents establish the basis for determining the bid price and influence the willingness of the prospective bidder to bid or enter into a contract. It is important, therefore, that the bid package documents accurately reflect the project to be constructed and the contract administration intentions of this owner or of the owner's representative. Any changes in detail, additions, corrections, and contract conditions that arise before bids are opened that are intended to become part of the bid package and the basis for bidding are incorporated into the bid package through *addenda*.

An *addendum* thus becomes part of the contract documents and provides the vehicle for the owner (or the owner's representative) to modify the scope and detail of a contract before it is finalized. It is important therefore that addenda details be rapidly communicated to all potential bidders prior to bid submission. Since addenda serve notice on the prospective bidder of changes in the scope or interpretation of the proposed contract, steps must be taken to ensure that all bidders have received all issued addenda. Consequently, addenda delivery is either documented through certified mail receipts or confirmed on bid submission through the bidder's submission of a signed document listing receipt of each duly identified addendum.

Once a contract has been signed, future changes in the scope or details of a contract may form the basis for a new financial relationship between contracting parties. The original

American Concrete Institute

ACI 211.1-91 Standard Practice for Selecting Proportions for Normal, Heavyweight, and Mass Concrete

ACI 211.2-98 Standard Practice for Selecting Proportions for Structural Lightweight Concrete

ACI 211.4R-93 Guide for Selecting Proportions for High-Strength Concrete with Portland Cement and Fly Ash

ACI 214-77 Recommended Practice for Evaluation of Strength Test Results of Concrete

ACI 304R-00 Guide for Measuring, Mixing, Transporting, and Placing Concrete

ACI 305R-99 Hot Weather Concreting

ACI 306R-88 Cold Weather Concreting

ACI SP-2-99 ACI Manual of Concrete Inspection

ACI 318-02/318R-02 Building Code Requirements for Structural Concrete and Commentary

ACI 421.1R-99 Shear Reinforcement for Slabs

ACI 550R-96 Design Recommendations for Precast Concrete Structures

American Society for Testing and Materials

ASTM A36/A36M-04 Standard Specification for Carbon Structural Steel

ASTM A82-02 Standard Specification for Steel Wire, Plain, for Concrete Reinforcement

ASTM A185-02 Standard Specification for Steel Welded Wire Reinforcement, Plain, for Concrete

ASTM A496-02 Standard Specification for Steel Wire, Deformed, for Concrete Reinforcement

ASTM A996/A996M-04 Standard Specification for Rail-Steel and Axle-Steel Deformed Bars for Concrete Reinforcement

ASTM A706/706M-04a Standard Specification for Low-Alloy Steel Deformed and Plain Bars for Concrete Reinforcement

ASTM C33-03 Standard Specification for Concrete Aggregates

ASTM C39/C39M-03 Standard Test Method for Compressive Strength of Cylindrical Concrete Specimens

ASTM C42/C42M-03 Standard Test Method for Obtaining and Testing Drilled Cores and Sawed Beams of Concrete

ASTM C55-03 Standard Specification for Concrete Brick

ASTM C90-03 Standard Specification for Loadbearing Concrete Masonry Units

ASTM C94/C94M-04 Standard Specification for Ready-Mixed Concrete

ASTM C109/C109M-02 Standard Test Method for Compressive Strength of Hydraulic Cement Mortars (Using 2-in. or [50-mm] Cube Specimens)

ASTM C140-03 Standard Methods for Sampling and Testing Concrete Masonry Units and Related Units

ASTM C270-03b Standard Specification for Mortar for Unit Masonry

ASTM D2000-03a Standard Classification System for Rubber Products in Automotive Applications

American Welding Society

AWS D1.4-98 Structural Welding Code— Reinforcing Steel

American Society of Civil Engineers

ASCE 7-02 Minimum Design Loads for Buildings and Other Structures

Building Seismic Safety Council

NEHRP Recommended Provisions for Seismic Regulations for New Building and Other Structures (2003 Edition)
Part One: Provisions
Part Two: Commentary

National Ready Mix Concrete Association

Concrete Plant Standards of Concrete Plant Manufacturers Bureau (12th Revision, 2000)

Truck Mixer, Agitator and Front Discharge Concrete Carrier (15th Revision, 2001)

Concrete Reinforcing Steel Institute

Reinforcement Anchorages and Splices (4th Edition, 1997)

Figure 2.11 Typical references to structural inspection and testing standards.

contract, in such cases, can no longer be accepted as forming the basis for a full description of the project. Such changes are referred to as *change orders* (see Section 3.5).

2.12   DECISION TO BID

After investigating the plans and specifications at the architect's office or a plans room, the contractor must make a major decision-whether or not to bid the job. This is a major financial decision, since it implies incurring substantial cost that may not be recovered. Bidding the job requires a commitment of man-hours by the contractor for the development of the estimate.

Estimating is the process of looking into the future and trying to predict project costs and various resource requirements. The key to this entire process is the fact that these predictions are made based on past experiences and the ability of the estimator to sense potential trouble spots that will affect field costs. The accuracy of the result is a direct function of the skill of the estimator and the accuracy and suitability of the method by which these past experiences were recorded. Since the estimate is the basis for determining the bid price of a project, it is important that the estimate be carefully prepared. Studies reveal the fact that the most frequent causes of contractor failure are incorrect and unrealistic estimating and bidding practices.

The quantities of materials must be developed from the drawings by an expert in quantity takeoff. The process of determining the required material quantities on a job is referred to as quantity takeoff or quantity surveying. Once quantities are established, estimators who have access to pricing information use these quantities and their knowledge of construction methods and productivities to establish estimates of the *direct costs* of performing each construction task. They then add to the totaled project direct costs those *indirect costs* that cannot be assigned directly to a particular estimating item (e.g., project supervising costs). Finally, the bid price is established by adding the management and overhead costs, allowances for contingencies, and a suitable profit margin. Appendix A gives typical considerations affecting the decision to bid.

The cost of the time and effort expended to develop a total bid price and submit a proposal is only recovered in the event the contractor receives the contract. A common rule of thumb states that the contractor's estimating cost will be approximately 0.25% of the total bid price. This varies, of course, based on the complexity of the job. Based on this rule, an estimating cost of $25,000 can be anticipated for a job with a total bid cost in the vicinity of $10 million. Expending this amount of money to prepare and submit a bid with only a probability of being awarded the work is a major monetary decision. Therefore, most contractors consider it carefully. In order to recover bidding costs for jobs not awarded, contractors place a charge in all bids to cover bid preparation costs. This charge is based on their frequency of contract award. That is, if a contractor, on the average, is the selected bidder on one in four of the contracts bid, he will adjust the bid cost included in each proposal to recover costs for the three in four jobs not awarded. In addition to the direct costs of bid preparation, the contractor may be required by the architect to pay a deposit for or purchase each set of plans and specifications used. Other costs include telephone charges related to obtaining quotations from subcontractors and vendors and the administrative costs of getting these quotations in writing. A small fee must be paid for a *bid bond* (see Section 2.15) and the administrative aspects of submitting the proposal in conformance with the *instructions to bidders*.

2.13   PREQUALIFICATION

In some cases the complexity of the work dictates that the owner must be certain that the selected contractor is capable of performing the work described. Therefore, before considering a bid, the owner may decide to prequalify all bidders. This is announced in the instructions to bidders. Each contractor interested in preparing and submitting a bid is asked to submit documents that establish his firm's expertise and capability in accomplishing similar types of construction. In effect, the owner asks the firm to submit its "resume" for consideration. If the owner has doubts regarding the contractor's ability to successfully complete the work, the owner can simply withhold qualification.

This is helpful to both parties. The contractor does not prepare a bid and incur the inherent cost unless he can qualify. On the other hand, the owner does not find himself in the position of being under pressure to accept a low bid from a firm he feels cannot perform the work. In the extreme case, a small firm with experience only on single-family residential housing may bid low on a complex radar-tracking station. If the owner feels the contractor will not be able to successfully pursue the work, he can fail to prequalify the contractor.

2.14   SUBCONTRACTOR AND VENDOR QUOTATIONS/CONTRACTS

As already noted, estimating section personnel will establish costs directly for those items to be constructed by the prime contractor with in-house forces. For specialty areas such as electrical work, interior finish, and roofing, the prime contractor solicits quotations from subcontractors with whom he has successfully worked in the past. Material price quotations are also developed from vendors. These quotations are normally taken telephonically and included in the bid. It is good business practice to use a subcontractor/vendor bid quotation form that includes a legal signature as well as the bid quotation. The contractor soliciting the quotation should request that the bid be faxed with the quotation and a signature binding the bidder to the quoted price. Alternatively, signed documentation of the quotation can be submitted as an email attachment.

The contractor integrates these quotations into the total bid price. Until the contractor has a firm subcontract or purchase order, the only protection against the vendor or subcontractor reneging on or changing the quoted price is the price submitted by fax or signed email attachment quotation form. Prior to the availability of electronically transmitted methods of submission, the contractor had only a telephone quote and the subcontractor's word that he would sign a formal agreement at the stated price. The use of electronically transmitted documents and signed quotations has greatly improved this situation and eliminated potential misunderstandings.

Following award of the contract, the prime contractor has his purchasing or procurement group move immediately to establish subcontracts with the appropriate specialty firms. Both the Associated General Contractors (AGC) and the EJCDC have standard forms for this purpose.

2.15   BID BOND

Various defaults are possible in the relationship between owner and contractor. The concept of a bond allows one party to protect itself against default in a relationship with a second party. A third party, referred to as the surety, provides protection such that, if a default between two parties occurs that results in damage (e.g., loss of money or other value), the surety protects the damaged party. This protection is typically in the form of offsetting or covering the damage involved. A construction bond, therefore, involves a relationship between three parties-the principal or party who might default, the obligee or party who may be damaged or lose some advantage, and the surety who will offset the damage or loss of advantage. For this reason, a bond is referred to as a three-party instrument establishing a three-parry relationship (see Fig. 2.12).

A similar relationship exists if a person of limited means attempts to borrow money at the bank. If the bank is concerned about the ability of the borrower to pay back the borrowed

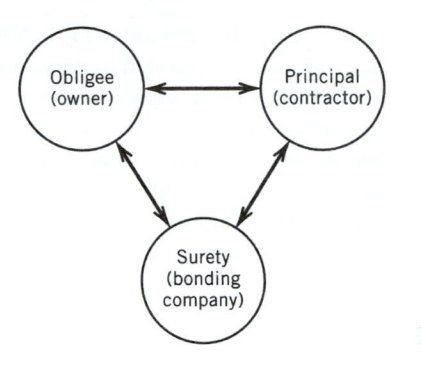

Figure 2.12 Bonding relationship (three-party).

money, it may require that a separate individual cosign the note or instrument of the loan. The cosigner is required to pay back the borrowed money in the event that the primary borrower defaults in repayment. In this situation, the bank is in the position of the obligee, the borrower is the principal, and the cosigner acts as the surety.

During the bidding process, the owner usually requires a *bid security*. The security is required to offset a "damage" occurring in the event that the firm selected fails to begin the project as directed. This may occur in the event that the selected bidder realizes that he has underbid the project and that pursuing the work will result in a financial loss.

In such cases, the owner would incur a damage, since he would be forced to contract with the next lowest bidder. If, for instance, a contractor bidding $3 million refuses to enter into contract, and the next low bid is $3,080,000, the owner is damaged in the amount of $80,000. A typical bid bond form is shown in Figure 2.13. Notice that the responsibility of the surety is indemnified (covered) by the principal. If the principal fails to enter into contract,

> the Principal shall pay to the Obligee the difference not to exceed the penalty hereof between the amount specified in said bid and such larger amount for which the Obligee may in good faith contract with another party to perform the work.

If the principal is unable to pay this amount, the surety must step in and cover the damage.

In most cases, the surety firm will not issue a bid bond unless it is very sure the assets of the principal will offset any default occurring due to failure to enter into a contract. Therefore, if issued at all, the bid bond is issued for a small administrative fee. From the bonding company's point of view, the importance of the bid bond is not the fee paid by the contractor for its issuance, but instead its implication that if the contract is awarded, the surety will issue performance and payment bonds. The bid bond is a "lead parachute," which pulls these two bonds out of the main pack. Typical performance and payment bonds are shown in Appendix B. The performance bond protects the owner against default on the contractor's part in completing the project in accordance with the contract documents. The payment bond protects the owner against failure on the part of the prime contractor to pay all subcontractors or vendors having outstanding charges against the project. If the surety fails to issue these bonds (required in the contract documents), the contractor is prevented from entering into contract and the surety could be forced to cover damages resulting from this default.

As an alternative to a bid bond, the owner will sometimes specify (in the notice to bidders or the proposal) acceptance of a cashier's check in a specified amount made out to the owner to secure the bid. If the contractor fails to enter into the contract, he forfeits this check and the owner can use it to defray the cost of entering into contract with the second lowest bidder at a higher bid price. This method of bid security is indicated in the notice to

2.15   Bid Bond   **43**

Commonwealth of Puerto Rico
Department of Transportation and Public Works
HIGHWAY AUTHORITY

BID BOND

KNOW ALL MEN BY THESE PRESENTS, that we
> (Here insert full name and address or legal
> title of Contractor)

as Principal, hereinafter called the Principal, and
> (Here insert full name and address or legal
> title of Surety)

a corporation duly organized under the laws of the State of
as Surety, hereinafter called the Surety, are held and firmly bound unto the Executive Director of the Puerto Rico Highway Authority as Obligee, hereinafter called the Obligee, in the sum of

Dollars ($),

for the payment of which sum well and truly to be made, the said Principal and the said Surety, bind ourselves, our heirs, executors, administrators, successors and assigns, jointly and severally, firmly by these presents.

WHEREAS, the Principal has submitted a bid for
> (Here insert full name, address and
> description of project)

NOW, THEREFORE, if the Obligee shall accept the bid of the Principal and the Principal shall enter into a Contract with the Obligee in accordance with the terms of such bid, and give such bond or bonds as may be specified in the bidding or Contract Documents with good and sufficient surety for the faithful performance of such Contract and for the prompt payment of labor and material furnished in the prosecution thereof, or in the event of the failure of the Principal to enter such Contract and give such bond or bonds, if the Principal shall pay to the Obligee the difference not to exceed the penalty hereof between the amount specified in said bid and such larger amount for which the Obligee may in good faith contract with another party to perform the Work covered by said bid, then this obligation shall be null and void, otherwise to remain in full force and effect.

Signed and sealed this   day of   19

	_____	_____
	(Principal)	(Seal)
_____		
(Witness)	_____	
	(Title)	
	_____	_____
	(Surety)	(Seal)
_____		
(Witness)	_____	
	(Title)	

Figure 2.13 Typical bid bond.

bidders in Figure 2.5. The notice to bidders states: "All Proposals must be accompanied by a *certified check*, or a bid bond of a reputable bonding company authorized to do business in the State of Georgia, in an amount equal to at least five (5%) percent of the total amount of the bid."

This procedure is further explained in the proposal form (Figure 2.9): "The bidder further agrees that, in case of failure on his part to execute the Construction Agreement and the Bonds[2] within ten (10) consecutive days after written notice being given of the award of the Contract, the check or bid bond accompanying this bid, and monies payable thereon, shall be paid into the funds of the {owner}."

All government construction contracts require a bid bond that is normally for 20% of the bid price. Private construction agencies for which bid bonds are required generally designate that the bid bond be for 5 or 10% of the bid price. For this reason, residential and commercial construction contractors are different from public construction contractors to a surety. A contractor failing to enter into contract after acceptance of his low bid in public construction places a surety at greater risk because of the larger bid bond for government contracts.

2.16 PERFORMANCE AND PAYMENTS BONDS

If the contractor is awarded the contract, performance and payment bonds are issued. A performance bond is issued to a contractor to guarantee the owner that the contract work will be completed and that it will comply with project specifications. In other words, a performance bond protects the owner against default on the part of the contractor in performing the project as required. If the contractor fails to perform the work as required, the surety must provide for completion of the project in compliance with the plans and specifications at the price originally quoted by the defaulting contractor.

A payment bond is issued to guarantee the owner protection against any liens or charges against the project that are unpaid as a result of the contractor's default. That is, if the contractor fails to pay outstanding liens and charges against the project occurring as a result of the construction work, the surety will pay these debts. If the contractor does not pay subcontracts or suppliers, the surety must protect the owner from their claims. Typical bonding forms for performance and payment bonds are shown in Appendix B.

Because of the potential cost and trouble involved in taking over the work of a contractor about to default, the surety may elect to negotiate short-term financing for a contractor who has current liquidity problems. The surety may grant loans directly or assist the contractor in getting additional loans for the construction. In the event of default, there is no surety payment until the contractor's funds are completely exhausted. Then the surety will normally rebid the job, at a cost to itself and delay to the owner. For these reasons, a surety will often seek to assist a contractor overcome temporary cash shortages.

Surety companies typically have a list of troubleshooters who have a proven record of quickly taking over projects that are in trouble and bringing them to successful completion. In certain cases, the surety will replace the defaulting contractor's management team with a team of troubleshooters and attempt to complete the job with the existing work force. In other cases, the surety may negotiate with a second contractor to complete the work at a fixed price that is acceptable under the circumstances. The owner's interest is that the loss of time and disruption occurring because of default is minimized.

2.17 COST AND REQUIREMENTS FOR BONDS

Performance and payment bonds are issued for a service charge. The common rate is 1% or $10 per $1000 on the first $200,000 of contract cost. At higher contract costs, the rate

---

[2] Performance and Payment bonds.

is reduced incrementally. Based on the size of the project and past performance of the contractor, rates fluctuate between 0.5% and 3%. Normally, the surety is not at any great risk, since the bond includes an indemnity agreement on the part of the contractor. In other words, the contracting corporation, partnership, or proprietorship must pledge to pay back any monies expended by the surety on its behalf. Since the contractor or principal in a construction bond is required to indemnify the surety against any loss as a result of the bond, the protection of personal wealth and assets of corporate stockholders typical of closely held corporations is not in force. Key personnel may be required to sign that they will back the bond with their personal wealth in the case of closely held corporations or limited partnerships.

The Miller Act (enacted in 1935) establishes the level of bonding required for federally funded projects. Performance bonds must cover 100% of the contractor amount while payment bonds are required based on a sliding scale as follows:

50% if the contract is $1,000,000 or less

40% if the contract is between $1,000,000 and $5,000,000

Fixed amount of $2,500,000 if the contract is greater than $5,000,000

A surety seeks to keep itself well informed of a contractor's progress on bonded projects and with the contractor's changing business and financial status. In order to help with this, the contractor makes periodic reports on the work in progress with particular emphasis on costs, payments, and disputes associated with uncompleted work. Based on these reports, the contractor's bonding capacity can be determined. This is calculated as a multiple of the net quick assets of the contractor as reflected in the company balance sheet. The net quick assets are the contractor's assets that can quickly be converted to cash or negotiable instruments to cover the cost of default. Such items as cash on hand, demand deposits, accounts receivable, and similar highly liquidable assets are available to the surety in case of a default.

The multiple to determine bonding capacity is based on the contractor's performance over the years. New contractors with no track record will have a low multiple such as 5 or 6. Old and reliable contractors may have a multiple of 40 or greater. Based on a multiple of 40, a reliable firm with *net quick assets* of $140,000 would have a bonding capacity of $5,600,000. In this case, the surety would provide bonding on work in progress (new jobs plus remaining value in jobs under way) up to the amount of $5,600,000. In other words, the firm will be able to pursue work for which bonds are required up to a total amount of $5,600,000.

Bonding companies have experienced larger than normal losses since the year 2000, and have raised rates and reduced the dollar amount available for writing new bonds. Many surety companies have raised premium rates by 10 to 30% since 2002 on performance and payment bonds. This has made it much more difficult for construction firms to find bonding. This has impacted operations and bid costs for firms working in markets which require bonds.

REVIEW QUESTIONS AND EXERCISES

2.1 What are the three major types of construction bonds? Why are they required? Name three items that affect bonding capacity.

2.2 In what major section of the contract is the time duration of the project normally specified?

2.3 Who are the three basic parties involved in any construction bonding arrangement?

2.4 What type of bond guarantees that if a contractor goes broke on a project the surety will pay the necessary amount to complete the job?

2.5 What is the purpose of the following documents in a construction contract?

    a. General conditions

    b. Special conditions

c. Addenda

d. Technical specifications

2.6 Why is the contractor normally required to submit a bid bond when making a proposal to an owner on a competitively bid contract?

2.7 What is the Miller Act and what does it specify regarding government contracts?

2.8 What is the purpose of the notice to bidders?

2.9 List the various specialty groups that are normally involved in the design of a high-rise building project.

2.10 How much money is the contractor investing in an advertised project available for bid at the time of:

a. Going to the architect/engineer's office to look at the plans and specifications?

b. Deciding to take the drawings to his home office for further consideration?

c. Deciding to make initial quantity take-off?

d. Full preparation of bid for submittal?

2.11 What are the major parameters to be considered in the prequalification assessment of a contractor? Investigate the local criteria used in the pre-qualification of both small housing and general contractors.

2.12 Obtain sample specification clauses relating to the quality of finish of an item such as face brick, exterior concrete, or paint surfaces. Who has the major responsibility for the definition, achievement in the field, and paid acceptance?

2.13 Read those clauses of the general conditions of the contract for construction that refer to the owner, architect, contractor, and subcontractor. Then list the major responsibilities of these agents with respect to the following:

a. The definition, or attention to, the scope of the project

b. The financial transactions on the project

c. The finished quality of the work

Chapter 3

Issues During Construction Phase

Project Rework Reduction

The Need

Rework, and particularly field rework, continues to be one of the major sources of unplanned cost growth on industrial construction projects. Rework occurs when the installed work does not comply with or meet required specifications. According to Construction Industry Institute research, if field rework alone can be significantly reduced, or even eliminated, as much as 10% of overall project costs can be saved. The savings are expected to be substantially greater across an entire project cycle, which includes engineering and procurement.

Field rework is not caused solely by construction site activities in isolation. In order for the field rework to be reduced, a substantial effort must be made to improve the effectiveness of the prior project phases with a view to preventing all to frequent 'catch-up' work during the site construction and commissioning phases. A number of computer-based tools have been developed to combat project-wide reworks.

3.1 ACCEPTANCE PERIOD/WITHDRAWAL

In formal competitive bid situations, the timing of various activities has legal implications. The issuance of the notice to bidders opens the bidding period. The date and time at which the bid opening is to take place mark the formal end of the bidding period. Usually a bid box is established at some central location. Bids that have not been received at the bid box by the appointed date and hour are late and are normally disqualified. Prior to the close of the bidding (i.e., bid opening), contractors are free to withdraw their bids without penalty. If they have noted a mistake, they can also submit a correction to their original bid. Once bid opening has commenced, these prerogatives are no longer available. If bids have been opened and the low bidder declares a mistake in bid, procedures are available to reconcile this problem. If it can be clearly established that a mathematical error has occurred, the owner usually will reject the bid. However, if the mistake appears contrived to establish a basis for withdrawal of the bid, the owner will not reject. Then the contractor must enter into contract or forfeit his bid security. The chronology of the bid procedure is shown in Figure 3.1.

The bid security protects the owner from failure by the contractor to enter into a formal construction agreement. The contractor is protected by the acceptance period. The notice to bidders specifies a period following bid opening during which the proposed bids are to remain in force. The indication is that, if the owner does not act in this period to accept one

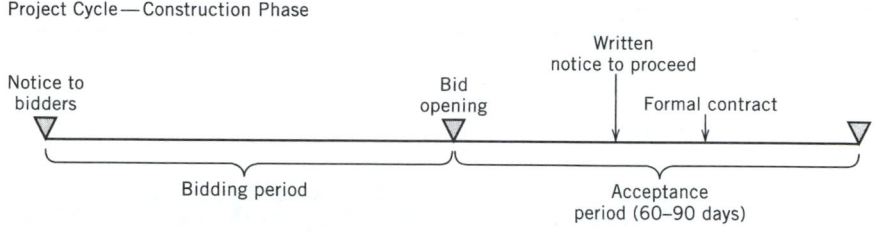

Figure 3.1 Chronology of bid procedure.

of the bids, then the contractors can withdraw or adjust their bids. This is indicated in the notice to bidders (Fig. 2.5) as follows:

Withdrawal of Bids   *No submitted bid may be withdrawn for a period of sixty (60) days after the scheduled closing time for the receipt of bid.*

This is designed to protect the bidder, since otherwise the owner could hold the contractors to their bids for an unspecified period. If the expected financing or appropriation for the project does not materialize, the owner could, in theory, say "Wait until next year, and I will enter into contract with you at this price." This, of course, would be potentially disadvantageous to the bidder. Therefore, the owner must send written notice of award (e.g., notice to proceed) to the selected contractor during the acceptance period or the bidders are released from their original proposals.

3.2   AWARD OF CONTRACT/NOTICE TO PROCEED

Notification of award of contract is normally accomplished by a letter indicating selection and directing the contractor to proceed with the work. This notice to proceed consummates the contractual relationship from a legal viewpoint despite the fact that a formal agreement has not been signed. The proposal (offer)–acceptance protocol of contractual law is satisfied by the issuance of this letter. The letter also implies that the site is free of encumbrances and that the contractor can occupy the site for work purposes. Provisions of the contract usually direct that selected bidders commence work on the site within a specified period of time, such as 10 days.

The notice to proceed has an additional significance. The date of the notice to proceed establishes the reference date from which the beginning of the project is calculated. Therefore, based on the stipulated duration of the project as specified in the supplementary conditions, the projected end of the project can be established. As will be discussed later, time extensions may increase the duration of the project, but the end of project beyond which damages will be assessed for failure to complete the project on time is referenced to the date of the notice to proceed. This might be specified as follows:

Work shall be completed not later than one thousand fifty (1050) calendar days after the date of receipt by the Contractor of Notice to Proceed.

Calendar days are used since they simplify the calculation of the end-of-project date. In certain cases, the duration of the project is specified in working days. The general conditions normally specify working days as Monday through Friday. Therefore, each week contains five working days.

In some projects, all encumbrances to entry of the construction site have not been reconciled. Therefore, the owner cannot issue a notice to proceed since he cannot authorize the contractor to enter the site. In such cases, in order to indicate selection and acceptance of a proposal, the owner may send the selected bidder a *letter of intent*. This letter will indicate the nature of encumbrance and establish the owner's intent to enter into contracts as soon as barriers to the site availability have been removed.

3.3   CONTRACT AGREEMENT

Although the issuance of the notice to proceed establishes the elements of a contract, this is formalized by the signing of a *contract agreement*. In a legal sense, the formal contract agreement is the single document that binds the parties and by reference describes the work to be performed for a consideration. It pulls together under one cover all documents to include (1) the drawings, (2) the general conditions, (3) the supplementary conditions, (4) the technical specifications, and (5) any addenda describing changes published to these original contract documents. As with other bid package components, standard forms for the contract agreement for a variety of contractual formats are available from various professional organizations. Forms for the *stipulated (lump) sum* and *negotiated* (cost of work plus a fee) type contract are given in Appendixes C and D.

3.4   TIME EXTENSIONS

Once the formal contract has been signed, certain aspects of the contractor's activity during construction must be considered. Often circumstances beyond the contractor's control, which could not have been reasonably anticipated at the time of bidding, lead to delays. These delays make it difficult or impossible to meet the project completion date. In such cases the contractor will request an extension of time to offset the delay. These time extensions, if granted, act to increase the duration of the project. Procedures for dealing with time extensions are established in the general conditions of the contract. Claims for extension of time must be based on delays that are caused by the owner or the owner's agents or on delays due to acts of God. Delays that result from design errors or changes are typical of owner-assignable delays and are not uncommon. A study of delay sources on government contracts indicated that a large percent of all delays can be traced to the reconciliation of design-related problems (see Table 3.1). Weather delays are typical of the so-called act of God type delay. Normal weather, however, is not justification for the granting of a time extension. Most general conditions state specifically that only "adverse weather conditions not reasonably anticipatable" qualify as a basis for time extensions. This means that a contractor working in Minnesota in January who requests a 15-day time extension due to frozen ground that could not be excavated will probably not be granted a time extension. Since frozen ground is typical of Minnesota in January, the contractor should have "reasonably anticipated" this condition and scheduled around it. Weather is a continuing question of debate, and many

Table 3.1   Average Percent Extension by Extension Type

Facility	Design problem	Owner modification	Weather	Strike	Late delivery	Other
Airfield paving/lighting	7.2	1.3	2.3	0.0	10.5	4.9
Airfield buildings	12.1	2.3	3.7	3.2	0.8	29.9
Training facilities	6.2	20.8	2.9	0.0	0.6	4.6
Aircraft maintenance facilities	12.0	2.0	8.4	1.0	2.2	0.2
Automotive maintenance facilities	12.9	2.3	3.4	1.4	0.7	0.4
Hospital buildings	16.0	3.4	2.6	0.6	0.6	0.9
Community facilities	6.7	5.4	2.3	1.7	1.5	0.3

Source: From D. W. Halpin and R. D. Neathammer, "Construction Time Overruns," Technical Report P-16, Construction Engineering Research Laboratory, Champaign, IL, August 1973.

contractors will submit a request for time extension automatically each month with their progress pay request, if the weather is the least bit out of the ordinary.

Time extensions are added to the original duration so that if 62 days of time extension are granted to an original duration of 1050 days, the project must be completed by 1112 calendar days after notice to proceed. If the contractor exceeds this duration, liquidated damages (see Section 3.9) are assessed on a daily basis for each day of overrun. The question of what constitutes completion can be answered as follows:

> *The Date of Substantial Completion of the Work or designated portion thereof is the Date certified by the owner's representative when construction is sufficiently complete, in accordance with the Contract Documents, so the Owner can occupy, or utilize the Work or designated portion thereof for the use for which it is intended.*

This is often referred to as the beneficial occupancy date, or BOD. Once the owner occupies the facility, he relinquishes a large portion of the legal leverage he has in making the contractor complete outstanding deficiencies. Usually, a mutually acceptable date is established when substantial completion appears to have been reached. On this date an inspection of the facility is conducted. The owner's representative (normally the architect/engineer) and the contractor conduct this inspection recording deficiencies that exist and noting items for correction. Correction of these items will satisfy the owner's requirement for substantial completion. This deficiency list is referred to in the industry as the punch list. Theoretically, once the contractor satisfactorily corrects the deficiencies noted on the punch list, the owner will accept the facility as complete. If the rapport between owner and contractor is good, this phase of the work is accomplished smoothly. If not, this turnover phase can lead to claims for damages on both sides.

An indication of the amounts of time extension granted for various reasons on some typical government projects is given in Table 3.1. The types of delay sources categorized were due to (1) design problems, (2) owner modification, (3) weather, (4) strike, (5) late delivery, and (6) other. The percentages presented were calculated as % extension = (no. of days of time extension granted ÷ originally specified project duration) × 100.

3.5   CHANGE ORDERS

Since the contract documents are included by reference in the formal agreement, the lines on the drawings, the words in the technical specifications, and all other aspects of the contract documentation are legally binding. Any alteration of these documents constitutes an alteration of the contract. As will be discussed in Chapter 4 certain contractual formats such as the unit-price contract have a degree of flexibility. However, the stipulated or lump-sum contract has virtually no leeway for change or interpretation. At the time it is presented to the bidders for consideration (i.e., is advertised), it represents a statement of the project scope and design as precise as the final drawings for an airplane or a violin. Changes that are dictated, for any reason, during construction represent an alteration of a legal arrangement and, therefore, must be formally handled as a modification to the contract. These modifications to the original contract, which themselves are small augmenting contracts, are called *change orders*.

Procedures for implementing change orders are specified in the general conditions of the contract. Since change orders are minicontracts, their implementation has many of the elements of the original contract bid cycle. The major difference is that there is no competition, since the contractor has already been selected. Normally, a formal communication of the change to include scope and supporting technical documents is sent to the contractor. The contractor responds with a price quotation for performing the work, which constitutes his offer. The owner can accept the offer or attempt to negotiate (i.e., make a counteroffer). This is, of course, the classical contractual cycle. Usually, the contractor is justified in increasing the price to recover costs due to disruption of the work and possible loss of job rhythm. If the original contract documents were poorly scoped and prepared, the project can turn into a patchwork of change orders. This can lead to a sharpening of the adversary roles of the contractor and the owner and can substantially disrupt job activities.

3.6 CHANGED CONDITIONS

Engineering designs are based on the project site conditions as they are perceived by the architect/engineer or designer. For structural and finish items as well as mechanical and electrical systems above ground, the conditions are constant and easily determined. Variation in wind patterns leading to deviation from original design criteria may pose a problem. But normally, elements of the superstructure of a facility are constructed in a highly predictable environment.

This is not the case when designing the subsurface and site topographical portions of the project. Since the designer's ability to look below the surface of the site is limited, he relies on approximations that indicate the general nature of the soil and rock conditions below grade. His "eyes" in establishing the design environment are the reports from subsurface investigations. These reports indicate the strata of soil and rock below the site based on a series of bore holes. These holes are generally located on a grid and attempt to establish the profile of soil and rock. The ability of the below-grade area to support weight may be established by a grid of test piles. The money available for this design activity (i.e., subsurface investigation) varies, and an inadequate set of bore logs or test piles may lead to an erroneous picture of subsurface characteristics. The engineer uses the information provided by the subsurface investigation to design the foundation of the facility. If the investigation is not extensive enough, the design can be inadequate.

The information provided from the subsurface investigation is also the contractor's basis for making the estimate of the excavation and foundation work to be accomplished. Again, if the investigation does not adequately represent the site conditions, the contractor's estimate will be affected. The topographic survey of the site is also a basis for estimate and, if in error, will impact the estimate and price quoted by the contractor. If the contractor feels the work conditions as reflected in the original investigation made available to him for bidding purposes are not representative of the conditions "as found," he can claim a changed condition. For instance, based on the boring logs, a reasonable estimate may indicate 2000 cu yd of soil excavation and 500 cu yd of rock. After work commences, the site may be found to contain 1500 cu yd of rock and only 1000 cu yd of soil. This, obviously, substantially affects the price of excavation and would be the basis for claiming a changed condition.

In some cases, a condition may not be detected during design, and the assumption is that it does not exist. For instance, an underground river or flow of water may go undetected. This condition requires dewatering and a major temporary-construction structure to coffer the site and to construct the foundation. If this condition could not reasonably have been foreseen by the contractor, there would be no allowance for it in his bid. The failure of the bid documents to reflect this situation would cause the contractor to claim a changed condition.

If the owner accepts the changed condition, the extended scope of work represented will be included in the contract as a change order. If the owner does not accept the changed condition claim, the validity of the claim must be established by litigation or arbitration.

3.7 VALUE ENGINEERING

Value Engineering (VE) was developed during World War II in the United States. It began as a search for alternative product components due to a shortage of critical items during the war. Innovation was required. It was discovered that a process of "function analysis" produced low-cost products without impacting functional characteristics or reducing quality. This initiative showed that innovation can yield products which cost less but maintain the expected levels of performance. In this case, "necessity was the mother of invention."

In the early 1960s, this concept of value was introduced in the construction industry through directives from the Navy and Army Corps of Engineers relating to facility procurement. Other major government agencies (e.g., Public Building Service) joined this movement by introducing incentive clauses in facility procurement (construction) contracts which provided rewards to contractors for value proposals which led to reduced construction costs while maintaining the functionality of the completed facility. These clauses are structured generally as follows:

> *The Contractor is encouraged to develop, prepare, and submit value engineering change proposals (VECP's) voluntarily. The Contractor shall share in any net acquisition savings realized from accepted VECP's, in accordance with the incentive sharing rates specified in the contract.*

In Army Corps of Engineers contracts, the VE incentive clause allows the construction contactor to share 50% or more of the net savings in firm fixed-price contracts. For example, if a contractor is constructing bridge towers supported (in the original design) by drilled pile foundations, and the contractor can re-design the foundations as spread mat footers with a savings of $400,000, a portion of the savings (usually 50%) is distributed to the contractor. The construction contractor must prepare a value engineering change proposal (VECP) which will be reviewed and then accepted or rejected by the owner. A potential reward to the contractor (in this case) of $200,000 is available if the proposal is accepted.

The VECP procedure allows the owner to harvest new and innovative ideas from the construction contractor. This overcomes, to some degree, the factors which obstruct the transfer of information from the contractor to the designer in classical Design-Bid-Build (DBB) contracts. In such contracts, information flow is impeded by the "friendly enemy" attitude, which is often characteristic of the relationship between the design professional and the contractor in competitively bid contracts. Also, due to the sequence of design and construction in the DBB format, contractors seldom have input to the design process.

Construction contractors are typically more knowledgeable about field conditions and construction methods than design engineers. The construction methods used to realize a given design in the field have a great impact on cost. Contractors are in a better position to know what materials are easiest to install and which designs are most constructable. This knowledge can greatly influence cost. The VECP process allows this expertise to be transferred to the owner yielding a cost saving.

The idea behind value engineering is the improvement of design by encouraging the contractor to make suggestions during construction. This is in contrast to the implementation of VE during the design phase which involves the designer or design professional in a systematic program of "value analysis." If at any point following selection, the contractor feels that a proposal to improve cost effectiveness of the design as transmitted at the time of bidding is appropriate, there is a monetary incentive to submit such a proposal. Again, if the contactor makes a suggestions that cuts the cost of the air-conditioning system by $60,000 and an equal sharing VE clause is in the contract, $30,000 is received for this VECP if accepted by the owner. The guiding principle in making the suggestion is that the cost is reduced while the functionality is maintained or improved.

VE can also be implemented during the design phase of project development. This aspect of VE uses various procedures such as brainstorming, prioritization, research, matrix analysis , and scoring systems to evaluate design alternatives. Criteria Evaluation can be used to assess multiple factors such as aesthetics, performance, safety, etc. A weighted analysis is used to do the final analysis. In the weighting process, a criteria matrix such as that shown in Figure 3.2 is used. All criteria to be considered are listed

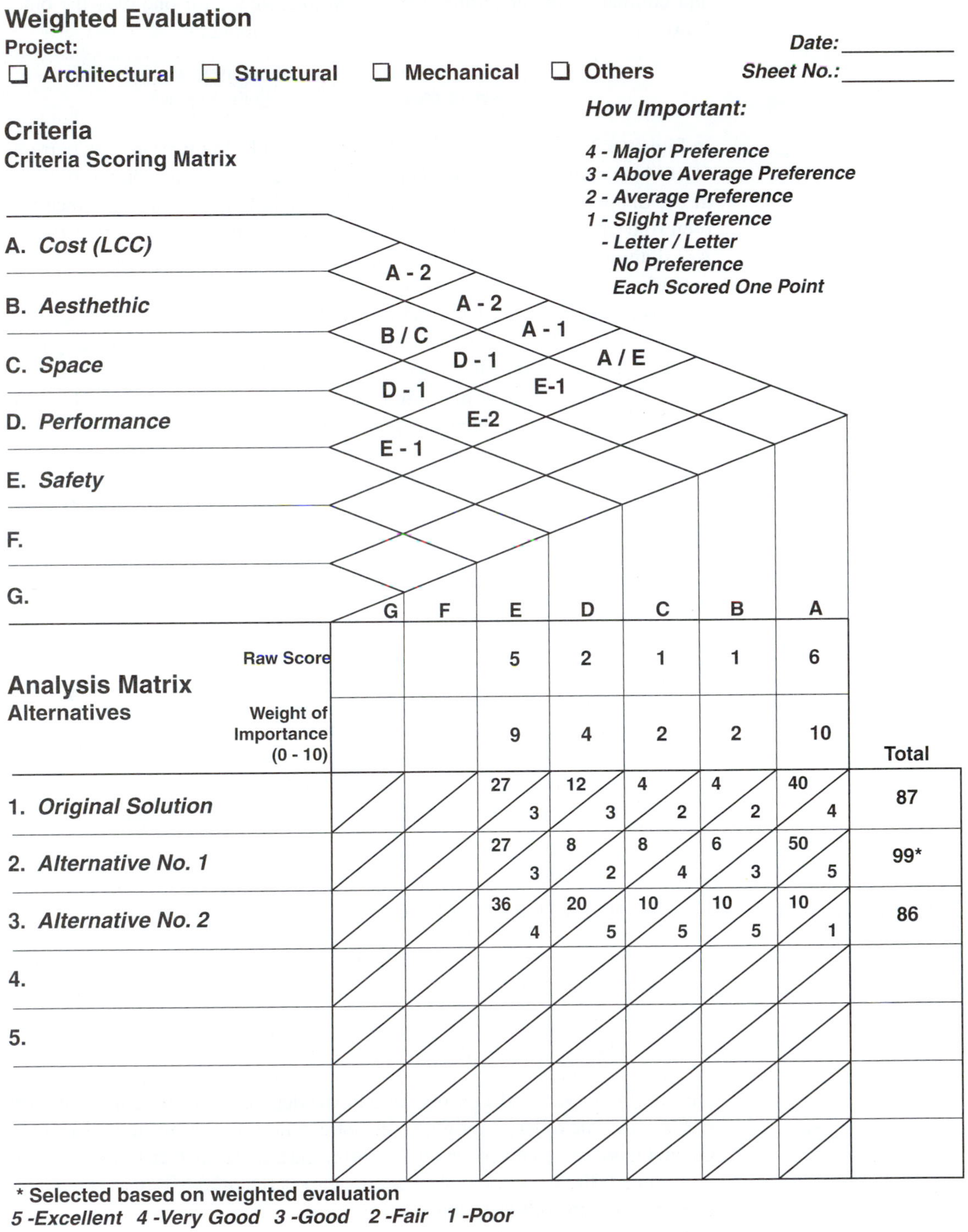

Weighted Evaluation

Project: Date: _____

☐ Architectural   ☐ Structural   ☐ Mechanical   ☐ Others   Sheet No.: _____

Criteria
Criteria Scoring Matrix

How Important:

4 - Major Preference
3 - Above Average Preference
2 - Average Preference
1 - Slight Preference
  - Letter / Letter
    No Preference
    Each Scored One Point

A. *Cost (LCC)*

A - 2

B. *Aesthethic*

A - 2

B / C   A - 1

C. *Space*

D - 1   A / E

D - 1   E-1

D. *Performance*

E-2

E. *Safety*

E - 1

F.

G.

Analysis Matrix
Alternatives

		G	F	E	D	C	B	A	
	Raw Score			5	2	1	1	6	
	Weight of Importance (0 - 10)			9	4	2	2	10	**Total**
1. *Original Solution*				27 / 3	12 / 3	4 / 2	4 / 2	40 / 4	87
2. *Alternative No. 1*				27 / 3	8 / 2	8 / 4	6 / 3	50 / 5	99*
3. *Alternative No. 2*				36 / 4	20 / 5	10 / 5	10 / 5	10 / 1	86
4.									
5.									

* Selected based on weighted evaluation
5 -Excellent   4 -Very Good   3 -Good   2 -Fair   1 -Poor

Figure 3.2 Criteria Evaluation Matrix.

and compared. In comparing two criteria, preference for one over the other is scored as follows:

1 = slight preference	3 = above average preference
2 = average preference	4 = major preference

In Figure 3.2, five criteria are listed as A through E. In the triangular portion of the matrix Cost (A) is compared with Aesthetic (B). Cost has an "average preference" so A-2 is placed in the diamond linking these two criteria. B and C are compared, since they are equally preferred (no preference) B/C is entered in the diamond linking these two criteria. Now the A-2 diamond is compared to the B/C diamond and the preference of A-2 prevails and is noted in the diamond linking these two composite preferences (A to B and B to C). This process is continued until all criteria have been compared. The right most diamond shows that the most preferred criteria are A and E. They are effectively equally preferred.

In the lower part of the matrix however, E is given a raw score of 5 while A receives a slightly higher score of 6. These raw scores are converted to Weights of Importance as shown. Two alternatives are compared with the original design solution. Each of the factors (A to E) are given a score between 1 (poor) and 5 (excellent). These scores are multiplied by the Weight of Importance. For the original solution, factor E receives a score of 3 times 9 or 27. The original solution score is $27 + 12 + 4 + 4 + 40$ or 87. This compares with scores for Alternative 1 and 2 of 99 and 86 respectively. Based on this weighted analysis, Alternative 1 would be selected.

This structured approach to evaluating alternatives during design helps to formalize consideration of various alternatives. A more comprehensive discussion of this topic is beyond the scope of this section. A basic reference on the use of VE is provided by Dell'Isola (Dell'Isola 1997).

Government agencies have led the way in the in the use of value engineering incentive clauses in construction contracts. This allows the owner (i.e., the government) to enhance the flow of cost saving information between the designer and the contractor during the construction phase. Success in the use of VECP clauses by public agencies has led to wide spread implementation of such incentive clauses in construction contracts used by private owners.

3.8 SUSPENSION, DELAY, OR INTERRUPTION

The standard general conditions utilized for many government contracts provide that:

> *The Contracting officer may order the Contractor in writing to suspend, delay, or interrupt all or any part of the work for such period of time as he may determine to be appropriate for the convenience of the government.*

Interrupting or suspending work for an extended period of time may be costly to the contractor, since he must go through a demobilization-remobilization cycle and may confront inflated labor and materials at the time of restarting. In such cases, within the provisions of the contract, the owner (i.e., the government) is required to pay an adjustment for "unreasonable" suspensions as follows:

> *An adjustment shall be made for any increase in cost of performance of this contract (excluding profit) necessarily caused by such unreasonable suspension, delay, or interruption and the contract modified in writing accordingly.*

The amount of this adjustment is often contested by the contractor and can lead to lengthy litigation. Normally, the owner will attempt to avoid interruptions. Difficulties in obtaining continuing funding, however, are a common cause for these suspensions.

3.9   LIQUIDATED DAMAGES

Projects vary in their purpose and function. Some projects are built to exploit a developing commercial opportunity (e.g., a fertilizer plant in Singapore) while others are government funded for the good and safety of the public (e.g., roads, bridges, etc.). In any case, the purpose and function of a project are often based on the completion of the project by a certain point in time. To this end, a project duration is specified in the contract document. This duration is tied to the date the project is needed for occupancy and utilization. If the project is not completed on this date, the owner may incur certain damages due to the nonavailability of the facility.

For instance, assume an entrepreneur is building a shopping center. The project is to be complete for occupancy by 1 October. The projected monthly rental value of the project is $30,000. If all space is rented for occupancy on 1 October and the contractor fails to complete the project until 15 October, the space cannot be occupied and half a month's rental has been lost. The entrepreneur has been "damaged" in the amount of $15,000 and could sue the contractor for the amount of the damage. Contracts provide a more immediate means of recourse in liquidating or recovering the damage. The special conditions allow the owner under the contractual relationship to charge the contractor for damages for each day the contractor overruns the date of completion. The amount of the liquidated damage to be paid per day is given in the special or supplementary conditions of the contract. The clause SC-2 of the Special Conditions in Figure 2.10 is typical of the language used, and reads:

> ***Liquidated Damages***   *In case of failure on the part of the Contractor to complete the work within the time fixed in the contract or any extensions thereof, the Contractor shall pay the owner as liquidated damages the sum of $3000 for each calendar day of delay until the work is completed or accepted.*

The amount of the liquidated damage to be recovered per day is not arbitrary and must be a just reflection of the actual damage incurred. The owner who is damaged must be able, if challenged, to establish the basis of the figure used. In the rental example given, the basis of the liquidated damage might be as follows:

$$\text{Rental loss: }\$30,000\text{ rent/month} \div 30\text{ days} = \$1000/\text{day}$$
$$\text{Cost of administration and supervision of contract: } = \underline{\$\ \ 200/\text{day}}$$
$$\$1200/\text{day}$$

If a project overruns, the owner not only incurs costs due to lost revenues but also must maintain a staff to control and supervise the contract. This is the $200 cost for supervision. The point is that an owner cannot specify an arbitrarily high figure such as $20,000 per day to scare the contractor into completion without a justification. The courts have ruled that such unsupported high charges are in fact not the liquidation of a damage but instead a penalty charge. The legal precedent established is that if the owner desires to specify a penalty for overrun (rather than liquidated damages), he must offer a bonus in the same amount for every day the contractor brings the project in early. That is, if the contractor completes the project three days late he must pay a penalty of $60,000 (based on the figure above). On the other hand, if the contractor completes the project three days early, he would be entitled to a bonus of $60,000. This has discouraged the use of such penalty-bonus clauses except in unusual situations.

Establishing the level of liquidated damages for government projects is difficult, and in most cases the amount of damage is limited to the cost of maintaining a resident staff and main office liaison personnel on the project beyond the original date of completion. In a claims court, it is difficult to establish the social loss in dollars of, for instance, the failure to complete a bridge or large dam by the specified completion date.

3.10 PROGRESS PAYMENTS AND RETAINAGE

During the construction period, the contractor is reimbursed on a periodic basis. Normally, at the end of each month, the owner's representative (e.g., project or resident engineer) and the contractor make an estimate of the work performed during the month, and the owner agrees to pay a progress payment to cover the contractor's expenditures and fee or markup for the portion of the work performed. The method of making progress payments is implemented in language as follows:

> *At least ten days before the date for each progress payment established in the Owner Contractor Agreement, the Contractor shall submit to the owner's representative an itemized Application for Payment, notarized if required, supported by such data substantiating the Contractor's right to payment as the Owner may require, and reflecting retainage, if any, as provided elsewhere in the Contract Documents.*

Retainage is considered in greater detail in Chapter 9 in discussing cash flow. The owner typically retains or holds back a portion of the monies due the contractor as an incentive for the contractor to properly complete the project. The philosophy of retainage is that if the project is nearing completion and the contractor has received virtually all of the bid price, he will not be motivated to do the small closing-out tasks that inevitably are required to complete the project. By withholding or escrowing a certain portion of the monies due the contractor as retainage, the owner has a "carrot," which can be used at the end of a project. He can say essentially, "Until you have completed the project to my satisfaction, I will not release the retainage." Retainage amounts are fairly substantial, and therefore, the contractor has a strong incentive to complete small finish items at the end of the project.

The amount of retainage is stated in the contract documents (e.g., general conditions) in the following fashion:

> *In making progress payments, there shall be retained 10 percent of the estimated amount until final completion and acceptance of the work.*

Various retainage formulas can be used, based on the owner's experience and policy. If work is progressing satisfactorily at the 50% completion point, the owner may decide to drop the retainage requirement as follows:

> *If the owner's representative (architect/engineer) at any time after 50 percent of the work has been completed finds that satisfactory progress is being made, he may authorize any of the remaining progress payments to be made in full.*

If a project has been awarded at a price of $1,500,000 and 10% retainage is withheld throughout the first half of the job, the retained amount is $75,000. This is a formidable incentive and motivates the contractor to complete the details of the job in a timely fashion.

3.11 PROGRESS REPORTING

Contracts require the prime contractor to submit a schedule of activity and periodically update the schedule reflecting actual progress. This requirement is normally stated in the general conditions as follows:

> *Progress Charts   The contractor shall within 5 days or within such time as determined by the owner's representative, after the date of commencement of work, prepare and submit to the owner's representative for approval a practicable schedule, showing the order in which the contractor proposes to carry on the work, the date on which he will start the several salient features (including procurement of materials, plant, and equipment) and the contemplated dates for completing the same. The schedule shall be in the form of a progress chart of suitable scale to indicate appropriately the percentage of work scheduled for completion at*

any time. The contractor shall revise the schedule as necessary to keep it current, shall enter on the chart the actual progress at the end of each week or at such intervals as directed by the owner's representative, and shall immediately deliver to the owner's representative three copies thereof. If the contractor fails to submit a progress schedule within the time herein prescribed, the owner's representative may withhold approval of progress payment estimates until such time as the contractor submits the required progress schedule.

This provision is fairly broad and could well be interpreted to require only grossly defined S-curves or bar charts. These bar charts may be based either on activities or percentage completion of the various work categories such as concrete, structural, electrical, and mechanical work. These reports are used at the time of developing the monthly progress payments and to ensure the contractor is making satisfactory progress. Figures 3.3 and

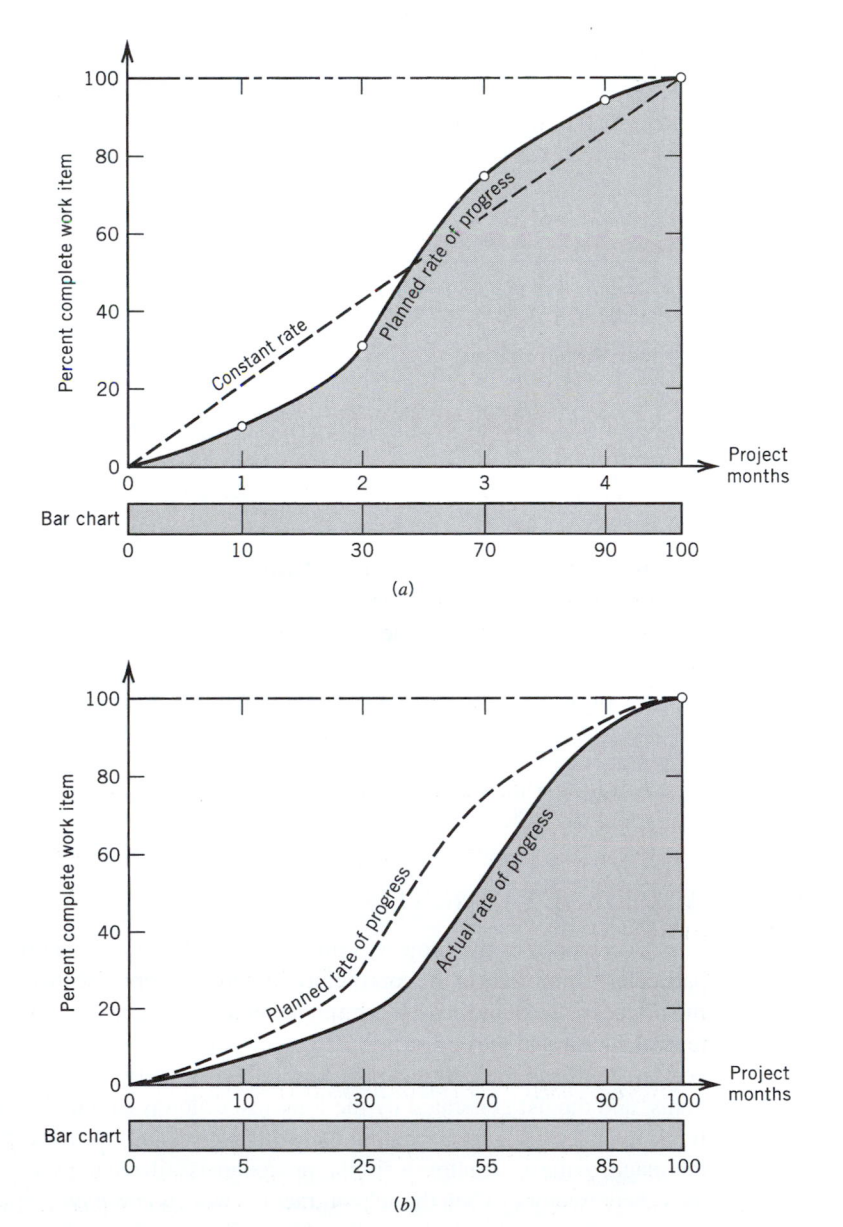

Figure 3.3 Bar chart planning and control models: (*a*) planned rate of progress and (*b*) actual rate of progress.

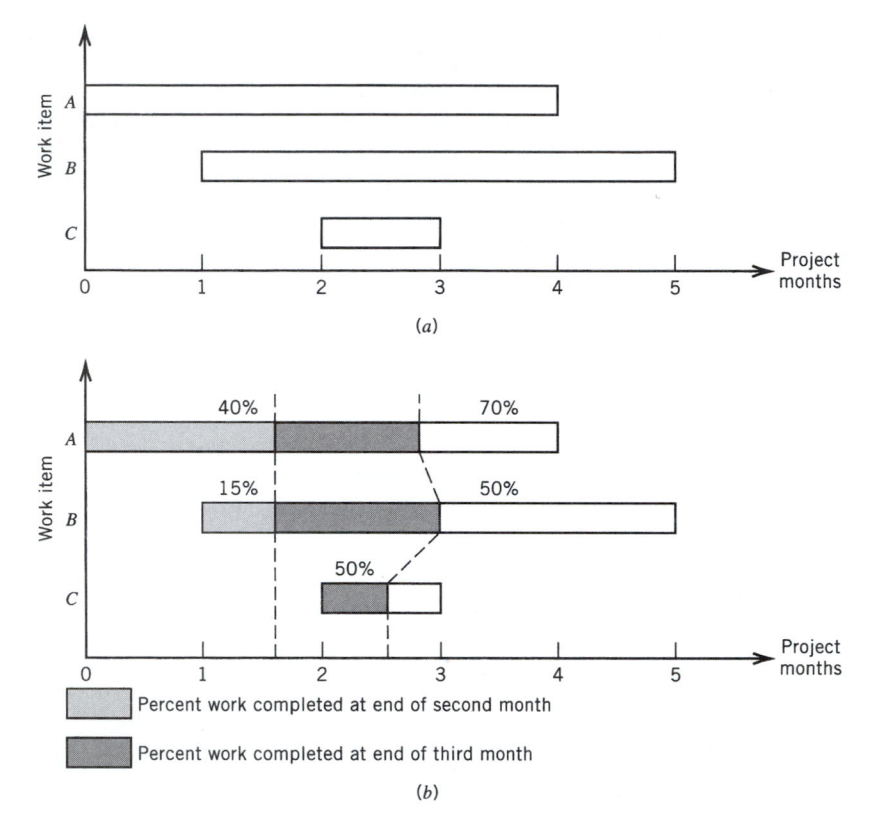

Figure 3.4 Bar chart project models: (*a*) bar chart schedule (plan focus) and (*b*) bar chart updating (control focus).

3.4 indicate sample reporting methods involving bar charts for work activities and S-curves of overall percentages complete.

Network methods provide greater detail and have the advantage during planning and scheduling of being oriented to individual activities and their logical sequence. From the owner's viewpoint they allow a more precise review of logic and progress during construction and acceptance periods. If the contractor is behind schedule on critical activities, a simple bar chart or S-curve will not highlight this. The network approach provides greater early warning of the impact of delays on total project completion.

3.12 ACCEPTANCE AND FINAL PAYMENT

Final acceptance of the project is important to all parties concerned. As noted above, it is particularly important to the contractor, since final acceptance means the release of retainage. Final acceptance of the project is implemented by a joint inspection on the part of the owner's representative and the contractor. The owner's representative notes deficiencies that should be corrected, and the contractor makes note of the deficiencies. These are generally detail items, and the list generated by the joint inspection is called the deficiency, or punch, list. It becomes the basis for accepting the work as final and releasing final payment (to include retainage) to the contractor. A similar procedure is utilized between the prime contractor and the subcontractors. When the subcontractor's work is complete, representatives of the prime and subcontractor "walk the job" and compile the deficiencies list for final acceptance of

```
                    May 20, 2XXX
            Punch List Items Acme Plastering Co.
                    Barfield-400 Project


Larger Building
    1. Caulking required between stucco and brick on the lower level.
    2. Streaks and cracking on stucco must be remedied.

Smaller Building
    1. Very noticeable line of stucco in rear of building.
    2. Streaks and cracking on stucco must be repaired.
    3. Exterior bridge entrances: patch stucco must be made uniform.
    4. Areas of excess spalling must be corrected.
```

Figure 3.5 Typical punch list.

subcontract work. An example of a punch list between prime contractor and subcontractor is shown in Figure 3.5.

3.13 SUMMARY

This chapter has presented an overview of the cycle of activity that moves a project from the bid award stage through construction to acceptance by the owner or client. It is necessarily brief but provides a general frame of reference indicating how the contractor receives the project and some of the contractual considerations he must be aware of during construction. The competitively bid type of contract and the bid sequence particular to this contractual format have been used as the basis for presentation. Other forms of contract will be discussed in Chapter 4. However, the basic chronology of events is the same. Having established this general mapping of the construction process, the following chapters develop the details of the contractor's role in the construction team.

REVIEW QUESTIONS AND EXERCISES

3.1 What is the difference between liquidated damages and a penalty for late completion of the contract?

3.2 What is the purpose of retainage?

3.3 During what period can a contractor withdraw the bid without penalty?

3.4 As a contractor you have built a 100-unit apartment complex that rents for $450 per unit a month. For late completion you were assessed $2000 per day. Would you call the assessment liquidated damages or a penalty? If the contract had included a bonus of $500 per day for early completion, would you expect to gain any assessment from court action? Why?

3.5 Describe the procedures to be followed for the receiving and opening of bids. If possible attend a bid opening and determine the number of bids that were submitted. For several unsuccessful bids determine the dollar amounts by which they exceeded the winning bid. Then calculate (relative to the winning bid) the percentages by which they exceed the winning bid. What do these figures tell about the strength of the current estimating and market environments? How much did the winning bidder "leave on the table"?

3.6 Scan a typical stipulated sum contract and identify those clauses that either prescribe, modify, or are related to time considerations. Then develop a time strip map (similar to Fig. 3.1) for the contract that locates the times (or time zones) for which each of the clauses are relevant. Which clauses rigorously fix time constraints for the contract and which are dependent on acts of God or the owner for relevance?

3.7 Describe the procedure to be followed by the contractor who wishes to claim a time extension. What sort of documentation do you think is necessary to either refute or defend a time extension claim due to unusual weather? What sort of records do local contractors keep of weather conditions?

3.8 Must a contractor accept and perform all the work involved in each contract change order? Is there a limit to the number or magnitude of change orders that can be applied to a contract? When can a contractor refuse to accept a change order?

3.9 List the common causes of changed conditions in a building contract. What typical contract clauses bear on the problems caused by changed conditions? Suppose separate contracts are let for the building foundations and all remaining work. If you are the second contractor and you find that the foundations are incorrectly located, either in plan or elevation, would you be able to claim a changed condition?

3.10 Prepare a punch list of deficiencies or repairs that you consider necessary for your room, garage, or classroom. Can

any of these items be related back to the original acceptance of the facility?

3.11 How would you go about either documenting a claim for a contractor's progress payment or its verification by the contract administrator for a typical building project in your locality?

Chapter 4

Construction Contracts

Web-based Contracts

The Need

Web-based contracting provides an environment supporting scheduling, controlling, regulating, analyzing, and auditing the procurement and delivery of materials and services for construction in an electronic format. By enabling online competitive bidding and improving the record-keeping associated with the purchasing process, buyers will be able to quickly and easily compare product offerings from different manufacturers, as well as solicit pricing and availability. This will put buyers in a position to make optimal price- and time-of-delivery decisions.

The Technology

A number of companies are developing network-based contracting services. These networks are adopting such solutions to improve their subcontract bidding process. With this system, construction companies can easily submit bid documents and specifications to solicit competitive bids for subcontracted work. They will be able to route RFQs/RFPs to approved contractors, or search for matching contractors according to attributes such as CSI (Construction Specifications Institute) classification, geographic location, specialty, minority status, licensing, and bonding. Eligible contractors and suppliers are notified of pending project bids and may then respond electronically. Their responses are automatically organized into bid summary spreadsheets for review and award. Buyers can anticipate an immediate return on investment from reduced costs associated with the processing of purchase orders, more competitive prices, and overall supply-chain efficiencies. Sellers will find new opportunities to increase sales by expanding their customer base and effectively communicating their product line and pricing. These online services help construction companies manage their complex supply chains, while enabling the project owner to make trade-off decisions about construction costs versus lifetime operational costs.

4.1   CONTRACT ENVIRONMENT

Construction is a product oriented activity that has many dimensions. One of these dimensions is the business side of construction. The business world is structured by contractual relationships, and the business aspects of construction require the establishment of legally binding relationships with a wide range of parties. The central role played by contracts is reflected by the fact that construction firms are referred to as "contractors." In addition to the contractual relationship with the owner/client, construction managers supervise contracts with subcontractors, specialty firms such as scheduling services, labor unions, as well as equipment and materials vendors. Insurance and bonds as well as the documents establishing the legal structure of a company have the elements of contractual requirements. In this chapter, we investigate the major contractual forms used to establish contracts for the construction of projects.

An agreement between two or more parties to do something for a consideration establishes the basis for a contract. "A contract is a promise or a set of promises for the breach of which the law recognizes duty. This amounts to saying that a contract is a legally enforceable promise" (Jackson, 1973). The courts are often called upon to determine:

1. Who are the parties to a contract?
2. What are their promises?
3. Other aspects of the contractual agreement.

A whole body of law has grown up around the many facets of contractual relationships. Because these issues remain constant for most construction situations, contract language in the construction industry has been normalized over many years and a variety of standard contract forms have developed.

4.2   PROCESS OF PURCHASING CONSTRUCTION

Construction contracts structure the way in which construction is "purchased." It is interesting to compare the construction purchasing system with the way in which we would buy a new lawn mower or a set of living room furniture. Consumers who need to purchase something go to a store, look at the range of product choices, and then pay a single supplier (the store owner) for the item of interest. If, for instance, we need a refrigerator, we go to an appliance store, inspect the various models, check prices, select one for purchase, pay for it, and the store owner sends it to our house or apartment within the next few days.

Two major aspects of this process contrast with the way in which we "buy" construction.

1. We have the finished product available for our inspection, and we can decide whether it meets our requirements. That is, the manufactured product is available for our inspection prior to purchase.
2. Since the final product is available, we purchase it from a single individual or source.

In construction the facility is purchased before it is "manufactured" based on a set of drawings and work descriptors. Also the end item requires the purchaser to coordinate many entities to include designer(s), contractor(s), specialty subcontractors, and vendors. It is as if to buy a refrigerator we must develop a drawing of the refrigerator, purchase the materials required, and then coordinate 10 different entities who build it for us. Typically, none of the entities building the end item will warrant the proper operation of the refrigerator. They will only warrant the work that they provide.

In the building of the Brooklyn Bridge referred to in Chapter 1, Washington Roebling ordered a gigantic wooden box (e.g., a caisson) that was built by a local shipyard. The shipyard required payment in advance and would only warrant that the box was built according to plans that Roebling provided. They would not guarantee that it would perform adequately as a caisson because they did not know what a caisson was or how it was to be used.

Ideally, we would like to go to a single source and purchase the construction project as a finished unit. In construction this is seldom possible. Traditional contract formats address this problem by focusing on the purchase of the design from a single entity (e.g., the design professional) and the construction of the facility by a general contractor who purchases the needed materials and services and coordinates the work of all the entities building the facility. As has been noted, even this three-party purchasing relationship (e.g., owner, designer, and constructor) can lead to an adversarial relationship between the parties.

Owners would, in general, like to work with a single source and be able to purchase the facility as-built (e.g., fully constructed and ready for occupancy). That single source would warrant the operation of the facility and act as a single point of contact to reconcile all problems with the product. This approach is used in single-family housing to a large extent. The home builder builds a house on a speculative basis without a buyer on hand. The house is presented to the public as a finished product, and the contractor is a single point of contact for purchase and warranty.

Tunnels, bridges, and most large construction projects are not first built and then presented to the public for purchase. These are not built "on spec" (i.e., on a speculative basis) simply because the risk of building a "white elephant" that cannot be sold is too great. For this reason, the ability to present a finished product to a prospective buyer of construction is not feasible for most construction projects (one exception being single-family housing).

Project delivery systems have been developed to provide the construction buyer (i.e., the client) with a single point of contact or source of purchase. These contract formats have gained popularity over the past 20 years and are still evolving. The two major varieties of contract formats designed to provide the client with the construction equivalent of "one-stop shopping" are (1) design-build contracts and (2) construction management contracts. Before discussing these more recent developments in construction contracting, it is important to understand the contract types that have been used most widely over the past 50 years.

4.3   MAJOR CONSTRUCTION CONTRACT TYPES

The most widely used format of contract is the competitively bid contract. For a number of reasons, almost all contracts that involve public funds are awarded using competitively bid contracts. A competitively bid contract is used since it yields a low and competitive price that ensures taxpayers that their monies are being equitably and cost-effectively disbursed. The basic sequence of events associated with this type of contract has been described in Chapters 2 and 3. The two main categories of competitively bid contracts are (1) the lump, or stipulated, sum contract and (2) the unit-price contract. The names of both of these contract formats refer to the method in which the price for the work is quoted.

The second most widely used contract format is the negotiated contract. This form of contract is also referred to as a *cost-plus* contract although this refers to the method of payment rather than the nature of the selection process. The contractor is reimbursed for the cost of doing the work plus a fee. In this type of contract, the contractor risk is greatly reduced since the requirement of completing the work at a fixed price is not present. The owner has the flexibility to select the contractor based on considerations other than lowest price quotation. The method of selection involves the identification of a group of qualified contractors who are invited to prepare proposals based on the project documentation available. The proposals present the credentials of the firm and an approximate estimate of cost based on the project data available. The estimate includes not only the "bricks-and-mortar" direct costs but also estimates of the cost of supervision by the contractor's personnel and the level of fee requested. The proposal is often presented in a semiformal interview framework in which the contractor meets with the client and his/her representatives. Selection of the contractor is based on the preferences of the owner and the strengths and weaknesses of the contractor's proposal. This format is not well suited to public projects since favoritism can play a major part in determining which contractor is selected.

4.4   COMPETITIVELY BID CONTRACTS

The mechanism by which competitively bid contracts are advertised and awarded has been described in Chapters 2 and 3. Essentially, the owner invites a quote for the work to be performed based on complete plans and specifications. The award of contract is generally made to the lowest *responsible* bidder. The word responsible is very important since the contractor submitting the lowest bid may not, in fact, be competent to carry out the work. Once bids have been opened and read publicly (at the time and place announced in the notice to bidders), an "apparent" low bidder is announced. The owner then immediately reviews the qualifications of the bidders in ascending order from lowest to highest. If the lowest bidder can be considered responsible based on his or her capability for carrying out the work, then further review is unnecessary.

The factors that affect whether a contractor can be considered responsible are the same as those used in considering a contractor for prequalification:

1. Technical competence and experience
2. Current financial position based on the firm's balance sheet and income statement
3. Bonding capacity
4. Current amount of work under way
5. Past history of claims litigation
6. Defaults on previous contracts

Because of shortcomings in any of these areas, a contractor can be considered a risk and, therefore, not responsible. Owners normally verify the bidder's financial status by consulting the Dun and Bradstreet *Credit Reports* (Building Construction Division) or similar credit reporting system to verify the financial picture presented in the bid documents.

Generally, the advantages that derive from the use of competitively bid contracts are twofold. First, because of the competitive nature of the award, selection of the low bidder ensures that the lowest responsible price is obtained. This is only theoretically true, however, since change orders and modifications to the contract tend to offset or negate this advantage and increase the contract price. Some contractors, upon finding a set of poorly defined plans and specifications will purposely bid low (i.e., zero or negative profit) knowing that many change orders will be necessary and will yield a handsome profit. That is, they will bid low to get the award and then negotiate high prices on the many change orders that are issued.

The major advantage, which is essential for public work, is that all bidders are treated equally and there are no favorites. This is very important since in the public sector political influence and other pressures could bias the selection of the contractor. Presently, public design contracts are not awarded by competitive bidding. The practice of negotiating design contracts is traditional and supported by engineering professional societies (e.g., the American Society of Civil Engineers and the National Society of Professional Engineers). Nevertheless, it has been challenged by the U.S. Department of Justice.

The competitive method of awarding construction contracts has several inherent disadvantages. First, the plans and specifications must be totally complete prior to bid advertisement. This leads to a sequentiality of design followed by construction and breaks down feedback from the field regarding the appropriateness of the design. Also, it tends to extend the total design-build time frame since the shortening of time available by designing and constructing in parallel is not possible. In many cases, the owner wants to commence construction as quickly as possible to achieve an early completion and avoid the escalating prices of labor and materials. The requirement that all design must be complete before construction commences preempts any opportunity for commencing construction while design is still under way.

4.5 STIPULATED-SUM CONTRACTS

A lump-sum, or stipulated-sum, contract is one in which the contractor quotes one price, which covers all work and services required by the contract plans and specifications. In this format, the owner goes to a set of firms with a complete set of plans and specifications and asks for a single quoted price for the entire job. This is like a client going to a marine company with the plans for a sailboat or catamaran and requesting a price. The price quoted by the boat builder is the total cost of building the vessel and is a lump-sum price. Thus the lump sum must include not only the contractor's direct costs for labor, machines, and so forth but also all indirect costs such as field and front office supervision, secretarial support, and equipment maintenance and support costs. It must also include profit.

In stipulated-sum contracts the price quoted is a guaranteed price for the work specified in the plans and supporting documents. This is helpful for the owner since he knows the exact amount of money that must be budgeted for the project, barring any contingencies or change of contractual documents (i.e., change orders).

In addition, the contractor receives monthly progress payments based on the estimated percent of the total job that has been completed. In other contract forms, precise field measurement of the quantity of work placed (e.g., cubic yards of concrete, etc.) must be made continuously since the contractor is paid based on the units placed rather than on the percent of job completed. Since the percent of the total contract completed is an estimate, the accuracy of the field measurements of quantities placed need only be accurate enough to establish the estimated percent of the project completed. This means that the number and quality of field teams performing field quantity measurements for the owner can be reduced. The total payout by the owner cannot exceed the fixed or stipulated price for the total job. Therefore, rough field measurements and observations, together with some "Kentucky windage," are sufficient support for establishing the amount of progress payment to be awarded.

In addition to the disadvantage already noted (i.e., the requirement to have detailed plans and specifications complete before bidding and construction can begin), the difficulties involved in changing design or modifying the contract based on changed conditions are an important disadvantage. The flexibility of this contract form is very limited. Any deviation from the original plans and specifications to accommodate a change must be handled as a change order (see Section 3.5). This leads to the potential for litigation and considerable wrangling over the cost of contract changes and heightens the adversary relationship between owner and contractor.

The stipulated-sum form of contract is used primarily in building construction where detailed plans and specifications requiring little or no modification can be developed. Contracts with large quantities of earthwork or subsurface work are not normally handled on a lump-sum basis since such contracts must be flexible enough to handle the imponderables of working below grade. Public contracts for buildings and housing are typical candidates for lump-sum competitively bid contracts.

4.6   UNIT-PRICE CONTRACTS

In contrast to the lump-sum, or fixed-price, type of contract, the unit-price contract allows some flexibility in meeting variations in the amount and quantity of work encountered during construction. In this type of contract, the project is broken down into work items that can be characterized by units such as cubic yards, linear and square feet, and piece numbers (e.g., 16 window frames). The contractor quotes the price by units rather than as a single total contract price. For instance, he quotes a price per cubic yard for concrete, machine excavation, square foot of masonry wall, and so forth. The contract proposal contains a list of all work items to be defined for payment. Items 1 to 4 in Section 1 of Figure 2.9 provide a typical listing for unit-price quotation. This section is reprinted for reference:

Item number	Quantity	Unit	Description	Unit price	Total amount
1	550	cubic yard (cu yd)	Rock excavation (for structures and pipes only)	$_____	$_____
2	50	linear foot (lin ft)	8-in. C.I. force main	$_____	$_____
3	20	cubic yard (cu yd)	Trench excavation for pipes	$_____	$_____
4	200	square yard (sq yd)	Paving	$_____	$_____

Four items of unit-price work are listed. A guide quantity is given for each work item. The estimated amount of rock excavation, for example, is 550 cu yd. Based on this quantity of work, the contractor quotes a unit price. The total price is computed by multiplying the unit price by the guide quantity. The low bidder is determined by summing the total amount for each of the work items to obtain a grand total. The bidder with the lowest grand total is considered the low bidder. In true unit-price contracts, the entire contract is divided into unit-price work items. Those items that are not easily expressed in units such as cubic yards are expressed in the unit column as "one job."

Unit-price quotations are based on the guide quantity specified. If a small quantity is specified, the price will normally be higher to offset mobilization and demobilization costs. Larger quantities allow *economies of scale*, which reduce the price per unit. That is, if 100 sq ft of masonry brick wall is to be installed, the cost per square foot would normally be higher than the cost for 5000 sq ft. Mobilization and demobilization costs are spread over only 100 units in the first case, whereas in the second case these costs are distributed over 5000 units, reducing the individual unit cost.

Most unit-price contracts provide for a price renegotiation in the event that the actual field quantity placed deviates significantly from the guide quantity specified. If the deviation exceeds 10%, the unit price is normally renegotiated. If the field quantity is over 10% greater than the specified guide quantity, the owner or the owner's representative will request a price reduction based on economies possible due to the larger placement quantity. If the field quantity underruns the guide quantity by more than 10%, the contractor will usually ask to increase the unit price. He will argue that he must recover his mobilization, demobilization, and overhead costs since the original quote was based on the guide quantity. That is, there are fewer units across which to recover these costs and, therefore, the unit price must be adjusted upward.

In developing the unit-price quotation, the contractor must include not only direct costs for the unit but also indirect costs such as field and office overheads as well as a provision for profit.

In unit-price contracts, the progress payments for the contractor are based on precise measurement of the field quantities placed. Therefore, the owner should have a good indication of the total cost of the project based on the grand total price submitted. However, deviations between field-measured quantities and the guide quantities will lead to deviations in overall job price. Therefore, one disadvantage of the unit price contract form is that the owner does not have a precise final price for the work until the project is complete. In other words, allowances in the budget for deviations must be made. In addition, the precision of field measurement of quantities is much more critical than with the lump-sum contract. The measured field quantities must be exact since they are, in fact, the payment quantities. Therefore, the owner's quantity measurement teams must be more careful and precise

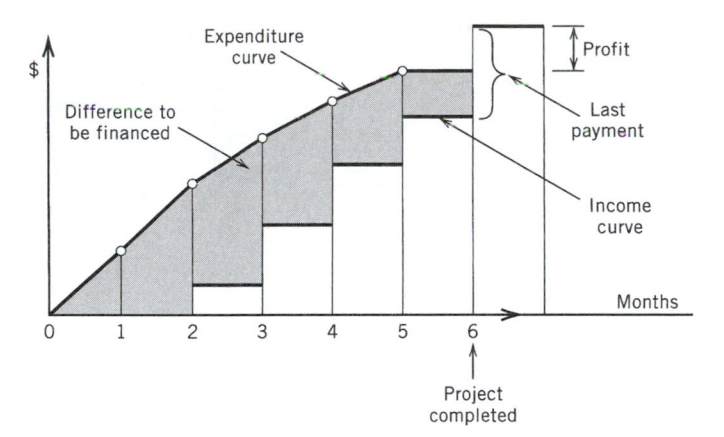

Figure 4.1 Project expense/income curves.

in their assessments since their quantity determinations establish the actual cost of the project.

Unit-price contracts can also be manipulated using the technique called unbalancing the bid. The relationship between the contractor's expenditures and income across the life of a typical project is shown schematically in Figure 4.1. Because of delays in payment and retainage as described in Section 3.10, the income curve lags behind the expenditure curve and leads the contractor to borrow money to finance the difference. The nature and amount of this financing is discussed in detail in Chapter 9.

The shaded area in Figure 4.1 gives an approximate indication of the amount of overdraft the contractor must support at the bank pending reimbursement from the client. In order to reduce this financing as much as possible, the contractor would like to move the income curve as far to the left as possible.

One way to achieve this is to unbalance the bid. Essentially, for those items that occur early in the construction, inflated unit prices are quoted. For example, hand excavation that in fact costs $50 per cubic yard will be quoted at $75 per cubic yard. Foundation piles that cost $40 per linear foot will be quoted at $60 per linear foot. Since these items are overpriced, in order to remain competitive, the contractor must reduce the quoted prices for latter bid items. "Close-out" items such as landscaping and paving will be quoted at lower-than-cost prices. This has the effect of moving reimbursement for the work forward in the project construction period. It unbalances the cost of the bid items leading to front-end loading.

The amount of overdraft financing is reduced, as shown by the income and expense profiles in Figure 4.2. Owners using the unit-price contract format are usually sensitive to this practice by bidders. If the level of unbalancing the quotations for early project bid items versus later ones is too blatant, the owner may ask the contractor to justify his price or even reject the bid.

Some contracts obviate the need to unbalance the bid by allowing the contractor to quote a "mobilization" bid item. This essentially allows the bidder to request front money from the owner. The mobilization item moves the income curve to the left of the expense curve (see Fig. 4.3). The contractor in the normal situation (Fig. 4.1) will bid the cost (e.g., interest paid) of financing the income/expense difference into his prices. Therefore, the owner ultimately pays the cost of financing the delay in payments of income. If the owner's borrowing (i.e., interest) rate at the bank is better than that of the contractor, money can be saved by providing a mobilization item, thereby offsetting the contractor's charge for interim financing. Large owners, for instance, are often able to borrow at the prime rate (e.g., 8 or 9%), while contractors must pay several percent above the prime rate (e.g., 11

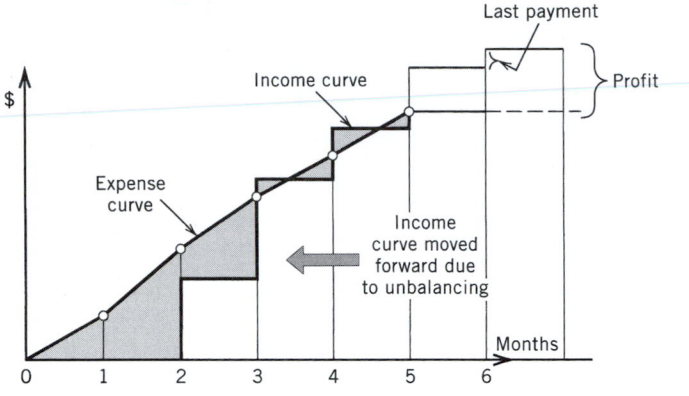

Figure 4.2 Unbalanced bid income profile.

or 12%). By providing a mobilization item, the owner essentially assumes the overdraft financing at his rate, rather than having the contractor charge for financing at the higher rate.

In addition to the flexibility in accommodating the variation in field quantities, the unit-price contract has the added advantage to the contractor that quantity estimates developed as part of the bidding process need only verify the guide quantities given in the bid item list (i.e., schedule). Therefore, the precision of the quantity takeoff developed for the contractor's estimate prior to construction need not be as exact as that developed for a fixed-price (lump-sum) contract. The leeway for quantity deviation about the specified guide quantities also normally reduces the number of change orders due to the automatic allowance for deviation.

Because of its flexibility, the unit-price contract is almost always used on heavy and highway construction contracts where earthwork and foundation work predominate. Industrial rehab work can also be contracted using the unit-price contract format with bid item list for price quotation. Major industrial facilities are typically bid using the negotiated contract format.

4.7   NEGOTIATED CONTRACTS

An owner can enter into contract with a constructor by negotiating the price and method of reimbursement. A number of formats of contract can be concluded based on negotiation between owner and contractors. It is possible, for example, to enter into a fixed-price or unit-price contract after a period of negotiation. In some cases, public owners will negotiate with the three low bidders on prices, materials, and schedule.

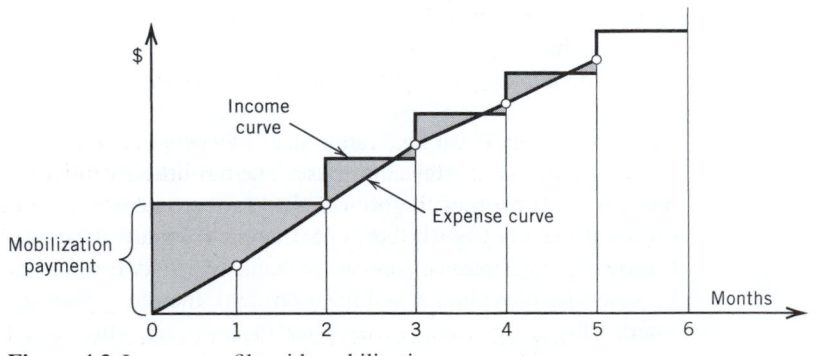

Figure 4.3 Income profile with mobilization payment.

The concept of negotiation pertains primarily to the method by which the contractor is selected. It implies flexibility on the part of the owner to select the contractor on a basis other than low bid. Therefore, a contractor competing for award of contract in the negotiated format cannot expect to be selected solely on the basis of low bid. This affects the bid cycle and the completeness of plans and specifications that must be available at the time of contractor selection. The owner invites selected contractors to review the project documentation available at the time of negotiation. This documentation may be total and complete design documentation as in the case of competitively bid contracts or only concept-level documentation. Based on the documentation provided, the contractor is invited to present his qualifications to perform the work and to indicate his projected costs and fee for completing the work. Since the level of the design documentation can vary from total detail to preliminary concept drawings, the accuracy of the cost projections will also vary. Within this presentation format, the owner evaluates the experience, reputation, facilities, staff available, charge rates, and fee structures of the various bidders participating. Based on this evaluation, the field is reduced to two or three contractors, and negotiations are opened regarding actual contract form and methods of reimbursement.

Since in most cases, the design documentation is not complete at the time of negotiation, the most common form of contract concluded is the COST + FEE. In this type of contract, the contractor is reimbursed for expenses incurred in the construction of the contracted facility. The contract describes in detail the nature of the expenses that are reimbursable. Normally, all direct expenses for labor, equipment, and materials as well as overhead charges required to properly manage the job are reimbursable. In addition, the contractor receives a fee for his expertise and the use of his plant in support of the job. The fee is essentially a profit or markup in addition to the cost reimbursement. The level and amount of fee in addition to the charge schedule to be used in reimbursement of the direct costs are major items of discussion during negotiation. Various formulas are used for calculating the fee and strongly influence the profitability of the job from the contractor's standpoint.

As in the case of competitively bid contracts, the contractor does the financing of the project and is reimbursed by periodic (e.g., monthly) progress payments. Both parties to the contract must agree to and clearly define the items that are reimbursable. Agreement regarding the accounting procedures to be used is essential. Areas of cost that are particularly sensitive and must be clearly established are those relating to home office overhead charges.[1] If the owner is not careful, he may be surprised to find out he has agreed to pay for the contractor's new computer network. Other activities that must be clearly defined for purposes of reimbursement are those pertaining to award and control of subcontracts as well as the charges for equipment used on the project.

Four types of fee structure are common. They lead to the following cost-plus types of reimbursement schemes:

1. Cost + percent of cost
2. Cost + fixed fee
3. Cost + fixed fee + profit-sharing clause
4. Cost + sliding fee

The oldest form of fee structure is the *percent of cost* form. This form is very lucrative for the contractor but is subject to abuse. There is little incentive to be efficient and economical in the construction of the project. Just to the contrary, the larger the cost of the job, the higher the amount of fee that is paid by the owner. If the cost of the job is $40 million and the fee is 2%, then the contractor's fee is $800,000. If the costs increase to $42 million the

---

[1] Overhead charges pertain to costs that cannot be linked to a specific work item such as concrete placement but are required for proper supervision and control of the project.

contractor's fee increases by $40,000. Abuse of this form of contract has been referred to as the 'killing of the goose that laid the golden egg.' Therefore this form of reimbursement is seldom used.

In order to offset this flaw in the percent-cost approach, the fixed-fee formula was developed. In this case, a fixed amount of fee is paid regardless of the fluctuation of the reimbursable cost component. This is usually established as a percent of an originally estimated total cost figure. This form is commonly used on large multiyear industrial plant projects. If the projected cost of the plant is $500 million, a fixed fee of 1% of that figure is specified and does not change due to variation from the original estimated cost. Therefore, the contractor's fee is fixed at $5 million. This form gives the contractor an incentive to get the job done as quickly as possible in order to recover his fee over the shortest time frame. Because of the desire to move the job as quickly as possible, however, the contractor may tend to use expensive reimbursable materials and methods to expedite completion of the project.

The fixed-fee plus profit-sharing formula provides a reward to the contractor who controls costs, keeping them at a minimum. In this formula it is common to specify a target price for the total contract. If the contractor brings the job in under the target, the savings are divided or shared between owner and contractor. A common sharing formula provides that the contractor shares by getting 25% of this underrun of the target. If, for instance, the target is $15 million and the contractor completes the job for $14.5 million, he receives a bonus of $125,000. The projection of this underrun of the target and the percent bonus to be awarded the contractor are used by some construction firms as a measure of the job's profitability. If the contractor exceeds the target, there is no profit to be shared.

In some cases, the target value is used to define a guaranteed maximum price (GMP). This is a price that the contractor guarantees will not be exceeded. In this situation, any overrun of the GMP must be absorbed by the contractor. The GMP may be defined as the target plus some fraction of the target value. In the example above, if the target is $15 million, a GMP of $16 million might be specified.

In this form, a good estimate of the target is necessary. Therefore, the plans and concept drawings and specifications must be sufficiently detailed to allow determination of a reasonable target. The incentive to save money below the target provides an additional positive factor to the contractor. The owner tends to be more ready to compromise regarding acceptance of the project as complete if the job is under target. Additional work on punch list costs the contractor 25 cents, but it costs the owner 75 cents. The quibbling that is often present at the time the punch list is developed is greatly reduced to the contractor's advantage.

A variation of the profit-sharing approach is the *sliding fee*, which not only provides a bonus for underrun but also penalizes the contractor for overrunning the target value. The amount of the fee increases as the contractor falls below the target and decreases as he overruns the target value. One formula for calculating the contractor's fee based on a sliding scale is

$$Fee = R(2T - A)$$

where T = target price

R = base percent value

A = actual cost of the construction

Negotiated contracts are most commonly used in the private sector, where the owner wants to exercise a selection criterion other than low price alone. The negotiated contract is used only in special situations in the public sector since it is open to abuse in cases where favoritism is a factor. Private owners are also partial to the negotiated format of contracting because it allows the use of phased construction (see Fig. 4.4) in which design

CM–phased construction versus lump-sum bidding

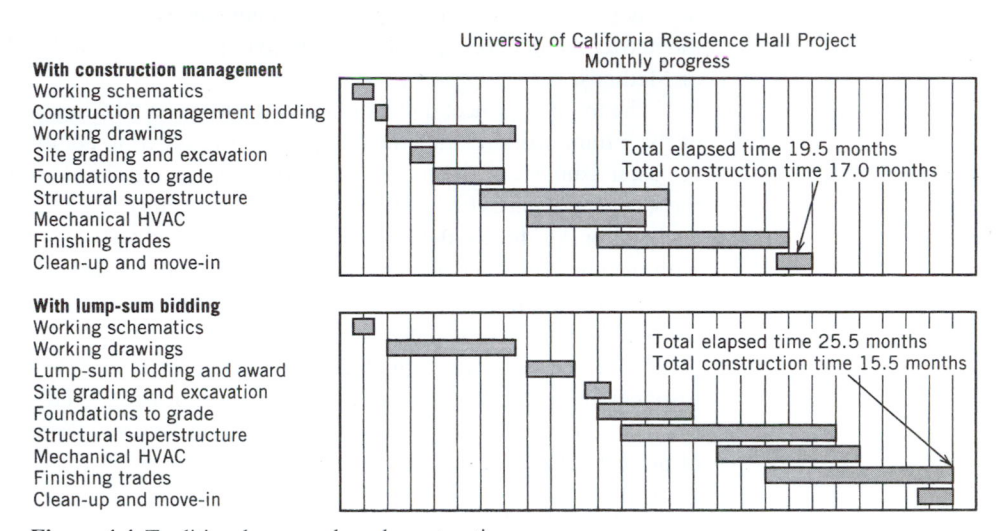

Figure 4.4 Traditional versus phased construction.

and construction proceed simultaneously. This allows compression of the classical "design first-then construct or design-bid-build" sequence. Since time is literally money, every day saved in occupying the facility or putting it into operation represents a potentially large dollar saving. The cost of interest alone on the construction financing of a large hotel complex can run as high as $50,000 a day. Financial costs generated by delays on complex industrial facilities are estimated at between $250,000 and $500,000 a day. Quite obviously, any compression of the design-build sequence is extremely important.

Large and complex projects have durations of anywhere from to 2 to 3 up to 10 years. For such cases, cost-plus contracts are the only feasible way to proceed. Contractors will not bid fixed prices for projects that continue over many years. It is impossible to forecast the price fluctuations in labor, material, equipment, and fuel costs. Therefore, negotiated cost-plus-fee contracts are used almost exclusively for such complex long-duration projects.

4.8   PROJECT DELIVERY METHODS

In ancient times, great structures were constructed by "master builders" who developed the project concept, designed the appearance and technical details of the finished building or monument, and mobilized the resources needed to realize the final structure. This classical approach was used to build the pyramids, the great castles and churches of the Middle Ages, and the civil engineering infrastructure of the industrial revolution. Master builders designed and constructed facilities acting as a single point of contact for the client.

Over the past 100 years, the processes of designing and building were gradually separated. Design and construction were viewed as separate endeavors. A design professional prepared the project plans, and a separate firm was contracted to perform the actual construction of a facility. This separation of activities also led to a sequencing of activities in which design was completed before construction commenced. This became the "traditional" sequence and is now referred to as Design-Bid-Build or DBB. This contracting procedure has been the basis for our discussions in Chapters 2 and 3. In contrast, master builders conducted design and construction simultaneously.

Over the past 30 years, a number of new concepts for project delivery have been developed to compress the time required to realize a constructed facility. It has been recognized that the DBB method of project delivery, with its sequential emphasis, leads to longer-than-necessary project time frames. It is advantageous from a time perspective to have design and construction proceed simultaneously. This has led to a reconsideration of the master builder concept and a discussion of what is meant by "Project Delivery Systems or Methods."

The topic of project delivery methods addresses "the organization or the development of the a framework relating the organizations required to complete or deliver a project and the establishment of the formal (i.e., contractual) and the informal relationships between these organizations." In the DBB approach, for instance, the owner holds a contract with the designer or A/E for the development of the plans and specifications and a separate contract with the construction contractor for the building of the facility. In other delivery systems, the owner contracts with a single group or entity for both the design and construction of a facility. Another accepted definition is as follows:

"A project delivery method is the comprehensive process of assigning the contractual responsibilities for designing and constructing a project. (AGC, 2004)."

Based on guidance given by the Associated General Contractors (AGC), the concept of project delivery addresses two critical issues:

1. Is the responsibility to the owner/client for project design and construction tied to a single entity (e.g., a performing group) or multiple entities? In other words, does the owner deal with a single entity or multiple entities when pursuing design and construction of a project?

2. Is the criteria for award based on lowest cost or on other criteria?

In the framework of this discussion, competitively bid contracts require multiple entities and the contract award is based on lowest quotation or cost to the owner. Negotiated contracts also involve multiple entities (e.g., architects, design professionals, construction contractors, etc.) and may be based on lowest cost, although, in certain cases other considerations take precedence over cost (e.g., expertise, previous performance, ability to react quickly, etc.). Two relatively new contract formats have been introduced which focus on simplifying the project delivery process. Low cost is not as central to these formats as in the competitive or negotiated types of contracts. The emphasis is on optimizing parameters other than cost (e.g., quality, time of completion, meeting market needs, safety, etc.). Design Build (DB) and Construction Management (CM) contracts differ from the traditional DBB format in terms of how they address the two critical issues of project delivery methods stated above. They also facilitate the use of "phased construction" or "fast-tracking" based on design and construction occurring in parallel (i.e., at the same time) in contrast to the sequential nature of the DBB approach.

4.9 DESIGN-BUILD CONTRACTS

As noted in Section 4.2, it is advantageous from the client's point of view to have a single contractor provide the entire project as a single contract package. In the 1970s, large firms began to offer both design and construction services in order to provide the client with a single source for project delivery. This approach of providing both design and construction services can be viewed as a natural evolutionary step beyond the negotiated contract and is often referred to as *integrated* design build. It has been common practice in industrial construction to use the design-build approach for complex projects that have tight time requirements. In such cases, it is advantageous for the client to have a single firm providing both design and construction services.

This system has the advantage that differences or disputes between the design team or group and the construction force are matters internal to a single company. This eliminates the development of an adversary relationship between two or more firms involved in realizing the project (i.e., it eliminates disputes between designer and constructor). Normally, the management of the design-build contractor is motivated to reconcile disputes or differences between design and construction in an expeditious and efficient manner. If such problems are not addressed, they can lead to loss of profit and potential dismissal of the contractor for poor performance.

Coordination between design and construction is also enhanced by having both functions within the same firm. This system improves the communication between designers and the field construction force and assists in designing a facility that is not only functional but is also efficient to construct. The firm is driven by the profit motive to optimize the design both for the functional life-cycle use of the building as well as to design for construction thus enhancing the efficiency of the construction process. This can be compared to the manufacturing design of the refrigerator referred to in Section 4.2. If a firm manufactures a refrigerator, it designs the appliance to be efficient for use in the home. It also designs the item so that it can be assembled in the most cost-efficient and timely manner so as to reduce production costs.

Design-build contracts also have the advantage that design and construction can be done concurrently. That means that work can be started in the field before a complete design is available. This allows for "phased construction," or a "fast track" approach as described above and a compression of the schedule since design must not be totally complete prior to commencement of construction. This compression of schedule is illustrated in Figure 4.4 in which design and construction proceed simultaneously.

During the 1970s this type of contract was used mainly on large and complex projects (e.g., petrochemical plants, industrial complexes, power plants, etc.) to improve the flow of information between the design team and the construction people in the field. Usually, only firms with large design and construction capabilities were able to provide design-build services. Projects built with a single design-build contractor were often referred to as "turn key" projects since the owner dealt with only one contractor and that contractor was charged with the completion of the facility so that the project was ready to be placed in operation at the "turn of a key." That is, this owner signed a single contract and said "Call me when you have the project complete and you want me to turn the key to start it up."

4.10   DESIGN-BUILD IN A CONSORTIUM FORMAT

In the past decade, the use of design-build contracts has become more common in the building construction sector. A number of firms have marketed this project delivery approach to private entrepreneurs (e.g., owners building hotels, apartment and office structures, etc.) in the building sector as a way to receive the best product in the most timely way at the best price. Since most building contractors do not have an in-house design capability, lead contractors typically form a team or consortium of designers and specialty contractors who work together to meet the needs of the client. The owner/client contracts with the consortium as a single group providing the total project package (e.g., design, construction, procurement, etc.).

Each member of the consortium is at risk and is motivated to work with other members to minimize delays and disputes. In effect, a group of designers and constructors form a consortium to build a project based on conceptual documentation provided by the owner. They agree to work together to achieve the project and, therefore, implicitly agree to avoid developing an adversarial relationship between one another.

The attraction of this consortium-based approach is the fact that the owner/client is given a stipulated-sum price for the project after 30 to 40% of the design of the project is completed. Barring major changes to the project, the consortium locks in the final price at the end of the preliminary design phase. This is very attractive to the owner since financing for the overall project can be lined up based on a definitive cost figure developed early in the design development process. This reduces the need for contingency funds and is attractive to the lender since the cost and, therefore, the amount of the borrowing is locked in.

The members of the consortium are motivated to be innovative and avoid disputes since failure to achieve the stipulated price quoted at the end of preliminary design will result in a loss to all of the members of the consortium. Again, the adversarial relationship typical of designer and constructors is largely eliminated since bickering and lack of cooperation among members of the team can lead to significant losses. Incentive to avoid disputes and to develop innovative solutions to field problems is inherent in this type of contract.

This consortium-based design–build contract has gained such wide acceptance in the private building construction community that it is now being used by the federal government on selected building projects. A number of large Internal Revenue Service (IRS) facilities have been built using design-build contracts. In its application in the private sector, stiff competition among a number of consortia for the same project has not been the normal case. In most cases, the owner/entrepreneur has worked with one or two lead contractors who form a design–build team to meet the customized requirements of the client. Competition among competing consortia in the private building construction sector has not been a major issue.

With the advent of the use of this method in the public sector, competition has been a major factor. Selection of the winning consortium is based on competitive review of proposals from each consortium. In this format, the nonselected consortia incur substantial losses based on the cost of organizing and developing a competitive proposal.

4.11   CONSTRUCTION MANAGEMENT (CM) CONTRACTS

The Construction Management format became popular in the 1970s as a method which provided a single point of coordination to owners. In this format, a single firm or entity called the Construction Manager (CM) is retained to coordinate all activities from concept design through acceptance of the facility. This firm represents the owner in all construction management activities. The CM coordinates the selection of all design and construction entities (firms) and supervises and controls the pre-design, design, pre-construction, and construction activities related to the project on behalf of the owner. In effect, the CM acts as the "agent" of the owner. For this reason, this contract format is often called agency construction management (CMa). In this type of contract, construction management is defined as "a group of management activities related to a construction program, carried out during the pre-design, design, and construction phases, that contributes to the control of time and cost in the construction of a new facility." The construction management firm's position in the classical relationship linking owner, contractor, and architect engineer is as shown in Figure 4.5.

One item of importance with the CMa approach is that although the construction manager works on behalf of the owner in managing and coordinating all aspects of the project, all contracts between design and construction firms involved in realizing the project are signed by the owner. Therefore, the CM is not "at risk." The CM is a coordinator and is responsible only to provide management services that are consistent with the norms of the industry established for other firms performing construction manager services.

In this coordination function, the CMa firm acts as a traffic cop monitoring and controlling the flows of information among all parties active on the project. The CM establishes the procedures for award of all contracts to architect/engineers, principal vendors, and the so-called trade or specialty contractors. Once contractual relationships are established, the CM controls not only the prime or major contractor but all subcontractors as well as major vendors and off-site fabricators. Major and minor contractors on the site are referred

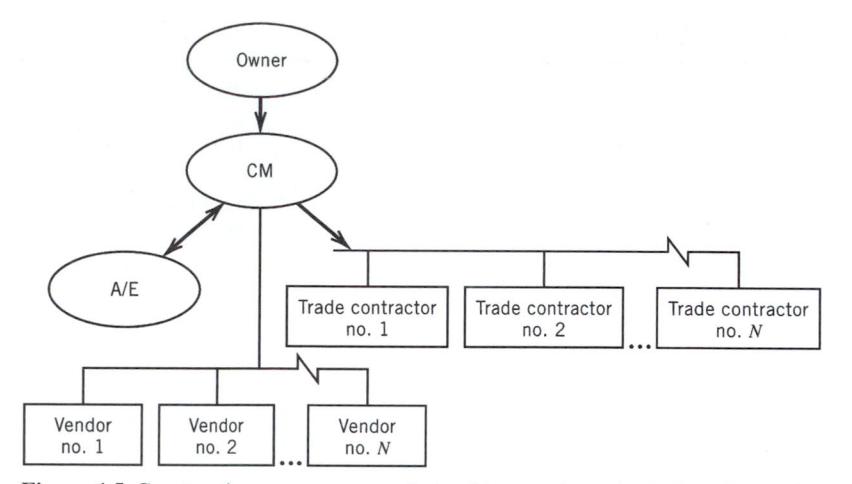

Figure 4.5 Construction management relationships to other principal parties on the project.

to as trade or specialty contractors. In this control or management function, the CM firm utilizes the project schedule as a road map or flight plan to keep things moving forward in a timely and cost-effective manner. The major functions carried out by the CM firm vary depending on whether the project is in the (1) predesign, (2) design, or (3) construction phase.

Agency construction management contracts are particularly attractive for organizations that periodically build complex structures (e.g., hospital authorities, municipalities, transit authorities, etc.) but do not desire to maintain a full-time construction staff to supervise projects on a recurring basis. In such cases an owner can retain a CMa firm to plan, develop, and coordinate the activities of one or more design professionals, trade contractors, vendors, and other interested parties such as licensing and control bodies.

4.12 CONSTRUCTION MANAGEMENT AT-RISK

Agency CMs are coordinators working on behalf of the client and are not contractually liable for the successful completion of the work. In addition, problems arising out of decisions made as a part of their oversight of the project often lead to legal issues which are difficult to resolve. For instance, if the CMa implements safety procedures on behalf of the owner which are found to be faulty and lead to unsafe construction operations, the owner, not the CMa, is normally viewed as liable for legal suites or claims arising out of these procedures. Therefore, the client, through lack of experience in dealing with construction problems and protocols, may be put at risk by poor decisions made by the CMa while supervising and controlling the project.

To close this loop hole, a version of the CM contract format has become popular in which the construction manager not only coordinates the project, but also assumes responsibility for the construction phase of the work. In the CM at-Risk contract, the CM assumes the same risk that a construction contractor in the DBB format would assume for the successful completion of the project. In this situation, the CM-at Risk signs all contracts related to the construction phase of the work. The design and other pre-constructions contracts and responsibilities are signed by or remain with the owner. Prior to the commencement of construction, the CM at-Risk provides services similar to those provided by a CMa firm. The CM at-Risk contract is similar to the DBB format in that design and construction are separate contracts. It differs from the DBB format, however, in that the selection criteria are based on issues other than lowest total construction cost.

Table 4.1   Project Delivery Method (PDM)

	Contract Type	Single or Multiple Contracts to Owners	Selection Criteria	Phased Construction
DBB	Competitively Bid	Multiple-Design Contract & Construction Contract	Low Construction Cost	No
	Negotiated	Multiple Design & Construction	Low Cost or Other	Possible
DB	DB	Single Contract with DB Firm	Usually not Low Cost— Based on Performance	Yes
CM	CMa	Contracts held by Owner—CM, Design, Construction, and Vendors	Based on Performance Expectations	Yes
	CM at Risk	Same as CMa above except CM and Construction Contracts are Combined	Based on Performance Expectations	Yes

4.13   COMPARING PROJECT DELIVERY METHODS

From the project delivery perspective, competitively bid contracts are required to be design-bid-build (DBB) contracts. Negotiated contracts can be viewed as DBB contracts although it is not unusual for design and construction to proceed in parallel (i.e., simultaneously) in a given situation. In other words, it is possible to use "phased construction" when working with negotiated construction contracts. A constructor involved in construction of a hotel building on a cost reimbursable basis can begin construction of the site excavation and sub-basements while the roof-top restaurant is still being designed. For both competitively bid and negotiated contracts, the owner holds separate contracts with the designer or design group and with the construction contractor.

In the Design Bid (DB) format, the owner enters into contract with a single entity—the Design Builder. The basis of selection of the Design Build firm or consortium is normally on the basis of considerations other than least cost. Fast tracking or phased construction is typical of DB contracts.

In both forms of the CM format, the owner holds multiple contracts. In the case of the *Agency Construction Management* format, the owner signs a management contract with the CMa, but holds contracts directly with the design and construction firms involved. Selection of the Agency CM firm is based on issues other than total construction cost (e.g., quality, schedule performance, etc.) Fast track construction is usual when using this format.

The CM at-Risk format requires separate contracts for the design team and the CM at-Risk firm (similar to the DBB format). Low total construction cost is not the basis for selection of the Construction Manager at-Risk. Fast track construction is possible when using this format. The major types of Project Delivery Methods are summarized in Table 4.1.

REVIEW QUESTIONS AND EXERCISES

4.1   Name and briefly describe each of the two basic types of competitively bid construction contracts. Which type would be most likely used for building the piers to support a large suspension bridge? Why?

4.2   If you were asked to perform an excavation contract competitively with limited boring data, what type of contract would you want and why?

4.3   Name three ways the construction contract can be terminated.

4.4   Name two types of negotiated contracts and describe the method of payment and incentive concept.

4.5   What is meant by unbalancing a bid? What type of contract is implied? Give an example of how a bid is unbalanced.

4.6   Why is cost plus a percentage of cost type of contract not used to a great extent?

4.7   Under what circumstances is a cost-plus contract favorable to both owner and contractor?

4.8   Valid contracts require an offer, an acceptance, and a consideration. Identify these elements in the following cases:

    a. The purchase of an item at the store

    b. The hiring of labor

    c. A paid bus ride

    d. A construction contract

    e. The position of staff member in a firm

4.9   Suppose you are a small local building contractor responsible for the construction of the small gas station in Appendix I. List the specialty items that you would subcontract.

4.10   Visit a local building site and ascertain the number and type of subcontracts that are involved. How many subcontracts do you think may be needed for a downtown high-rise building? Why would there be more subcontractors in a building job as opposed to a heavy construction job?

4.11   From the point of view of the owner's contract administrator, each different type of contract places different demands on supervision. List the significant differences that would impact the complement (number) of field personnel required to monitor the contract.

4.12   Visit a local contractor and determine the proportion of contracts that are negotiated against those that are competitively bid awards. Is this percentage likely to change significantly with small building contractors? Is there a difference between building contractors and heavy construction contractors?

Chapter 5

Legal Structure

Hoover Dam (http://www.usbr.gov)

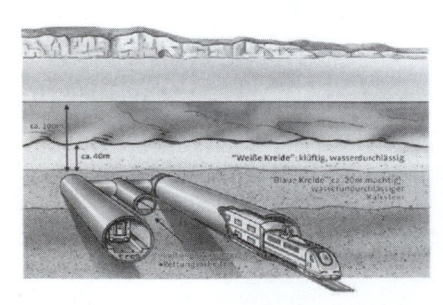

Channel Tunnel

Joint Venturing

The Need

Larger and complex projects often exceed the capability of a single firm or contractor to do the work. In such cases, a team or group of contractors will combine their resources to bid and execute the work. A famous example of this situation is the construction of the Hoover Dam during the 1930s, in which a group of six of the largest contractors in the U.S. banded together to build this project. The approach of a group of firms or professionals establishing a team to complete a project is referred to as "a joint venture." Joint ventures are also referred to as consortia (e.g., a single team or group is a consortium).

There are many reasons why firms will decide to legally combine for a specific period of time to pursue a given project. A given project may be so large that the financial resources of a number of companies are required to bond the project. For instance, the construction of an addition to the McCormick Exhibition Center in Chicago was so large that bonding companies were not prepared to provide performance and payment bonds to a single company. Being prudent, the bonding companies wanted to spread their risk.

This was also a consideration in the financial structure for the construction of the Channel Tunnel which led to the establishment of a large consortium of companies (both French and English) who allied for the specific purpose of construction the "Chunnel." From a financial, political and technical perspective, it would not have been possible for one company to build this epic project.

The Approach

A joint venture is a business relationship undertaken by two or more companies to form a legal entity for the purpose of performing a specific work item or, in the case of construction, a given project. A team of firms may be involved in both the design and construction of a project or only (as in the Hoover Dam) the construction phase of a project. In any case, the owner contracts with the joint venture.

Therefore, the joint venture must be legally established in a rigorous fashion so that the contract required is binding. A major benefit to the owner/client when working with a joint venture is that the owner deals with single entity as opposed to a number of different companies. From a management point of view, the owner has a single point of contact or a single contractor made up of many sub-entities.

Joint ventures differ from proprietorships, partnerships, and corporations in that they exist for a fixed period of time defined by the duration of the project being undertaken. Therefore, as a legal entity they exist to perform a given objective and are then dissolved.

The partners in a joint venture must each bring important contributions or capabilities to the undertaking. Each firm in the consortium brings special abilities which may include technical expertise, financial resources, or special knowledge—all of which are key to successful completion of the project being pursued. In the Design Build contract format discussed in Chapter 4, some of the partners are involved with the design aspects of the project while others are focused purely on the construction phase of the work. The legal aspects of joint venture formation and operation are typically unique to each project and will vary based on special aspects of the team partners (e.g., is the consortium multinational or not, etc.) and the location and nature of the project being constructed.

5.1   TYPES OF ORGANIZATION

One of the first problems confronting an entrepreneur who has decided to become a construction contractor is that of deciding how best to organize the firm to achieve the goals of profitability and control of business as well as technical functions. When organizing a company, two organizational questions are of interest. One relates to *the legal organization* of the company, and the second focuses on the *management organization*. The legal structure of a firm in any commercial undertaking, be it construction or dairy farming, is extremely important since it influences or even dictates how the firm will be taxed, the distribution of liability in the event the firm fails, the state, city, and federal laws that govern the firm's operation, and the firm's ability to raise capital. Management structure establishes areas and levels of responsibility in accomplishing the goals of the company and is the road map that determines how members of the firm communicate with one another on questions of common interest. The types of company legal organization will be considered in this chapter.

5.2   LEGAL STRUCTURE

At the time an entrepreneur decides to establish a company, one of the first questions to be resolved is which type of legal structure will be used. The nature of the business activity may point to a logical or obvious legal structure. For instance, if the entrepreneur owns a truck and decides to act as a free agent in hauling materials by contracting with various customers, the entrepreneur is acting alone and is the proprietor of his own business. In situations where a single person owns and operates a business activity and makes all of the major decisions regarding the company's activity, the company is referred to as a *proprietorship*. If the business prospers, the entrepreneur may buy additional trucks and hire drivers to expand his fleet, thereby increasing business. The firm, however, remains a proprietorship even if he has 1000 employees so long as the individual retains ownership and sole control of the firm.

If a young engineer with management experience and a job superintendent with field experience decide to start a company together, this firm is referred to as a partnership. The size of a *partnership* is not limited to two persons and may consist of any number of partners. Law firms as well as other professional companies (e.g., accounting firms) are often organized as partnerships consisting of as many as 10, 12, or more partners. If two or three individuals decide to form a partnership, the division of ownership is decided by the initial contribution to the formation of the company on the part of each partner. The division of ownership may be based solely on the monetary or capital assets contributed

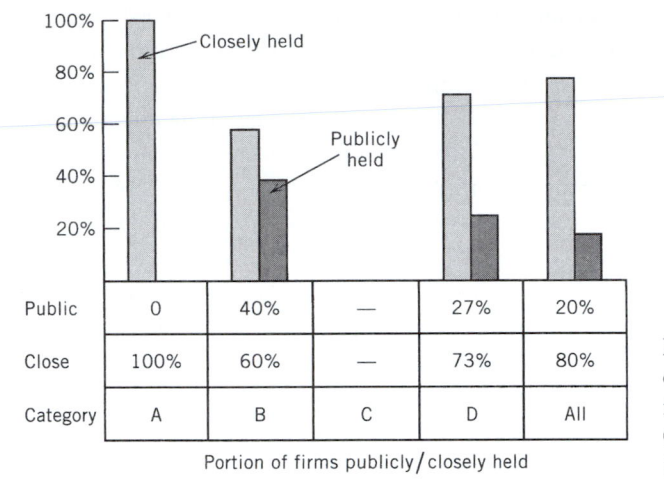

Public	0	40%	—	27%	20%
Close	100%	60%	—	73%	80%
Category	A	B	C	D	All

Portion of firms publicly/closely held

Figure 5.1 Forms of legal ownership in the construction industry (study by T. Gibb, Georgia Institute of Technology, 1975).

by each partner. Therefore, if three individuals form a partnership with two contributing $20,000 and a third contributing $10,000, the division of ownership among the partners is 40, 40, and 20%. In other cases, one of the partners may bring a level of expertise that is recognized in the division of ownership. For instance, in the example just cited, if the partner contributing the $10,000 were the expert in the area of business activity to be pursued, his expertise could be valued at the nominal level of $10,000, making his overall contribution to the firm $20,000. Therefore, ownership would be equally divided among the three partners. The actual division of ownership is usually specified in the charter of the partnership. If no written charter exists, and the partnership was concluded by verbal mutual agreement only, the assumption is that the division of ownership among the partners is equal.

In some business activity the risk of failure or exposure to damage claims may be such that a corporate structure is deemed appropriate. This form of ownership recognizes the company itself as a legal entity and makes only those assets that belong to the firm attachable for settlement of claims in the event of bankruptcy or damage claims. This allows principals or stockholders in a corporation to protect their personal and private assets from being called in to settle debts or claims arising out of the firm's operation or insolvency. Therefore, if a stockholder in a corporation has private assets of $1 million and the corporation declares bankruptcy, the $1 million cannot be attached to settle debts of the corporation.[1] Other desirable features of corporate structure that cause firms to select this legal structure will be discussed later in this chapter.

Two types of corporations are commonly encountered. Corporations in which a small number of persons hold all of the stock in the firm are referred to as *close* or *closely held* corporations. This form of ownership is very common in the construction industry since it offers risk protection and also allows a small group of principals to control company policies and functions. A *public* corporation, in contrast to a *closely held* corporation, allows its stock to be bought and sold freely. The actual ownership of the stock varies daily as the stock is traded by brokers, in the case of large corporations, on the stock market. Figure 5.1 gives a graphical indication of the forms of legal ownership utilized by a set of building construction companies located throughout the southeastern United States. In this example, the companies have been grouped according to the volume of work done using

---

[1] In certain situations, stockholders may by ancillary agreement, such as bond, waive some of the protection offered by the corporate structure and find that their personal assets are subject to attachment.

fixed-price contracts versus that done using negotiated contracts. The groups were defined as follows:

Group A. Contractors doing 25% or less of their volume in negotiated contract format.

Group B. Contractors doing between 25 and 50% negotiated work.

Group C. Contractors doing between 50 and 75% negotiated work.

Group D. Contractors doing more than 75% of their work in negotiated contract format.

The figure indicates that the close corporation format is very popular.

Another form of organization that has legal implications is the joint Venture. This is not a form of ownership but a temporary grouping of existing firms defined for a given period to accomplish a given task or project. A joint-venture organizational structure is used when a very large project is to be constructed and requires the pooling of resources or expertise from several companies. Typically the companies establish a basis for division of responsibility on the job and cooperate toward the end of successfully completing the project. They are bound together for a period of cooperation by a legal agreement that defines the nature of the relationship. Joint venturing first became popular during the construction of large dams such as the Grand Coulee and Hoover Dam in the western United States and has since been used for a wide variety of large construction tasks.

5.3   PROPRIETORSHIP

The simplest form of legal structure is the proprietorship. In this form of business ownership, an individual owns and operates the firm, retaining personal control. The proprietor makes all decisions regarding the affairs of the firm. The assets of the firm are held totally by one individual and augment the individual's personal worth. All revenue to the firm is personal cash revenue to the proprietor, and all losses or expenses incurred by the firm are personal expenses to the proprietor. The proprietor is, therefore, taxed as an individual and there is not separate taxation of the firm. Consider Uncle Fudd, who has a small contracting business. The firm generated $187,000 in total volume during the calendar year. The firm has $100,000 in expenses, so that the before-tax income of the firm is $87,000. Uncle Fudd declares this income on his personal income tax return. Assuming this is his total income (i.e., he received no further income from other sources) and that he has $17,000 in deductions and exemptions, his taxable income is $70,000.

Since the owner's capital and that of the firm are one and the same, the credit that the firm can obtain and its ability to generate new capital are limited by the personal assets of the proprietor. Furthermore, any losses incurred by the firm must be covered from the personal assets of the proprietor. Any liabilities incurred by the firm are the owner's liability, and he must cover them from his personal fortune. Therefore, bankruptcy of the firm is personal bankruptcy. Since there is no limitation of liability, high-risk businesses do not normally use the proprietorship form of structure.

The life of the proprietorship corresponds to that of the owner. Upon the death of the owner, the proprietorship ceases to exist. Assets of the proprietorship are normally divided among the heirs to the proprietor's estate.

5.4   PARTNERSHIP

The partnership is similar to the proprietorship in the sense that liabilities of the firm are directly transmitted to the partners. That is, there is no limitation of liability. However, in this case, since there are two or more partners, the liability is spread among several principals. The reason for forming a partnership is based on the principle of division of risk and pooling of management and financial resources. The ownership of the firm is shared among the partners to a degree defined in the initial charter of the partnership. Since several persons come together to form a partnership, the capital base of the firm is broadened to include the personal assets of the partners involved. This increase in assets increases the line of credit available to a partnership as opposed to a proprietorship. Control of the firm, however, is divided among the principals, who are called *general partners*. Partners share the profits and losses of the firm according to their degree of ownership as defined in the partnership agreement, but since the liability of each of the partners is not limited, one partner may carry more liability in the case of a major loss. Assume that Carol, Joan, and Bob are partners in a small contracting business. The personal fortunes and percent ownership of the three principals are as follows:

Carol	$1,400,000	40% ownership
Joan	800,000	30% ownership
Bob	100,000	30% ownership

The firm loses $1,000,000 and must pay this amount to creditors. The proportionate shares of this loss are:

Carol	$400,000
Joan	300,000
Bob	300,000

However, since Bob can only cover $100,000, the remaining $900,000 must be carried by Carol and Joan in proportion to their ownership share.

A *limited partnership*, as the term implies, provides a limit to the liability that is carried by some partners. This concept allows the general partners to attract capital resources to the firm. The *limited partner* is liable only to the extent of his or her investment. Assume that Tom comes into the partnership described above as a limited partner. He makes $200,000 available for the capitalization of the firm. The percentages of ownership are redefined to provide Tom with 15% ownership. He, therefore, shares in the profit and loss of the firm in this proportion. Nevertheless, his level of loss is limited to the $200,000 he has invested. No amount beyond this investment can be attached from his personal fortune to defray claims against the firm. This provides the general partners with a mechanism to attract wealthy investors who desire liability limitation but profit participation. Limited partners have the position of a stockholder in a corporation in that loss is limited to the amount of their investment.

Limited partners have no voice in the management of the firm. Therefore, the *general partners* retain the same level of control but increase the capital and credit bases of the firm by bringing in limited partners. There must be at least one general partner in any partnership. The limited form of partnership (i.e., a partnership that includes limited partners) is more difficult to establish and subject to more regulation by state chartering bodies (usually the Office of the Secretary of State of the state in which the partnership is chartered). This is because limited partnerships realize some of the advantages available in the corporate legal structure. Corporations are subject to close control by state chartering bodies.

The contribution made by the limited partner must be tangible. That is, the limited partner cannot contribute a patent, copyright, or similar instrument. The contribution must have a tangible asset value (i.e., equipment, cash, notes. shares of stock in a corporation, etc.).

Any partnership is terminated in the event of the death of one of the partners. However, arrangements can be made to provide for the continuity of the partnership should one of the partners die. An agreement can be made among the partners that in the event of the death of a partner the remaining principals will purchase the ownership share of the deceased partner. Usually a formula that recognizes the fluctuating worth of the partnership is adopted in this agreement. The remaining partners pay this amount to the estate of the deceased partner.

General partners who are actively involved in the day-to-day management of the firm may decide to pay themselves a salary. In this way, the time and level of expertise contributed to the operation of the partnership are recognized. This level of day-to-day participation may be different from the level of initial contribution made in capitalizing the firm. In the case of Carol, Joan, and Bob, the levels of ownership were 40, 30, and 30%, respectively. If Bob is most active in the management of the partnership, he may be paid a full-time salary to recognize his commitment. Carol and Joan being active only on a part-time basis will be paid proportionately smaller or part-time salaries. Taxation, in any case, will be on both salary and earnings deriving from the operation of the partnership.

The action of one partner is binding on all partners. For instance, in the partnership described, if Joan enters into a contract to construct a building for the client, this agreement binds Bob and Carol as well. In this sense, a partnership is a "marriage," and any partner must be able to live with any commitment made on behalf of the partnership by another partner. On the other hand, it is not proper for a partner to sell or mortgage an asset of the partnership without the consent of the other partners. If the partner sells the asset, the income accrues to the partnership. If the partner utilizes a partnership asset to secure a personal note or loan, the other partners could advise the noteholder that they contest the use of this asset as security.

5.5   CORPORATION

A corporation is a separate legal entity and is created as such under the law of a state in which it it chartered. In most states, corporations are established by applying to the office of the secretary of state or similar official. This office issues a chartering document and approves the initial issuance of shares of stock in the corporation to establish the level of ownership of initial stockholders. As in the case of a partnership, the initial stockholders contribute financial capital and expertise as well as other intangible assets such as patents and royalty rights. The level of contribution is recognized by the number of shares of stock issued to each of the founding stockholders. If, in the partnership just described, Carol, Joan, and Bob decided to incorporate and the level of ownership was to remain the same, shares in the proper proportion would be issued to each principal. The number of shares and the share value defined at the initialization of a corporation are arbitrary and are selected to facilitate the recognition of ownership rather than actual value of the corporate assets. If the Carol-Joan-Bob (CJB) Corporation is established by the issuance of 1000 share of stock, Carol would receive 400 shares (40%), and Joan and Bob would receive 300 shares each (30%). For simplicity, each share could have a par value of one dollar. This assignment of one dollar per share simplifies the unit (i.e., share value) used to recognize ownership. On the other hand, the initial capital contributed to the formation of the corporation might have been $100,000. Therefore, the book value of each share of stock would be $100 per share. The book value of each share of a corporation is the net worth of the corporation divided by the number of shares issued. In this case, 1000 shares are issued and the asset value is $100,000. Therefore, each share has a book value of $100.

In addition to the par and book values associated with a share of stock in a corporation, each share has a traded or market value. This is the value that is listed on stock exchanges for those publicly traded corporation shares and that is printed in the newspaper. It indicates what the general public or stock traders are willing to pay for a share of ownership in the corporation. If the future looks good, traders will anticipate an increase in the value of the corporation's stock and will pay to own a stock that is increasing in value. If the corporation is about to experience a loss, the market price of the stock may indicate this by declining in value. To illustrate, if CJB, Incorporated, wins a contract that promises to net the corporation an after-tax profit of $100,000, the market price of the stock will tend to move up. In fact, as already noted, most construction firms hold their stock closely and do not trade it publicly. Therefore, the market value of the stock is of interest primarily to the giant construction firms that are publicly traded.

Because of the legal procedures required, the corporation is the most complicated form of ownership to establish. A lawyer is normally retained to prepare the proper documents, fees must be paid to cover actions by the chartering body (e.g., Office of the Secretary of State), printed stock is prepared, and formal meetings by the principals are required. Since the corporation can sell further stock to raise capital, it has an advantage in this respect over the proprietorship and the partnership. This power to sell stock can be and has been abused. Once a corporation is established, it may sell stock to unsuspecting buyers based on an idea or concept that is not properly presented or explained. For this reason and others, the corporation is closely controlled by the chartering agency in regard to its issuance and sale of additional stock. Federal law also dictates certain aspects of the presentation of corporate stock for sale.

The most desirable aspect of the corporate structure to businesses that are exposed to high risk such as the construction industry is its limitation of liability. Since the corporation is a legal entity of itself, only the assets of the corporation are subject to attachment in the settling of claims against and losses incurred by the corporation. This means that stockholders in a corporation can lose the value of their investment in stock, but that is the limit of their potential loss. Other assets that they own outside of the corporation cannot be impounded to offset debts against the corporation.

One disadvantage associated with the corporation is the double-taxation feature. Since the corporation is a legal entity, it is subject to taxation. The same profit that is taxed within the corporation is taxed again when it is distributed to stockholders as a dividend. This distributed profit becomes taxable as personal income to the individual stockholders. Assume that CJB Corporation has a before-tax profit (e.g., revenue − expenses) of $100,000 during the corporation's first year of operation. Let us assume the CJB Corporation is taxed by the IRS at the rate of 34% of profit provided income is in excess of $75,000. The corporation would be taxed $34,000 for $100,000 of before-tax profit.[2] The after-tax income would be $66,000. Assume the CJB decides to distribute $30,000 to the three stockholders. That is, Carol, Joan, and Bob as directors of their closely held corporation distribute $30,000 to themselves and retain $36,000 of these earnings within the corporation as working capital. In this case Carol, the major stockholder, receives a dividend of $12,000. Joan and Bob would receive $9000 each. If we assume that each stockholder pays approximately 25% on personal taxable income, Carol will pay $3000 in tax on this dividend, and Joan and Bob will pay $2250. In other words, the federal tax at the corporation and stockholder levels combined will be $34,000 plus $7500, or $41,500.

The double-taxation feature does not always prove to be a disadvantage. Returning to the situation of Uncle Fudd who is organized as a proprietorship, assume his before-tax income with the proprietorship is $147,000.[3] Assume that Uncle Fudd decides to incorporate his proprietorship and become Fudd Associates, Inc. As president of this corporation, Uncle Fudd pays himself a salary of $85,000. At this salary level, Uncle Fudd is taxed at 21% of his taxable income (i.e., his gross income minus deductions and exemptions). In the proprietorship format, his tax would be 25% of $147,000 minus $12,000 in deductions and exemptions.[4] He will pay 25% of $135,000, or $33,750 in tax. In the corporate format,

---

[2] Corporate taxation levels vary over time due to changes in Federal and State legislation.

[3] This example is different from the previous situation in which the taxable income was $87,000.

[4] The corporate rate for less than $75,000 taxable income is assumed to be 25%.

Uncle Fudd's tax will be:

$147,000

−85,000   Fudds Salary = expense

───────

$62,000   Gross income of corporation

Corporate tax $= 0.25(62,000) = \$15,500$ (See footnote 4, page 84)

Personal tax $= (0.21)\{\$85,000 - 12,000 \text{ (deductions and exemptions)}\}$

$= \$15,330$

Therefore, Uncle Fudd's tax in the corporate format will be $15,500 + $15,330 = $30,830. In this case, the corporate form of ownership yields a lower tax payment despite the double taxation. For this reason, a good tax consultant is a very valuable advisor when deciding which form of ownership is most appropriate.

Certain states provide for a special corporate structure that avoids the double-taxation feature of a normal corporation but retains the protection of limited liability. This is referred to as a subchapter "S" corporation. In a subchapter S corporation, the principals are taxed as if they were members of a partnership. That is, corporate income is taxed only once as personal income. The corporate shareholders are, however, still protected and their loss is limited to the value of the stock they possess.

As noted earlier, the corporation is very advantageous when attraction of additional capital is of interest. Figure 5.2 shows a typical stock certificate as issued at the time of incorporation. The certificate indicates that 250 shares of stock are represented. In addition, the corporation has authority to issue a total of 50,000 shares. Therefore, the directors of a corporation can decide to raise money for capital expansion by selling stock rather than borrowing money. This provides for the generation of additional capital by distributing ownership. It has the advantage that the money generated is not subject to repayment and therefore is not a liability on the company balance sheet.

The corporation also has a continuity that is independent of the stockholders. Unlike the proprietorship or partnership in which the firm is terminated on the death of one of the principals, the corporation is perpetual. Unless the corporation is bankrupt or the corporate charter lapses, the corporation continues in existence until all stockholders agree to dissolve it. In most states, clauses can be included in the corporate charter that in effect allow the control of sale of stock outside of the circle of present stockholders. That is, any stockholder who wishes to sell a block of stock must first offer the stock for sale to the other stockholders. They have an option to purchase it before it is sold to others. This allows the closed nature of a closely held corporation to be maintained. If a stockholder should die, the stockholder's heirs are committed to offer it to the present stockholders before selling it to others. The heirs can, of course, decide simply to retain the stock.

Two disadvantages that are inherent in the corporate form of ownership are the reduced level of control exercised in management decision making and certain restrictions that can be placed on the corporation when operating outside of its state of incorporation. The larger a corporation becomes the more decentralized the ownership becomes. On questions of dividend levels, the issuance of stock to generate capital, and other critical operational decisions, agreement of all stockholders must be obtained. In large corporations, this leads to involved balloting to establish the consensus of the ownership. This process is cumbersome and greatly reduces the speed with which corporations can respond to developing situations. In small closely held corporations, however, this presents no more of a problem than it does in a partnership.

When a corporation operates in a state other than the one in which it is incorporated, it is referred to as a *foreign* corporation. For instance, a corporation incorporated in Delaware

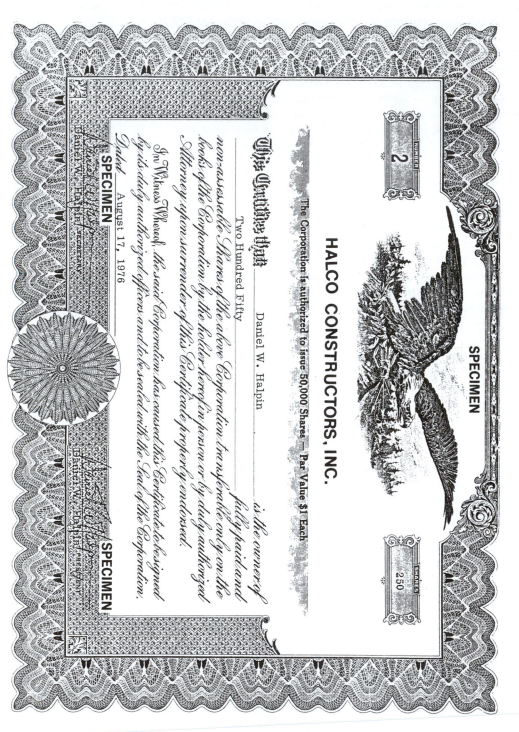

Figure 5.2 Typical stock certificate.

is considered a foreign corporation in Indiana. Corporations in certain industries may encounter restrictive laws when operating as a foreign corporation. They must establish legal representation in states in which they operate as foreign corporations. Restrictive legislation of this type cannot be applied to proprietorships and partnerships, since these entities consist of individuals who are legally recognized. The individual is protected by equal treatment under the Constitution, and what is a legal restriction when placed on a corporation is illegal when applied to a proprietorship or a partnership.

5.6   COMPARISON OF LEGAL STRUCTURES

The decision to choose a particular legal structure for the firm hinges on seven major considerations. The pluses and minuses of each type of structure are summarized in Table 5.1. These considerations have already been introduced in general form. Specifically, an owner contemplating a legal structure for the firm must consider:

1. Taxation
2. Costs associated with establishing the firm
3. Risk and liability
4. Continuity of the firm
5. Administrative flexibility and impact of structure on decision making

Table 5.1   Considerations in Choosing Legal Structure

	Proprietorship	Partnership	Corporation
Tax	Tax on personal income; tax on earnings whether or not they are withdrawn	Tax on personal salary and earnings	Lower taxes in some cases.[a] Dividends are not deductible; double taxing. Taxes on dividends, that is, money actually received
Costs and procedures in starting	No special legal procedure; apply for licenses; register with IRS	General: Easy—oral agreement Limited: More difficult—must closely adhere to state law	More complex and expensive. Meeting must be held
Size of risk	Personal liability	Personal liabilities: Extent of personal fortune. Limited: each partner is protected; loss of limited partner cannot exceed initial investment	Limited to assets of corporation
Continuity of the concern	No continuity on death of proprietor	Dissolution: No continuity on death of partner. Surviving partners can buy share if in agreement.	Perpetual (charter can expire)
Adaptability of administration	Simplicity of organization; direct control	Decisions and policies implemented by oral agreement	Directors—good if involved. Policy decisions predefined by by-laws
Influences of applicable laws	Laws are well defined; no limit on doing business in various states	Laws are also well defined; a license may be required	Foreign corporation status; requires legal counsel on permanent basis
Attraction of additional capital	Limited potential for capital expansion Borrowing; line of credit; personal fortune investment	Better; more capital; limited partner concept	Issue securities; collateral provided by corporate assets; issue stock

[a] See Fudd Associates, Inc., example in text.

6. Laws constraining operations

7. Attraction of capital

The question of taxes to be paid in each organizational format is mixed, and the anticipated balance sheet and cash flow of each firm must be studied to arrive at a "best" solution. The corporation has the disadvantage that the firm is taxed twice, once on corporate profit and a second time when the stockholders must pay tax on the dividends received as distributed income. The subchapter S type of firm circumvents this to a degree in that the stockholders are taxed as individuals as if the firm were a partnership. The normal proprietorship and partnerships have the disadvantage that all income is taxed whether or not it is withdrawn from the firm. Thus, as in the example of Uncle Fudd, incorporating yields a benefit despite the double-taxation feature.

Costs and procedures associated with establishing the firms are generally minimal for a proprietorship, slightly more involved for a simple partnership, and a major financial consideration for limited partnerships and corporations. Normally whatever costs and procedures are associated with local, state, and federal tax registration and the purchasing of a license are all that must be considered in establishing a proprietorship or simple partnership. These as well as significant legal costs ($2000 to $5000) must be considered in establishing limited partnerships and corporations. These costs may be justified, however, based on the limitation of liability achieved and the benefits of medical, health, and insurance plans that can be implemented in a corporate format.

Corporations and limited partnerships limit the level of loss in the event of a default or bankruptcy to the level of investment. That is, stockholders cannot lose more than the value of their shares. The loss of a limited partner cannot exceed the amount of his investment. If he initially invested $20,000, he can lose this amount, but his other assets cannot be attached in the event of bankruptcy or default. The assets of stockholders in a subchapter S corporation are similarly protected. Personal assets are used to pay creditors in the proprietorship or simple partnership form of ownership. This can lead to personal bankruptcy.

Proprietorships also have the disadvantage that they terminate when the proprietor dies. This may present a problem, particularly if the firm as an asset must be divided among several heirs. It can be circumvented in part by willing the firm and its market and "goodwill" to one heir (who will carry on the business) and providing that the heir will compensate the other heirs for their share. If a partner dies, the partnership is dissolved. Again, however, provisions in the partnership agreement can provide the means for surviving partners to purchase the deceased partner's share from his estate. Corporations are perpetual, and the stock certificates are transferred directly as assets to heirs of the estate.

Policy decisions are relatively simple in the proprietorship and partnership formats. Principals make all decisions. In the corporate format, certain decisions must be approved by the stockholders, which may impact the corporation's ability to react to a developing need or situation. In closely held corporations, however, this is no problem since the partners are able to call ad hoc board meetings to react quickly. Corporations with large numbers of shareholders are not as flexible in this regard. The chief operating office or president handles the day-to-day decision making. A board of directors is charged with intermediate-range and strategic planning and decision making. Major decisions, however, such as stock expansion and acquisition of other firms or major assets, must often be approved by all stockholders in a formal vote.

Local laws may encourage the formation of small and local businesses by placing restrictive constraints and burdensome additional cost on out-of-state, or foreign, corporations. These discriminating practices must be investigated when bidding construction work in a state other than the one in which the construction corporation is chartered. Special licenses and fees are sometimes required of foreign corporations. Proprietorships and partnerships that consist of individuals are protected against these discriminatory practices by the Constitution and enabling "equal rights" legislation.

In raising capital, proprietorships and simple partnerships must rely on the personal borrowing of the principals to generate capital for expansion. The unique feature of the corporation that permits it to sell stock allows corporate entities to attract new capital by further distributing ownership. The corporate assets as well as future projections of business provide a collateral basis to attract new stockholders. This mechanism is not always viable, however. From time to time, corporations may be unable to sell large issues of stock for capital expansion and will be forced to go to the commercial banks to borrow. In periods of economic uncertainty, sale of stock as a method of attracting capital may be limited.

Good information regarding the advantages and disadvantages of various legal forms of organization are contained in the Small Business Administration management guides available from the Government Printing Office.

REVIEW QUESTIONS AND EXERCISES

5.1 Name the three principle forms of business ownership in construction and state the liability limits of the owners in each case.

5.2 Which legal structure is most difficult to establish and why?

5.3 Name three types of partnerships.

5.4 Describe briefly two advantages and two disadvantages of a corporate form of business organization as compared to a partnership.

5.5 Jack Flubber, who owns Sons of Flubber Construction Co. and runs it as a proprietorship, had gross profits last year of $80,000. His personal and family expenses are $52,000 and he has $7000 in exemptions and deductions. He paid $17,000 in taxes. If he paid himself a salary of $55,000 taxed at 20%, would it be advantageous for him to incorporate as a closely held corporation? Explain.

5.6 What is meant by the term foreign corporation?

5.7 What would be the advantages of organizing as a subchapter S corporation?

5.8 Is it possible to characterize the legal structures of local contractors using the Yellow Pages as a guide?

5.9 What steps must be taken to set up a partnership? How can a partnership be dissolved?

5.10 In problem 5.5, what taxes would Jack pay if he organized as a closely held corporation (as described) and, after paying his salary, also issued himself a dividend of $10,000?

5.11 What is the difference between par and book values of corporate stock? If an incorporated construction company wins a large cost plus fixed-fee contract, what impact might this have on the market value of the company's stock?

5.12 Uncle Fudd has decided to sell his ownership in the Cougar Construction, Inc. to Cousin Elmer. How would the legal firm handling this transaction determine a fair price per share?

Chapter 6

Project Planning

Web-Based Project Management

The Need

Considering that 1 to 2% of project cost is simply paperwork, multimillion dollar amounts are expended to support communication and information transfer. Geographically dispersed team members need to share information, documents, drawings, and strategies. The likelihood of errors, miscommunication, or missed deadlines is reduced through web-based project management tools.

Most web-based project management tools enable project teams to share all relevant project information which includes directory, specification, correspondence, sketches, meeting notes, shop drawing logs, field reports, requests for information (RFIs), and CAD drawings, etc.

The Technology

With the Internet acting as the ultimate communications medium, web-based project management applications provide an instant, on-demand, secure online solution for all team members to communicate, share documents, and collaborate. The general features include document management, workflow analysis, schedule/calendar development and management, messaging, conferencing, discussion function, directory service, revision control, and project camera, etc.

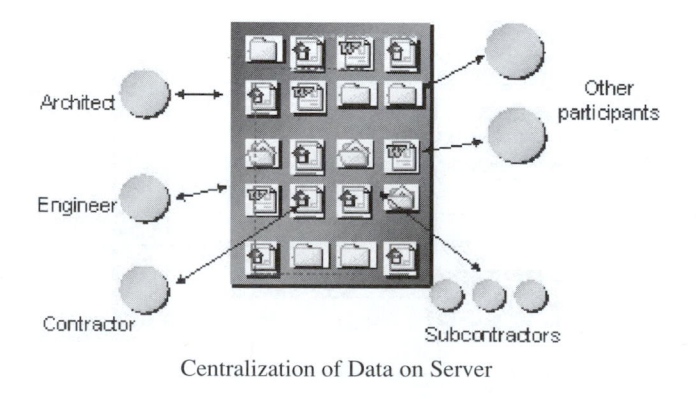

Centralization of Data on Server

6.1 INTRODUCTION

The planning of a project involves the concept of an objective or facility, and a *scope of work* defining the work product or deliverable. The bid package consisting of the plans and specifications establishes the scope of work to be performed. In order to be properly managed, the scope of work must be broken into components which define work elements or building blocks which need to be accomplished in order to realize the end objective. The assumption is that the project is the summation of its sub-elements.

Definition of the sub-elements is important since it determines how the project is to be realized in the field. The sub-elements are often referred to as work packages. The summation of the work packages can be shown in a hierarchical format called a work breakdown structure or WBS. Figure 6.1 is an example of a WBS for a small building.

The *building* defined at Level I is sub-divided into major sub-systems at Level II. The *foundation* work is again broken into major work activities at Level III. Similarly, the *pile cap* consists of work tasks such as forming, installation of reinforcement, etc.

Development of a WBS requires a thorough understanding of the project scope of work. Experience in building is key to establishing a functional WBS. The WBS and the hierarchy of work packages of which it is composed are used to determine the status of a project and to manage the project from a time, cost, and quality perspective. Mentally, building a WBS structures the work which must be physically accomplished to realize the project and its end objective.

Planning can be thought of as the definition and sequencing of the work packages within a given project. That is:

PLANNING = WORK BREAKDOWN + WORK SEQUENCING

Planning leads to a refinement of the Scope of Work as established in the contract documents. A good plan reduces uncertainty and improves efficiency.

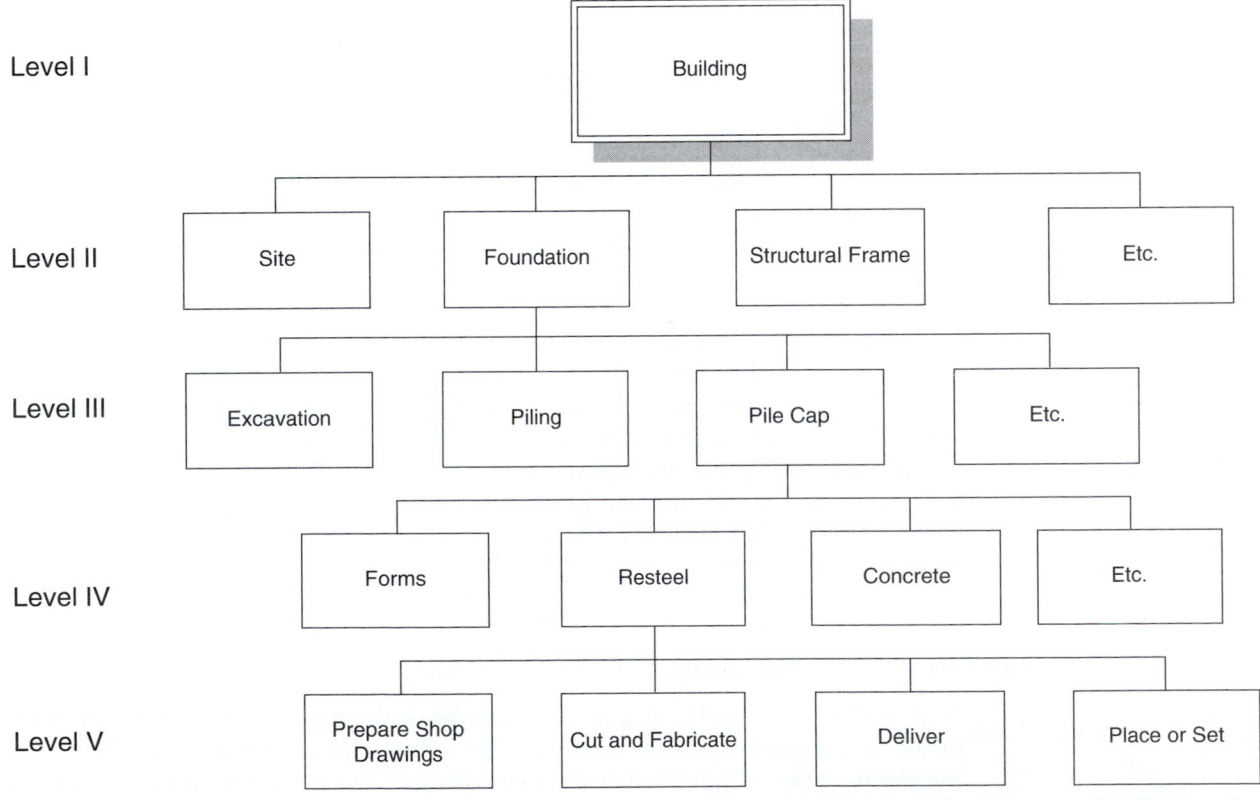

Figure 6.1 Work Breakdown Structure (WBS) Example.

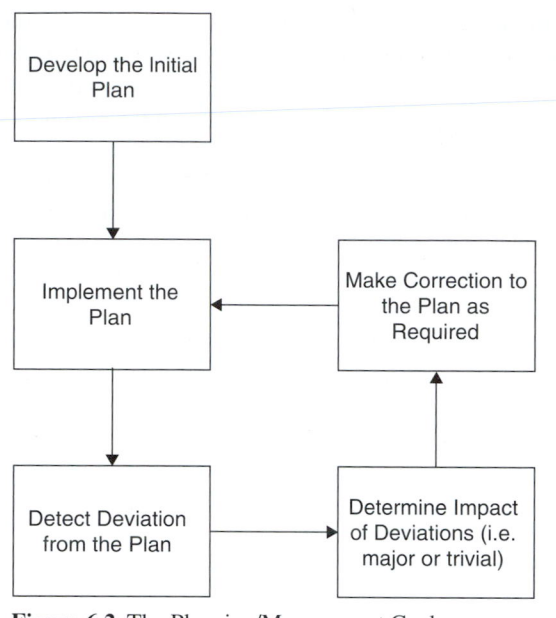

Figure 6.2 The Planning/Management Cycle.

The WBS also assists in determining the amount of planning needed. That is, it defines the level of planning required. For instance, if we are traveling to Washington, D.C., certain major elements of the travel should be planned. If we are traveling by air we need to book an airline ticket. We need to determine what to pack. Typically, we will need a place to stay, so we probably will reserve a hotel room unless we are staying with friends. At a lower level in our planning hierarchy, we must determine how to get from the airport to the hotel. If time is critical, we may reserve a limousine to meet us at the airport. On the other hand, we may decide to make a decision upon arrival as to whether to take a cab, a shuttle bus, or the Washington Metro MassTransit system. In effect, the development of the WBS and the definition of work packages is an exercise in "thinking things out ahead of time." It provides an orderly mechanism which facilitates planning. How detailed and comprehensive the WBS must be will vary with the situation and the complexity of the project.

Planning allows us to develop a framework for project execution, monitoring, and control. This minimizes uncertainty, clarifies sub-objectives within the overall objective, assists in establishing sequencing of activity, and helps to avoid crisis management.

Planning is, however, an ongoing task and continues throughout the life of the project—General Dwight D. Eisenhower once said "Plans are nothing, Planning is everything." Initial planning is inevitably impacted by events which cause changes to the plan. The success of a project is tied to the manager's agility in identifying deviations from the plan and solving the challenges precipitated by these deviations. Figure 6.2 reflects this cycle of planning in terms of a simple flow chart. Once the WBS has been developed, it provides the framework within which planning can proceed throughout the life of the project. It becomes the vehicle for identifying deviations, assessing their impact, and making appropriate corrections to the existing plan. Therefore, "planning is everything."

6.2 DEVELOPING THE WORK BREAKDOWN STRUCTURE

Work packages are the building blocks of the Work Breakdown Structure (WBS). They should be defined to assist the manager in determining the project status or the level of completion of the project. One definition of the WBS is "the progressive hierarchical breakdown of the project into smaller pieces to the lowest practical level to which cost is applied."

When monitoring and controlling a project, cost and time are areas of primary interest. The WBS is extremely useful in developing both cost and time (schedule) plans.

In establishing the WBS, the following guidelines need to be considered:

1. Work packages must be clearly distinguishable from other work packages.
2. Each work package must have unique starting and ending dates.
3. Each work package should have its own unique budget.
4. Work packages should be small enough that precise measurement of work progress is possible.

For example, in Figure 6.1 one work package at Level IV defines work associated with the installation of reinforcing steel in a pile cap. This work package is (1) clearly defined and separate from other work packages, (2) has a starting and ending date, and (3) has a cost budget which is unique and is small enough for accurate progress measurement. The work packages at Level V become more generic and more difficult to distinguish as unique. For instance, tasks such as cutting and setting steel are very short and assigning a unique budget to this level of work becomes difficult. Therefore, the Level V work packages in Figure 6.1 can be viewed as sub-tasks which are pro-rated to the more unique Level IV work packages.

6.3   A WORK BREAKDOWN EXAMPLE

In order to better understand the concept of work breakdown, consider the small gas station project for which simplified plans and concept drawings are shown in Appendix I.

In construction, the various aspects of the work that contribute to breakdown of the project into packages relate to:

1. Methods used to place work
2. Skills needed for the work
3. Craft workers involved
4. Critical Resources (e.g., cranes, crews, etc.)

The definition of work packages can be facilitated by using four categories which help in establishing a level of uniqueness. These categories are:

1. Location or Area within the Project (e.g., foundation – pile cap)
2. Material Type (e.g., concrete, resteel, etc.)
3. Method of placement (e.g., excavation)
4. Organizational Resources Required (e.g., labor and equipment needed)

In the gas station project, the construction requires a *foundation* for the load bearing walls. The foundation can be thought of as a location (as well as a structural support system). The LOCATION or AREA of the work is a physical part of the construction. That is, one can walk up and touch the location of the package. A work package defining the floor slab in Section A on the 3rd floor of an academic building is something we can locate and physically touch in the completed facility. The fact that the slab is concrete (i.e., rather than wood or metal) is another important parameter. It has implications from a procurement, as well as from a placement and work content point of view.

LOCATION and MATERIAL TYPE will influence the method of placement or installation. The METHOD OF INSTALLATION and the material type will determine the human skills and equipment resources needed for installation. The method of placement or installation dictates the TYPES OF RESOURCES required, thus differentiating one package from another. For instance, in one case we may be placing concrete using a concrete pump, where as in another situation the concrete may be transported using concrete buggies. In each case, labor and equipment resources, the budget, and the productivity of the concrete placement will be different.

6.4 WORK PACKAGES FOR THE GAS STATION PROJECT

Let's develop a set of work packages and a WBS for the gas station construction. First, locations which are work package related will be determined. As noted above, the building foundation can be considered a location. It would be important to know whether the scope of work includes the parking and service area surrounding the station. For the purposes of this exercise, we will assume it is within the scope of work. LOCATION work packages would be as follows:

1. Parking and Service Area
2. Foundation
3. Building Walls/Structural Panels
4. Building Roof
5. Interior Floors/Slabs (separate from the Foundation)
6. Interior Finishes
7. Exterior Finishes
8. Electrical Systems
9. Mechanical Systems

Adding the category of MATERIAL TYPE expands the number of work packages as shown in Table 6.1. Although this listing is, by no means, complete, it indicates the level of detail which must be considered even in a relatively small project (when only two levels of hierarchy are defined).

If mechanical work is expanded to cover location, material type, methods and resources the following partial list of work packages would be added to the hierarchy of the WBS.

1. Excavation of Waste Water System
2. Drainage Tile installation—Waste Water
3. Septic Tank installation
4. Fresh Water lines (piping)
5. Sinks, basins, toilets installation
6. Hot Water System installation
7. Pneumatic Air System installation

Table 6.1 Work Packages for the Gas Station Project

(1) Earthwork for Parking and Service (P&S) Area	(10) Interior Built-ins (e.g., Cabinets, etc.)
(2) Asphalt Paving for P&S Area	(11) Interior Painting
(3) Concrete Hardstands in P&S Area	(12) Interior Drywall
(4) Concrete Foundations	(13) Interior Doors, Frames, Hardware, etc.
(5) Walls–Masonry Bearing Walls	(14) Interior Floor Coverings (if required)
(6) Walls–Prefab Metal Sandwich Panels	(15) Exterior Brick Façade
(7) Interior Concrete Floors	(16) Exterior Glazing
(8) Built-up Roof	(17) Exterior Doors and Signage
(9) Roof Gutters/Drainage	(18) Mechanical Systems
	(19) Electrical Systems

Figure 6.3 shows a work breakdown structure based on this partial development of work packages.

6.5   DETERMINING SEQUENCE OF WORK PACKAGES

Having broken the work in to work packages, activities which facilitate time management and control can be defined and logically placed in sequence. The word ACTIVITY is generally used when discussing time control or scheduling to refer to the work elements which appear in the schedule in their expected sequence or logical order. As mentioned in Chapter 1, the word "technology" implies that operations have a logic or sequence. An understanding of this logic is a key to successful project management.

In arranging the work package sequence for time control, the criteria of (1) *location* (2) *material* (3) *method* and (4) *required resources* developed in Section 6.3 must be re-considered from the perspective of how these criteria impact the order or sequence of work activities. For instance, location can determine sequence. It is normal to complete the structural frame for the 1st floor of a building before beginning work on the structural frame for the 2nd floor. This could be considered a *physical* constraint since the 2nd floor frame cannot be supported until the first floor frame is completed. Such physical constraints or *physical logic* are common and characteristic of construction operations (e.g., the floor must be complete before installing the floor covering, etc.). Locational aspects of a work package may, therefore, determine its sequence in the overall project.

In some cases, physical requirements do not dictate order or sequence. For instance, in finish work relating to a rest room, fixtures such as sinks, commodes and partitions must be installed. Wall and floor finishes such as ceramic tiles must also be completed. Since there is no physical constraint, it is a management decision as to whether the walls and floors are completed first or the fixtures are installed first. In this case, the situation is controlled by a management decision (e.g., the fixture subcontractor is available first and is instructed to proceed) and the sequence is driven by *management logic*.

Again, consider the small gas station project. A preliminary sequencing of the work packages is shown in Figure 6.4. One activity is defined to address mobilizing men, material, and machines to the site and preparation of the site. Then the locational or area work packages are ordered in sequence starting with the foundation and completing with the interior and exterior finishes and the mechanical and electrical systems which can be worked on at the same time (i.e., in parallel). This preliminary sequencing provides the framework for a more detailed schedule development to be presented in Chapter 7.

Following the preparation of the site, the footers and pier foundations are to be poured. After the footers and pier foundations have reached sufficient strength, the building structure is erected. It should be noted that, in this case, the floor slabs of the service bays, the show room, and the office areas as well as the utility room and toilets are not poured until the building structure is erected. Since the foundation consists of a "ring" footing[1] and individual piers to support walls and columns, etc., the building shell and roof can be completed prior to casting the various floor slabs throughout the building. When the roof is completed and the building is "closed in," work inside the building can proceed without concern for the weather.

As a more detailed time plan (i.e., schedule) is developed, consideration must be given to other time consuming activities which are not necessarily identified using the location, material, method, and resource criteria.

1. Administrative actions such as inspections, permit issuance, noise constraints, etc. must be considered in developing the time schedule logic.

---

[1] A ring footing is a footing that supports the periphery of the building.

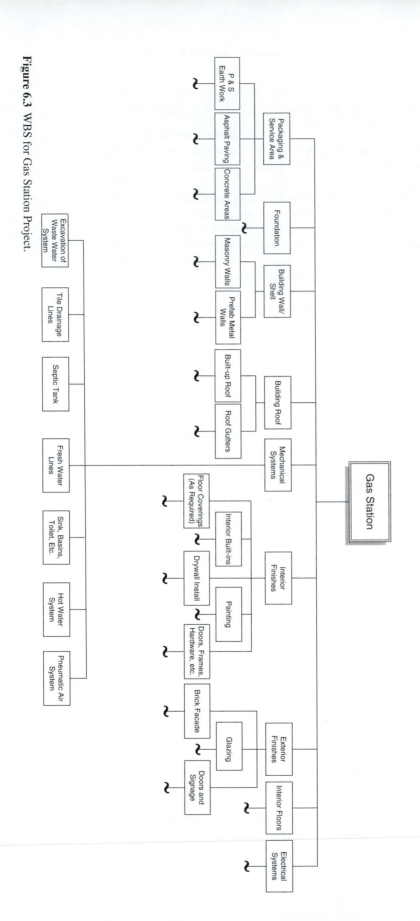

Figure 6.3 WBS for Gas Station Project.

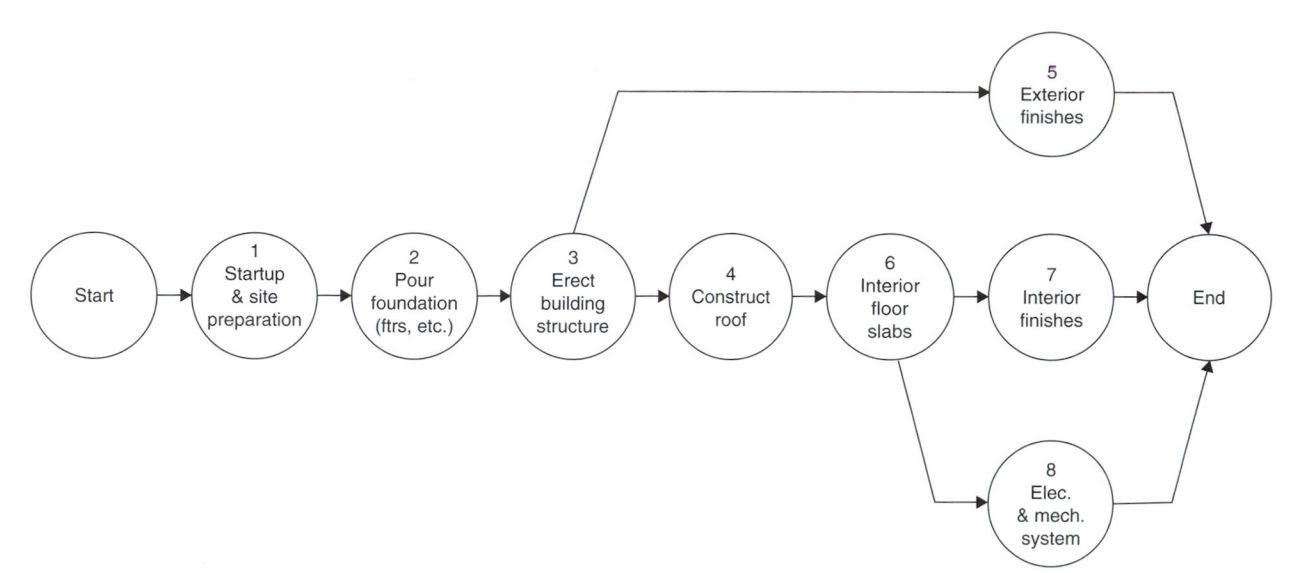

Figure 6.4 Preliminary project breakdown.

2. Deliveries of materials and similar logistically issues must also be factored into the schedule.

3. Finally, certain special activities tied to the physical properties of the materials or procedures required (e.g., curing of concrete, moisture content measures for soil compaction, etc.) must be included in the time schedule.

A well defined WBS facilitates the development of both preliminary and detailed schedules.

6.6   ESTIMATE DEVELOPMENT AND COST CONTROL RELATED TO THE WBS

A good WBS facilitates cost control during the life of the project.[2] Work packages are defined so that they have their own unique budgets. When referring to work packages in the context of cost control, the terminology "cost accounts" or "control accounts" is often used. During the bidding process, the contractor prepares an estimate of cost which becomes the basis for the bid price submitted for the proposed work. If the bid is accepted, the detail estimate used for bid submittal is converted to a *budget* which serves as a cost baseline to control spending during the life of the project. The concept of cost control is shown schematically in Figure 6.5.

Based on a refinement of the bid estimate, a control budget is prepared. The budget structure is tied to the breakdown of the project into major cost elements. For small and simple projects, such as the paving of a residential driveway the cost breakdown of the budget may consist of relatively few elements (e.g., labor, materials, equipment, special item, etc.). For larger and more complex projects however, the structure and level of detail of the cost breakdown is key to effective control of project spending. In the case of the gas station project, for instance, budgets for each of the work packages shown in Figure 6.3 would be developed. The summation on these individual work package budgets can then be used to track total project costs and determine an overall cost status for the project at any time during construction.

In large and complex projects (e.g., large buildings, manufacturing plants, etc.), a comprehensive WBS is required. Literally thousands of work packages must be defined and a consistent and reliable system of referring to these elements of the work breakdown

---

[2] If a WBS is used, the major items are work packages and the control accounts which are sub-elements of the work packages.

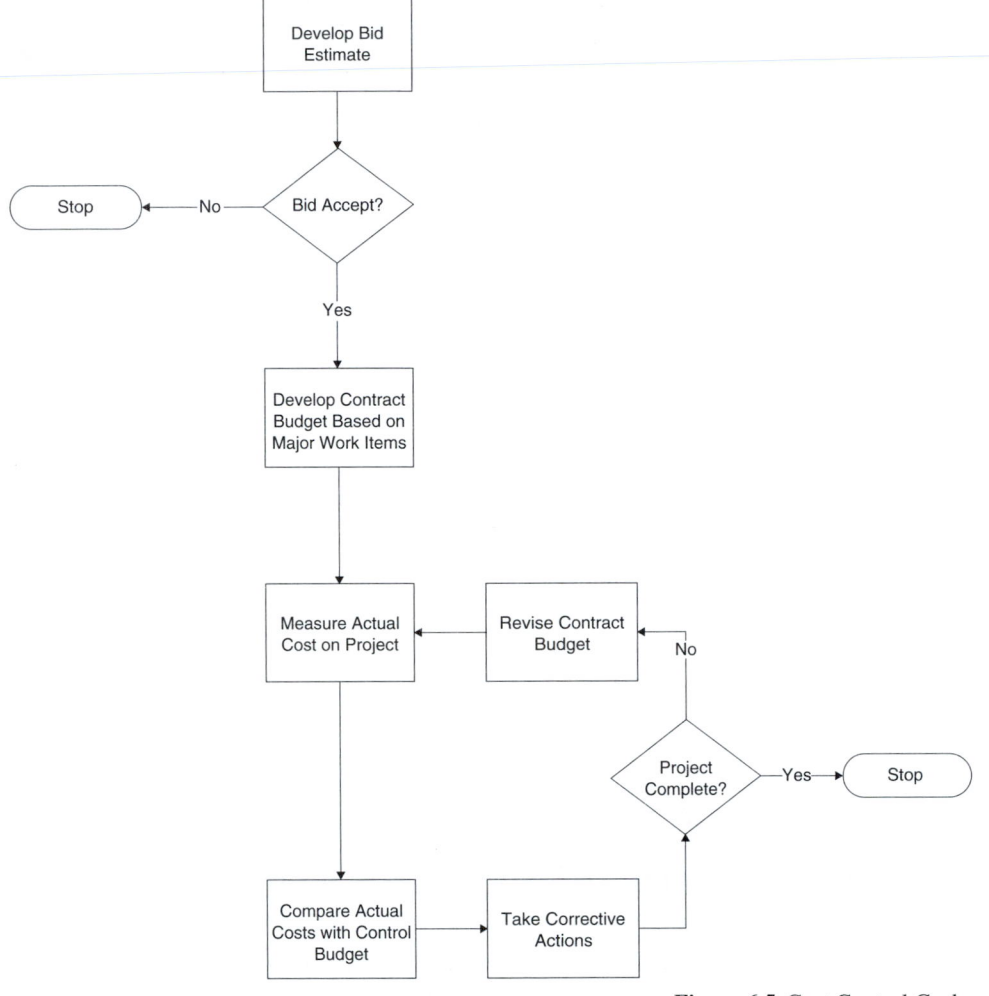

Figure 6.5 Cost Control Cycle.

is essential. To provide consistency and structure to the management of large cost control systems, a *code of accounts* is used as a template or guide in defining and cataloging the cost centers within a project. Several cost coding systems are presented in Chapter 15.

To be consistent with the WBS, such *cost coding systems* utilize a numerical or alphanumerical labeling which is multi-level or hierarchical in nature. This numerical labeling must be (1) suitable for computerization, (2) compatible with the company's financial accounting system, and (3) facilitate gathering and reporting of information for broad cost categories (e.g., cost of all slab concrete) and for highly detailed information (e.g., cost of labor for pump installation-pump 101 in the cooling tower). The cost accounting system should facilitate the aggregation or "rolling up" of costs. The hierarchical or multi-level nature of the WBS provides an excellent format for the collection and review of cost throughout the life of a complex project.

6.7 ROLE OF CODE OF ACCOUNTS

The development of cost coding systems is discussed in some detail in Chapter 15. In general, these labeling or coding systems focus on the multiple or hierarchical levels of cost spending typical of complex construction projects. Defining work packages, location of the work, material type, method of installation, and resources required help to establish the various levels of the breakdown. In cost code development, these identifiers can be

Table 6.2 Cost Code Structure (Example)

Level	Project	Area	Discipline	Trade
1	21300			
2		804		
3			724	
4				112

Cost Code $= 21300 - 804 - 724 - 112$

characterized in many ways. A typical cost code might define (1) the project designator, (2) area of work, (3) work discipline (e.g., civil, mechanical, electrical, etc.), and (4) the trade specialty required. Table 6.2 gives an example of how such a code might be developed. The code in Table 6.2. is 14 digits long. Therefore, computer databases which are used to store and retrieve cost information must be able to handle long numerical designators or labels such as the one shown here.

On complex or unique projects, preparation of work package budget control sheets may be appropriate. Such a sheet allows for collection of data regarding actual costs versus estimated cost for each work element or package. A typical sheet for the interior slab construction on the small gas station project is shown in Figure 6.6.

The sheet acts as documentation of the control budget for each work package. Locations for the base cost calculations for each resource type (e.g., materials, installed equipment, labor, and equipment) are provided. As work packages are completed, the "as-built" cost and productivity achieved are recorded. This is "back-up" material for the cost estimating database and can be used as a reference when estimating the cost of similar work in future. It also supports the comparison activity (e.g., Compare Actual Costs with Control Budget) in the flow chart of Figure 6.5.

Work Package Identification 06-123 Description: Concrete Placement Interior Floor Slabs						
	Materials					Actual Productivity
Resource Code	Description	Unit	Qty	Unit Cost	Extension	
101	Concrete, 2500psi	CY	30	40.00	1200.00	
	Installed Equipment					Notes
	Crew Labor	NR	Hours	Cost/Hr	Extension	Cost Summary
020	Foreman	1	8	30.00	240.00	Actual Cost
029	Laborer	4	8	15.00	480.00	Labor =
022	Finisher	1	8	20.00	160.00	Materials =
063	Pump Operator	1	8	25.00	200.00	Equipment =
				Total	1080.00	Variation from Budget
	Equipment Not Charged As Indirects					
		NR	Hours	Cost/Hr	Extension	
505	Vibrator	1	8	10.00	80.00	
517	Finisher	1	8	15.00	120.00	
308	Concrete Pump	1	8	150.00	1200.00	
				Total	1400.00	

Figure 6.6 Work Package Control Account Sheet.

6.8 SUMMARY

Construction project planning focuses heavily on time and cost control. Planning is a continuous task. Deviations from the original plan are the norm rather than the exception. Therefore, an organized approach to identifying change from the original plan is critical. It has been noted that:

> "In order to MANAGE, one should be able to CONTROL.
>
> In order to CONTROL, one should be able to MEASURE.
>
> In order to MEASURE, one should be able to DEFINE.
>
> In order to DEFINE, one should be able to QUANTIFY."
>
> D. Burchfield, 1970

The Work Break System (WBS) approach provides a rigorous way to quantify, define, measure, and control the elements or work packages of a given scope of work. Breaking a construction project into work elements to be managed is essential for both time and cost planning. In addition to time and cost planning, a number of other planning efforts are needed when constructing a facility. Decisions and supporting plans must be developed to address many dimensions of the project. The following is a partial list of other plans that must be developed:

1. Procurement Plan
2. Safety Plan
3. Subcontracting Plan
4. Quality Plan
5. Communication Plan
6. Organizational Plan
7. Completion and Start-up Plan

The effective manager is constantly involved in developing and updating plans. Planning is a never ending task. Therefore, one is continuously improving and perfecting the plan. "Planning is Everything."

REVIEW QUESTIONS AND EXERCISES

6.1 Develop a two level Work Breakdown Structure consisting of at least 16 work packages for the building of the Brooklyn Bridge as described in Chapter 1.

6.2 Develop a preliminary work breakdown structure (WBS) for a small one-story commercial building to be constructed on the site of an existing small-frame structure. It is 30 by 60 ft in plan (see illustration). The exterior and interior walls are of concrete block. The roof is constructed of bar joists covered with a steel roof deck, rigid insulation, and built-up roofing. The ceiling is suspended acoustical tile. The floor is a concrete slab on grade with an asphalt tile finish. Interior finish on all walls is paint.

    a. Show the first level of this structure for the total project (WBS is developed from top to the bottom).

    b. Select one work package (or building system) of the first level and develop the second-level structure for this work package.

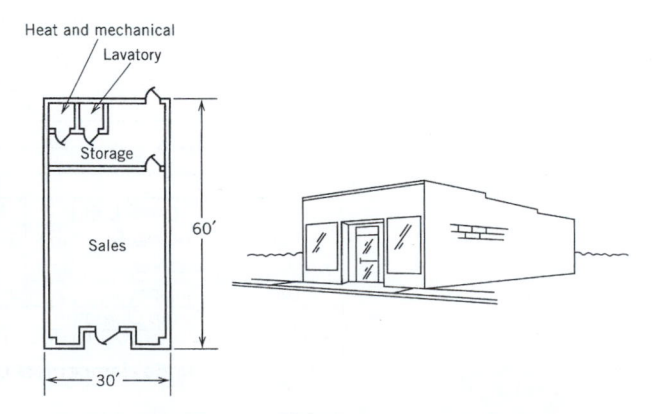

6.3 Describe how you would develop a subcontracting plan for the construction of a small child-care facility.

Chapter 7

Project Scheduling

4D Modeling

The Need

Traditional design and construction planning tools, such as 2D drawings and network diagrams, do not support the timely and integrated decision making necessary to move projects forward quickly. They do not provide the information modeling, visualization, and analysis environment necessary to support the rapid and integrated design and construction of facilities. Synthesis of construction schedules from design descriptions and integrated evaluation of design and schedule alternatives are still mainly manual tasks. Furthermore, the underlying representations of a design and a construction schedule are too abstract to allow the multiple stakeholders to visualize and understand the cross-disciplinary impacts of design and construction decisions.

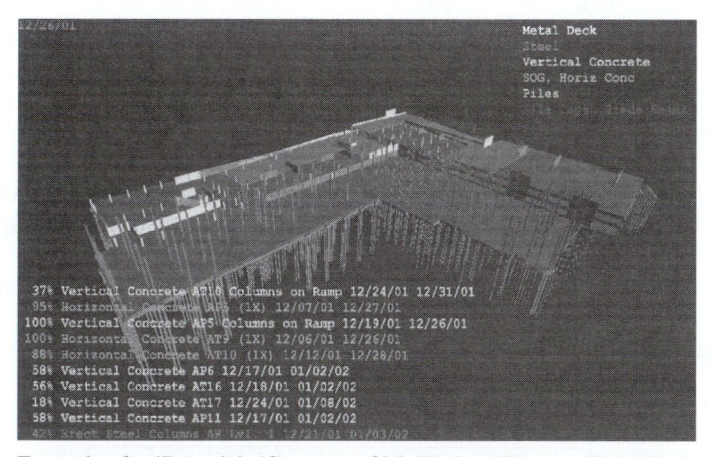

Example of a 4D model. (Courtesy of M. Fischer, Common Point Technologies, Inc. and DPR Construction, Inc.)

4D technologies are now being used by planners, designers, and, engineers to analyze and visualize many aspects of a construction project, from the 3D design of a project to the sequence of construction to the relationships among schedule, cost, and resource availability data. These intelligent 4D models support computer-based analysis of schedules with respect to cost, interference, safety, etc., and improve communication of design and schedule information.

The Technology

Extending the traditional planning tools, visual 4D models combine 3D CAD models with construction activities to display the progression of construction over time. However, 4D models are very time-consuming to generate manually and cannot currently support analysis programs. The difficulty and cost of creating and using such models are currently blocking their widespread adoption. The construction knowledge necessary to build 4D models has been formalized and developed by a methodology that guides project planners in generating 4D models from 3D product models. This formalized knowledge enables project managers to create and update realistic schedules rapidly and to integrate the temporal and spatial aspects of a schedule as intelligent 4D models.

7.1   INTRODUCTION

As noted in the previous chapter, time planning is among the most important aspects of successful project management. The concept of project scheduling addresses the issues associated with time planning and management. Early scheduling methods utilized simple bar charts or Gannt charts to achieve a very simple and straightforward representation of time and work activity sequencing. During the past 40 years network based scheduling methods have become the norm, and many contracts require the use of network schedules to reflect project progress to the owner/client. Simple barcharting concepts as well as network scheduling concepts will be introduced in this chapter.

7.2   BAR CHARTS

The basic modeling concept of the bar chart is the representation of a project work item or activity as a time scaled bar whose length represents the planned duration of the activity. Figure 7.1(a) shows a bar representation for a work item requiring four project time units (e.g., weeks). The bar is located on a time line to indicate the schedule for planned start, execution, and completion of the project work activity.

   In practice the scaled length of the bar is also used as a graphical base on which to plot actual performance toward completion of the project work item see Fig. 7.1(b). In this way the bar chart acts both as a planning-scheduling model and as a reporting-control model. In this use of the bar chart, the length of the bar has two different meanings:

1. The physical length of the bar represents the planned duration of the work item.

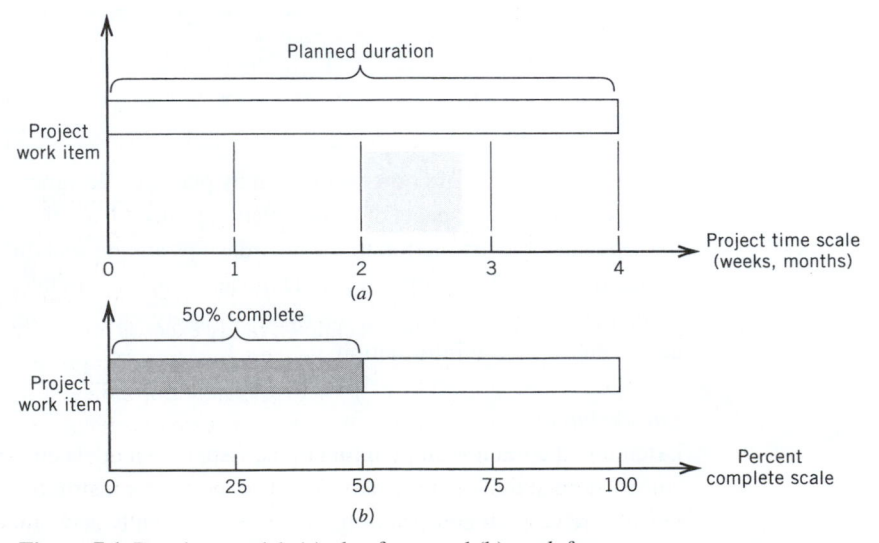

Figure 7.1 Bar chart model: (a) plan focus and (b) work focus.

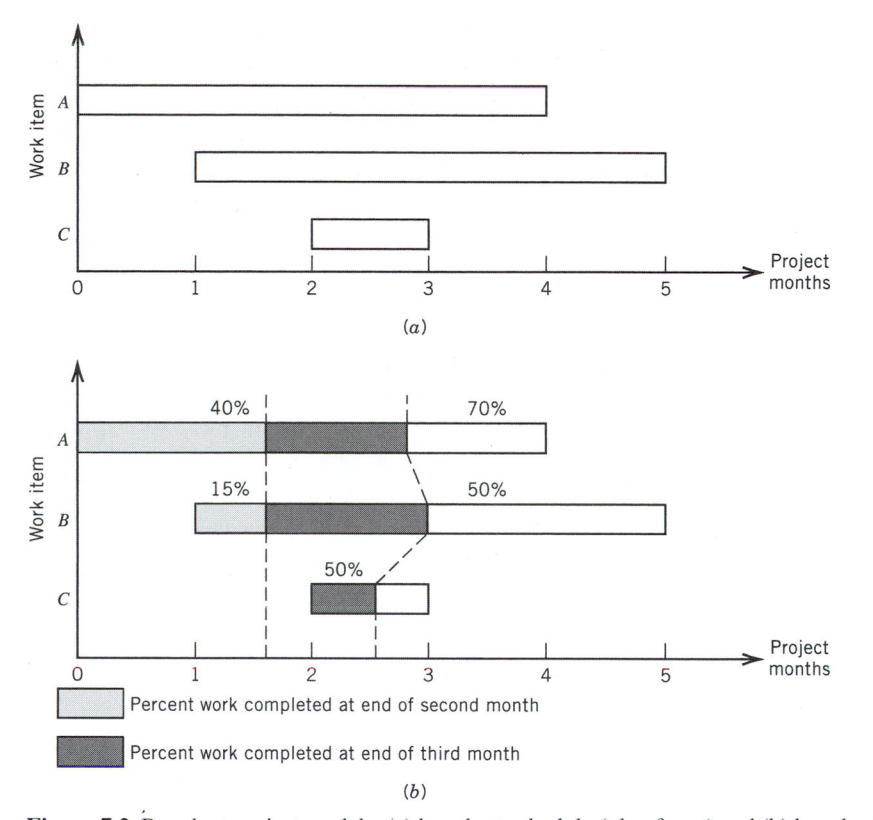

Figure 7.2 Bar chart project models: (a) bar chart schedule (plan foucs) and (b) bar chart updating (control focus).

2. It also provides a proportionally scaled baseline on which to plot at successive intervals of time, the correct percentage complete.

Figure 7.1(b) shows a bar for a project work item that has been half completed. In a situation where the work rate is constant and field conditions permit, this would occur in half the planned duration. If, however, actual work rates vary from time to time according to resource use and field conditions, then the work will be half completed sooner or later than half the planned duration. In this modeling concept actual work progress is modeled independently of the actual time scale of the bar chart.

Figure 7.2(a) shows a schedule for a project consisting of three activities. Activity A is to be carried out in the first four months, activity B in the last four months, and activity C in the third month. Actual progress in the project can be plotted from time to time on these bars as shown in Figure 7.2(b).

In this manner, project status contours can be superimposed on the bar chart as an aid to management control of the project. By using different shading patterns, the bar chart can indicate monthly progress toward physical completion of the activities.

Project bar chart models are developed by breaking down the project into a number of components. In practice the breakdown rarely exceeds 50 to 100 work activities and generally focuses on physical components of the project. If a project time frame is established, the relative positioning of the project work activities indicates the planned project schedule and the approximate sequence of the work. One disadvantage of the traditional bar chart is the lack of precision in establishing the exact sequence between activities. This problem can be addressed by using directional links or arrows connecting the bars to give a precise indication of logical order between activities. This connected diagram of bars is referred to as a bar-net.

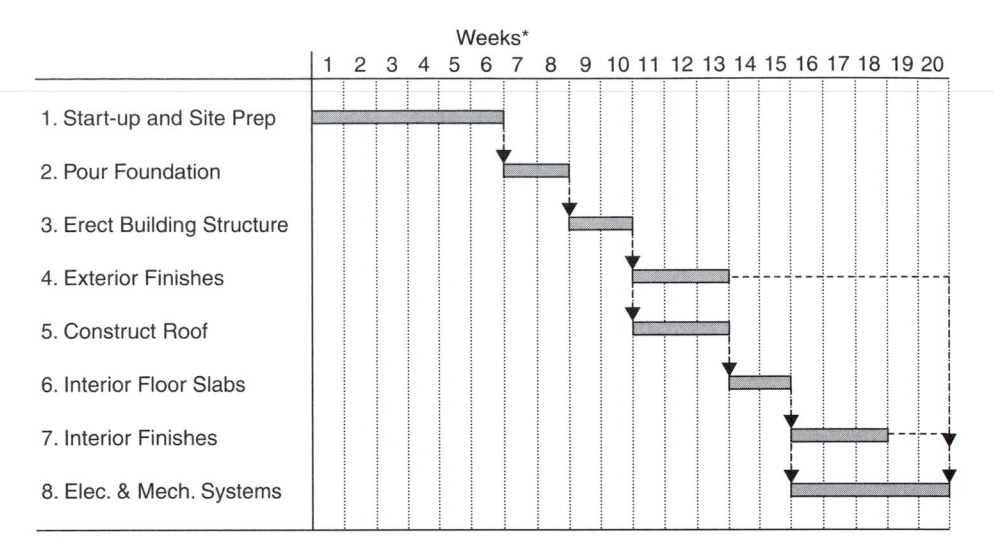

*Weeks are assumed to be working weeks consisting of 5 working days.

Figure 7.3 Preliminary Bar-Net Schedule for the Small Gas Station.

A bar-net showing the major activities defined in the preliminary project breakdown diagram for the small gas station (Figure 6.4) is shown in Figure 7.3. The bars are positioned in sequence against a time line. The sequence or logic between the bars is formalized by connecting the end of the preceding bar to the start of the following bar. For instance, the end of bar 3. Erect Building Structure is connected using a directional link or arrow to the two activities that follow it (activities 5 and 4). The use of directional arrows to connect preceding and following activities leads to the development of a preliminary scheduling document called a bar-net. This is a schedule that combines the graphical modeling features of the bar (e.g., length to indicate duration, and scaling to a time line) with the sequencing features of directional arrows. Positioning the eight activities as bars in their logical sequence using the arrow connectors against a time line plotted in weeks allows us to visually determine that the duration of the entire project is roughly 20 weeks.

This bar-net diagram also allows one to determine the expected progress on the project as of any given week. For example, as of week 11, activities 1, 2, & 3 should be completed. Activities 4 and 5 should be in progress. If we assume a linear rate of production (i.e., half of a two-week activity is completed after one week), we could assume that 1/3 of 4 and 5 will be completed as of the end of week 11.

A bar-net is a somewhat more sophisticated version of a bar chart which emphasizes the sequencing of the activities by using arrow connectors. Use of this arrow connection approach to show logical order will be a key element of developing network schedules to be discussed later in Section 7.5.

7.3   SCHEDULING LOGIC

In order to develop a schedule, the logical sequence or scheduling logic which relates the various activities to one another must be developed. In order to gain a better understanding of the role played by sequencing in developing a schedule, consider, a simple pier made up of two lines of piles with connecting headers and simply supported deck slabs.

A schematic view of a portion of the pier is shown in Figure 7.4(a). The various physical components of the pier have been identified and labeled. An exploded view of the pier is shown in Figure 7.4(b), which shows each physical component individually separated but

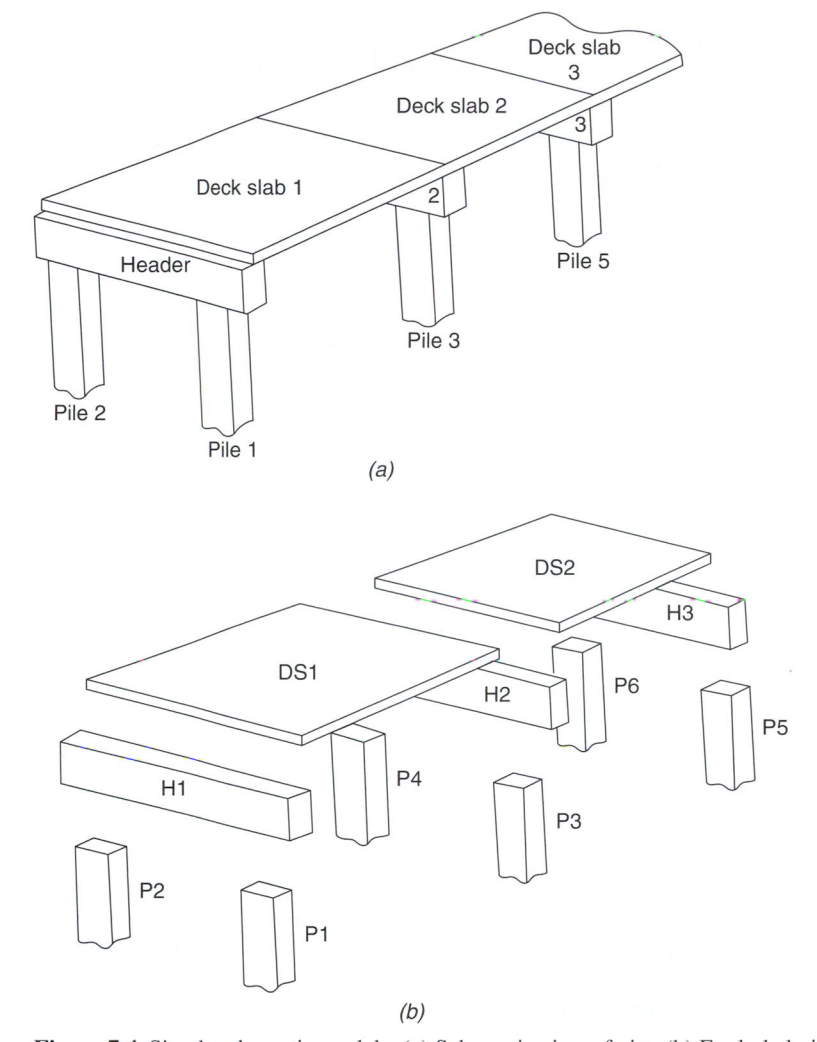

Figure 7.4 Simple schematic models. (a) Schematic view of pier. (b) Exploded view of pier. (Antill and Woodhead, 1982).

in the same relative positions. Notice that abbreviated labels have now been introduced. Clearly, these figures are schematic models (i.e., not physical models), but they have rather simple conceptual rules so that the physical relationship between the components of the structure is clear.

Now suppose that each component or element is represented by a labeled circle (or node). Figure 7.5 gives a "plan" view of the pier components shown in Figure 7.4. Such an abstraction or model can be used as the basis for portraying information about the physical makeup of the pier or about the order in which the physical components will actually appear on the site.

For example, an indication of the adjacency of physical components or the relational contact of physical components may be required. A model to portray these properties requires a modeling element (say a line) to indicate that the property exists. Assuming the modeling rationale of Figure 7.6(a), the various nodes of Figure 7.5 can be joined by a series of lines to develop a graph structure portraying the physical component adjacency or contact nature of the pier. If the idea of contact is expanded to indicate the order in which elements appear and physical contact is established, a directed line modeling rationale may be used, as shown in Figure 7.6(b). Using this conceptual modeling rule, Figure 7.7 can

P2   P4   P6

H1   S1   H2   S2   H3

P1   P3   P5

Figure 7.5 Conceptual Model of Pier Components.

be developed. This figure shows, for example, that header 1 (H1) can only appear (i.e., be built) after piles 1 and 2 (i.e., P1, P2) appear; in fact, header 1 is built around, on top of, and therefore in contact with piles 1 and 2. Finally, if the order of appearance of physical elements is to be modeled for all elements, whether or not in contact, a directional arrow such as that shown in Figure 7.6(c) may be necessary.

As an example of the above modeling techniques, consider the pier pile driving operation. A number of possible pile driving sequences are shown in Figures 7.8(a), (b), and (c). In Figure 7.8(a) it is assumed that the pile driving rig is swung by its mooring cables to drive the piles alternatively from one line to the next (i.e., P1, P2) before being relocated for the next set of piles (P3, P4), and so forth. In Figure 7.8(b) the pile driving sequence is along one line first (i.e., piles P1, P3, P5) and then along the other line (P2, P4, P6, etc.). Figure 7.8(c) shows a situation that may result if field events interrupt the planned sequence. In this case the figure indicates a situation where, for example, pile 2 (P2) is broken or lost during pile driving operations, so that to conserve time the pile driving rig moves on to drive piles P4 and P3 and then returns to re-drive a new pile P2 before resuming normal pile driving sequences. This situation is common in practice and indicates the major difficulty with scheduling models in relation to what actually happens in the field.

Figure 7.8(d) indicates the basic modeling rationale of bar charts wherein specific identification with individual piles is hidden. First, it implies the concept that each pile requires a certain time to appear in its driven position on site; second, it implies that the actual sequence of driving piles on the site is not absolutely fixed or essential to field management.

In order to schedule a project, the sequence of activities and their relationship to one another must be defined. The use of directional links or arrows provides a flexible modeling technique which establishes the scheduling logic or sequence of activities within a project.

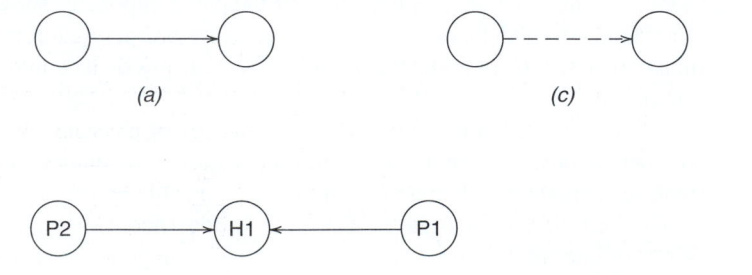

(a)   *(c)*

P2 ———→ H1 ←——— P1

(b)

Figure 7.6 Logical modeling rationales. (a) Adjacency of contact modeling. (b) Physical structure order modeling. (c) Physical construction order modeling.

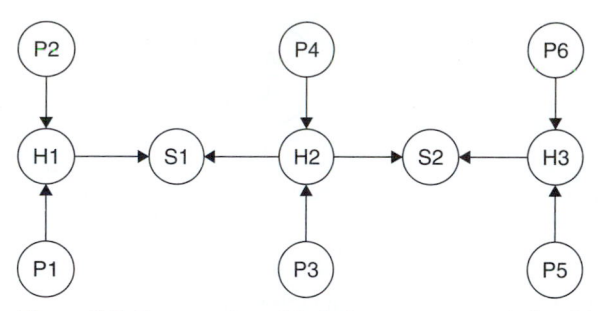

Figure 7.7 Conceptual model of pier component relationships.

7.4   SCHEDULING NETWORKS

A scheduling network consists of nodes and links. Depending on the notation, the nodes may represent events in time or activities. In Figure 7.9(a), nodes are used to represent events marking the beginning or ending of an activity. In Figure 7.9(b), a single larger node is used to represent an activity (e.g., "Activity A") with a duration of 10 days. Similarly, links may be used to represent activities or to indicate the logical sequence between activities. In order to show sequence, links become directional and are converted to arrows by the addition of an arrowhead. In Figure 7.9(a), the directional link represents an activity which begins on node i and ends on node j. The length of the link implies duration similar to the length of a bar in a bar chart. The arrow head implies that the activity begins on one node (i.e. "i") and ends on another. Figure 7.9(b) shows directional links which place activity "A" in a logical sequence relating it to other activities. The left-hand arrow links A to a preceding activity. The right-side arrow leads to a following activity.

Networks of activities can be constructed using the nodes and directional links or arrows. In Figure 7.10(a), a network of activities is shown in which nodes represent the activities and arrows represent the sequence or relationship of the various activities to one another. For instance, activity E is preceded by activities B and C. Activity E is followed by activity F. This network format is called *activity on node* or *precedence notation*.

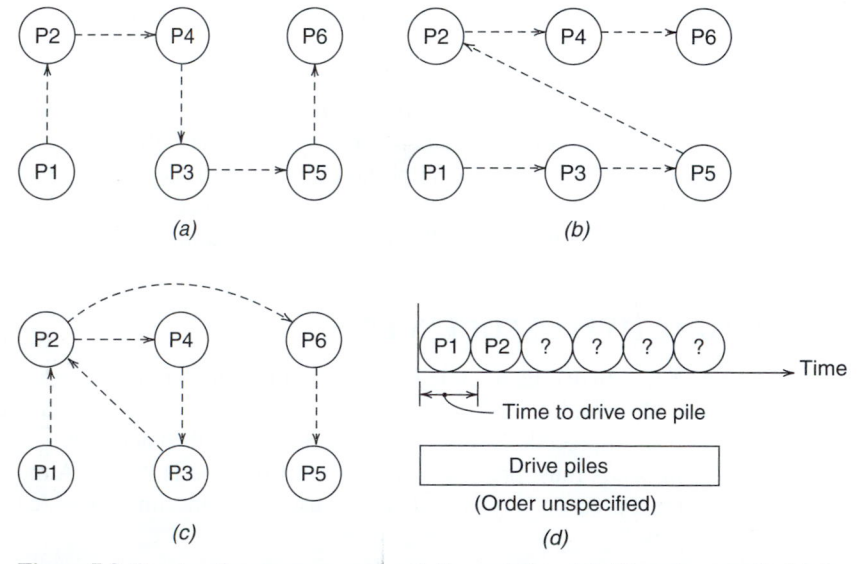

Figure 7.8 Construction sequence and activity modeling. (a) Alternate row pile driving. (b) Sequential row pile driving. (c) Field mishap alteration to pile driving sequence. (d) Bar chart model of pile driving operation. (Antill and Woodhead, 1982.)

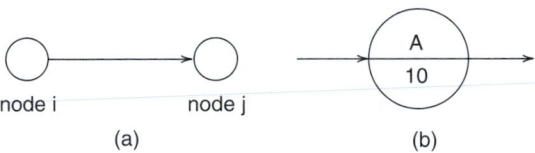

node i   node j

(a)   (b)

Figure 7.9 (a) node to represent an event. (b) node to represent an activity.

In Figure 7.10(b), the same group and sequence of activities are shown with the nodes representing events in time and the arrows representing activities. The nodes represent the beginnings and endings of activities (i.e., events in time). It will be noted that in order to indicate that activity E is preceded by both activities B and C, an extra activity called a "dummy" activity has been inserted between nodes 3 and 4. The reason the dummy activity is needed is to avoid a logical anomaly or mistake. If activity B ends directly on node 4 as shown in Figure 7.11, the sequence is incorrect. Activity B does precede both activities D and E (correct), but now the logic indicates that activity C precedes both activities D and E. We know, however, from Figure 7.10(a), that C only precedes E and has no connection with activity D. Therefore, the network in Figure 7.11 is not equivalent to that shown in Figure 7.10(a). To correct for this, the "dummy" activity (shown as a dashed arrow) is inserted. Now activity B is connected to activity E, but C has no connection with D. The use of the arrows to represent activities is called *Activity on Arrow*, *AOA* or *arrow notation*. It is sometimes referred to as "i-j" notation since, as noted in Figure 7.9(a), the activities are defined as beginning on a generic node i and end on a node j.

Activity networks have become the most common method for scheduling projects involving complex sequences of activities or activity paths. In the late 1950s, Kelly and Walker demonstrated that in activity networks a longest path of activities can be identified using simple mathematical methods.[1] The longest path or paths define a set of activities which control the total duration of the project represented by the network. The set of activities represented by the longest path can also be considered "*critical*," since if one or more of the critical activities are delayed, the total project duration will be extended. Kelly and Walker called their approach to scheduling the *Critical Path Method* or CPM.

At the time Kelly and Walker developed the basis for network scheduling and Critical Path calculations, they were involved in the management of complex industrial facilities being constructed by the DuPont Corporation. Methods such as bar charting could not capture the complex interlinking and interaction of scheduled activities. Knowledge of this interaction and its implication for meeting project completion dates required a more formal approach to include mathematical rigor and the ability to quickly update large and complex schedules. It was also important to be able to determine whether activity criticality was changed due to duration changes or scheduling delays. Within this framework, the iterative method we now call CPM or network scheduling was developed.

7.5 TYPICAL CPM ACTIVITY SEQUENCES

The network diagrams in Figure 7.12 illustrate some of the logical sequences used to develop CPM schedules using arrow notation. In part (1), it is obvious that A must precede B, and B must precede C. In (2), A must precede both B and C. In (3), A and B must precede C. In (4), A must precede C, and B must precede D. In (5), A must precede C and D, and B must precede D; this necessitates using a connecting arrow (called a dummy) to maintain

[1] Kelley, J. E. and M. R. Walker, *Critical Path Planning and Scheduling*, Proceedings of the Eastern Joint Computer Conference, December 1959, pp. 160–173.

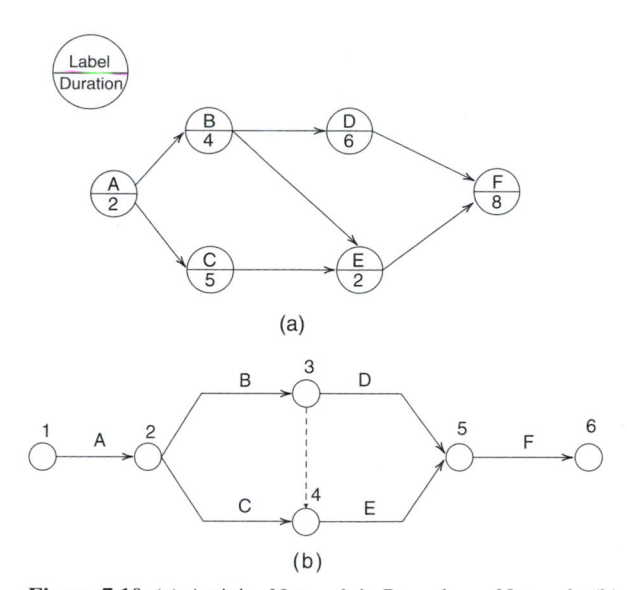

(a)

(b)

Figure 7.10 (a) Activity Network in Precedence Network. (b) Activity Network in Arrow Notation.

the logical sequence of events between A and D. Dummy activities have zero time duration; they are shown by broken arrows. Dummies may also be required to maintain specific activity identification between events, as shown in (6), where A must precede B and C, and B and C must precede D. Events and activities should be labeled, and they are usually numbered for computer identification of the network elements.

Figure 7.13 shows elements of a precedence notation network corresponding to the same activity sequences shown in Figure 7.12 in arrow notation.

Consider the simple construction of concrete footings, which involves earth excavation, reinforcement, formwork, and concreting. A preliminary listing of activities might be:

  A. Lay out foundations.

  B. Dig foundations.

  C. Place formwork.

  D. Place concrete.

  E. Obtain steel reinforcement.

  F. Cut and bend steel reinforcement.

  G. Place steel reinforcement.

  H. Obtain concrete.

Examination of the list of activities shows that some grouping is obvious (see Figure 7.14). Thus, considering physical constraints only, the following physical chains can be developed.

  1. From a consideration of the actual footings: A, B, C, G, D.

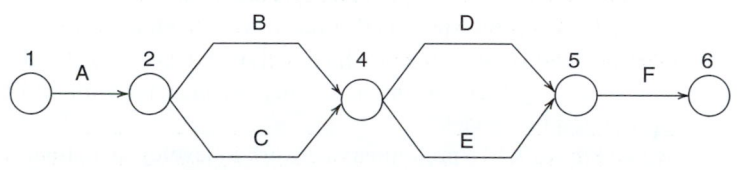

Figure 7.11 Mistake in Logical Sequence.

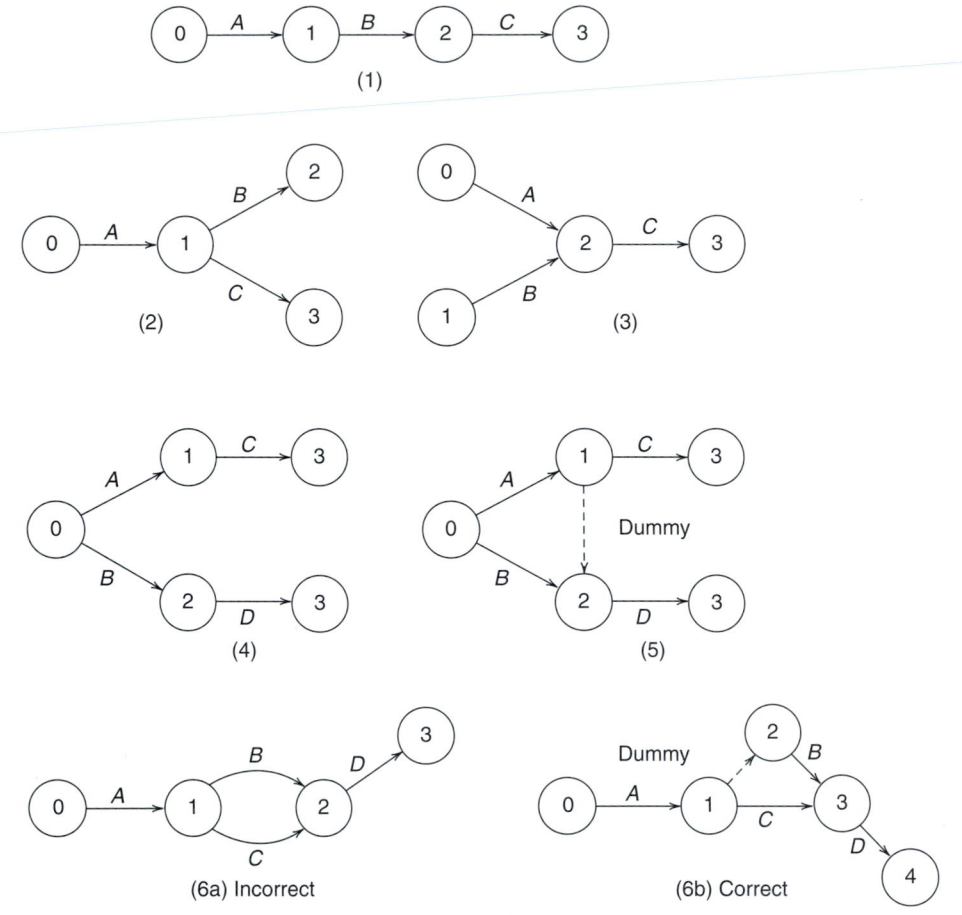

Figure 7.12 Elements of an arrow network. (After Antill and Woodhead, 1982).

2. From a consideration of the steel reinforcement: E, F, G, D.

3. From a consideration of the concrete only: H, D.

When the project is seen from these different viewpoints, individual chains of activities emerge; but, on viewing the job as a whole, it is obvious that interrelationships exist. For example, it is useless to pour concrete before the steel reinforcement is placed and the formwork is installed. Therefore, all the chains must merge before pouring the concrete. And if steps are to be taken to obtain the steel and the concrete immediately when work begins (this would be a management decision or constraint), then the chains all start at the same point or event with the laying out of the foundations.

The development of a preliminary network for the project is possible at this stage because first, a list of activities has been defined and second, a rough construction logic has emerged.

The actual representation and appearance of the network depend on the modeling form adopted and on the spatial locations of the symbols as drawn. As mentioned previously, there are two basic ways in which activities can be modeled: (1) when the activities are represented by "arrows" in an activity-oriented network and (2) when the activities are represented by "nodes."

In Figure 7.14 a preliminary network is developed, in both arrow and circle forms, from the above information.

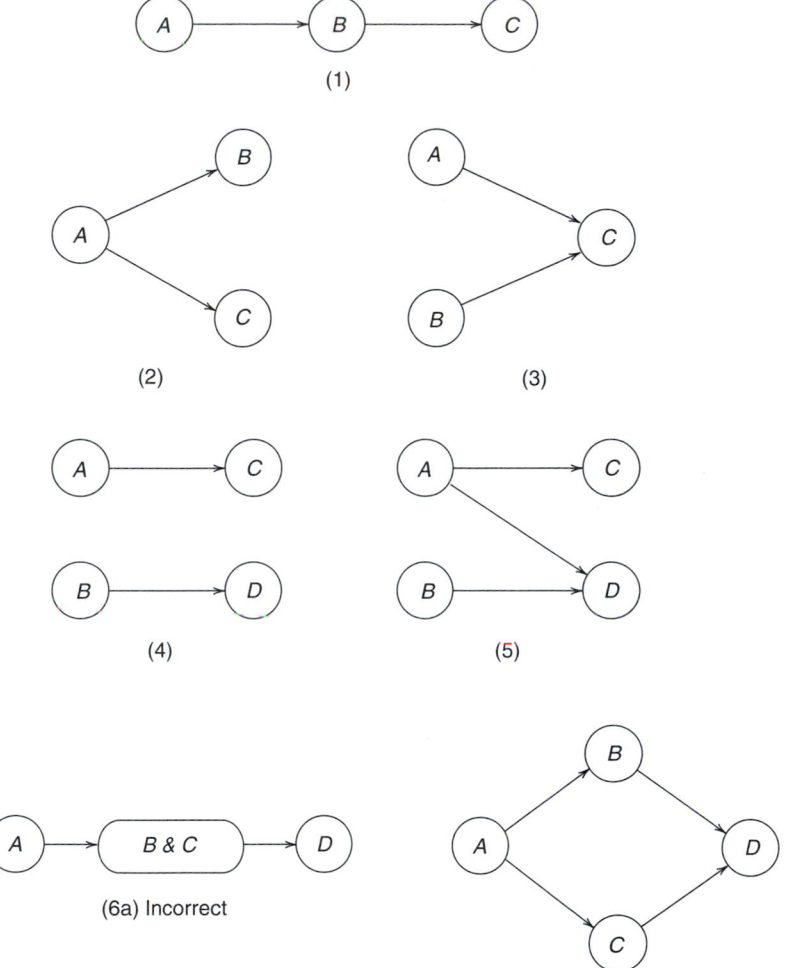

Figure 7.13 Elements of a precedence network. (After Antill and Woodhead, 1982).

7.6   NETWORK SCHEDULE ANALYSIS USING PRECEDENCE NOTATION[2]

The objective of analyzing a project network is to:

1. Find the critical set of activities that establishes the longest path and defines the minimum duration of the project.

2. Calculate the early start times for each activity.

3. Calculate the late start times for each activity.

4. Calculate the float, or time, available for delay for each activity.

By definition, critical activities cannot be delayed without extending the project duration. Therefore, the float or amount of delay time associated with critical activities is zero. Activities that are critical lie along the longest path through the network.

In order to determine which path is the longest, a variety of methods have been developed. In this chapter, a method based on the use of two algorithms is used to identify the

---

[2] A discussion of network scheduling calculations using arrow notation is presented in Appendix E.

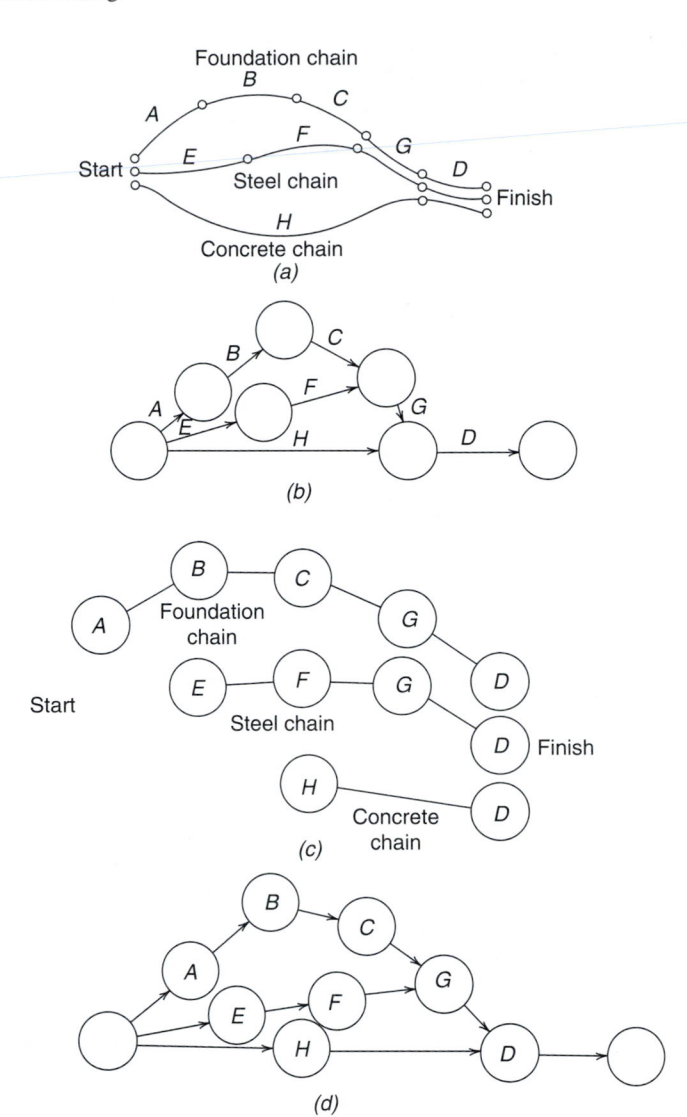

Figure 7.14 Preliminary network diagram. (a) Initial sketch, arrow notation. (b) First draft – arrow notation. (c) Initial sketch – precedence notation. (d) First draft – precedence notation. (After Antill and Woodhead, 1982).

longest and critical path. An algorithm is a formula, or recipe, that is repetitively applied to solve a problem in a stagewise manner. In solving for the critical path(s),[3] the *forward-pass algorithm* is used to calculate the earliest event times for each node. It also allows calculation of the minimum duration of the project. The *forward-pass algorithm* is not sufficient, however, to calculate the critical path consisting of the critical set of activities in the project.

In order to identify the critical path, a second algorithm, called the *backward-pass algorithm*, is needed. This algorithm allows for the calculation of the latest event times for each of the activities. Once the latest event times for each node have been determined, the critical activities and hence the critical path can be identified. Those activities for which the earliest and latest start times are the same are critical since they cannot be delayed without

---

[3] Sometimes there are multiple longest pathes which are equal in duration.

causing a delay in the total project completion time. Since the earliest and latest start times are the same, these activities have zero float.

Activities for which the earliest and latest start times are not the same can be delayed to a degree without delaying the total project. The amount (e.g., number of days) by which they can be delayed as noted above is called *float*. Such activities will have positive float and, therefore, are not critical. These activities can be delayed to the degree established by the float available without impacting the completion time of the total project.

When working in precedence notation, the forward-pass algorithm is applied to the first or source activity in the network. If there are multiple starting activities (i.e., nodes), then these multiple starts are connected to a single node called START to allow for a common source node to be used for commencing application of the algorithm. In effect, the calculations start at the leftmost or start activity node and the algorithm is applied in a "boot strapping" fashion until calculation of the late finish time of the final or END activity of the network has been calculated.

In order to demonstrate this method, consider the small precedence notation network shown below as Figure 7.15. The activities are shown as nodes or circles. The label for each activity is shown in the upper half of the circle, and the activity duration is shown in the lower half of the circle.

Four values will be calculated for each activity as shown in Figure 7.15 on the reference circle. These are:

$$EST(I) = \text{Early start time of activity I}$$
$$EFT(I) = \text{Early finish time of activity I}$$
$$LST(I) = \text{Late start time of activity I}$$
$$LFT(I) = \text{Late finish time of activity I}$$

These four values for each activity will be calculated by starting at activity A and proceeding from left to right in the network. This calculation moving from left to right in time is called a "forward pass." Two equations form the basis of the forward pass algorithm as follows:

$$\boxed{EFT(I) = EST(I) + DUR(I)} \tag{7.1}$$

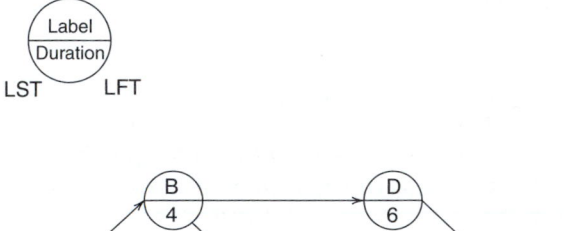

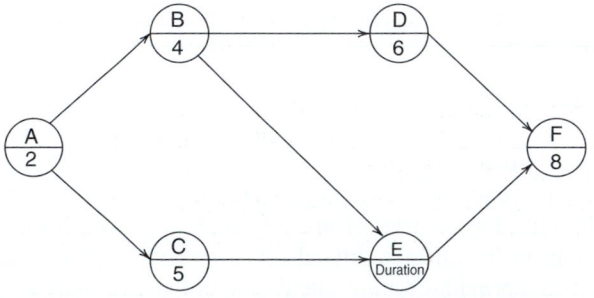

Figure 7.15 Precedence Notation Scheduling Network.

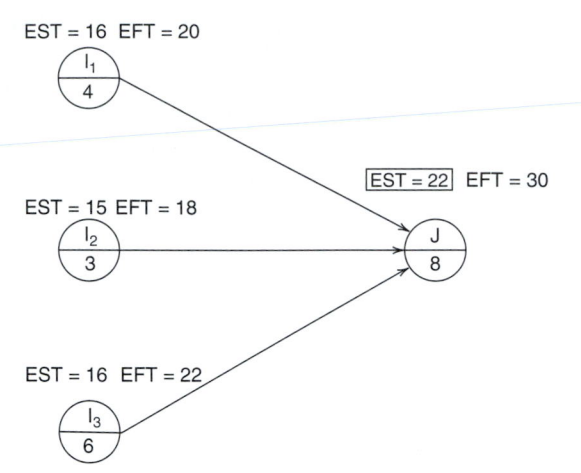

Figure 7.16 Calculation of EST(J).

where DUR(I) is the duration of the activity being considered. The early start is calculated based on the EFTs of the activities directly preceding the activity of interest as follows:

$$\text{EST}(J) = \max_{I \in M}^{\text{all } I} [\text{EFT}(I)] \qquad (7.2)$$

where I is a member activity of the set of M activities that precede activity J.

A graphical illustration example of this equation is shown in Figure 7.16.

The EFT values of the three preceders shown in Figure 7.16 are 20, 18, and 22. Since the largest of these values is 22, the EST(J) = 22. The EFT values for each of the preceders were calculated using EFT(I) = EST(I) + DUR(I). Equations 7.1 and 7.2 define the calculations needed to perform the forward-pass algorithm.

Applying the forward-pass algorithm to the network shown in Figure 7.15 yields the following results for activities A through F.

Activity	Calculation	
A	EST(A) = 0	EFT(A) = 2
B	EST(B) = max[EFT(A)] = 2	EFT(B) = 2 + 4 = 6
C	EST(C) = max[EFT(A)] = 2	EFT(C) = 2 + 5 = 7
D	EST(D) = max[EFT(B)] = 6	EFT(D) = 6 + 6 = 12
E	EST(E) = max[EFT(B), EFT(C)] = max[6, 7] = 7	EFT(E) = 7 + 2 = 9
F	EST(F) = max[EFT(D), EFT(E)] = max[12, 9] = 12	EFT(F) = 12 + 8 = 20

This set of equations (algorithmic calculations) tells us that the earliest completion time for activity "F" is 20 days (assuming the durations given are in days). Therefore, the duration of the entire project is at least 20 days. In fact, the forward pass is sufficient to tell us that the duration of both the entire project and the critical or longest path is 20 days. The EST and EFT values for the network are shown plotted on each activity in Figure 7.17.

In order to identify the critical set of activities, a backward pass must be done. The backward pass provides for the calculation of the Late Start and Late Finish times for each activity. The critical set of activities and consequently the critical path within the network

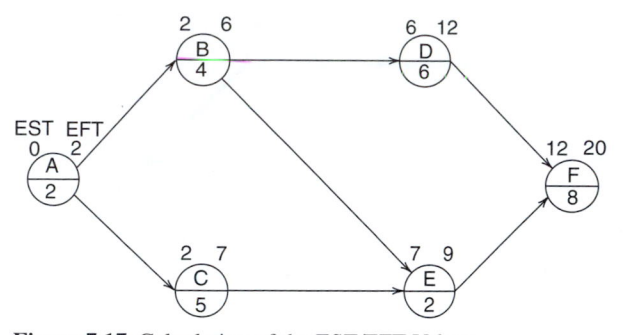

Figure 7.17 Calculation of the EST/EFT Values.

is defined by those activities which have LSTs and ESTs which are the same. (The LFTs and EFTs will also be the same for all critical activities). Since the early start and late start times are the same, any delay associated with a critical activity will cause the total project duration to be extended.

The backward-pass calculations utilize the following two equations.

$$LST(J) = LFT(J) - DUR(J) \qquad (7.3)$$

$$LFT(I) = \min_{J \in M}^{all\ J} [LST(J)] \qquad (7.4)$$

where J is a member activity of the set of activities M that follows activity I.

A graphical version of the application of Equation 7.4 is shown in Figure 7.18. The minimum of the LSTs for the three following activities is 28. Therefore the LFT(I) = 28.

Again, to get started on the backward pass, the LFT of the last activity F is assumed to be equal to the EFT of F. That is, LFT(F) = EFT(F) = 20. Calculations move from right to

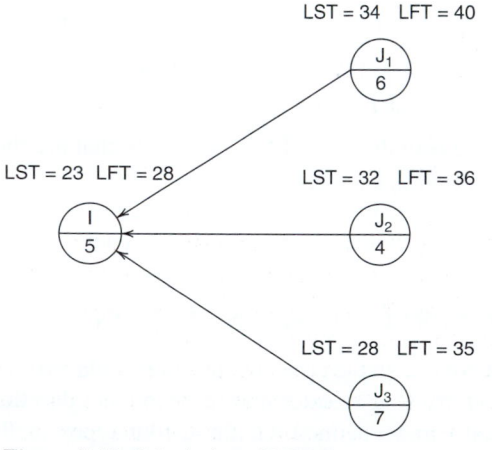

Figure 7.18 Calculation of LFT(I).

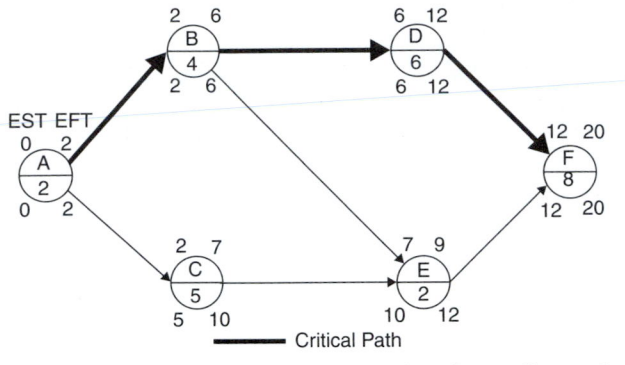

Figure 7.19 EST, EFT, and LST, LFT values for small precedence notation network.

left. Therefore,

$$LST(F) = LFT(F) - DUR(F) = 20 - 8 = 12$$

The other backward-pass calculations are as follows:

Activity	Calculation
E	$LFT(E) = min[LST(F)] = 12$ $LST(E) = LFT(E) - DUR (E) = 12 - 2 = 10$
D	$LFT(D) = min[LST(F)] = 12$ $LST(D) = LFT(D) - DUR(D) = 12 - 6 = 6$
C	$LFT(C) = min[LST(E)] = 10$ $LST(C) = LFT(C) - DUR(C) = 10 - 5 = 5$
B	$LFT(B) = min[LST(D), LST(E)]$ $= min(6, 10) = 6$ $LST(B) = LFT(B) - DUR(B) = 6 - 4 = 2$
A	$LFT(A) = min[LST(B) - LST(C)]$ $= min(2, 5) = 2$ $LST(A) = LFT(A) - DUR(A) = 2 - 2 = 0$

If the calculations are performed correctly, the EST and LST of the initial activity (e.g., Activity A) should be zero (0). That is, the forward- and backward-pass calculations should close to an EST of zero for the source activity.

The critical path activities will satisfy the following relationship:

$$LST(I) = EST(I) \quad \text{which is equivalent to}$$
$$LFT(I) = EFT(I)$$

Therefore, all of the activities with ESTs and LSTs that are the same will be critical. The critical activities are A, B, D, and F.

Figure 7.19 shows the EST, EFT, LST, LFT values for all activities (A to F) as well as the critical path for this simple precedence notation network.

7.7   FLOAT CALCULATIONS WITH PRECEDENCE NOTATION

All activities that are not on the critical path can be delayed a certain number of time units (e.g., days) without causing an extension of the project duration. For noncritical activities four types of float can be defined. Of these four types of float, three have a practical interpretation within the context of a construction project. The four types of float are shown schematically in Table 7.1. The table also includes the formula for calculating each type of float.

Total float is the total, or maximum, number of time units that an activity can be delayed without increasing the total project duration. In calculating the amount of float, it is useful to think in terms of the "window" of time available for each activity to occur. For a critical activity, the window available for it in the schedule is just large enough to fit the duration of the activity. Therefore, a critical activity has zero float. For non-critical activities, the window is larger and accommodates some amount of delay without impacting the total project duration. The window of time available is defined by the Early Start Time (EST) and the Late Finish Time (LFT) of the activity being considered. This opens the window of time available to its maximum position. Given this maximum open position of the available time window, the activity is inserted as shown in Table 7.1.

The bar representing the activity is placed so that its start is at the EST(I) position. Having positioned the activity bar in this manner, the amount of time between the LFT(I) and the EFT(I) is the total float. Written as a formula:

$$TF(I) = LFT(I) - EFT(I)$$

Consider Activity E in the network of Figure 7.19. Then $TF(E) = LFT(E) - EFT(E) = 12 - 9 = 3$. This implies the Activity E can be delayed 3 days (e.g., time units) without delaying the overall duration of the project. The same calculation for Activity D yields:

$$TF(D) = LFT(D) - EFT(D) = 12 - 12 = 0$$

This is what we would expect since D is a critical activity. Since the TF is zero, we cannot delay Activity D without causing an increase in the total project duration.

Use of all of the total float available to an activity may reduce the float available to activities that follow it in sequence. For instance, the $TF(C) = LFT(C) - EFT(C) = 10 - 7 = 3$. For each day that C is delayed, the float available to activity E will decrease. IF 2 days of float on C are used, this will decrease the total float available to E to one day. Therefore, total float cannot be used without potentially having an effect on following activities.

The second type of float that is important is *free float*. Free float is the amount of time (e.g., number of time units) that an activity can be delayed without impacting activities that follow it. In other words, the float is free in the sense that it can be used without reducing

Table 7.1   Four Types of Activity Float

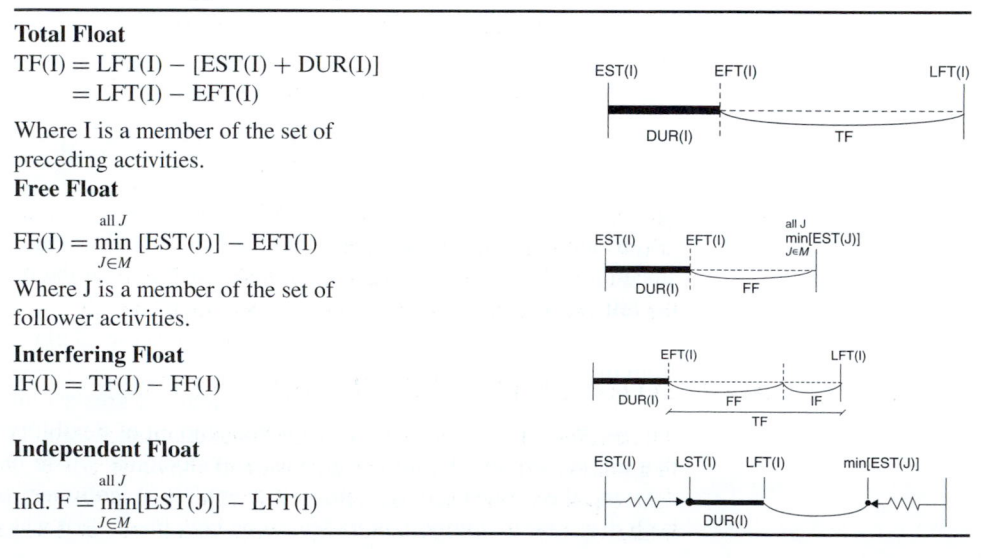

Total Float

$$TF(I) = LFT(I) - [EST(I) + DUR(I)]$$
$$= LFT(I) - EFT(I)$$

Where I is a member of the set of preceding activities.

Free Float

$$FF(I) = \min_{J \in M} [EST(J)] - EFT(I)$$

(all J)

Where J is a member of the set of follower activities.

Interfering Float

$$IF(I) = TF(I) - FF(I)$$

Independent Float

$$\text{Ind. F} = \min_{J \in M} [EST(J)] - LFT(I)$$

(all J)

the float available to following activities. Again, the idea of the window of available time is useful in determining the free float. The left side of the window remains as it was for total float. That is, the left side of the window is defined by the EST(I) as shown in Table 7.1. The right side of the "window," however, closes to the point defined by the minimum EST of the following activities. This may reduce the overall "width" of the window. Therefore, the formula for free float is:

$$FF(I) = \min_{\substack{J \in M}}^{\text{all } J}[EST(J)] - EFT(I)$$

where M is the set of activities which follow I and J is a member of that set. In the next section, this formula will be used to determine whether there is FF available on any of the activities in a more complex network schedule. Applying this formula to activities C and E we find:

$$FF(C) = \min [EST(E)] - EFT(C) = 7 - 7 = 0$$
$$FF(E) = \min [EST(F)] - EFT(E) = 12 - 9 = 3$$

This implies that any use of the TF(C) is not "free" and may, as we have discussed, take float away from activities which follow it. TF(E) and FF(E) are both 3. Therefore, using any float available to activity on E will not "rob" float available on following activities. In this sense, use of this float is free.

Interfering float is the amount of the total float utilized that interferes with the following activities. It is defined as:

$$IF(I) = TF(I) - FF(I)$$

The interfering float for activity C is:

$$IF(C) = TF(C) - FF(C) = 3 - 0 = 3$$

This implies that although 3 days of delay can occur on activity C without impacting the total project duration, each day of IF used in conjunction with activity C will "interfere" with the float available for following activities.

The last type of float to be considered here is called the *independent float*. In this case, the left side of the window shown in Table 7.1 is defined by the LST of the activity rather than the EST. The right side of the window is the same as for free float. The independent float is given by the formula:

$$\text{Ind. } F(I) = \min_{\substack{J \in M}}^{\text{all } J} [EST(J)] - (LST(I) + DUR(I))$$

$$= \min_{\substack{J \in M}}^{\text{all } J} [EST(J)] - [LFT(I)]$$

In some cases, the independent float will be negative. This can occur in some situations since the EFT(I) is later in time than the earliest EST of the following activities. In effect, the left side of the "window" is near or later than the right side. If the Ind. F is positive, this implies that float exists even if preceding activities use all of their float and no float is taken from the following activities.

As a practical matter the two floats that are of greatest interest are the *total float* and the *free float*. Floats not only indicate the amount of flexibility available for the activities that are not critical. Floats are also used in situations where one contractor is accused of delaying the project through failure to complete at the times specified in the schedule. In such disputes, it is important to determine how much float was available, who utilized the float, and who "owned" the float. The owner of the float is the party who is authorized to allocate the use of float. This may be a matter established in the contract documents or, failing contractual specification, may be determined by case law and legal precedent.

7.8   DEVELOPING A SCHEDULE FOR THE SMALL GAS STATION (PRECEDENCE NOTATION)

Using the bar-net shown in Figure 7.3 and information from the WBS given in Figure 6.3 for the small gas station, the original eight activities can be expanded to a set of 22 activities providing a more comprehensive scheduling framework for this project. The next step in scheduling is to assign durations to each of the activities. The scheduler must consult with field personnel and get the best estimate of the duration for each activity based on the methods selected for accomplishing the work, the resources available, and the experience the field management has in estimating productivity and time durations with these methods and resources.

For instance, based on the crew size, the equipment available, and the method of placement, the field superintendent can establish how long it will take to cast the concrete footers. The superintendent will know whether the concrete is to be mixed on site or brought in using a transit mix truck. He will also know the type of forms to be used, the nature of the reinforcement to be installed, the nature of any embedments and penetrations required, and the placement and number of anchor bolts required for the building structure. The estimate of duration for a given activity is given in working days. Table 7.2 gives a listing of the expanded list of activities together with estimated durations for each activity. An expanded

Table 7.2   Durations of Activities for the Small Gas Station

Activity	Title	Duration (Days)
1	Mobilize	10
2	Obtain permits	15
3	Site work	8
4	Exterior utilities	12
5	Excavate catch basin	2
6	Excavate footers	5
7	Excavate foundation piers	6
8	Pour footers, etc.	8
9	Erect bldg. frame	10
10	Exterior brick facade	14
11	Exterior fascia panels	4
12	Roof construction	15
13	Landscaping	12
14	Pour interior slabs	10
15	Glazing and doors	6
16	Interior walls	10
17	Elec. & mech. Systems	25
18	Shelves	3
19	Floor coverings	6
20	Interior finishes	8
21	Final inspection	1
22	Demobilization	3

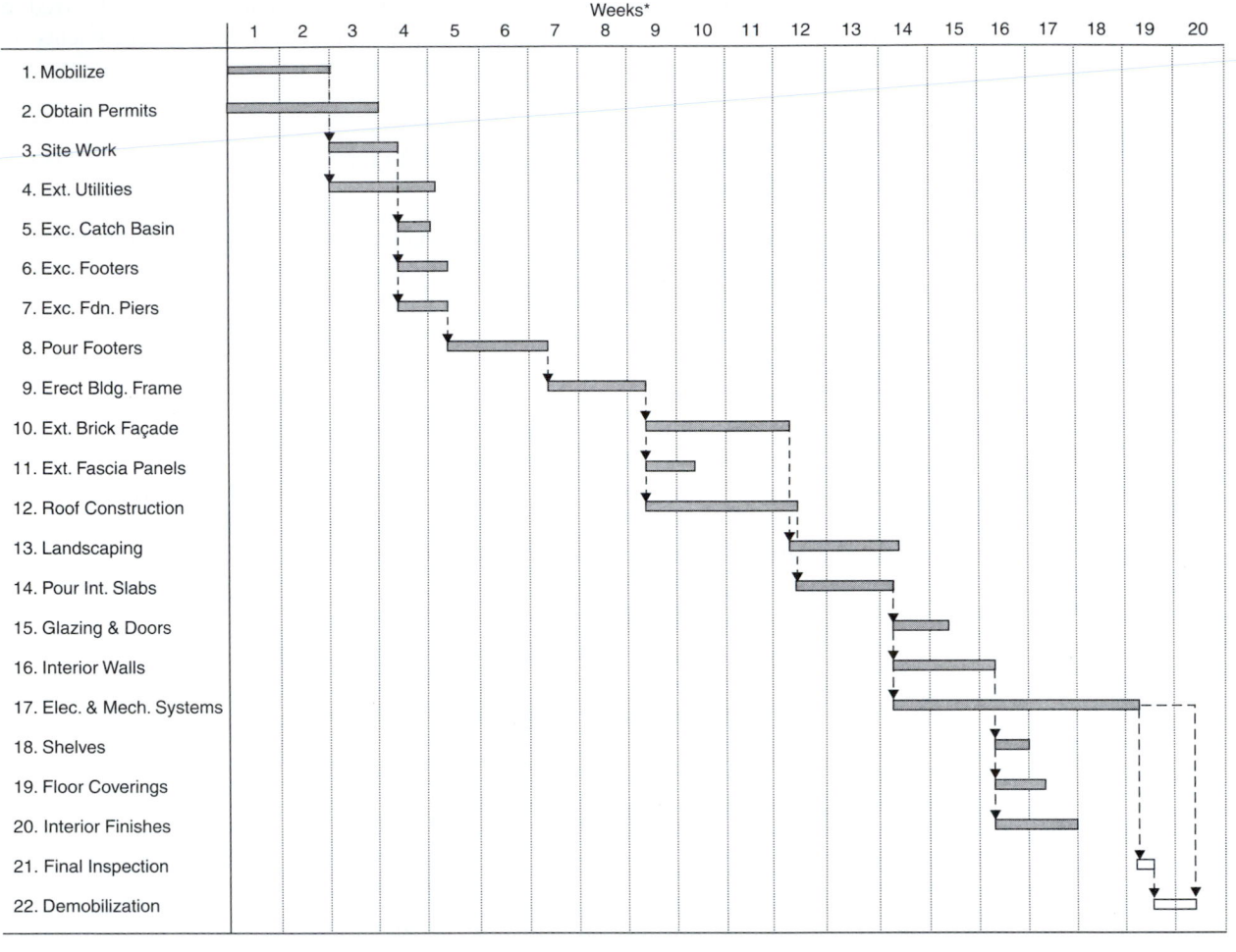

*Weeks are assumed to be work weeks consisting of 5 working days.

Figure 7.20 Expanded Bar-Net Schedule for the Small Gas Station.

bar-net based on Table 7.2 and including the logical sequencing of the 22 activities is shown in Figure 7.20.

The bar-net has been used to develop a precedence network model of the small gas station project. This network is shown in Figure 7.21. The values of the ESTs and EFTs are shown above the circle representing each activity. Similarly, the LSTs and LFTs are shown below the activity circle.

The forward-pass algorithm is applied repetitively starting at the source node labeled START. All events related to the START node are given the value zero. Two activities (e.g., 1 and 2) follow the START node. Calculations for all of the activities are shown in Table 7.3. Calculated EST and LST values are shown on each activity in Figure 7.13. Based on the calculations in Table 7.3, it can be seen that the duration of the longest and therefore critical path in the network is 96 working days.[4] The minimum project duration based on this critical path analysis is 96 working days or slightly over 19 working weeks. If we are plotting the project duration in calendar days, this will equate to $(19 \times 7) + 1$ or 134 calendar days. If

---

[4] Working days are differentiated from calendar days. There are typically 5 working days in a week as opposed to 7 calendar days.

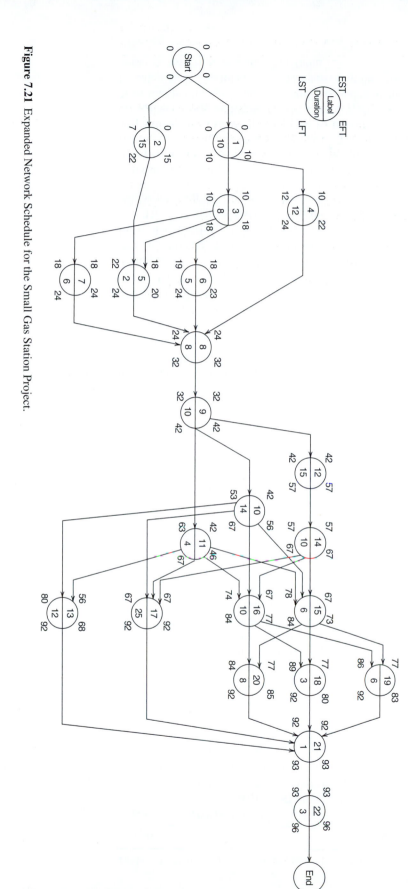

Figure 7.21 Expanded Network Schedule for the Small Gas Station Project.

Table 7.3   Forward-Pass Calculations for the Small Gas Station Project

Activity	Calculations	
Start	EST(START) = 0	EFT(START) = 0
1	EST(1) = max[EFT(START)] = 0	EFT(1) = 0 + 10 = 10
2	EST(2) = max[EFT(START)] = 0	EFT(2) = 0 + 15 = 15
3	EST(3) = max[EFT(1)] = 10	EFT(3) = 10 + 8 = 18
4	EST(4) = max[EFT(1)] = 10	EFT(4) = 10 + 12 = 22
5	EST(5) = max[EFT(2), EFT(3)] = 18	EFT(5) = 18 + 2 = 20
6	EST(6) = max[EFT(3)] = 18	EFT(6) = 18 + 5 = 23
7	EST(7) = max[EFT(3)] = 18	EFT(7) = 18 + 6 = 24
8	EST(8) = max[EFT(4), EFT(5), EFT(6), EFT(7)] = 24	EFT(8) = 24 + 8 = 32
9	EST(9) = max[EFT(8)] = 32	EFT(9) = 32 + 10 = 42
10	EST(10) = max[EFT(9)] = 42	EFT(10) = 42 + 14 = 56
11	EST(11) = max[EFT(9)] = 42	EFT(11) = 42 + 4 = 46
12	EST(12) = max[EFT(9)] = 42	EFT(12) = 42 + 15 = 57
13	EST(13) = max[EFT(10), EFT(11)] = 56	EFT(13) = 56 + 12 = 68
14	EST(14) = max[EFT(12)] = 57	EFT(14) = 57 + 10 = 67
15	EST(15) = max[EFT(10), EFT(11), EFT(14)] = 67	EFT(15) = 67 + 6 = 73
16	EST(16) = max[EFT(10), EFT(11), EFT(14)] = 67	EFT(16) = 67 + 10 = 77
17	EST(17) = max[EFT(10), EFT(11), EFT(14)] = 67	EFT(17) = 67 + 25 = 92
18	EST(18) = max[EFT(15), EFT(16)] = 77	EFT(18) = 77 + 3 = 80
19	EST(19) = max[EFT(15), EFT(16)] = 77	EFT(19) = 77 + 6 = 83
20	EST(20) = max[EFT(15), EFT(16)] = 77	EFT(20) = 77 + 8 = 85
21	EST(21) = max[EFT(13), EFT(17), EFT (18), EFT(19), EFT(20)] = 92	EFT(21) = 92 + 1 = 93
22	EST(22) = max[EFT(21)] = 93	EFT(22) = 93 + 3 = 96

the project is to begin on Monday March 1, the estimated project completion date will be July 12, 20xx.

The forward-pass calculations establish the minimum project duration (based on no delays to the critical activities). It does not, however, identify the critical path. In order to identify the critical set of activities that constrain the project to a minimum duration of 96 days, we must apply the backward-pass algorithm. To start the backward-pass algorithm, the LFT (END) is set to 96 days. This is equivalent to setting the LFT for activity 22 to 96 days. As we have just seen, based on our forward-pass calculations, the LFT of activity 22 cannot be less than 96 days. If we set LFT (22) to a duration greater than 96 days, the finish date for the project will be extended.

The calculations for the backward-pass are given in Table 7.4, and the LST and LFT values for each activity are shown in Figure 7.21.

7.9   FLOAT CALCULATIONS AND THE CRITICAL PATH

The critical set of activities can be identified as those which have zero float. Once the critical activities are identified, the critical path linking these activities can be established. Now that the forward-and backward-pass calculations have been completed, the floats can be calculated. The total, free, and interfering float for all activities in the small gas station network have been calculated in Table 7.5.

Table 7.4   Backward-Pass Calculations for the Small Gas Station Project

Activity	Calculations	
END	LFT(END) = 96	LST(END) = 96
22	LFT(22) = min[LST(END)] = 96	LST(22) = 96–3 = 93
21	LFT(21) = min[LST(22)] = 93	LST(21) = 93–1 = 92
20	LFT(20) = min[LST(21)] = 92	LST(20) = 92–8 = 84
19	LFT(19) = min[LST(21)] = 92	LST(19) = 92–6 = 86
18	LFT(18) = min[LST(21)] = 92	LST(18) = 92–3 = 89
17	LFT(17) = min[LST(21)] = 92	LST(17) = 92–25 = 67
16	LFT(16) = min[LST(18), LST(19), LST(20)] = 84	LST(16) = 84–10 = 74
15	LFT(15) = min[LST(18), LST(19), LST(20)] = 84	LST(15) = 84–6 = 78
14	LFT(14) = min[LST(15), LST(16), LST(17)] = 67	LST(14) = 67–10 = 57
13	LFT(13) = min[LST(21)] = 92	LST(13) = 92–12 = 80
12	LFT(12) = min[LST(14)] = 57	LST(12) = 57–15 = 42
11	LFT(11) = min[LST(13), LST(15), LST(16), LST(17) = 67	LST(11) = 67–4 = 63
10	LFT(10) = min[LST(13), LST(15), LST(16), LST(17)] = 67	LST(10) = 67–14 = 53
9	LFT(9) = min[LST(10), LST(11), LST(12)] = 42	LST(9) = 42–10 = 32
8	LFT(8) = min[LST(9)] = 32	LST(8) = 32–8 = 24
7	LFT(7) = min[LST(8)] = 24	LST(7) = 24–6 = 18
6	LFT(6) = min[LST(8)] = 24	LST(6) = 24–5 = 19
5	LFT(5) = min[LST(8)] = 24	LST(5) = 24–2 = 22
4	LFT(4) = min[LST(8)] = 24	LST(4) = 24–12 = 12
3	LFT(3) = min[LST(5), LST(6), LST(7)] = 18	LST(3) = 18–8 = 10
2	LFT(2) = min[LST(5)] = 22	LST(2) = 22–15 = 7
1	LFT(1) = min[LST(3), LST(4)] = 10	LST(1) = 10–10 = 0

Those activities with zero float are noted in Table 7.5 with an asterisk. The critical set of activities consists of activities 1, 3, 7, 8, 9, 12, 14, 17, 21, and 22. Based on the activities which have zero total float, the critical path can be identified in Figure 7.21. As a practice exercise verify that the duration the critical chain of activities is 96 days. Now take several paths through the network and verify that the total duration of each of these non-critical paths is less than 96 days.

To better understand the concepts of float in this project network consider the non-critical activity 15. This activity has 11 days of total float. The window of time available for installing glazing and doors allows for slippage of up to 11 days without extending the total duration of the project. Since the free float is 4 days, this means that the activity can be delayed 4 days from its EST without impacting any following activities. The interfering float is 7 days, so that after four days of delay, float used will impact activities which follow activity 15 (e.g., 18, 19, and 20).

7.10   SUMMARY

A number of techniques exist for time planning, scheduling, and control of construction projects. Each has strengths and weaknesses. In this chapter, the concepts of bar charting, bar-nets, and network scheduling have been introduced. These techniques are used widely, and critical path analysis or CPM is often required by contract when constructing even relatively small projects. For complex projects with numerous activities being worked

Table 7.5   Float Values for the Small Gas Station Project

Activity	Total Float	Free Float	Interfering Float
* 1	TF(1) = 10–10 = 0	FF(1) = 10–10 = 0	IF(1) = 0–0 = 0
2	TF(2) = 22–15 = 7	FF(2) = 18–15 = 3	IF(2) = 7– 3 = 4
* 3	TF(3) = 18–18 = 0	FF(3) = 18–18 = 0	IF(3) = 0–0 = 0
4	TF(4) = 24–22 = 2	FF(4) = 24–22 = 2	IF(4) = 2–2 = 0
5	TF(5) = 24–20 = 4	FF(5) = 24–20 = 4	IF(5) = 4–4 = 0
6	TF(6) = 24–23 = 1	FF(6) = 24–23 = 1	IF(6) = 1–1 = 0
* 7	TF(7) = 24–24 = 0	FF(7) = 24–24 = 0	IF(7) = 0–0 = 0
* 8	TF(8) = 32–32 = 0	FF(8) = 32–32 = 0	IF(8) = 0–0 = 0
* 9	TF(9) = 42–42 = 0	FF(9) = 42–42 = 0	IF(9) = 0–0 = 0
10	TF(10) = 67–56 = 11	FF(10) = 56–56 = 0	IF(10) = 11–0 = 11
11	TF(11) = 67–46 = 21	FF(11) = 56–46 = 10	IF(11) = 21–10 = 11
* 12	TF(12) = 57–57 = 0	FF(12) = 57–57 = 0	IF(12) = 0–0 = 0
13	TF(13) = 92–68 = 24	FF(13) = 92– 68 = 24	IF(13) = 24– 24 = 0
* 14	TF(14) = 67–67 = 0	FF(14) = 67–67 = 0	IF(14) = 0–0 = 0
15	TF(15) = 84–73 = 11	FF(15) = 77–73 = 4	IF(15) = 11–4 = 7
16	TF(16) = 84–77 = 7	FF(16) = 77–77 = 0	IF(16) = 7–0 = 7
* 17	TF(17) = 92–92 = 0	FF(17) = 0–0 = 0	IF(17) = 0–0 = 0
18	TF(18) = 92–80 = 12	FF(18) = 92–80 = 12	IF(18) = 12–12 = 0
19	TF(19) = 92–83 = 9	FF(19) = 92–83 = 9	IF(19) = 9–9 = 0
20	TF(20) = 92–85 = 7	FF(20) = 92–85 = 7	IF(20) = 7–7 = 0
* 21	TF(21) = 93–93 = 0	FF(21) = 0–0 = 0	IF(21) = 0–0 = 0
* 22	TF(22) = 96–96 = 0	FF(22) = 0–0 = 0	IF(22) = 0–0 = 0

simultaneously and complex sequences of procurement and installation, critical path network scheduling may by the only adequate means of planning and scheduling. When the number of activities in a network schedule exceeds 200 to 300, updating and control become extremely tedious. Such projects are almost always scheduled and controlled using specialty software based on the CPM algorithms discussed here.

As computers continue to make tedious calculations more tractable, greater complexity can be addressed to include the impact of available resources (e.g., crews, cranes, trucks, etc.) in determining a realistic sequence of work. Our discussion in this chapter assumes that all resources required are available and the logic developed makes the same assumption (i.e., the network as presented is not resource constrained). That is, the network is calculated as if an infinite pool of the resources required is available. This, of course, is not realistic.

Several activities scheduled for the same day may require trucks. If the total number of trucks needed is 12 and only 9 trucks are available, there is a resource conflict or constraint. If additional trucks cannot be found, this means that at least one and maybe more than one activity must be delayed. Scheduling techniques are beginning to focus on this aspect of project management to a much greater extent. Since the construction of a project is very sensitive to the number of resources available, one speaks of construction as being a resource-driven industry. As a practical matter, both time and resource availability are critical. Resource constrained project models can be considered to be advanced project scheduling systems.

REVIEW QUESTIONS AND EXERCISES

7.1 Develop the precedence diagram network for the following project and then calculate the total float, free float, and interfering float for each activity.

Activity	Duration	Immediately Following Activities
a	22	dj
b	10	cf
c	13	dj
d	8	—
e	15	cfg
f	17	hik
g	15	hik
h	6	dj
i	11	j
j	12	—
k	20	—

7.2 Make a clear and neat sketch of the network specified below using precedence notation.
On the precedence diagram calculate and show the early start (EST), early finish (EFT), late start (LST), late finish (LFT), and total free and interfering float in each activity using the notation shown below. Start calculations with day zero 0. Show the critical paths with colored pencil.

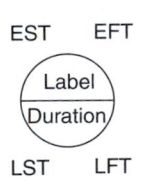

EST   EFT
Label
Duration
LST   LFT

The activity logic and durations are as follows:

Label	Duration	Must Follow Operations
A	2	—
B	4	A
C	7	A
D	3	A
E	5	A
F	7	B
G	6	B
H	7	F
I	5	G
J	3	G
K	8	C,G
L	9	H,I
M	4	F,J,K
N	7	D,K
O	8	E,K
P	6	M,N
Q	10	N,O
R	5	L,O,P
S	7	Q,R

7.3 From the following network data, determine the critical path, starting and finishing times, and total and free floats.

Activity	Description	Duration
1–2	Excavate stage 1	4
1–8	Order and deliver steelwork	7
2–3	Formwork stage 1	4
2–4	Excavate stage 2	5
3–4	Dummy	0
3–5	Concrete stage 1	8
4–6	Formwork stage 2	2
5–6	Dummy	0
5–9	Backfill stage 1	3
6–7	Concrete stage 2	8
7–8	Dummy	0
7–9	Dummy	0
8–10	Erect steel work	10
9–10	Backfill stage 2	5

Draw the network in precedence notation.

7.4 Using the information given in figures shown, develop a precedence notation CPM net-work for the bridge project

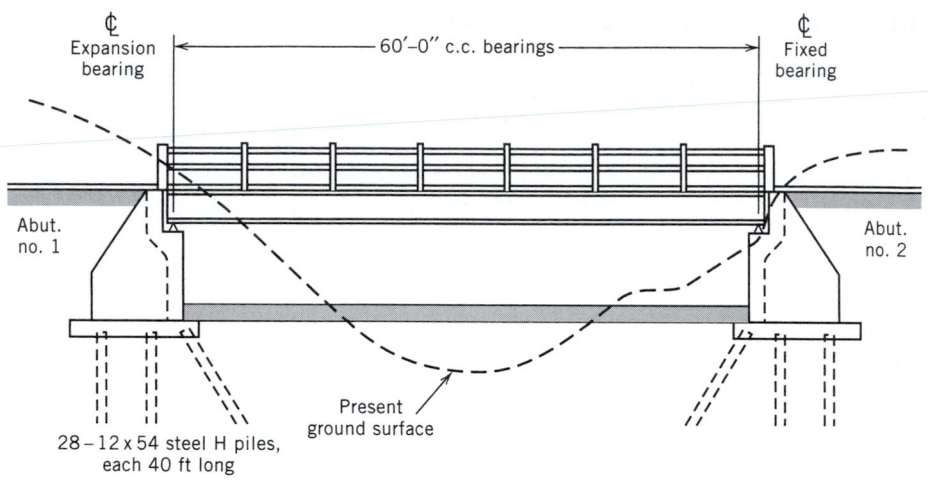

Example project, bridge profile.

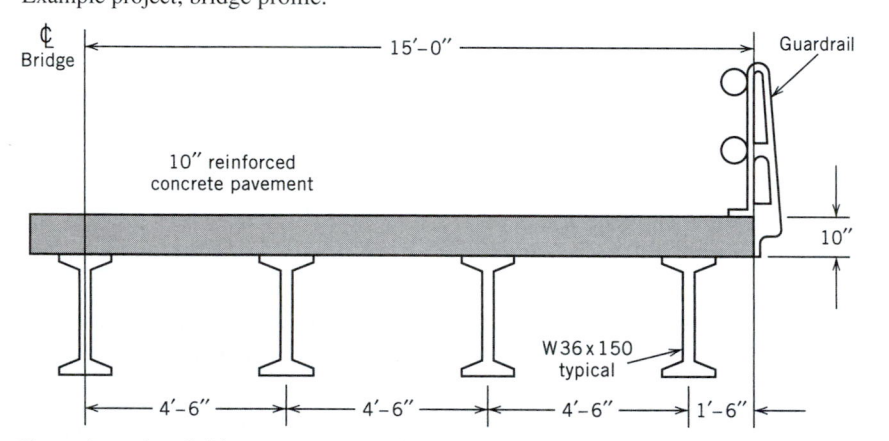

Example project, bridge transverse section for problem 7.4.

		June	July	August	Sept
Exc. abut. no. 1	8	17 ▭21			
Make abut. forms	7	17 ▭21			
Exc. abut. no. 2	10	22 ▭23			
Drive piles abut. no. 1	9		5 ▭8		
Forms & steel foot. no. 1	12		9 ▭12		
Drive piles abut. no. 2	13		9 ▭13		
Pour foot. no. 1	14		▪13		
Strip foot. no. 1	15		▪14		
Form & steel foot. no. 2	17		15 ▭16		
Pour foot. no. 2	19		▪19		
Form & steel abut. no. 1	16		20 ▭23		
Strip foot. no. 2	21		▪20		
Pour abut. no. 1	18		26 ▭27		
Strip & cure abut. no. 1	20		28 ▭30		
Back abut. no. 1	22			2 ▭4	
Rub. conc. abut. no. 1	30			2 ▭4	
Form & steel abut. no. 2	23			2 ▭5	
Pour abut. no. 2	24			6 ▭9	
Strip & cure abut. no. 2	25			10 ▭12	
Rub. conc. abut. no. 2	32			13 ▭17	
Back abut. no. 2	27			13 ▭17	
Set girders	28			13 ▭16	
Deck forms & steel	29			17 ▭20	
Pour & cure deck	31			23 ▭25	
Strip deck forms	33			26 ▭30	
Saw contraction joints	34			26 ▪	
Painting	35			31 ▭7	
Guardrail	36			31 ▭2	
Clean up	37				8 ▭10
Inspection	38				13 ▪

Bar chart schedule for bridge project for problem 7.4. (based on an example in Clough and Sears, *Construction Project Management*)

described. Certain logical relationships are implied by the bar chart that has been supplied to you by the field superintendent who has been chosen to run the job. According to contract specfications your company must submit a network scheduling (CPM) to the owner. Knowing that you are a CPM expert, your boss has given you the job of setting up the network. He also has asked you to calculate the project duration in working days. The duration of each activity can be developed from the dates given on the bar chart.

MOBILIZATION AND PROCUREMENT ACTIVITY ARE AS FOLLOWS:

Act	Duration	Description	Followers
1	10	Shop drawings, abutment, and deck steel	11
2	5	Shop drawings, foot steel	6
3	3	Move in	7,8
4	15	Deliver piles	9
5	10	Shop drawings, girders	26
1	15	Deliver abutment and deck steel	16
6	7	Deliver footer steel	12
6	25	Deliver girders	28

Hint

1. There is only one excavation crew.

2. There is one set of formwork material for footers and abutments.

3. Project duration should be approximately 65 days long.

4. General sequence of activities:

    a. Excavataion $\cdots$ piles driven $\cdots$ footer $\cdots$ abutment $\cdots$ deck $\cdots$

    b. Forming $\cdots$ pouring $\cdots$ curing $\cdots$ stripping $\cdots$

5. In order to calculate the work days from the bar chart, assume June 17 is a Thursday. Workdays are Monday through Friday.

7.5 A new road section with concrete pavement, shown in longitudinal section, is 11,600 ft long. It is to be constructed in accordance with the following conditions:

    a. The balanced earthworks from station 00 to 58(00) may be done at the same time as the balanced earthworks from station 58(00) to 116(00) using two separate independent crews.

    b. The double-box culvert will be built by one crew, and another crew will build the two small culverts. Concrete may be supplied either from the paving batch plant or from small independent mixers at the culvert sites, whichever is expedient.

    c. One small slip-form paver will do all the concrete paving work, and all the shouldering will then follow with one crew after the concrete pavement is cured.

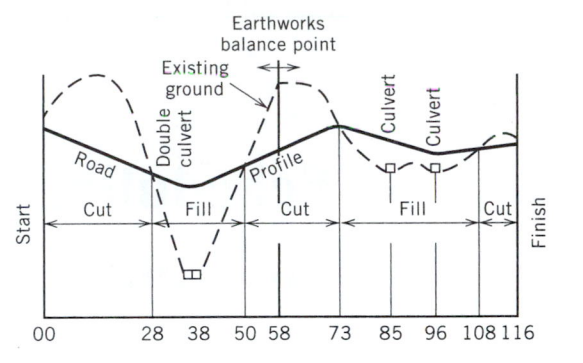

    d. Seeding the embankments with grass must be left as late as possible.

Prepare a network diagram and determine the minimum possible project duration.

If independent concrete mixers are used for the culverts, what is the latest day for delivery of the paving batch plant to the site, so that the paving crew may have continuity of work (no idle time at all)?

Activity Description	Duration
Deliver rebars—double-box culvert	10
Move in equipment	3
Deliver rebars—small culverts	10
Set up paving batch plant	8
Order and deliver paving mesh	10
Build and cure double-box culvert, station 38	40
Clear and grub, station 00–58	10
Clear and grub, station 58–116	8
Build small culvert, station 85	14
Move dirt, station 00–58	27
Move part dirt, station 58–116	16
Build small culvert, station 96	14
Cure small culvert, station 85	10
Cure small culvert, station 96	10
Move balance dirt, station 58–116	5
Place subbase, station 00–58	4
Place subbase, station 58–116	4
Order and stockpile paving materials	7
Pave, station 58–116	5
Cure pavement, station 58–116	10
Pave, station 00–58	5
Cure pavement, station 00–58	10
Shoulders, station 00–58	2
Shoulders, station 58–116	2
Guardrail on curves	3
Seeding embankments with grass	4
Move out and open road	3

Chapter 8

Scheduling – PERT Networks and Linear Operations

VRML Applications in Construction

The Need
Traditionally, construction process information is communicated with paper documents and 2D CAD drawings. Recently, the industry has embraced many kinds of web-based technologies, but construction still uses document-based models. It is believed that transition to model-based information can be done through web-based 3D user interfaces. Moreover, there is a need to easily model structures to be used in a web-based user interface.

VRML Model of the NIST Fire Research Facility Emissions Control System

The Technology
The applicability of the Virtual Reality Modeling Language (VRML) is being investigated for visualizing the activities at a construction site and creating an advanced web-based 3D user interface for construction process information. The Computer-Integrated Construction Group at the National Institute for Standards and Technology (NIST) in Gaithersburg, MD is developing this concept.

In principle, VRML is an open standard that offers the possibility of accessing many types of construction project data using readily available and well-accepted graphical user interfaces. These interfaces are based on web-based 3D visualizations of a model. In order to view the VRML world, the users should have a VRML browser, which can be a stand-alone application, a helper application, and/or a plug-in. Using this environment, models such as these pictured on this page can be readily developed.

VRML Excavator, Tower Crane, and Dump Truck

8.1 INTRODUCTION

Bar charts and critical path method (CPM) networks assume that all activity durations are constant or deterministic.[1] An estimate is made of the duration of each activity prior to the commencement of a project, and the activity duration is assumed to remain the same (e.g., a nonvariable value) throughout the life of the project. In fact, this assumption is not realistic. As soon as work begins, due to actual working conditions, the assumed durations

---
[1] Logic is also considered to be constant or invariable throughout the life of the project.

for each activity begin to vary. The variability of project activities is addressed in a method developed by the U.S. Navy at approximately the same time as CPM. This method was called the Program Evaluation and Review Technique. It is now widely known as the PERT scheduling method.

PERT incorporates uncertainty into the project by assuming that the activity durations of some or all of the project activities are variable. The variability is defined in terms of three estimates of the duration of each activity as follows:

1. Most pessimistic duration
2. Most optimistic duration
3. Most likely duration

Let's assume that a 20,000-sq ft slab on grade is to be cast in place. For scheduling purposes, the project superintendent is asked for three durations (i.e., most pessimistic, etc.) rather than for a single constant duration. The three estimates are used to calculate an expected activity duration. The calculations are loosely based on concepts from mathematical probability. The expected duration, t_e, is assumed to be the average value of a probability distribution defined by the three-estimate set. The expected duration, t_e, of each activity with variable characteristics is given as:

$$t_e = \frac{[t_a + 4t_m + t_b]}{6}$$

where t_a is the most optimistic duration estimate

t_m is the most likely duration estimate

t_b is the most pessimistic duration estimate

For instance, if for the slab pour, the three estimates from the superintendent are:

$$t_a = 5 \text{ days}$$
$$t_m = 8 \text{ days}$$
$$t_b = 12 \text{ days}$$

the expected activity duration is calculated as:

$$t_e = \frac{5 + 4(8) + 12}{6} = 49/6 = 8.17 \text{ days, say 9 working days}$$

The expected value for each activity with a constant value, k, is $t_e = k$.

Once the t_e values for each variable duration activity have been calculated, the longest path and project duration are determined using the same methods developed in CPM. The probability of completing the project within a predetermined time duration is calculated by assuming that the probability distribution of the total project duration is normally distributed with the longest path of t_e values as a mean value of the normal distribution.

The normal distribution is defined by its mean value $\bar{x}$ (i.e., in this case the value of the longest path through the network) and the value, σ, which is the so-called "standard deviation" of the distribution. The standard deviation is a measure of how widely about the mean value the actual observed values are spread or distributed. Another parameter called the variance is the square of the standard deviation or σ^2. It can be shown mathematically that 99.7% of the values of normally distributed variables will lie in a range defined by three

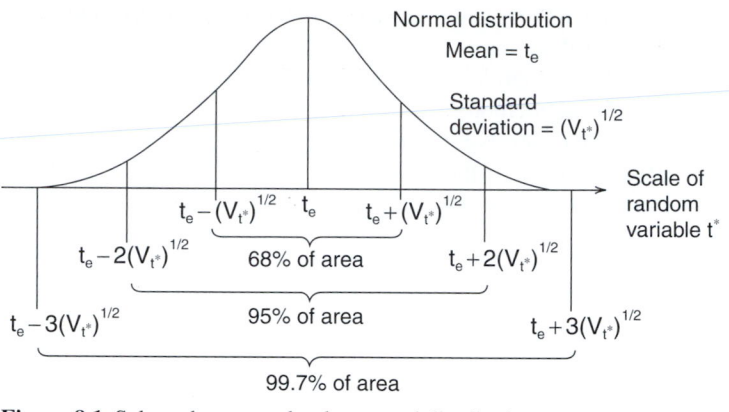

Figure 8.1 Selected areas under the normal distribution curve.

standard deviations below the mean and three standard deviations above the mean (see Fig. 8.1).

In PERT, the standard deviation[2] σ of the normal distribution for the total project duration is calculated using the variance of each activity on the critical path. The variance for each PERT activity is defined as:

$$\sigma^2 = \left[\frac{(t_b - t_a)}{6}\right]^2$$

If the variance of each activity on the longest path is summed, that value is assumed to be the variance of the normal distribution of the entire project duration values.

The fact that the Normal distribution is used to represent the probability distribution of the possible total project durations is based on a basic concept from probability theory called the Central Limit Theorem. This is explained by Moder and Phillips as follows:

"Theorem":

Suppose m independent tasks are to be performed in order; (one might think of these as the m tasks that lie on the critical path of a network). Let $t_1{}^*, t_2{}^*, \ldots t_m{}^*$ be the times at which these tasks are actually completed.

Note that these are random variables with true means $t_1, t_2, \ldots, t_m$, and true variances $V_{t1}{}^*, V_{t2}{}^*, \ldots V_{tm}{}^*$, and ... actual times are unknown until these specific tasks are actually performed. Now define T^* to be the sum:

$$T^* = t_i{}^* + t_2{}^* + \cdots + t_m{}^*$$

And note that T^* is also a random variable and thus has a distribution. The Central Limit Theorem states that if m is large, say four or more, the distribution of T^* is approximately normal with mean T and variance $V_T{}^*$ given by

$$T = t_1 + t_2 + \cdots + t_m$$
$$V_T{}^* = V_{t1}{}^* + V_{t2}{}^* + \cdots + V_{tm}{}^*$$

That is, the mean of the sum, is the sum of the means; the variance of the sum is the sum of the variances; and the distribution of the sum of activity times will be normal regardless of the shape of the distribution of actual activity performance times (Moder and Phillips, 1964)."

---

[2] The variance is the standard deviation squared.

Table 8.1   Three Estimate Values and Calculated Values for Each Activity

Activity	t_m	t_a	t_b	t_e	Var
1	3	1	5	3	0.44
2	6	3	9	6	1.00
3	13	10	19	13.5	2.25
4	9	3	12	8.5	2.25
5	3	1	8	3.5	1.36
6	9	8	16	10	1.23
7	7	4	13	7.5	2.25
8	6	3	9	6	1.00
9	3	1	8	3.5	1.36

8.2   AN EXAMPLE PERT NETWORK

To demonstrate the use of the PERT approach, consider the small arrow notation network shown in Figure 8.2. Three estimate durations for each activity in this activity network are given in Table 8.1.

The t_e values shown for each activity are calculated using the formula:

$$t_e = \frac{(t_a + t_m + t_b)}{6}$$

For instance, t_e for activity 7 is:

$$t_e = \frac{4 + 4(7) + 13}{6} = \frac{45}{6} = 7.5$$

The variance for each activity is approximated by the equation

$$\sigma^2 = \left[\frac{(t_b - t_a)}{6}\right]^2$$

For activity 7, the variance is var(7) =

$$\left[\frac{13 - 4}{6}\right]^2 = \left[\frac{9}{6}\right]^2 = 2.25$$

Using the forward and backward pass methods described in Chapter 7, two paths have an expected duration of 17.5 days. These pathes are shown below.

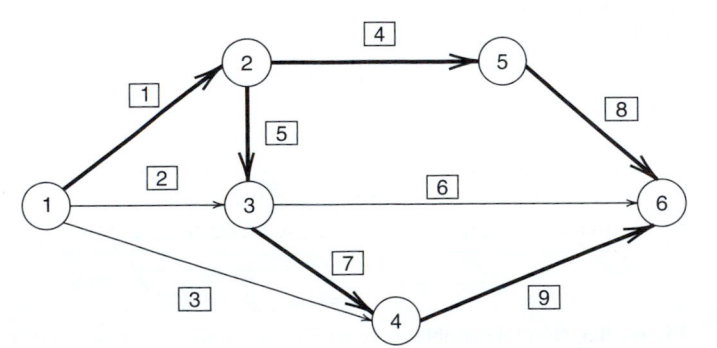

Figure 8.2 Small   PERT network.

	DURATION	VARIANCE
Path 1 (1-4-8)	$3 + 8.5 + 6 = 17.5$	$\text{Var} = .444 + 2.25 + 1.0 = 3.694$
Path 2 (1-5-7-9)	$3 + 3.5 + 7.5 + 3.5 = 17.5$	$\text{Var} = .444 + 1.361 + 2.25 + 1.361 = 5.416$

The mean of the normal distribution is therefore assumed to be 17.5 days. The variances of the two longest paths are calculated by adding variances of the individual activities in each path. The variance of path two (5.416) is greater than that of path one (3.694). Since this means a greater spread of the probable total project durations, the variance of path two is selected as the variance to be used for further PERT calculations. The PERT normal distribution for Total Project Duration is shown in Figure 8.3.

The normal distribution is symmetrical about the mean. The standard deviation will be $\sigma = \sqrt{5.416} = 2.327$. PERT answers the question: "What is the probability (given the variable durations of the activities) that the project can be completed in N days?" The probability of completing the project is given by the area under the normal distribution to the left of the value N selected for investigation. Since we know that 99.7% of the area (representing probability) under the normal distribution is in the range of 3 σ below the mean 3 σ above the mean, we can say that there is a better than 99.7% chance that the project can be completed in [x + 3 σ] or [17.5 + 3 (2.327)] = 24.5 days or less. That is, at least 99.7% of the area under the normal curve is to the left of the value 24.5 days in Figure 8.3. In other words, we can be almost 100% sure that the project can be completed in 25 days or less.

What if we want to know the probability of completing in 19 days?[2] Given the values of the mean and the variance, we can use a cumulative normal distribution function table such as that shown in Appendix K to calculate the area under the curve left of the value 19. First, we must calculate the Z value for a given value (e.g., 19):

$$Z = \frac{\text{Mean} - x}{\sqrt{\text{Variance}}} \quad \text{or} \quad Z = \frac{(\overline{X} - x)}{\sigma}$$

where σ is the standard deviation of the cumulative normal distribution.
In our case:

$$Z = \frac{|17.5 - 19|}{\sqrt{5.416}} = \frac{1.5}{2.327} = 0.644$$

Consulting the cumulative normal distribution function table given in Appendix K with a Z value of 0.644 yields a value of .7389 or 73.89% probability of completing the project in 19 days. What would be the probability of completing the project in 16 days?

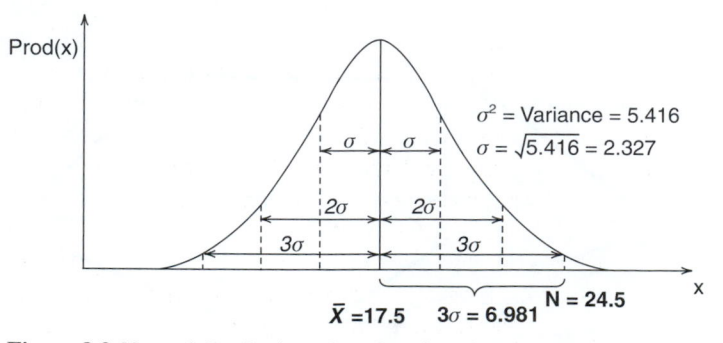

Figure 8.3 Normal distribution of total project durations for small PERT network.

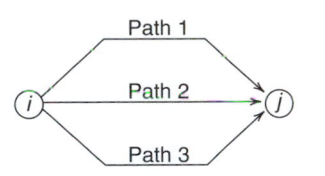

Figure 8.4 Merge event bias.

8.3   PERT SHORTCOMINGS

In fact, the PERT results are too optimistic. The method of using the t_e value to determine the longest path through the project network and assuming that the duration of this path is the most probable value for the total project duration is not totally accurate. Although PERT introduces elements of probability into the calculation of the project duration, it consistently underestimates the duration. The principal cause of this underestimation is a condition known as "merge event bias." Briefly, merge event bias occurs when several paths converge on a single node. Figure 8.4 is a simplified depiction of how several paths in a schedule network might converge on a single node.

PERT calculations give the early expected finish time of this node as the summation of times on the longest path leading to the node. This path then becomes part of the longest path through the network that determines expected project duration. However, since the duration of the activities on the paths are random variables, it is possible that some other path converging on the node could have an activity with a random duration longer than its expected (mean) duration. Thus, this longer path would determine the early finish time of the node. That this potential longer path is *not* taken into account in the PERT calculation leads to an underestimation of project duration.

Additionally, the PERT method assumes statistical independence between activities. This assumption allows the variance of activities along a path to be added, giving the variance of the duration of the project. The assumption of independence, however, may not always be appropriate. For instance, weather can create a positive correlation between activities, and a delay in one activity may create a negative correlation between activities.

One solution to the difficulties noted above is computer simulation. Because Monte Carlo simulation of schedule networks does not use a single number to represent activity durations, it avoids the merge event bias described above.

8.4   LINEAR CONSTRUCTION OPERATIONS

Often construction sites have linear properties that influence the production sequence. Road construction, for instance, is worked in sections which require that a set of work processes be completed in a particular sequence before the section is completed. The individual sections can be thought of as "processing through" a series of workstations.

For example, a road job may be subdivided into 14 sections that must be completed see Figure 8.5 (a). This type of breakdown is typically established based on centerline stationing (e.g., section 1 is defined as running from station 100 to station 254.3). Each of the 14 stations must undergo the following work activities: (1) rough grading, (2) finish grading, (3) aggregate base installation, (4) 5-in. concrete pavement, (5) 9-in. concrete pavement, and (6) curb installation.

Each of the 14 sections can be thought of as being processed by crews and equipment representing each of the six work processes. Since the site is linear, the normal way for the work to proceed would be to start with section 1, then go on to the second section, and so forth. This implies that the sections will first be rough graded, then finish graded, then

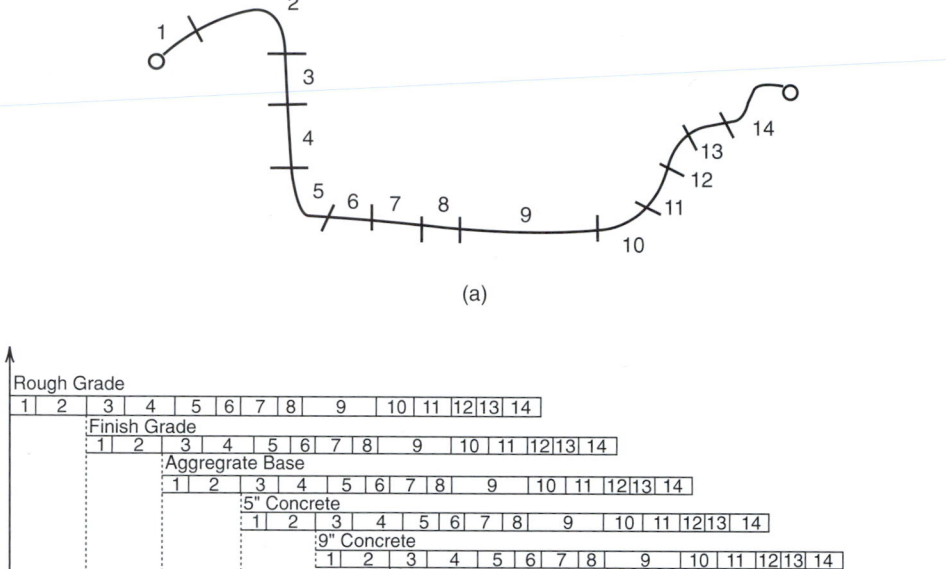

(a)

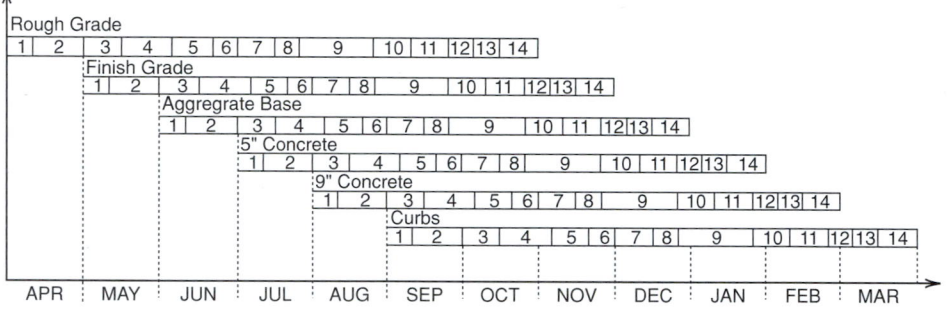

(b)

Figure 8.5 Road project divided into 14 sections.

aggregate base will be placed, and so on. A bar chart indicating this sequence of activity is shown in Figure 8.5(b).

The bar chart indicates that work activity overlaps such that several operations are in progress simultaneously during the middle of the job. The required sequentiality leads to a "train" effect. That is, a section must complete 5-in. concrete before preceding to 9-in. concrete. Therefore, the sections can be thought of as a "train" or "parade" of work that must pass each station represented by the six construction processes.

Many types of projects exhibit this kind of rigid work sequence. A high-rise building, for instance, requires each floor to pass through a set of operations. Each floor can be thought of as a "car" in the "train" of work to be completed. Construction processes such as erect formwork, install reinforcing steel and imbedments, and pour concrete can be viewed as workstations through which each floor must pass.

Tunnels are worked in sections in a fashion similar to road or pipeline work. Each section must be processed through work processes such as drill, blast, remove muck, and advance drilling shield. This again leads to a repetitive sequence that is rigidly sequenced.

8.5   PRODUCTION CURVES

Bar charts and network schedules provide only limited information when modeling linear operations and projects. They typically do not readily reflect the production rate or speed with which sections or units are being processed. Since the rate of production will vary across time, this has a major impact on the release of work for following work processes. Delays in achieving the first units of production occur as a result of mobilization requirements. As the operation nears completion, the rate of production typically declines because of demobilization or closeout considerations. The period of maximum production is during the midperiod of the process duration. This leads to a production curve with the shape of

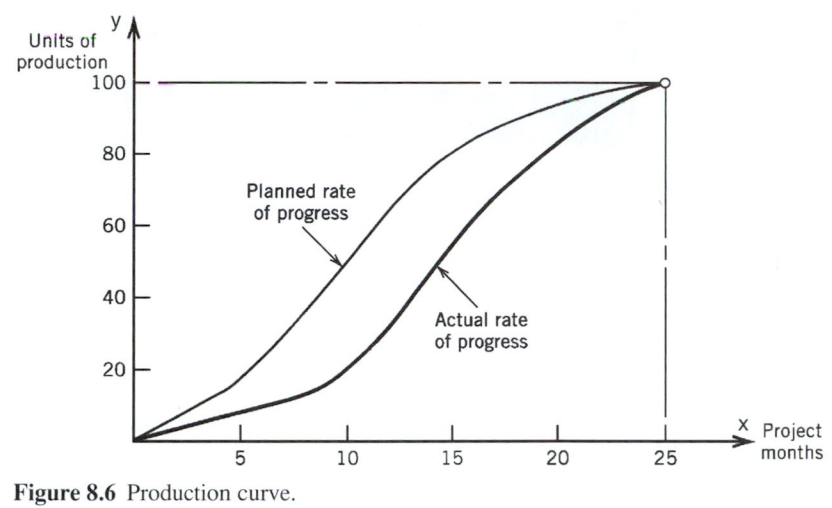

Figure 8.6 Production curve.

a "lazy S," as shown in Figure 8.6. The slope of the curve is flat at the beginning and the end, but steep in the midsection. The slope of the curve is the production rate.

These curves are also called time-distance, time-quantity, or velocity diagrams since they relate units of production (i.e., quantities or distance) on the y axis (vertical, ordinate) with time plotted on the x axis (horizontal, abscissa). The slope of the curve relates the increase in production units on the y axis with the increment of time as shown on the x axis. The slope of the curve, therefore, represents the number of units produced over a given time increment. This is the rate of production.

The production curves for a typical road job are shown in Figure 8.7. The curves indicate the beginning and ending points in time for each of the processes. The slope of each curve is the production rate for each process. The distance between the beginning points of each process establishes the lag between processes. The aggregate base operation begins in week 6 and lags the finish-grading operation by two weeks. This means that two weeks of work (i.e., completed finish-grade sections) are built up before the aggregate base operation is started.

Leading processes generate work area or availability so that follow-on processes have a "reservoir" of work from which to operate. Reservoirs of work are cascaded so that units of work must be available from an "upstream" process reservoir before work is available at a

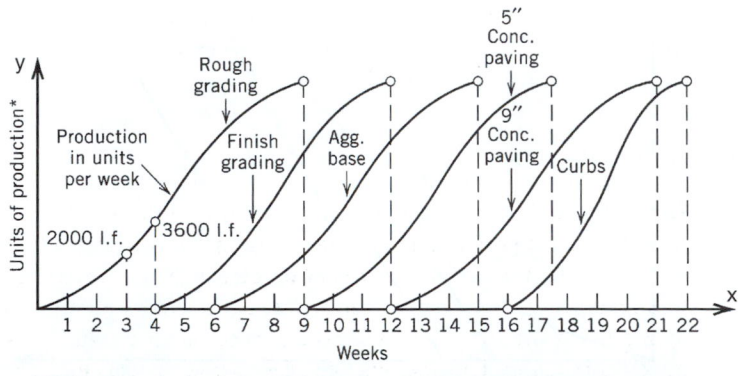

* Units of production (e.g., l.f. for rough and finish grading, tons of aggregate base, cu. yd. or sq. ft. of paving, etc.)

Figure 8.7 Velocity diagrams for a road construction project.

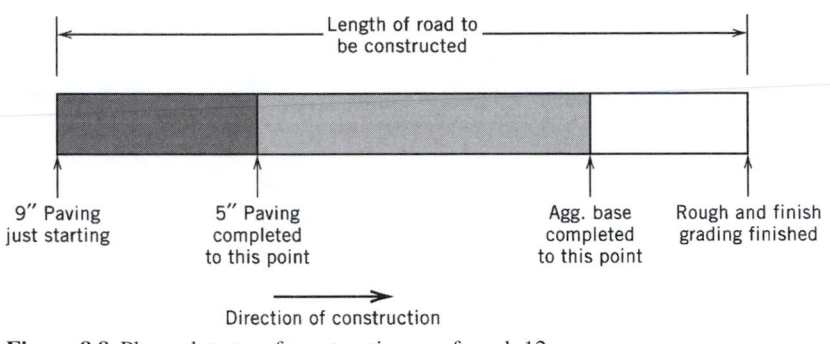

Figure 8.8 Planned status of construction as of week 12.

following process reservoir. This illustrates that workflow moves from leading to following processes.

In addition to indicating the rate of production, production curves or velocity diagrams are helpful in establishing the project status. The planned status of the job as of week 12 can be determined by simply drawing a vertical line at week 12 on the x axis of Figure 8.7. This will intersect the aggregate base and 5-in. concrete curves. It also represents the beginning of work on the 9-in. concrete pavement (overlaying the 5-in. base concrete). It can be readily determined that:

1. Both rough and finish grading should be completed.

2. Approximately 80% of the aggregate base has been placed.

3. Placement of the 5-in. concrete base is approximately 30% complete.

4. Placement of 9-in. concrete is just commencing.

The planned status of construction as of week 12 is shown in Figure 8.8.

There is a definite advantage in balancing the production rates between processes. Balancing rates means ensuring that the slopes of the production curves which interact are roughly parallel and do not intersect. If rates are not balanced, the situation shown in Figure 8.9 can develop. In this example, the slope (production rate) of process *B* is so steep that it catches or intersects the process *A* curve at time *M*. This requires a shutdown of process *B* until more work units can be made available from *A*. Again, at time *L*, process *B* overtakes

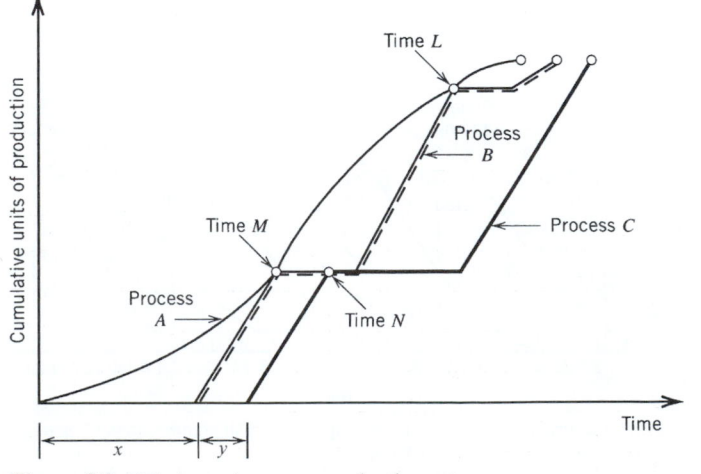

Figure 8.9 Unbalanced process production rates.

the production in process *A*, resulting in a work stoppage. This is clearly inefficient since it requires the demobilization and restarting of process *B*. The stoppage of process *B* at *M* also causes a "ripple" effect since this causes a shutdown of process *C* at time *N*.

It should be clear that these stoppages are undesirable. Thus, processes should be coordinated so as to avoid intersections of production curves (e.g., times *M*, *N*, and *L*). Obviously, one way to avoid this is to control production in each process so that the slopes of the curves are parallel. This implies the need to design each process so that the resources utilized result in production rates that are roughly the same for all interacting construction processes. Since the six curves for the road job are roughly parallel, we can assume that the production rates have been coordinated to avoid one process overtaking its leading or preceding process.

8.6   LINE-OF-BALANCE CONCEPTS

Line of balance (LOB) is a graphical method for production control integrating barcharting and production curve concepts. It focuses on the planned versus actual progress for individual activities and provides a visual display depicting differences between the two. Indication of these discrepancies enables management to provide accurate control in determining priorities for reallocation of labor resources. Those activities indicated ahead of schedule can be slowed by directing part or all of their labor crews to individual activities that lag behind schedule. This obviously assumes that resources are interchangeable. This can present a limitation to the application of this procedure in construction.

The LOB method serves two fundamental purposes. The first is to control production and the second is to act as a project management aid. Each objective is interrelated through development and analysis of four LOB elements. These elements provide the basis for progress study on critical operations throughout the project duration. The four elements are:

1. The objective chart
2. The program chart
3. The progress chart
4. The comparison

The *objective chart* is a segmental curve showing cumulative end products to be produced over a calendar time period. The number of end products may be specified in the contract. Assume that the units being considered in this example are precast panels for the exterior of a high-rise building. A typical objective chart is shown in Figure 8.10.

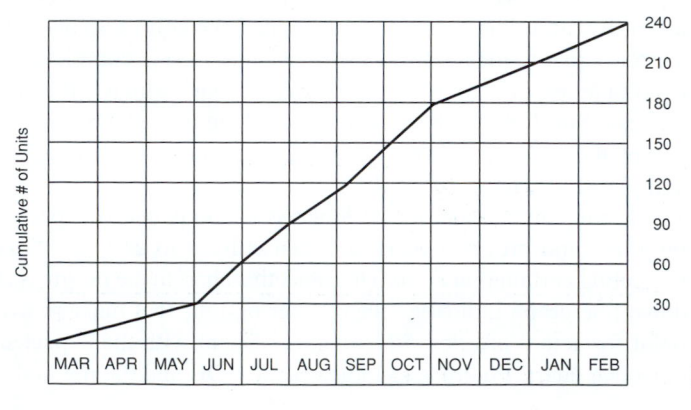

Figure 8.10 Objective chart.

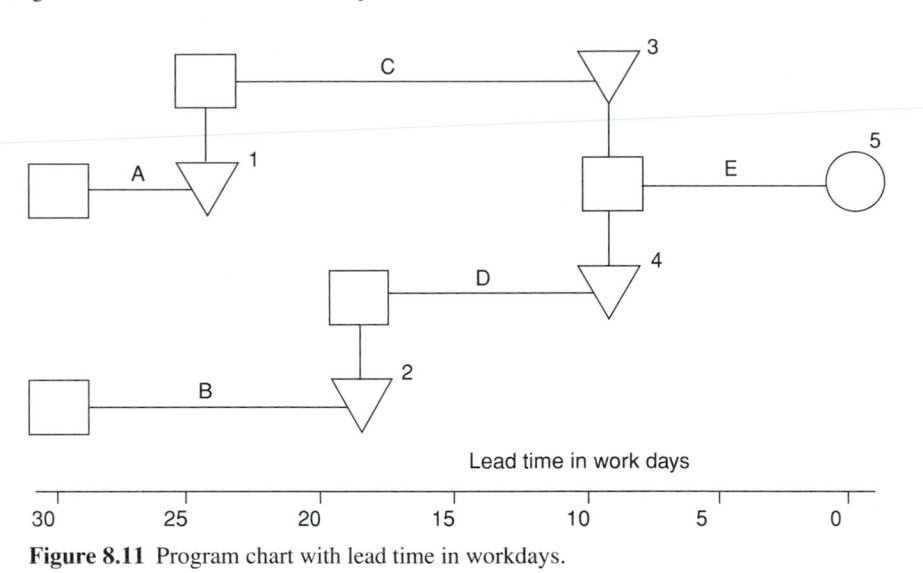

Figure 8.11 Program chart with lead time in workdays.

This example indicates a total of 30 units to be delivered or completed by June 1, 60 units to be delivered and completed by July 1, 180 units to be delivered and completed by November 1, and a total of 240 units to be delivered and completed by February 28. The contract start date is shown as March 1.

The *program chart* is the basic unit of the LOB system. It is a flow process chart of all major activities, illustrating their planned, sequenced interrelationships on a "lead-time" basis. Three aspects to consider in development of the program chart are determination of (1) operations to be performed, (2) the sequence of operations, and (3) processing and assembly lead time.

The program chart indicated in Figure 8.11 describes the production process for the 240 units mentioned in the objective chart. Each activity (A through E) has associated with it a lead time (latest start time) signified by an event starting symbol (□) and an event coordination symbol (△) signifying its end or completion. These event coordination symbols, referred to as *progress monitoring points*, are labeled from top to bottom and from left to right. All five activities must be completed before one unit can be ready for delivery. This takes 30 working days as shown on the program chart's lead-time scale.

The *progress chart* is drawn to the same vertical scale as the objective chart and has a horizontal axis corresponding to the progress monitoring points labeled in chronologic order. Vertical bars represent the cumulative progress or status of actual performance at each monitoring point, usually based on visiting the site and measuring actual progress (e.g., assessing status of completion).

The progress chart of Figure 8.12 indicates that on a given day when inventory was taken, 120 units had passed through monitoring point 5. In other words, the vertical height of bar 5 is equal to the number of units actually having completed station 5. This corresponds to completion of activity *E* in the program chart, which is the last activity in the production process. Similarly, activity *D* (bar 4) had completed 120 units and activity *C* (bar 3) had completed 130 units; activity *B* (bar 2) had completed 150 units; and activity *A* (bar 1) had completed 180 units.

In the comparison, actual progress is compared to expected progress. The objective, program, and progress charts are then utilized to draw the "line of balance" or LOB by projecting certain points from the objective chart to the progress chart. This results in a step-down line graph indicating the number of units that must be available at each monitoring point for actual progress to remain consistent with the expected progress as given by the objective chart. Figure 8.13 indicates the LOB and the method used to project it from the

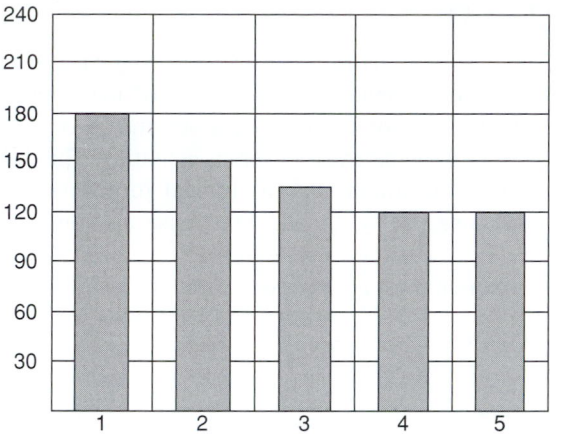

Figure 8.12 Progress Chart

objective chart to the progress chart. The procedure for striking the line of balance is as follows:

1. Plot the balance quantity for each control point.

    a. Starting with the study date (e.g., Sept 1) on the horizontal axis of the objective chart, mark off to the right the number of working days (or weeks or months, as appropriate) of lead time for that control point. This information is obtained from the program chart.

    b. Draw a vertical line from that point on the horizontal axis to the cumulative objective curve.

    c. From that point draw a horizontal line to the corresponding bar on the progress chart. This is the balance quantity for that bar.

2. Join the balance quantities to form one stair-step-type line across the progress chart.

Analysis of the LOB reveals that activities 2 and 5 are right on schedule while activities 3 and 4 show deficit units. Activity 1 shows surplus. This surplus is the difference between the 180 units actually completed by activity 1 and the 157 units indicated as necessary by the LOB. On the other hand, activities 3 and 4 are lagging by 5 and 15 units, respectively. The LOB display enables management to begin corrective action on activities 3 and 4 to ensure that they do not impede the progress rate of the remaining units.

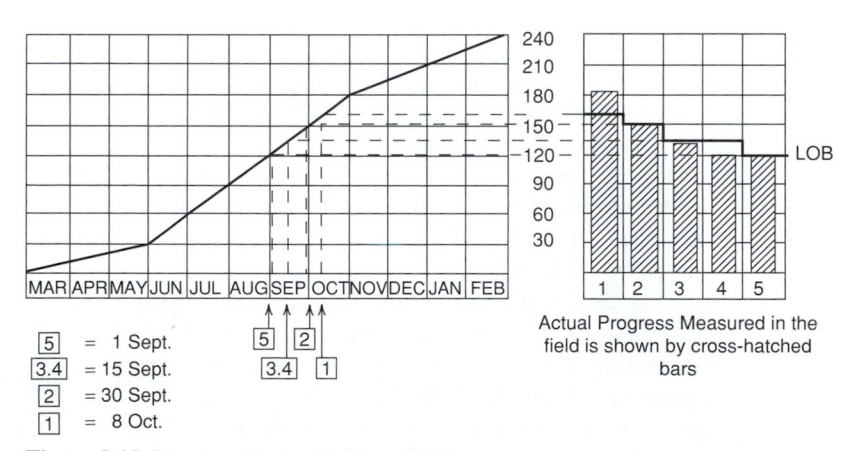

Figure 8.13 Progress Chart with Line of Balance.

8.7   LOB APPLIED TO CONSTRUCTION

To illustrate the use of LOB in a construction context, consider a high-rise building in which repetitive activity sequences are a part of the floor-to-floor operation. In order to ensure a smooth flow of production, a schedule would be necessary that accounts for the interrelationships between different activities. This becomes even more obvious when an additional constraint such as limited formwork is involved. Each floor consists of four sections (A, B, C, and D). These sections can be viewed as processed units.

Each floor section must be processed through the following work activities:

1. Erect Forms
2. Place Reinforcing Steel
3. Place Concrete
4. Dismantle Forms
5. Place Curtain Wall (Exterior Façade)
6. Place Windows

Figure 8.14 shows a schematic of the status of activities at a given point in time. At the time illustrated, work is proceeding as follows:

1. Erect forms section A, floor N +5
2. Place reinforcing steel, section D, floor N +4
3. Place concrete section C, floor N + 4
4. Dismantle forms sections B, floor N +1
5. Place curtain wall section D, floor N
6. Place windows section D, floor N −1

Crews proceed from section A to B to C to D.

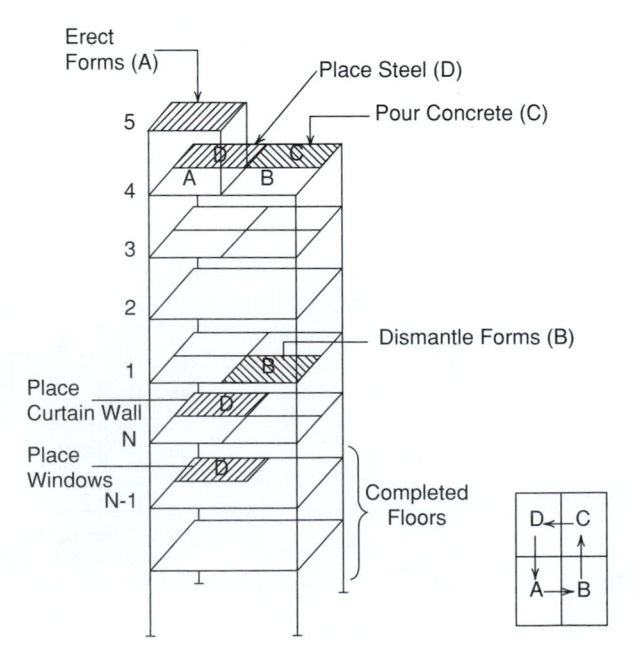

Figure 8.14 Schematic of floor cycle work tasks.

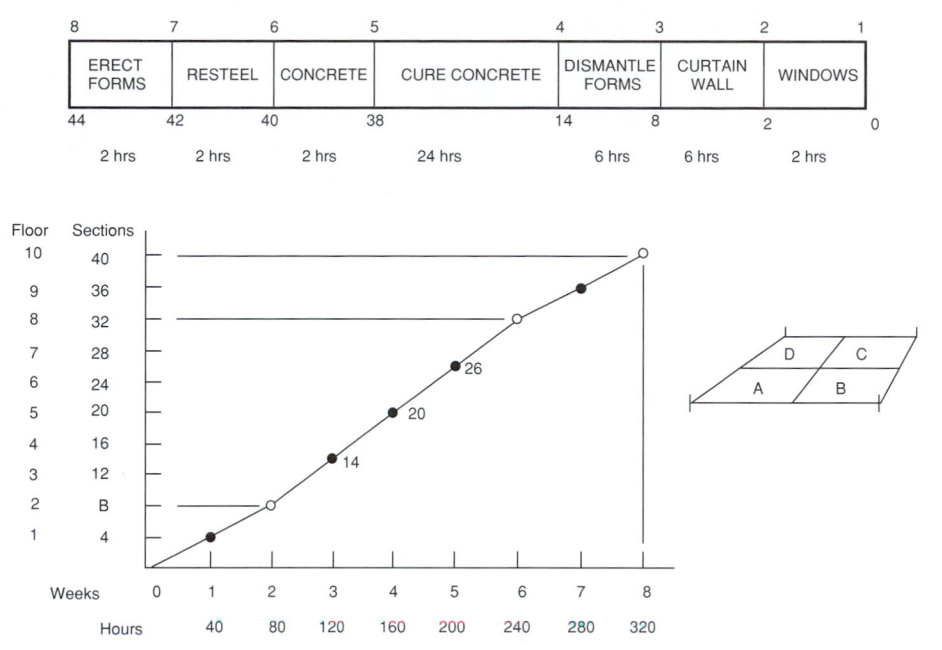

Figure 8.15 Program chart and objective.

The diagram in Figure 8.15 shows the LOB objective chart for a 10-story building. The program chart for a typical section is shown above the objective chart. During the first two weeks the floor cycle required is one floor (four sections) per week. For weeks 2 through 6 the rate of floor production is 1.5 floors (six sections) per week. That is, six floors must be completed in the four-week period from week 2 to week 6. In the last two weeks, the rate is reduced to one floor per week. The lead times required for various activities are shown on the bar program chart above the objective. To strike a line of balance for the beginning of week 5, the lead times are projected as described in Section 8.6. A diagram of this projection is shown in Figure 8.16. The LOB values can be calculated by determining the slope relating horizontal distance (lead time) to vertical distance (required sections). The slope of the objective during weeks 5–6 is six sections (1.5 floors) per 40 hr (one week) or 6/40 sections per hour.

During the remaining weeks, the slope is four sections (one floor) per 40 hr or 1/10 section per hour. The LOB for control point I is given as:

$$\text{LOB(I)} = \text{Section completed as for week 5}$$
$$+[(\text{slope}) \times (\text{lead time of control point I})]$$

The number of sections to be completed as of week 5 is 6.5 or 26 sections (6.5 × 4). Therefore,

$$\text{LOB(1)} = 26$$
$$\text{LOB(2)} = 26 + (6/40)2 = 26.3$$
$$\text{LOB(3)} = 26 + (6/40)8 = 27.2$$
$$\text{LOB(4)} = 26 + (6/40)14 = 28.1$$
$$\text{LOB(5)} = 26 + (6/40)38 = 31.7$$
$$\text{LOB(6)} = 26 + (6/40)40 = 32$$

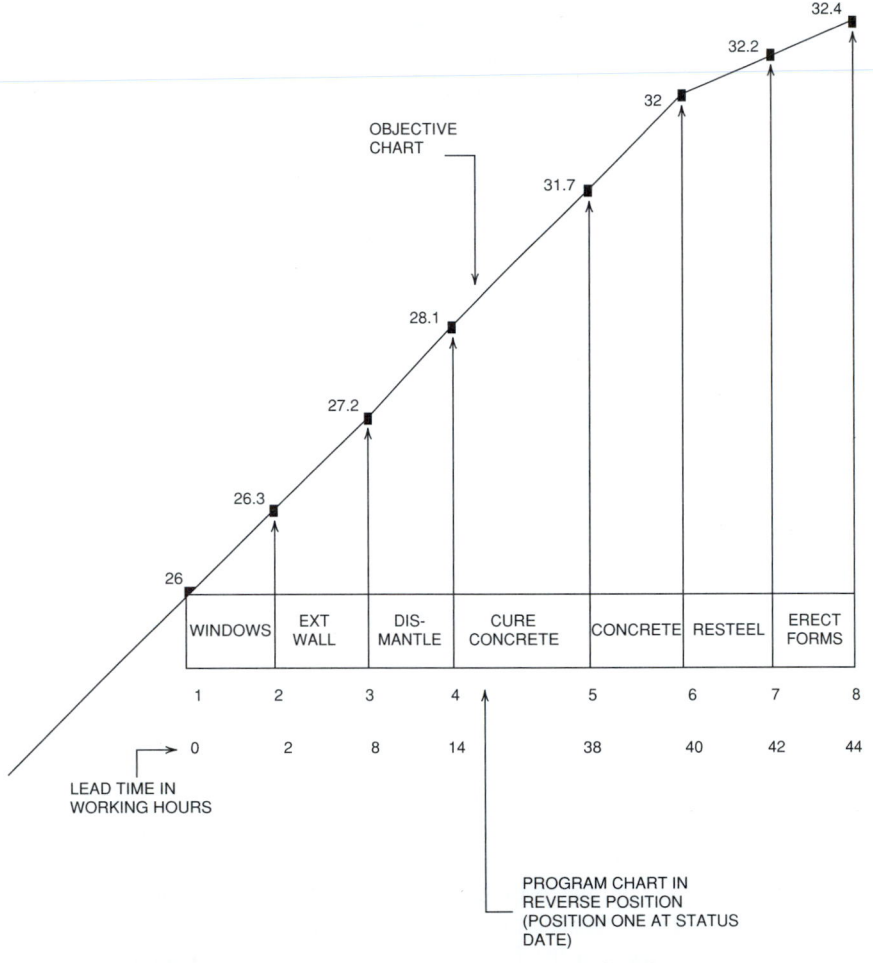

Figure 8.16 Enlarged projection of program chart onto objective chart.

Control points 7 and 8 plot to the flatter portion of the objective:

$$\text{LOB}(7) = 32 + (1/10)(42 - 40) = 32.2$$
$$\text{LOB}(8) = 32 + (1/10)(44 - 40) = 32.4$$

The line of balance for week 5 is shown in Figure 8.17. Field reports would be utilized to establish actual progress, and a comparison will determine whether actual progress is consistent with expected progress.

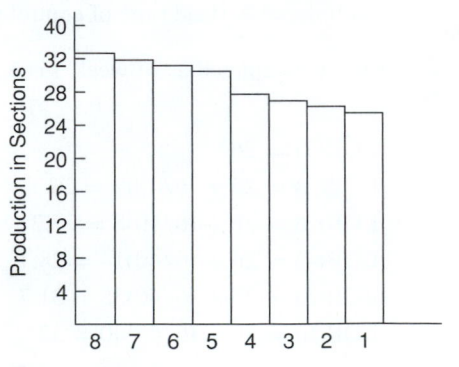

Figure 8.17 Line of balance for week 5.

REVIEW QUESTIONS AND EXERCISES

8.1 **(a)** Using PERT, calculate the expected project duration and determine the critical path in the network defined by the data below.

(b) What is the probability of completing this project in 32 weeks?

Activity #	Activity	Type	Duration (Weeks)	Followed by Act #
10	Prefab Wall Forms	constant	2	40
20	Excavate Cols and Walls	constant	3	50, 60, 70
30	Let Elec and Mech Subcontract	t_a, t_m, t_b	3, 4, 8	60, 70
40	Deliver wall Forms	constant	4	80, 90, 100
50	Forms, Pour & Cure Wall & Col Fig	t_a, t_m, t_b	6, 7, 8	80, 90, 100
60	Rough-in Plumbing	t_a, t_m, t_b	5, 7, 10	110
70	Install Conduit	$t_a, t_m\ t_b$	9, 11, 15	110
80	Erect Wall Forms & Steel	constant	9	110
90	Fabricate & Set Interior Column Forms	constant	6	120
100	Erect Temporary Roof	t_a, t_m, t_b	12, 16, 18	140
110	Pour, Cure & Strip Walls	constant	10	130
120	Pour, Cure & Strip Int. Walls	constant	6	140
130	Backfill for Slab on Grade	constant	1	140
140	Grade & Pour Floor Slab	constant	5	END

8.2 **(a)** Given the data below for a small pipeline project, based on a PERT analysis what is the expected project duration?

(b) What is the probability of completing this project in 120 days?

Activity	Description	t_a	t_m	t_b	Followed by Activity
1	Start	0	0	0	2
2	Lead Time	10	10	10	3, 4, 5
3	Move to Site	18	20	22	6
4	Obtain Pipes	20	30	100	8, 9, 10
5	Obtain Valves	18	20	70	11
6	Lay Out Pipeline	6	7	14	7
7	Dig Trench	20	25	60	8, 10
8	Prepare Valve Chambers	17	18	31	11
9	Cut Specials	7	9	17	11
10	Lay Pipes	18	20	46	12
11	Fit Valves	8	10	12	13, 14
12	Concrete Anchors	11	12	13	13, 14, 15
13	Finish Valve Chambers	8	8	8	17
14	Test Pipeline	5	6	7	16
15	Backfill	8	10	20	16
16	Clean Up	2	3	10	17
17	Leave Site	3	4	5	18
18	End				

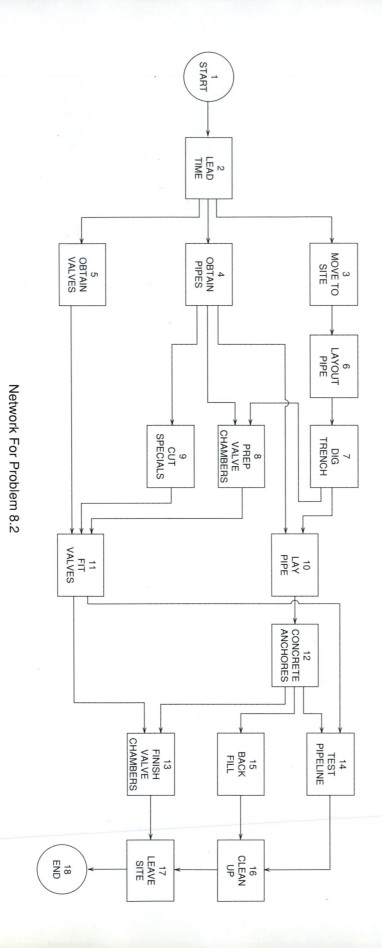

Network For Problem 8.2

8.3 Consider the road job described in Figure 8.5, which consists of 14 road sections to be completed. The objective chart for this job is given below. Assume that each month consists of 20 working days on the average. The program chart for this process is also shown below. Calculate the line of balance for the study representing the beginning of month 9 (i.e., day to right of number 8).

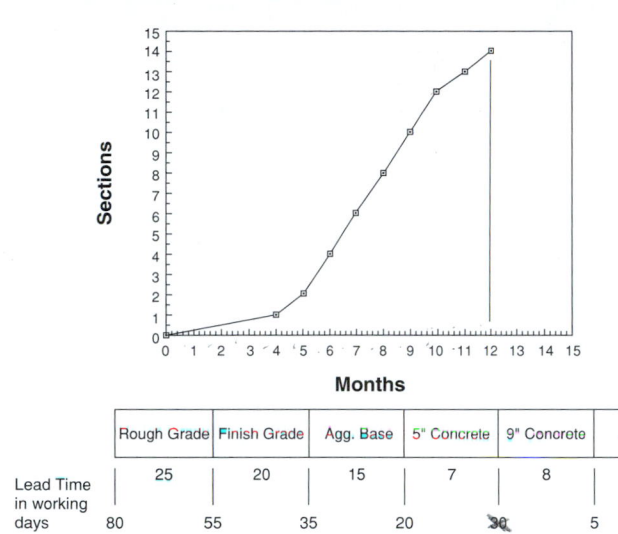

Months

	Rough Grade	Finish Grade	Agg. Base	5" Concrete	9" Concrete	Curbs
Lead Time in working days	25	20	15	7	8	5
	80	55	35	20	30 ~~13~~	5	0

Program Chart

8.4 The objective chart for a 10-story building is show below. Each floor is divided into four sections (A, B, C, D). The production for a typical section is shown below the objective chart, Calculate the LOB values for control points 1–8 for week 5 (200 hr). Determine the LOB values in numbers of floor sections.

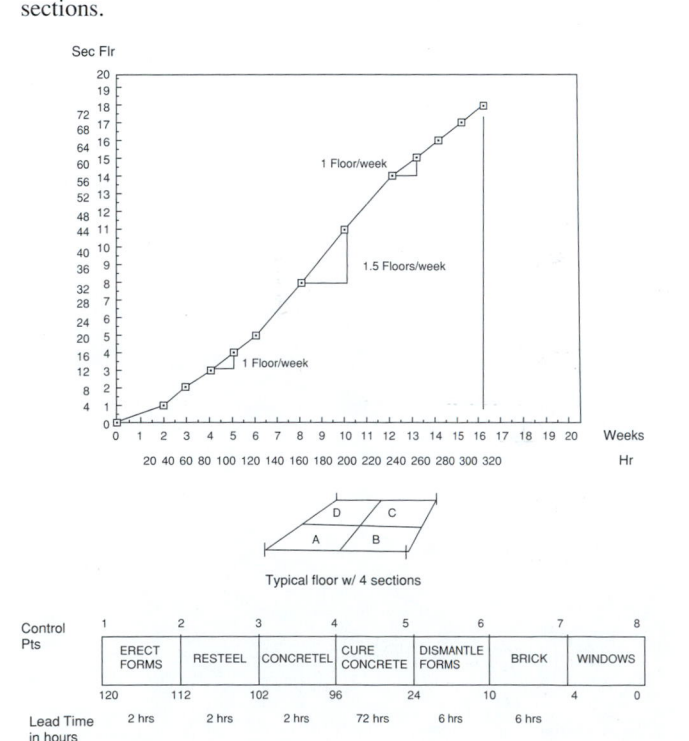

Typical floor w/ 4 sections

Control Pts	1	2	3	4	5	6	7	8
	ERECT FORMS	RESTEEL	CONCRETEL	CURE CONCRETE	DISMANTLE FORMS	BRICK	WINDOWS	
	120	112	102	96	24	10	4	0
Lead Time in hours	2 hrs	2 hrs	2 hrs	72 hrs	6 hrs	6 hrs		

8.5   Given the following charts, calculate the LOB quantities for July 1. What are these charts called?

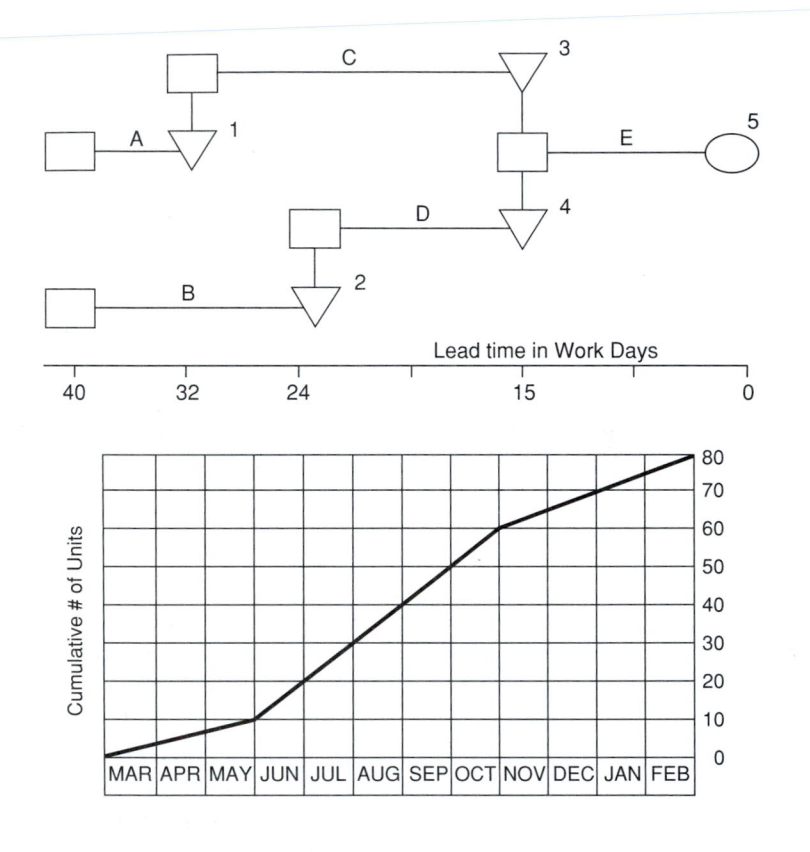

Chapter 9

Project Cash Flow

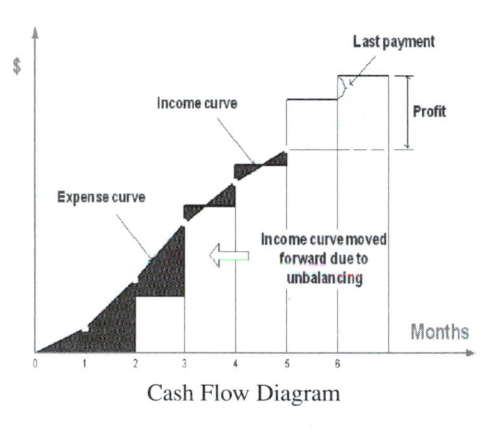

Cash Flow Diagram

Scenario Testing

The Need

Construction company operations are project based. Cash flows can be estimated by attempting to assess flows from (1) projects in progress, (2) projects under contract but not yet begun, and (3) potential projects which will start during the coming financial accounting period. These sources of income can be viewed as (1) Birds in the hand, (2) Birds in the bush, and (3) Birds flying in the sky. In other words, cash flows can be projected from projects in progress and projects which may, with some probability, start in the coming period for which forecasts are being made. The advent of spreadsheet analysis and high speed computing has led to "scenario testing" of future cash flow expectations.

The Technology

Spreadsheets allow managers to run probabilistic cash flow projections that take into account the factors noted above. More advanced analysis can also factor in historical evidence of payment trends and the potential impact of macroeconomic factors. These techniques go beyond the typical best-, expected-, and worst-case scenario modeling and may rely on Monte Carlo simulation, Markov modeling, or the use of "fuzzy" data sets to build up statistically valid outcomes. At the most advanced level, when future cash flows are tied to a multitude of unknowns, probabilistic techniques may be employed in combination with real-options theory to gain an improved view of the impact of a financial decision (e.g., accepting or declining a project or changing market strategy) on value creation for a company. This level of analysis used to be in the economist's realm, but is now commonplace in the finance and business development groups of corporations.

9.1 CASH FLOW PROJECTION

The projection of income and expense during the life of a project can be developed from several time-scheduling aids used by the contractor. The sophistication of the method adopted usually depends on the complexity of the project. In many contracts (e.g., public contracts such as those used by state agencies), the owner requires the contractor to provide an S-curve of estimated progress and costs across the life of the project. The contractor develops this by constructing a simple bar chart of the project, assigning costs to the bars, and smoothly connecting the projected amounts of expenditures over time.

Consider the highly simplified project (Fig. 9.1) in which four major activities are scheduled across a four-month time span. Bars representing the activities are positioned along a time scale indicating start and finish times. The direct costs associated with each activity are shown above each bar. It is assumed that the monthly cost of indirect charges (i.e., site office costs, telephone, heat, light, and supervisory salaries, which cannot be

147

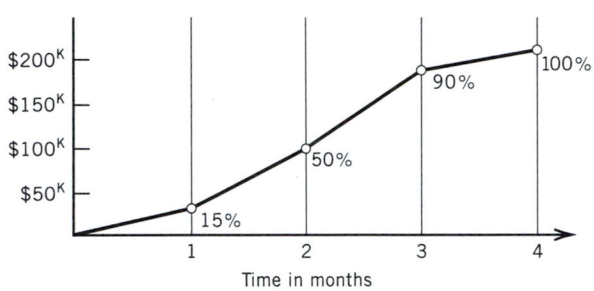

Monthly direct costs	$25,000	$ 65,000	$ 75,000	$ 15,000
Monthly indirect costs	$ 5,000	$ 5,000	$ 5,000	$ 5,000
Total monthly costs	$30,000	$ 70,000	$ 80,000	$ 20,000
Cumulative monthly costs	$30,000	$100,000	$180,000	$200,000

The letter "K" is used to indicate thousands of dollars.

Figure 9.1 Development of the S-curve.

charged directly to an activity) is $5,000. Assuming for simplicity that the direct costs are evenly distributed across the duration of the activity, the monthly direct costs can be readily calculated and are shown below the time line. The direct charges in the second month, for example, derive from activities A, B, and C, all of which have a portion in the period. The direct charge is simply calculated based on the portion of the activity scheduled in the second month as:

$$\text{Activity } A: \frac{1}{2} \times 50{,}000 = \$25{,}000$$

$$\text{Activity } B: \frac{1}{2} \times 40{,}000 = \$20{,}000$$

$$\text{Activity } C: \frac{1}{3} \times 60{,}000 = \underline{\$20{,}000}$$

$$\$65{,}000$$

The figure shows the total monthly and cumulative monthly expenditures across the life of the project. The S-curve is nothing more than a graphical presentation of the cumulative expenditures over time. A curve is plotted below the time-scaled bars through the points of cumulative expenditure. As activities come on-line, the level of expenditures increases and the curve has a steeper middle section. Toward the end of a project, activities are winding down and expenditures flatten again. The points are connected by a smooth curve since the assumption is that the expenditures are relatively evenly distributed over each time period. This curve is essentially a graphical portrayal of the outflow of monies (i.e., expense flow) for both direct and indirect costs.

9.2   CASH FLOW TO THE CONTRACTOR

The flow of money from the owner to the contractor is in the form of progress payments. As already noted, estimates of work completed are made by the contractor periodically

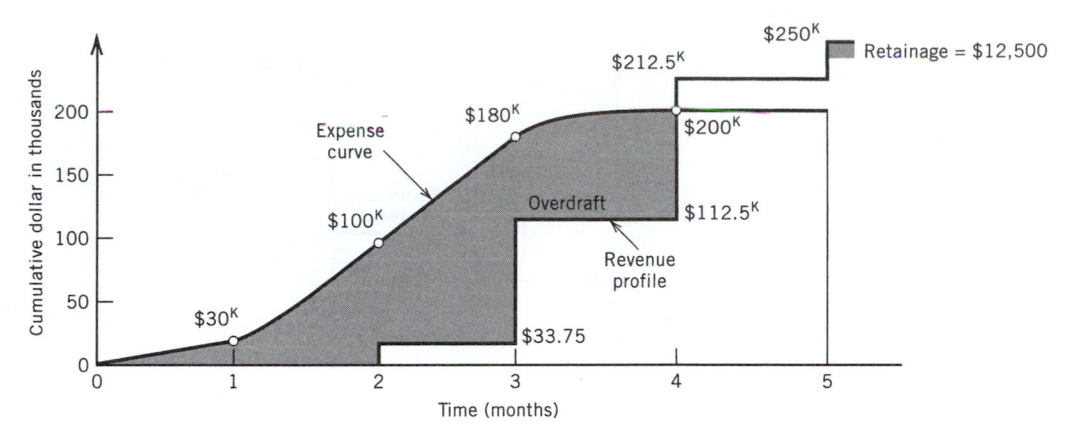

Figure 9.2 Expenses and income profiles.

(usually monthly) and are verified by the owner's representative. Depending on the type of contract (e.g., lump sum, unit price, etc.), the estimates are based on evaluations of the percentage of total contract completion or actual field measurements of quantities placed. This process is best demonstrated by further consideration of the four-activity example just described. Assume that the, contractor originally included a profit or markup in his bid of $50,000 (i.e., 25%) so that the total bid price was $250,000. The owner retains 10% of all validated progress payment claims until one-half of the contract value (i.e., $125,000) has been built and approved as an incentive for the contractor to complete the contract. The retainage will be deducted from the progress payments on the first $125,000 and eventually paid to the contractor on satisfactory completion of the contract. The progress payments will be billed at the end of the month, and the owner will transfer the billed amount minus any retainage to the contractor's account 30 days later. The amount of each progress payment can be calculated as:

$$\text{Pay} = 1.25(\text{indirect expense} + \text{direct expense})$$
$$-0.10[1.25(\text{indirect expense} + \text{direct expense})]$$

The minus term for retainage drops out of the equation when 50% of the contract has been completed. Because of the delay in payment of billings by the owner and the retainage withheld, the revenue profile lags behind the expense S-curve as shown in Figure 9.2.

The revenue profile has a stair-step appearance since the progress payments are transferred in discrete amounts based on the preceding equation. The shaded area in Figure 9.2 between the revenue and expense profiles indicates the need on the part of the contractor to finance part of the construction until such time as he is reimbursed by the owner. This difference between revenue and expense makes it necessary for the contractor to obtain temporary financing. Usually, a bank extends a line of credit against which the contractor can draw to buy materials, make payments, and pay other expenses while waiting for reimbursement. This is similar to the procedure used by major credit card companies in which they allow credit card holders to charge expenses and carry an outstanding balance for payment. Interest is charged by the bank (or credit card company) on the amount of the outstanding balance or overdraft[1]. It is, of course, good policy to try to minimize the amount of the overdraft and, therefore, the interest payments. The amount of the overdraft is

---

[1] Similar examples of this type of inventory financing can be found in many cyclic commercial undertakings. Automobile dealers, for instance, typically borrow money to finance the purchase of inventories of new car models and then repay the lender as cars are sold. Clothing stores buy large inventories of spring or fall fashions with borrowed money and then repay the lender as sales are made.

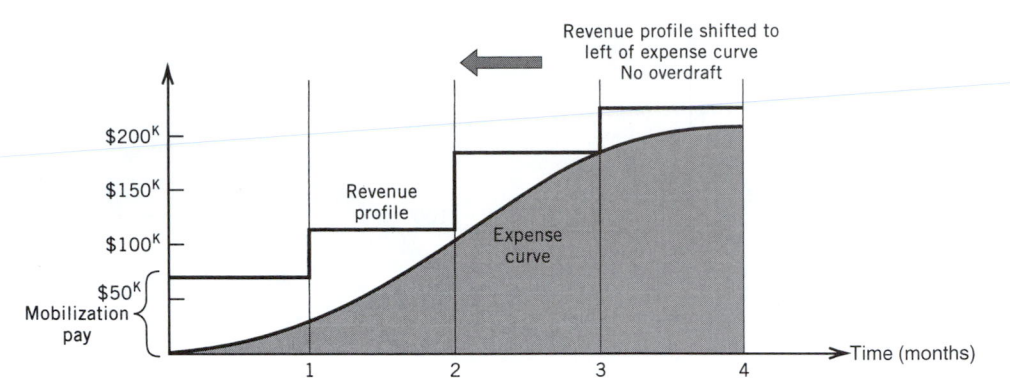

Figure 9.3 Influence of front, or mobilization, payment on expense and income profiles.

influenced by a number of factors, including the amount of markup or profit the contractor has in his bid, the amount of retainage withheld by the owner, and the delay between billing and payment by the owner.

Interest on this type of financing is usually quoted in relationship to the prime rate. The *prime rate* is the interest rate charged preferred customers who are rated as very reliable and who represent an extremely small risk of default (e.g., General Motors, Exxon, etc.). The amount of interest is quoted in the number of points (i.e., the number of percentage points) above the prime rate. The higher-risk customers must pay more points than more risky borrowers. Construction contractors are normally considered high-risk borrowers; if they default, the loan is secured only by some materials inventories and partially completed construction. In the event that a manufacturer of household appliances defaults, the inventory of appliances is available to cover part of the loss to the lender. Additionally, since construction contractors have a historically high rate of bankruptcy, they are more liable to be charged higher interest rates in most of their financial borrowings.

Some contractors offset the overdraft borrowing requirement by requesting front, or mobilization, money from the owner. This shifts the position of the revenue profile so that a reduced, or zero, overdraft occurs (Fig. 9.3). Since the owner is normally considered less of a risk than the contractor, he can borrow short-term money at a lower interest rate. If the owner agrees to this approach, he essentially takes on the interim financing requirement normally carried by the contractor. This can occur on cost-reimbursable contracts where the owner has great confidence in the contractor's ability to complete the project. In such cases it represents an overall cost savings to the owner, since otherwise he will ultimately be back-billed for the contractor's higher financing rate if the contractor must carry the overdraft.

9.3   OVERDRAFT REQUIREMENTS

In order to know how much credit must be made available at the bank, the contractor needs to know what the maximum overdraft will be during the life of the project. With the information given regarding the four-activity project, the overdraft profile can be calculated and plotted. For purposes of illustration, the interest rate applied to the overdraft will be assumed to be one percent per month. That is, the contractor must pay the bank 1% per month for the amount of the overdraft at the end of the month. More commonly, daily interest factors may be employed for the purpose of calculating this interest service charge. Month-end balances might otherwise be manipulated by profitable short-term borrowings at the end of the month. The calculations required to define the overdraft profile are summarized in Table 9.1.

Table 9.1 Overdraft Calculations

				Month		
	1	2	3	4	5	6
Direct cost	$25,000	$65,000	$75,000	$15,000		
Indirect cost	5,000	5,000	5,000	5,000		
Subtotal	30,000	70,000	80,000	20,000		
Markup (25%)	7,500	17,500	20,000	5,000		
Total billed	37,500	87,500	100,000	25,000		
Retainage withheld (10%)	3,750	8,750	0	0		
Payment received		$33,750	$78,750	$100,000	$37,500	
Total cost to date	30,000	100,000	180,000	200,000		
Total amount billed to date	37,500	125,000	225,000	250,000		
Total paid to date		$33,750	112,000	212,500	250,000	
Overdraft end of month	30,000	100,300	147,553	90,279	(8,818)[b]	(46,318)[b]
Interest on overdraft balance[a]	300	1,003	1,476	903	0	0
Total amount financed	30,300	101,303	149,029	91,182	(8,818)	0

[a] A simple illustration only. Most lenders would calculate interest charges more precisely on the amount/time involved employing daily interest factors.
[b] Parentheses indicate a positive balance in this case.

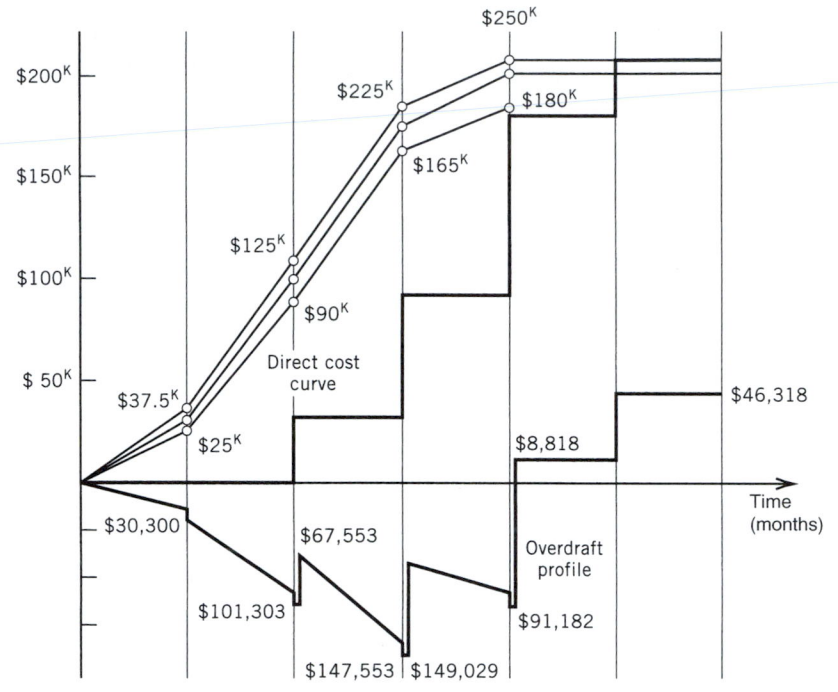

Figure 9.4 Plot of maximum overdraft.

The table indicates that the payment by the owner occurs at the end of a month based on the billing at the end of the previous month. It is assumed that the interest is calculated on the overdraft and added to obtain the amount financed. This amount is then reduced by the amount received from the owner for previous billings. To illustrate: The overdraft at the bank at the end of the second month is $100,300. The interest on this amount is $1,003 and is added to the overdraft to obtain the total amount financed ($101,303). To obtain the overdraft at the end of the third month, the progress payment of $33,750 is applied to reduce the overdraft at the beginning of the third month to $67,553. The overdraft at the end of the period is, then, $67,553 plus the costs for the period. Therefore, the overdraft is $67,553 plus $80,000, or $147,553. The information in the table is plotted in Figure 9.4. The overdraft profile appears as a sawtooth curve plotted below the baseline. This profile shows that the maximum required is $149,029. Therefore, for this project the contractor must have a line of credit that will provide at least $150,000 at the bank plus a margin for safety, say $175,000 overall to cover expenses.

Requirements for other projects are added to the overdraft for this project to get a total overdraft or cash commitment profile. The timing of all projects presently under construction by the contractor leads to overlapping overdraft profiles that must be considered to find the maximum overdraft envelope for a given period of time. Bids submitted that may be accepted must also be considered in the projection of total overdraft requirement. The plot of total overdraft requirements for a set of projects is shown in Figure 9.5.

Cash flow management involves all of the techniques described in this chapter—and very much more. It is fairly true to say, for example, that you cannot budget the other fellow's payments! That is, cash flows are affected by a significant degree of uncertainty. A cash flow management model of a relatively simple kind involves making provision for a set of at least 50 variables and requires a computer program to secure sufficient, timely, and usable cash management decision-making information.

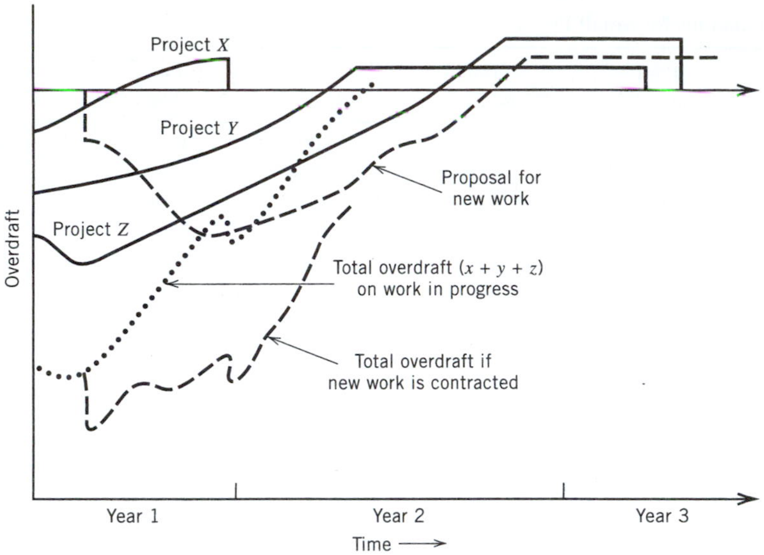

Figure 9.5 Composite overdraft profiles.

9.4   COMPARISON OF PAYMENT SCHEMES

Rate-of-return (ROR) analysis is helpful in comparing the economic value to a contractor of varying payment schemes. This technique utilizes engineering economy to evaluate the value of economic plans and strategies based on the time value of money. It is assumed in this section that the reader is familiar with the concepts of engineering economy. This subject is discussed in a number of textbooks (e.g., Collier and Ledbetter, *Engineering Economics and Cost Analysis*, 1988). It provides a vehicle for examining the economic impact of (1) varying retainage policies, (2) delay in payment strategies, and (3) the payment of a mobilization item to the contractor.

Consider the small four-activity project of Figure 9.1. The owner will consider the payment of a mobilization item at the end of the first period. This will be deducted from the final payment to the contractor. The amount of the payment will be $20,000. To determine the impact of this payment, the rate of return on the original payment sequence will be compared with the rate of return given the mobilization payment. Figure 9.6 shows a diagram of the original payment and expenditure sequence. Expenditures as taken from Table 9.1 are shown above the baseline in the figure. Revenues are shown below the line.

In order to determine the rate of return for a given sequence of payments and expenditures, a value for the interest rate must be found which satisfies the following relationship.

$$\sum^{all\ I} PW[REV(I)] - \sum^{all\ I} PW[EXP(I)] = 0$$

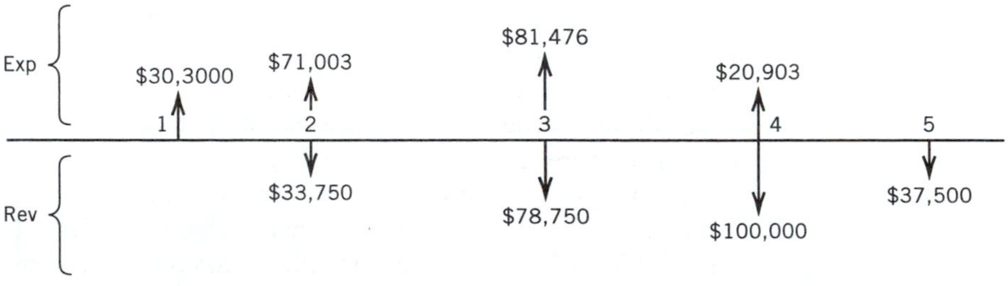

Figure 9.6 ROR for small bar chart problem.

Table 9.2 ROR Calculations for Small Project

N	NET[a]	PWF[b] @ 20%	Total @ 20%	PWF @ 25%	Total @ 25%	PWF @ 22%	Total @ 22%
1	−30300	.8333	−25249	.8000	−24240	.8196	−24834
2	−37253	.6944	−25868	.6400	−23842	.6719	−25030
3	−2726	.5787	−1577	.5120	−1396	.5507	−1501
4	79097	.4822	38140	.4096	32398	.4514	35704
5	37500	.4019	15071	.3277	12289	.3700	13875
			$\sum = +517$		$\sum = -4971$		$\sum = -1786$

$$\frac{X}{2\%} = \frac{517}{(1786 + 517)}$$

$$X = 0.45\%$$

$$ROR = 20\% + 0.45\%$$

$$= 20.45\%$$

[a] A negative net value indicates expenses exceed revenue for this period.
[b] PWF = Present Worth Factor.

$$\text{where} \qquad REV(I) = \text{revenue for period } I$$

$$EXP(I) = \text{expenditure for period } I$$

$$PW = \text{present worth of these values}$$

Since the difference between revenues and expenditures for a given period is REV(*I*) − EXP(*I*) = NET(*I*), the equation can be reduced to $\sum$PW[NET(*I*)] = 0. In other words, the sum of the present worth values for all period NET values must equal to zero. This assumes that all expenditures and all revenues are recognized at the end of each period, *I*. The effective ROR will be the value of interest that satisfies this equation or, in effect, causes the present value of all expenditures (recognized at the end of each month) to equal the present value of all revenues (again recognized at the end of the month). Since this method does not consider a company's cost of capital, it is often referred to as the Internal Rate of Return (IRR).

We do not have a closed-form mathematical expression that allows for the determination of the correct interest value *i*. Therefore, an iterative approach must be used to bracket the proper value of *i*. Values for the present value of revenues and expenditures or net values in each period are calculated using an assumed value of interest, *i*. The summation of the net values at present worth must equal zero. If the value of the summation of NET(*I*) changes sign (from minus to plus) between two different values of *i*, then the value of *i* that satisfies the equation is contained between those two values. Table 9.2 summarizes the calculations required to determine the rate of return for the original payment scheme given in Table 9.1. The net values [NET(*I*)] for the five periods are shown in the first column. A value of *i* of 20% is selected. The summation of the net values is calculated to be +517. The *i* value [present worth factor (PWF)] is increased from 20 to 22%, and the summation of net values

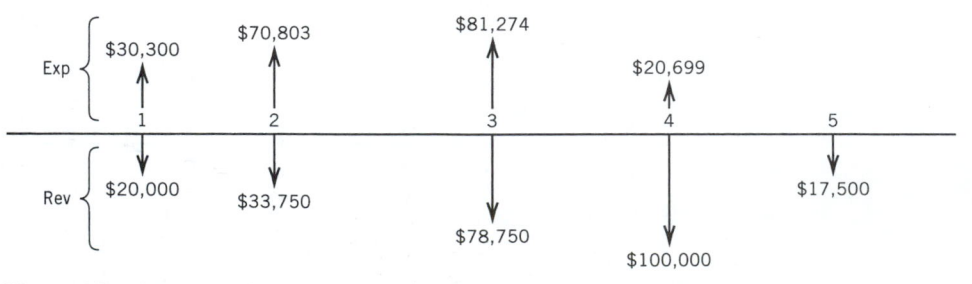

Figure 9.7 ROR for small bar chart problem with mobilization payment.

Table 9.3 Overdraft Calculations with Mobilization Payment

			Month			
	1	2	3	4	5	6
Direct cost	$25,000	$65,000	$75,000	$15,000		
Indirect cost	5,000	5,000	5,000	5,000		
Subtotal	30,000	70,000	80,000	20,000		
Markup (25%)	7,500	17,500	20,000	5,000		
Total billed	37,500	87,500	100,000	25,000		
Retainage withheld (10%)	3,750	8,750	0	0		
Payment received		$20,000	$33,750	$78,750	$100,000	$17,500
Total cost to date	30,000	100,000	180,000	200,000	200,000	
Total amount billed to date	37,500	125,000	225,000	250,000	250,000	
Total paid to date		$20,000	$53,750	$132,000	$232,500	$250,000
Overdraft end of month	30,000	80,300	127,353	69,877	(29,424)[a]	(46,924)[a]
Interest on overdraft balance	300	803	1,274	699	0	0
Total amount financed	$30,300	$81,103	$128,627	$70,576	(29,424)	0

[a]Parentheses indicate a positive balance.

Table 9.4 ROR Calculations to Include Mobilization Payment

N	Net[a]	PWF[b] 30%	Total @ 30%	PWF 32%	Total @ 32%	PWF 34%	Total @ 34%
1	−10300	.7692	−7923	.7575	−7802	.7463	−7687
2	−37053	.5917	−21925	.5739	−21265	.5569	−20635
3	−2524	.4552	−1149	.4348	−1097	.4156	−1049
4	79301	.3501	27765	.3294	26122	.3101	24591
5	17500	.2693	4713	.2495	4366	.2315	4051
			$\sum = 1482$		$\sum = 324$		$\sum = -729$

$$\frac{X}{2\%} = \frac{324}{(324 + 729)}$$

$$X = 0.62$$

$$\text{ROR} = [32 + .62]\%$$

$$= 32.62\%$$

[a] A negative net value indicates expenses exceed revenue for this period.
[b] PWF = Present Worth Factor

becomes −1786. The value of i that satisfies the equality of $\text{PW} \sum [\text{NET}(I)] = 0$ must be between 20 and 22%. The correct ROR value is found by interpolation to be 20.45%.

The alternative sequence of payment including the $20,000 mobilization payment is shown in Figure 9.7. This sequence pays the contractor $20,000 at the end of the first work period and deducts the prepayment from the final payment. The final payment is reduced from $37,500 to $17,500. The cash flow calculations for this sequence are shown in Table 9.3. It can be seen that the mobilization payment causes the revenue profile to move closer to the expense curve, thus reducing the area between the two. This also reduces the overdraft and peak financial requirement.

The calculations to determine the ROR of this payment sequence are given in Table 9.4. The correct value is bracketed between 32 and 34%. The final ROR is 32.6%. This indicates that payment of the mobilization payment at the end of the first period increases the rate of return on this project to the contractor in the amount of approximately 12%. This is due in part to the reduction in the amount of inventory financing that must be carried by the contractor. What would be the impact of paying a $30,000 mobilization payment immediately upon commencement of the job? This is left as an exercise for the reader to determine the change in the rate of return.

REVIEW QUESTIONS AND EXERCISES

9.1 Given the following cost expenditures for a small warehouse project (to include direct and indirect charges), calculate the peak financial requirement, the average overdraft, and the rate of return on invested money. Sketch a diagram of the overdraft profile.

Assume 12% markup
Retainage 10% throughout project
Finance charge = 1.5% month
Payments are billed at end of month and received one month later

Month	1	2	3	4
Indirect + Direct Cost ($)	$69,000	$21,800	$17,800	$40,900

9.2 The table and graph on page 157 represent a contractor's overdraft requirements for a project. Complete the table shown for costs, markup, total worth, retainage, and pay received. Retainage is 10%, markup is 10%, and interest is 1% per month.

The client is billed at the end of the month. Payment is received the end of the next month, to be deposited in the bank the first of the following month.

Overdraft	−50,000	−120,500	−82,205	−13,727	+10,336
Interest	−500	−1,205	−822	−137	—
Cumulative	−50,500	−121,705	−83,027	−13,864	+10,336

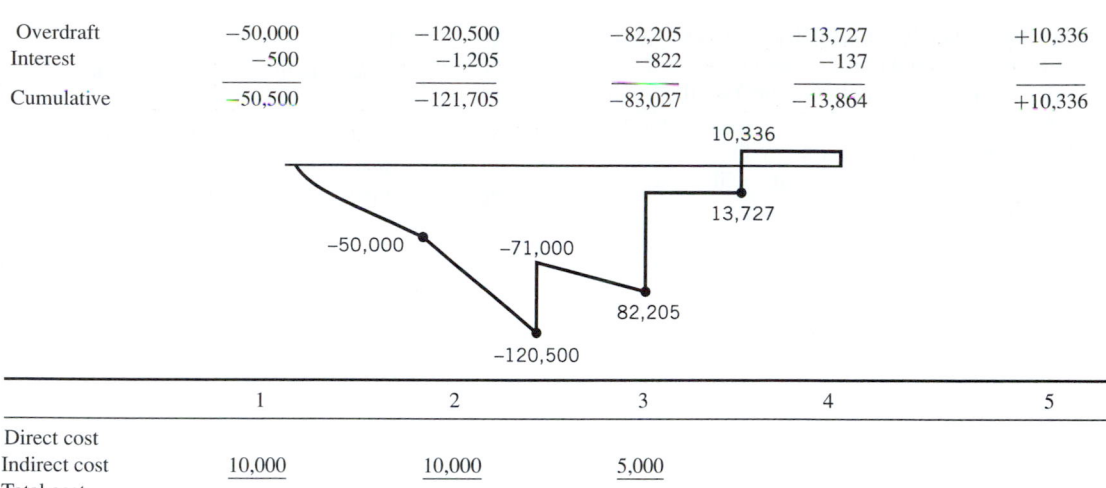

	1	2	3	4	5
Direct cost					
Indirect cost	10,000	10,000	5,000		
Total cost					
Markup					
Total worth					
Retainage					
Pay received					

9.3 Given the bar chart in Figure 9.1 with the direct costs for each activity as shown, calculate the rate of return of the contractor. Assume that (a) the markup is 15%; (b) retainage is 5% on the first 50% of worth, and 0% thereafter; (c) payment requests are submitted at the end of each month, and payments are received one month later; and (d) the finance charge is 1% per month of the amount of the overdraft at the end of the month.

Timing and allocation	$25,000	$65,000	$75,000	$15,000	
			Total direct costs		$180,000
Indirect costs $5000/month	5,000	5,000	5,000	5,000	
			Total indirect costs		$20,000
	$30,000	$70,000	$80,000	$20,000	$200,000

9.4 A contractor is preparing to bid for a project. He has made his cost estimate together with the schedule of work. His expected expenses and their time occurrence are as shown in the following table. For simplicity of analysis he assumed that all expenses are recognized at the end of the month in which they occur.

Month	Mobilization Demobilization	Subcontractors	Materials	Payroll	Equipment	Field Overhead
0	$40,000	$0	$0	$0	$0	$0
1	0	10,000	10,000	10,000	20,000	1,000
2	0	30,000	20,000	15,000	10,000	5,000
3	0	30,000	30,000	20,000	20,000	6,000
4	0	40,000	30,000	20,000	30,000	6,000
5	0	50,000	40,000	40,000	20,000	6,000
6	0	50,000	40,000	40,000	15,000	6,000
7	0	40,000	30,000	40,000	10,000	6,000
8	0	40,000	10,000	20,000	10,000	6,000
9	0	70,000	10,000	10,000	10,000	6,000
10	0	30,000	5,000	5,000	10,000	6,000
11	0	30,000	5,000	5,000	5,000	6,000
12	20,000	50,000	0	5,000	5,000	5,000
Total	$60,000	$470,000	$230,000	$230,000	$165,000	$65,000

Total cost = $60,000 + $470,000 + $230,000 + $230,000 + $165,000 + $65,000 = $1,220,000
Profits + overhead @ 10% = $122,000
Bid price = $1,342,000

(a) The contractor is planning to add 10% to his estimated expenses to cover profits and office expenses. The total will be his bid price. He is also planning to submit for this progress payment at the end of each month. Upon approval, the owner will subtract 5% for retainage and pay the contractor one month later. The accumulated retainage will be paid to the contractor with the last payment (i.e., end of month 13).

(i) Develop the cash flow diagram.

(ii) What is the peak financial requirement and when does it occur?

(b) Assume the same as in part (a), except that the owner will retain 10% instead of 5%. Plot the cash flow diagram and calculate the peak financial requirement.

Chapter 10

Project Funding

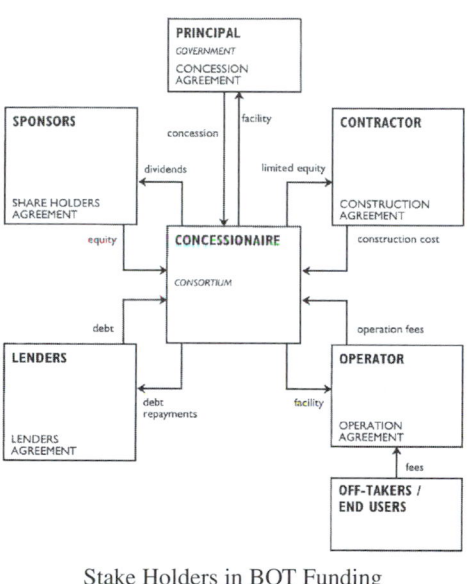

Stake Holders in BOT Funding

Build Operate and Transfer (BOT)

Concept

Pollalis (1996) defines Build Operation and Transfer as follows:

"In the BOT approach, a private party or concessionaire retains a concession for a fixed period from a public party, called principal (client), for the development and operation of a public facility. The development consists of the financing, design, and construction of the facility, managing and maintaining the facility adequately, and making it sufficiently profitable. The concessionaire secures return of investment by operating the facility and, during the concession period, the concessionaire acts as owner. At the end of the concession period, the concessionaire transfers the ownership of the facility free of liens to the principals at no cost."

The modern implementation of BOT concepts is generally accredited to the Turkish government under the leadership of Prime Minister Ozal.

This method of funding and constructing a large infrastructure program was initiated in Turkey starting in 1984. In fact, the financing and construction of the Suez Canal by the French in the nineteenth century was done using a system which would be considered BOT by today's standards (Levy 1996).

Confederation Bridge Crossing – Prince Edwards Island – Canada

Advantages

Traditionally, highways, dams, public buildings (e.g., jails), tunnels, etc. have been constructed using funds which were generated from taxes levied by public entities (e.g., federal, state, municipal government). In many cases, tax payers have rebelled and failed to support the issuance of bonds and similar borrowing instruments to allow construction of critically needed public facilities. In developing countries with relatively weak economies, modest tax revenues have led to a delay in developing infrastructure to support national development.

The concept of privatization as defined by the BOT approach became popular in the early 1980s. Privatization addresses the problems of developing infrastructure projects by utilizing private funds to finance and construct public projects. For instance, if a bridge is needed to connect two political entities separated by a river or a strait, a private consortium can raise the funds and construct the bridge recovering the cost and effort involved by charging tolls to users of the bridge. The Confederation Bridge linking New Brunswick with Prince Edwards Island in Canada is an example of a BOT infrastructure project.

10.1   MONEY: A BASIC RESOURCE

The essential resource ingredients that must be considered in the construction of a project are usually referred to as the four Ms. These basic construction resources are (1) money, (2) machines, (3) manpower, and (4) materials. They are presented in this order since this is the sequence in which they will be examined in the next few chapters. Here, the first of these resources to be encountered in the construction process, money, is considered. Money (i.e., actual cash or its equivalent in monetary or financial transactions) is a cascading resource that is encountered at various levels within the project structure. The owner or developer must have money available to initiate construction. The contractor must have cash reserves available to maintain continuity of operations during the time he is awaiting payment from the owner. The major agents involved in the flow of cash in the construction process are shown in simple schematic format in Figure 10.1.

Rising construction costs have increased the pressure on the construction industry to carefully monitor and control the flow of money at all levels. As a result, more emphasis is being placed on cash flow and cost control functions in construction management than ever before. In the planning phases, more thorough investigations and more accurate cost estimates are being required for those seeking financial backing. To remain competitive, contractors are being forced to monitor their cost accounts more closely and to know where losses are occurring. In this chapter, the methods by which the owner/entrepreneur acquires project funding will be considered. The relationship between the flow of money from owner to contractor and its impact on the contractor's project financing has been discussed in Chapter 9.

10.2   CONSTRUCTION FINANCING PROCESS

The owner's financing of any significant undertaking typically requires two types of funding: *short-term* (construction) funding and *long-term* (mortgage) funding. The short-term funding is usually in the form of a construction loan, whereas the long-term financing involves a mortgage loan over a term ranging from 10 to 30 years.

The short-term loans may provide funds for items such as facility construction, land purchases, or land development. Typically these short-term loans extend over the construction period of the project. For large and complex projects, this can be a period of 6 to 8 years as in the case of utility power plants. A short-term loan is provided by a lending

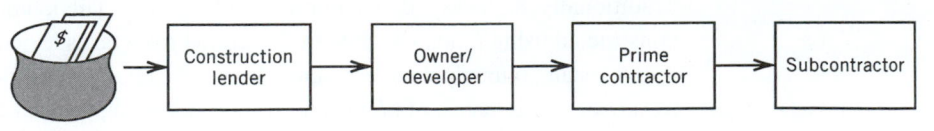

Figure 10.1 Project money flow.

institution, based on the assurance that it will be repaid with interest, by some other loan. This subsequent mortgage loan constitutes the long-term financing. Therefore, the first objective of any entrepreneur is to seek a commitment for long-term, or permanent, financing from a mortgage lender. Regardless of the type of project, this commitment will permit the construction loan, and any other funding required, to be obtained with relative ease or, at least, more easily.

Unless he is in a position to raise the funds required directly by the issue of his own securities, the entrepreneur will seek to obtain a commitment from one of several alternate sources, including real estate investment trusts (REITs), investment or merchant banks, commercial banks, savings and loan associations, insurance companies, governmental agencies (VA, FHA), or, in special cases, from one of the international development banks. Public institutions often raise project construction funds by the sale of bonds. The choice of lender depends on the type and size of project. The choice of the form of security employed depends on a number of factors such as relative cost, the time period for which the funds will be available, and the degree of flexibility involved (the freedom to pay out or refinance) as to whether there are any restrictions involved and whether there is any sacrifice of control to the lender. The funding of some larger projects may be handled by a consortium of international bankers (e.g., the Channel Tunnel).

Lending institutions are cautious; they are not interested in financing failures or in owning partially completed projects. Therefore, they will undertake a great deal of research and evaluation prior to providing a commitment for funding. At a minimum, an entrepreneur will be expected to provide the following as part of the loan application:

1. A set of financial statements for the firm
2. Personal financial statements from the principals of the firm
3. Proof of clear title to the land for the project and that it has an appropriate zoning
4. Preliminary floor plans and elevations for the project
5. Preliminary cost estimates
6. A market research study to verify expected income
7. A detailed pro forma indicating projected income and expense throughout the life of the mortgage loan

---

EXAMPLE

An example of a long-term finance pro forma for a venture involving the construction and leasing of a 75-unit apartment complex is shown in Figure 10.2a. This document indicates that the annual income from the proposed apartment complex project will be $306,830. The requested loan is $2,422,000, and the annual debt service (i.e., interest) on this amount is $236,145, realizing an income after debt service of approximately $70,000. The ratio between income and debt service is 1.3. Lenders normally wish this ratio to be below 1.3. The basis for the loan amount is given in Figure 10.2b. Items 1 to 34 are construction-related items and are developed from standard references [e.g., R. S. Means, Co., *Building Construction Cost Data* (published annually)] based on unit measures such as square footage. The lender normally has a unit-price guide for use in verifying these figures. Items 35 to 46 in Figure 10.2b cover nonconstruction costs that are incurred by the entrepreneur. It should be noticed that the interest for the construction loan is included in the costs carried forward to the long-term financing.

The method used to calculate the actual dollar amount of the loan is of great interest to the entrepreneur. The interest the developer pays for the use of the borrowed money is an expense, and it is generally considered prudent business policy to minimize expenses. One way to minimize the interest expense would be to borrow as little as possible. This is not, however, the way the developer moves toward his objective. The developer seeks primarily to protect his own personal assets (or those of his company) in his efforts to complete the project. The more he invests, the more he stands to lose if the project fails. With this consideration in mind, the developer may seek to minimize his own investments. That is, the developer tries to expand his own small initial asset input into a large amount of usable money. This is called *leverage*. He takes a small amount and levers, or amplifies, it into a large amount.

The amount of the mortgage loan should be a happy medium between too much and too little. If the mortgage is too small, there will not be enough to cover the project. On the other hand, if the mortgage is too large, the developer will find that the individual mortgage payments will exceed his available revenue, and he may be unable to meet all of his obligations.

The amount the lender is willing to lend as long-term funding is derived from two concepts: the economic value of the project and the capitalization rate (cap rate). The economic value of the project is a measure of the project's ability to earn money. One method of predicting the economic value is called the *income approach* to value and is the method shown in Figure 10.2a. Simply stated, it is the result of an estimated income statement of the project in operation. Like any income statement, it shows the various types of revenue and their sum. These are matched against the predicted sums of the different expenses. Although the predicted net income is a function of many estimated numbers, commonly a fairly reasonable degree of accuracy is achieved. The expected net income divided by the cap rate produces the economic value of the project. The cap rate used in Figure 10.2a is 9.5%. The capitalized economic value of the project is obtained by dividing the net income ($306,830) by the cap rate factor (0.095). This yields an economic value of $3,229,789.

How is the cap rate obtained? First, a lender generally provides a mortgage that is about 75% of the estimated economic value of the project. This is done because 25% of the value, or thereabouts, must be invested by the developer and will serve as an incentive for his making the project a success. That is, the lender furnishes 75% and the developer furnishes 25%. The lender must then decide what the interest rate will be and takes up the developer's rate of return. The sum of these numbers, times their respective portions, gives the cap rate.

As an example, suppose that the lender decides that the interest rate will be 8.5% and that the developer's planned rate of return will be 12%. Then, the cap rate is obtained as 8.5% times 75% plus 12% times 25%, which gives 9.375% or 0.09375 as the cap rate factor. Obviously, the value of the cap rate can be adjusted by the values that the lender places on his interest rate and the developer's rate of return. These numbers are a function of the existing

Market rent for subject property (unfurnished) 55 two-bedroom A, B, or C units—1167 sq ft @ 41.0 cents/sq ft = $478.47/mo or $480 × 55 $ 26,400.00
20 three-bedroom A, B units—1555 sq ft @ 37.3 cents/sq ft = $580.00/mo $580 × 20   11,600.00

Total estimated monthly income $ 38,000.00
Other income: Coin laundry, vending machine   150.00

                                                                           $ 38,150.00
× 12 = annual total   457,800.00
Less vacancy factor of 5% (based on historical data)   −22,890.00

Adjusted gross annual income   434,910.00
Less estimated expenses @ 29.45%   −128,080.00

Net income before debt service $ 306,830.00

Capitalized value @ 9.5% = $3,229,789.00 = $\frac{306,830.00}{(0.095)}$

Requested loan value   = $2,422,000.00
Loan/value ratio   = 75% (high) governed by law
Long-term debt service @ 9.75% constant   = $236,145.00
Debt service coverage ratio   = 1.3
Loan per unit   = $32293.33
Loan per square foot   = $25.42

Figure 10.2a Pro forma for 75 apartment units.

	1. Excavation and grading	$ 67,500
	2. Storm sewers	48,000
	3. Sanitary sewers	84,030
	4. Water lines	28,000
	5. Electric lines	14,000
	6. Foundations	31,000
	7. Slabs	96,000
	8. Lumber and sheathing	185,000
	9. Rough carpentry	185,000
	10. Finish carpentry	81,362
	11. Roofing and labor	20,035
	12. Drywall and plaster	70,000
	13. Insulation	28,888
	14. Millwork	140,556
	15. Hardware	8,813
	16. Plumbing	165,000
Construction	17. Heating and air conditioning	95,025
Related	18. Electrical	90,350
Costs	19. Linoleum and tile	17,752
	20. Carpeting	101,881
	21. Kitchen cabinets	62,075
	22. Painting and decorating	107,000
	23. Masonry, block	20,680
	24. Masonry, brick	100,200
	25. Ranges and hoods	29,638
	26. Disposals	3,139
	27. Exhaust fans	1,022
	28. Refrigerator	35,040
	29. Paving	20,915
	30. Walks and curbs	20,792
	31. Landscaping	30,000
	32. Fence and walls	36,792
	33. Fireplace	51,100
	34. Cleanup	29,200
	35. Lender's fee	32,000
	36. Surveyor's fee	1,000
	37. Architect's fee	12,500
	38. Land cost	80,000
	39. Attorney's fee	7,500
Non-	40. Title insurance premium	5,762
Construction	41. Other closing costs	150
Costs	42. Hazard insurance premium	4,780
	43. Construction loan interest	120,000
	44. Appraisal	750
	45. Building permit	1,500
	46. Tax	50,000
	Total	$2,422,000

Figure 10.2b Construction cost breakdown for 75 apartment units.

economic conditions and thus fluctuate with the state of the economy. The lender, therefore, cannot exert as much influence on their values as might at first be expected. In addition, the lender is in business to lend and wisely will not price himself out of the competition. He will attempt to establish a rate that is conservative but attractive. The expected income divided by the cap rate yields the economic value. The mortgage value may then be on the order of 75% of the calculated economic value. Not every lender will follow this type of formula approach; some, for example, may have a policy of lending a fixed proportion of their own assessed valuation, which may not be based on the economic value but instead on their estimate of the market value of the property.

The mortgage loan may be the critical financial foundation of the entire project and may also involve protracted and complex negotiations. For this reason, the project developing company may exercise its right to hire a professional mortgage broker whose business it is to find a source of funds and service mortgage loan dealings. The broker's reputation is based on his ability to obtain the correct size mortgage at the best rate that is also fair to his client. The broker acts as an advisor to his client, keeping him apprised of all details of the proposal financing in advance of actually entering into the commitment. For this service, the mortgage broker receives a fee of about 2% of the mortgage loan, although the rate and amount will vary with the size of the loan.

10.3   MORTGAGE LOAN COMMITMENT

Once the lending institution has reviewed the venture and the loan committee of the lender has approved the loan, a preliminary commitment is issued. Most institutions reserve their final commitment approval until they have reviewed and approved the final construction plans and specifications.

The commitment issued is later embodied in a formal contract between the lender and borrower, with the borrower pledging to construct the project following the approved plans, and the lender agreeing that upon construction completion, and the achievement of target occupancy, he will provide the funds agreed upon at the stated interest rate for the stated period of time. As noted earlier, the actual amount of funds provided generally is less than the entire amount needed for the venture. This difference, called *owner's equity*, must be furnished from the entrepreneur's own funds or from some other source. The formal commitment will define the floor and ceiling amounts of the long-term loan.

During the construction period, no money flows from the long-term lender to the borrower. Funds necessary for construction must be provided by the entrepreneur or obtained from a short-term construction lender. Typically the lender of the long-term financing will pay off the short-term loan in full, at the time of construction completion, thereby canceling the construction loan and leaving the borrower with a long-term debt to the mortgage lender.

10.4   CONSTRUCTION LOAN

Once the long-term financing commitment has been obtained, the negotiation of a construction loan is possible. Very often commercial banks make construction loans because they have some guarantee in knowing the loan will be repaid from the long-term financing. However, even in these situations, there are definite risks involved for the short-term lender. These risks relate to the possibility that the entrepreneur or contractor may, during construction, find themselves in financial difficulties. If this occurs, it may not be possible for the entrepreneur/contractor to complete the project, in which case the construction leader may have to take over the job and initiate action for its completion. This risk is offset by a discount (1–2%), which is deducted from the loan before any money is disbursed. For example, if the amount of construction money desired is $1,000,000, the borrower signs a note that he will pay back $1,020,000. The borrower, in effect, pays immediately an interest of $20,000. This is referred to as a discount and may be viewed as an additional interest rate for the construction loan. The current trend to minimize these risks is to require the borrower to designate his intended contractor and design architect. The lender may also require that all contactors involved in the construction be bonded as well. Some commercial banks evaluate and seek to approve the owner's intended contractor, his prime subcontractors, and the owner's architect, as a prerequisite to approving the construction loan. This evaluation extends to an evaluation of their financial positions, technical capabilities, and current workloads.

To minimize the risks involved, the banks will also base their construction loans on the floor of the mortgage loan, and only 75 to 80% of this floor will be lent. Of course, the developer may need additional funds to cover construction costs. One way to ensure this is to finance the gap between the floor and ceiling of the long-term mortgage loan. The entrepreneur goes to a lender specializing in this type of financing and obtains a standby commitment to cover the difference or gap between what the long-term lender provides and the ceiling of the long-term mortgage. Then, if the entrepreneur fails to achieve the breakeven rent roll, he still is ensured of the ceiling amount. In this situation, the construction lender will provide 75 to 80% of the ceiling rather than the floor. If the floor of the loan is $2,700,000 and the ceiling is $3,000,000, the financing of the gap can lead to an additional $240,000 for construction (i.e., 80% of $300,000). Financing of the gap is usually expensive, requiring a prepaid amount of as much as 5% to the gap lender. In the above example, this would be $15,000 paid for money that may not be required if the rent roll is achieved. Nevertheless, the additional $240,000 of construction funding may be critical to completion of the project and, therefore, the $15,000 is well spent in ensuring that the construction loan will include this gap funding.

Once the construction loan has been approved, the lender sets up a draw schedule for the builder or contractor. This draw schedule allows the release of funds in a defined pattern, depending on the site and length of the project. Smaller projects, such as single-unit residential housing, will be set up for partial payments based on completion of various stages of construction (i.e., foundation, framing, roofing, and interior), corresponding to the work of the various subcontractors who must be paid (see Fig. 10.3). For larger projects, the draw schedule is based on monthly payments. The contractor will invoice the owner each month for the work he has put in place that month. This request for funds is usually sent to the owner's representative or architect who certifies the quantities and value of work in place. Once approved by the architect and representative, the bank will issue payment for the invoice, less an owner's retainage (see Chapter 3).

The owner's retainage is a provision written into the contract as an incentive for the contractor to continue his efforts, as well as a reserve fund to cover defective work that must be made good by the contractor before the retainage is released. Typically this retainage is 10%, although various decreasing formulas are also used. When the project is completed, approved, cleared, and taken over by the owner, these retainages are released to the builder.

In addition to the funds mentioned, the developer should be aware that some front money is usually required. These funds are needed to make a good-faith deposit on the loan to cover architectural, legal, and surveying fees and for the typical closing costs.

10.5   OWNER FINANCING USING BONDS

Large corporations and public institutions commonly use the procedure of issuing bonds[1] to raise money for construction projects. A bond is a kind of formal IOU issued by the borrower promising to pay back a sum of money at a future point in time. Sometimes this proviso is supported by the pledging of some form of property by way of security in case of default by the borrower. A *series* of bonds or debentures, issued on the basis of a prospectus, are the general type of security issued by corporations, cities, or other institutions, but not by individual owner-borrowers. In this discussion, owner financing means financing arrangements made by those corporations or institutions that are the owner

[1] In this case, bonds refer to financial borrowing instruments.

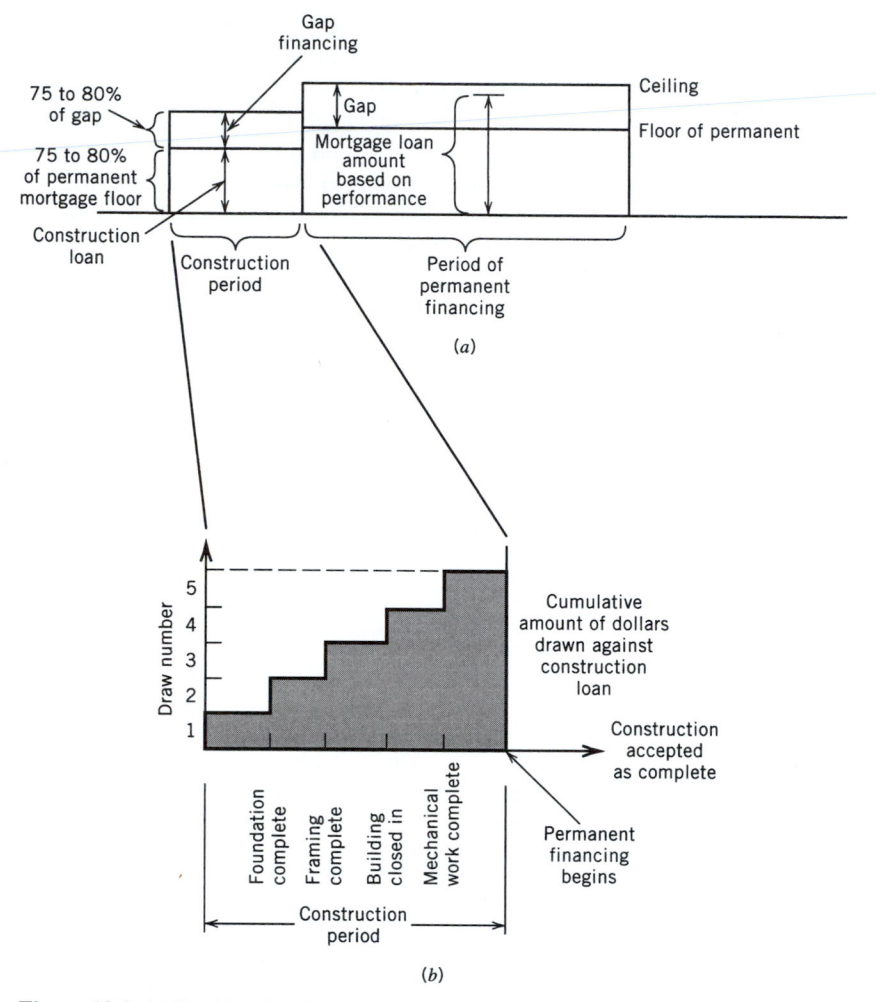

Figure 10.3 (*a*) Profile of project financing by the entrepreneur and (*b*) draw schedule.

of the project property. In the illustrative material that follows, "Joe" stands as a surrogate for "any borrower" ("Joan" would have served as well!). During the period in which he has use of the money, the borrower promises to pay an amount of interest at regular intervals. For instance, Joe borrows $1000 and agrees to pay back the $1,000 (referred to as the principal) in full at the end of 10 years. He pays an annual interest of 8% (at the end of each year). That is, he, in effect, pays a rent of $80.00 per year on the principal sum of $1000 for 10 years and then pays back the amount borrowed. The rent is payable at the end of each year. The sequence of payments for this situation would be as appears in Figure 10.4. When a series of bonds is issued, there may be a commitment to pay the interest due in quarterly installments rather than in one amount at the end of the year. A bond, as a long-term promissory note, may take any one of a variety of forms depending on the circumstances; mortgage bonds involve the pledging of real property, such as land and buildings; debentures do not involve the pledging of specific property. Apart from the security offered, there is the question of interest rates and the arrangements to be made for the repayment of the principal sum. Sometimes a sinking fund may be set up to provide for the separate investment, at interest, of capital installments that will provide for the orderly retirement of the bond issue. Investors find this type of arrangement an attractive condition in a bond issue.

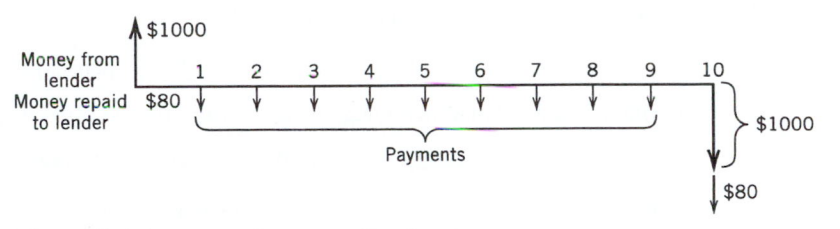

Figure 10.4 Sequence of payments for a bond.

In preparing for a bond or debenture issue, financial statements must be drawn up and sometimes a special audit may be required. A prospectus for the issue may need to be drawn up, and this will involve settling the terms of issue and of repayment, the interest rates payable, and the series of promises or conditions related to the issue, such as its relative status in terms of priority of repayment, limitations on borrowing, the relative value of the security, and the nomination of a trustee to watch the interests of bond or debenture holders. These details are usually settled with the aid of specialists such as a CPA firm or the mortgage broker.

Public bodies may need the approval of some local regulatory authority, and corporations may have to file and have approved a prospectus for the proposed bond issue. Charters or other constitutional documents must, of course, confer on the public body or corporation the power to borrow money in this way; this power is exercised by the council or by the board of directors or governors. For public offerings that are particularly attractive, banks bid for the opportunity to handle the placement of the bonds. The banks recover their expense and profit by offering to provide a sum of money slightly less than the amount to be repaid. As noted above, this is called *discounting the loan*. The fact that more will be repaid by the borrower than is lent by the lender leads to a change in the actual interest rate. This is established through competitive bidding by the banks wishing to provide the amount of the bond issue. The bank that offers the lowest effective rate is normally selected; this represents the basic cost incurred for the use of the money.

Consider the following situation in which a city that has just received a baseball franchise decides to build a multipurpose sports stadium. The design has been completed, and the architect's estimate of cost is $40.5 million. The stadium building authority has been authorized to issue $42 million in bonds to fund the construction and ancillary costs. The bonds will be redeemable at the end of 50 years with annual interest paid at 5% of the bond principal. Neither the term nor its rate purport to be representative of current market conditions. At this time the term for any bond issue would tend to be shorter and its rate higher. In some commercial dealings "index number" escalation clauses are also occasionally seen. The banks bid the amounts for which they are willing to secure payment support. Suppose the highest bid received is $41 million.

In order to determine the effective rate of interest, a rate-of-return analysis may be used. The profile of income and expense is shown in Figure 10.5. The effective rate of interest is that rate for which the present worth of the expenses is equal to the present worth (PW) of

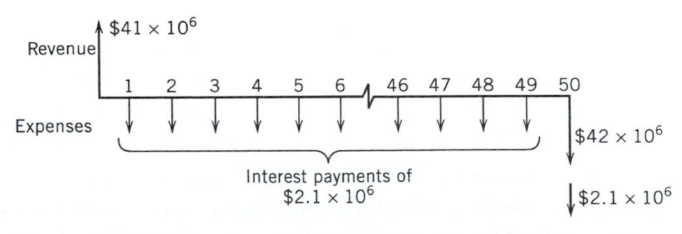

Figure 10.5 Profile of revenue-expense for a bond issue.

the revenue (in this case, 41×10). That is,

$$PW \text{ (revenue)} = PW \text{ (expenses)}$$

Utilizing the information above, this expression for the bond issue problem becomes

$$\$41 \times 106 = \$2,100,000 \,(PWUS, i, 50)^2 + \$42,000,000 \,(PWSP, i, 50)^3$$

The annual interest is $2.1 million, and this is a uniform series of payments for 50 years. The $42 million must be repaid as a single payment at the end of 50 years. The notation used in the equation is consistent with that used in a number of standard engineering economy textbooks.

In making this approach to a solution, a trial-and-error method (similar to that used in Section 9.4) must be employed to solve the equation. That is, values of i must be assumed and the equation solved to see if the relationship [e.g., PW(revenue) − PW(expenses) = 0] is satisfied. In this case, two initial candidates for consideration are $i = 0.05$ and $i = 0.06$. Consulting appropriate tables for the present worth factors, the right side of the equation becomes

$$i = 0.05 \quad PW = \$2.1 \times 10^6 \,(18.256) + \$42 \times 10^6 \,(0.0872)$$
$$= \$42 \times 10^6 \quad \text{difference} = +1.0 \times 10^6$$
$$i = 0.06 \quad PW = \$2.1 \times 10^6 \,(15.762) + \$42 \times 10^6 \,(0.0543)$$
$$= \$35.38 \times 10^6 6 \quad \text{difference} = -\$5.62 \times 10^6$$

Since the equation balance goes from plus to minus, the value satisfying the relationship is between 5 and 6%. Using linear interpolation, the effective interest rate is found to be

$$i = 0.05 + (0.06 - 0.05) \times \frac{1.0 \times 10^6}{(1.0 + 5.62) \times 10^6} = 0.515$$

or 5.15% as an approximation.

REVIEW QUESTIONS AND EXERCISES

10.1   What is the present level of the prime rate? How does this rate relate to the current financing and overdraft charges for new building construction in your locality? (How many points above the prime is this rate?) Does this overdraft rate vary with the magnitude of the monies involved?

10.2   Referring to the example in Figure 10.2aA, suppose the market rent for a two-bedroom unit is $550 per month and for a three-bedroom unit is $650 per month. If the going cap rate is 10%, rework the pro forma calculations for the apartment project of Figure 8.2. Then determine the lender's interest rate. What is the new breakeven vacancy factor?

10.3   What determines the number of draws a builder can make in completing his facility? What is the existing policy regarding number of draws in your locality?

10.4   Suppose that for the multipurpose sports stadium example considered in the text the bond issue was for 40 years and the annual interest rate is 9% of the bond principal. Using the architect's estimate of cost of $41 million, determine the effective interest rate.

10.5   Suppose in the preceding problem, that the bonds are financed by a bank that discounts the bond issue to $40 milion. What is the new effective interest rate?

---

[2](PWUS, i, 50) is the Present Worth Uniform Series Factor for an interest rate, i, over a period of 50 years.

[3](PWSP, i, 50) refers to the Present Worth Single Payment Factor.

Chapter 11

Equipment Ownership

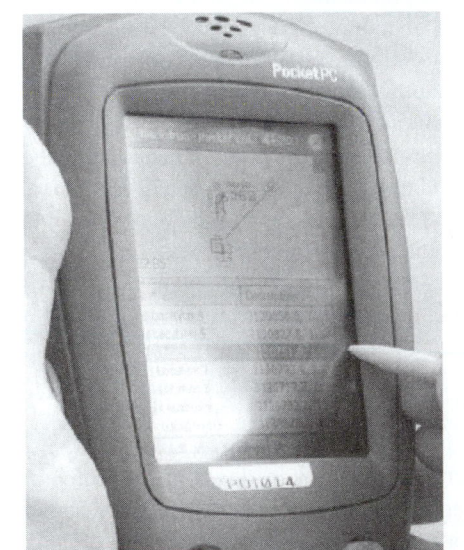

Hand held RFID device

RFID Application in Construction

The Need

For tracing and identifying construction equipment on construction sites, electronic identification tags are becoming widely used. With Radio Frequency Identification (RFID) technology, no line of sight or direct contact is required between the reader and the tag. Since RFID does not rely on optics, it is ideal for dirty, oily, wet, or harsh environments. RFID is an automatic identification technology, similar to bar code technology, with positive identification and automatic data transfer between a tagged object and a reader. Since the RFID tags are read by low wattage radio waves, instead of light waves (as with bar-codes), they will communicate through non-metallic materials such as paint, plastic, grease, and dirt, and are impervious to vibration, light, water, and heat up to 100°C in most cases.

The Technology

A RFID system consists of two major components (reader and the tag) which work together to provide the user with a non-contact solution to uniquely identify people, assets, and locations. The reader performs several functions, one of which is to produce a low-level radio frequency magnetic field. The RF magnetic field serves as a "carrier" of power from the reader to the passive (no battery required) RFID tag. When a tag is brought into the magnetic field produced by the reader, the recovered energy powers the integrated circuit in the tag and the memory contents are transmitted back to the reader. Once the reader has checked for errors and validated the received data, the data are decoded and restructured for transmission to a user in the format required by the host computer system. The RFID tags used are both readable and writable. This capability enables information to be written back to the tag for enhanced asset management. RFID tags do not require a line of sight for identification, and readability is not affected by bright lighting situations.

11.1   GENERAL

Equipment resources play a major role in any construction activity. Decisions regarding equipment type and combination can have a major impact on the profitability of a job. In this respect, the manager's goal is to select the equipment combination that yields the maximum production at the best or most reasonable price. Quite obviously, the manager must have a basic understanding of the costs associated with a particular piece of equipment. He must also be capable of calculating the rate of production of the piece or combination

169

of equipment. The cost and the rate of production combine to yield the cost per unit of production. For example, if it is estimated that the cost of a particular fleet of haulers and loaders is $500 per hour and the production rate is 750 cu yd/hr, the unit price can be easily calculated as $0.66 per cubic yard.

Construction equipment can be divided into two major categories. *Productive equipment* describes units that alone or in combination lead to an end product that is recognized as a unit for payment. *Support equipment* is required for operations related to the placement of construction such as movement of personnel and materials and activities that influence the placement environment. Typical production units are pavers, haulers, loaders, rollers, and entrenchers. Hoists, lighting sets, vibrators, scaffolds, and heaters represent typical classes of support equipment. In most cases, equipment units are involved either in handling construction materials at some point in the process of placing a definable piece of construction (e.g., crane lifting a boiler, pavers spreading concrete or asphalt into lifts on a base course) or in controlling the environment in which a piece of construction is realized (e.g., heaters controlling ambient temperature, prefabricated forms controlling the location of concrete in a frame or floor slab).

In heavy construction, large quantities of fluid or semifluid materials such as earth, concrete, and asphalt are handled and placed, leading to the use of machines. The equipment mix in such cases has a major impact on production, and the labor component controls production rates only in terms of the skill required to operate machines. Therefore, heavy construction operations are referred to as being equipment intensive. Heavy construction contractors normally have a considerable amount of money tied up in fixed equipment assets, since capitalizing a heavy construction firm is a relatively expensive operation.

Building and industrial construction require handwork on the part of skilled labor at the point of placement and are therefore normally not as equipment intensive.

Equipment is required to move materials and manpower to the point of installation and to support the assembly process. Emphasis is on hand tools; and, although heavy equipment pieces are important, the building and industrial contractors tend to have less of their capital tied up in equipment. Also because of the variability of equipment needs from project to project, the building contractor relies heavily on the renting of equipment. The heavy construction contractor, because of the repetitive use of many major equipment units, often finds it more cost effective to own this equipment.

11.2   EQUIPMENT OWNING AND OPERATING COSTS

The costs associated with construction equipment can be broken down into two major categories. Certain costs (e.g., depreciation, insurance, and interest charges) accrue whether the piece of equipment is in a productive state or not. These costs are fixed and directly related to the length of time the equipment is owned. Therefore, these costs are called *fixed*, or *ownership*, costs. The term *fixed* indicates that these costs are time dependent and can be calculated based on a fixed formula or a constant rate basis. On the other hand the operation of a machine leads to operating costs that occur only during the period of operation. Some of these costs accrue because of the consumption of supplies, such as tires, gas, and oil, and the widespread practice of including the operator's wages in the operating costs. Other costs occur as a result of the need to set aside moneys for both routine and unscheduled maintenance. Thus operating costs are *variable costs*.

The total of owning and operating costs for items of equipment such as tractors, shovels, scrapers, dozers, loaders, and backhoes is typically expressed on an hourly basis. These two categories of cost accrue in different ways. *Ownership costs* are usually arrived at by relating the estimated total service life in hours to the total of those costs. If the equipment is idle for some of those hours, the relevant costs would be taken up as part of general operating overhead; when the equipment is in use, the hourly costs are charged to the job or project.

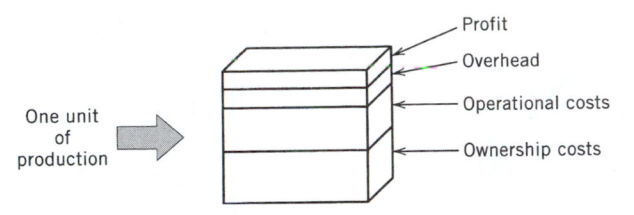

Figure 11.1 Cost components in a production unit.

Operating costs are variable in total amount, being a function of the number of operating hours, but these hourly costs are found to be relatively constant.

The *hourly charge* for a piece of equipment is made up of four elements. An allowance for estimated hourly overhead costs is added to the ownership and operating costs. The fourth element is an amount for income or profit. A schematic illustration of this breakdown of the hourly charge for a piece of equipment is shown in Figure 11.1.

Ownership costs are composed of two elements: first, an estimate for depreciation on the cost of using the equipment itself. Each piece of equipment represents an estimated number of hours of useful service life and the depreciable value, the major part of its original cost, is divided by the total hours to yield a charging rate for this element of equipment costs. The second component of ownership costs consists of estimates of allowance for interest, insurance, and taxes.

Operating costs cover a broader range of items, the principal elements being: fuel, oils and lubricants, hydraulics fluid, grease, filters, and other supplies; maintenance, general overhauls, and repairs; and parts replacement (cutting edges, blades, buckets), tire replacements, and the like. Also included here are the direct labor costs— the operator's wages—including all of the expense loadings for holidays, sick leave, and insurance.

To the direct operating costs just enumerated are added allowances for general overhead expenses and the indirect costs of supervisory labor. This total establishes the total hourly cost of owning and operating a unit of equipment. A percentage markup is added to provide for an income or profit element.

Some of these costs are incurred and paid for concurrently with the operation of the equipment, but the allowances or estimates included for items such as repairs and maintenance are provisions for costs that will have to be paid at some future time.

General administrative costs (e.g., overhead), including items such as telephones, stationery, postage, heat, light and power, and the costs of idle equipment in general are aggregated together as general overhead expense, an allowance that forms part of the hourly charging rate.

11.3 DEPRECIATION OF EQUIPMENT

The method by which depreciation is calculated for tax purposes must conform to standards established by the Internal Revenue Service (IRS). Federal law has introduced the use of fixed percentages as given in published tables to calculate the amount of depreciation for various classes of equipment and depreciable property. The tables have replaced accelerated methods referred to as the declining balance and sum-of-years-digits (SOYD) methods, which were used prior to 1981. Since the methods used under pre-1981 legislation are still relevant in understanding the tables presently used and are still required in some situations, they will be described briefly.

The four most commonly used methods of calculating depreciation on equipment prior to 1981 are:

1. Straight line
2. Declining balance

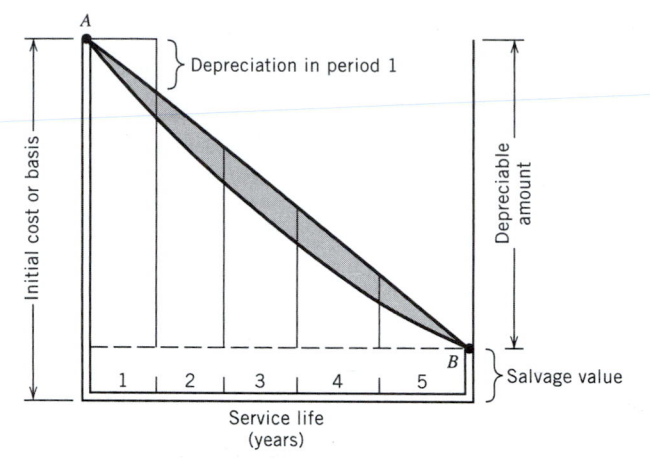

Figure 11.2 Factors in depreciation.

3. Sum of years digits

4. Production

Declining balance and sum of years are referred to as accelerated methods since they allow larger amounts of depreciation to be taken in the early years of the life of the asset. (Only the declining-balance method will be described in this chapter.) The contractor usually selects a method that offsets or reduces the reported profit for tax purposes as much as possible. In effect, for companies paying taxes at the corporate rate (assume 34%), each dollar of depreciation reduces the amount of tax paid by 34 cents (assuming that the depreciation does not reduce revenue below zero).

Most heavy construction contractors assume that each machine in the fleet is a small "profit center" and will attempt to apply any depreciation associated with a piece of equipment to offset the profit generated by that machine. The major factors to be considered in calculating the depreciation of an asset are shown in Figure 11.2. The three major factors form the three sides of the depreciation "box" that are linked by the method of depreciation selected. They are:

1. Initial cost or basis in dollars

2. Service life in years or hours

3. Salvage value in dollars

The amount that can be depreciated or claimed by way of a tax deduction is the difference between the initial net value of the asset and its residual or salvage value. This is referred to as the depreciable amount and establishes the maximum number of depreciation dollars available in the asset during its service life.

The declared initial cost of the asset must be acceptable in terms of the IRS definition of depreciable cost. For instance, suppose a $75,000 scraper is purchased. The tires on the scraper cost $15,000. These tires are considered a current period expense and therefore are not depreciable. That is, they are not part of the capital asset for purpose of depreciation. The tires are considered consumables and have a service life different from that of the asset. In this case, the initial value of the scraper for depreciation purposes is $60,000.

The initial depreciable cost or basis is often referred to as the net first cost. In addition to the purchase price minus major expenses, items such as tires, freight costs, and taxes are included in the net first cost and are part of the amount depreciable. If we have purchased a rubber-tired wheeled tractor, the net first cost for purposes of depreciation would be arrived

Table 11.1   Estimated Service Life Table (Caterpillar Tractor Co.)

Type of equipment	Excellent conditions: hours	Average conditions: hours	Severe conditions: hours
Track-type tractors Traxcavators Wheeled loaders Wheeled tractors Scrapers	12,000	10,000	8,000
Motor graders	15,000	12,000	10,000

To determine the cost per hour due to depreciation, the above information may be used as follows:

$$\text{Depreciation cost per hour} = \frac{\text{Purchase price} - \text{Tire value}}{\text{Estimated service life in hours}}$$

at as follows:

Purchase price	$84,000 (FOB[1] at factory)
Less tires	$14,000
	$80,000
Plus tax at 5%	$ 4,000
Plus freight	$ 2,800
Net first cost	$86,800

The depreciable basis for the calculation of depreciation allowances is this first cost of $86,800.

The concept of salvage value implies that there is some residual value in the piece of equipment (i.e., scrap value) at the end of its life. Unless this value exceeds 10% of the first cost of the equipment, this value is neglected and the entire first cost is considered to be available for depreciation. In the case cited, if the salvage value is less than $8,680, the entire first cost will be considered as depreciable and the piece of equipment will yield tax payment reductions in the amount of $29,512 (i.e., $86,800 × 0.34) across its service life.

The IRS publishes tables indicating the appropriate service life values. Most construction equipment items fall into the 3-, 5-, or 7-year service life categories. Manufacturers typically publish tables such as that shown in Table 11.1 indicating a variable service life based on operating conditions. Service life is defined by the IRS tables, and the only question has to do with the category or class of property to which an equipment type is to be assigned.

Given the present highly defined system of depreciation based on fixed tabular percentages, decisions regarding depreciation are simplified as to whether an accelerated or linearly prorated system of depreciation is to be used. To better understand the concepts behind the tables and the prorated system, two of the basic methods of calculating depreciation will be discussed in the following sections.

---

[1]FOB is discussed in Section 16.2 of Chapter 16. In this case it indicates the cost of the equipment at the factory prior to shipment.

11.4   STRAIGHT-LINE METHOD

An accountant (and the IRS) would describe the straight-line method of calculating allowable depreciation as being based on the assumption that the depreciation, or the loss in value through use, is uniform during the useful life of the property. In other words, the net first cost or other basis for the calculation, less the estimated salvage value, is deductible in equal annual amounts over the estimated useful life of the equipment. An engineer would call this a linear method. This simply means that the depreciable amount is linearly prorated or distributed over the service life of the asset. Let us assume that we have a piece of equipment that has an initial cost or base value of $16,000 and a salvage value of $1000. The service life is 5 years and the depreciable amount is $15,000 (initial cost minus salvage value). If we linearly distribute the $15,000 over the 5-year service life (i.e., take equal amounts each year), we are using the straight-line method of depreciation. The amount of depreciation claimed each year is $3000. This is illustrated in Figure 11.3.

The remaining value of the piece of equipment for depreciation purposes can be determined by consulting the stepwise curve of declining value. During the third year of the asset's service life, for example, the remaining base value, or book value, of the asset is $10,000. If we connect the points representing the book value at the end of each year (following subtraction of the depreciation), we have the "straight line."

The concept of the base value, or book value, has further tax implications. For instance, if we sell this asset in the third year for $13,000, we are receiving more from the buyer than the book value of $10,000. We are gaining $3000 more than the depreciated book value of the asset. The $3000 constitutes a *capital gain*. The reasoning is that we have claimed depreciation up to this point of $6000 and we have declared that as part of the cost of doing business. Now the market has allowed us to sell at $3000 over the previously declared value, demonstrating that the depreciation was actually less than was claimed. We have profited and, therefore, have received taxable income. Prior to the 1986 tax law, a capital gain was not taxed at the full rate but at approximately half of the tax rate for normal income. Presently, capital gains are taxed as normal income (i.e., 34%). Business entities have been pressing for the reinstatement of the alternate capital gains tax rate.

The base value for depreciation is affected if we modify substantially the piece of equipment. Assume in the above example, that in the third year we perform a major modification on the engine of the machine at a total cost of $3000. Since this is a capital improvement, the term *basis* is used to refer to the depreciation base. The modification increases the base

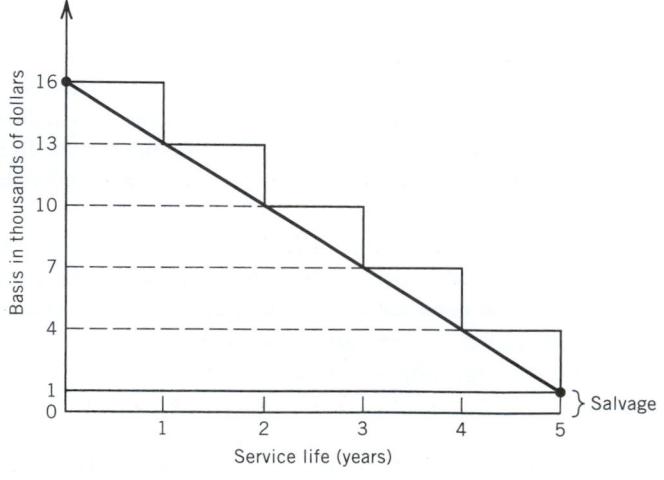

Figure 11.3 Straight-line depreciation.

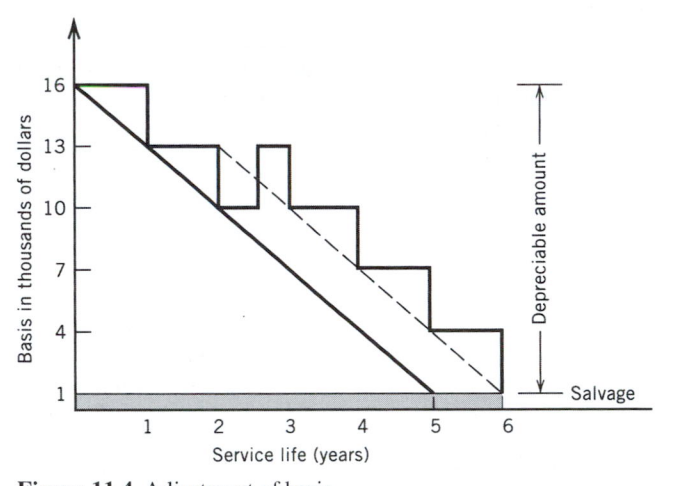

Figure 11.4 Adjustment of basis.

value of the unit by $3000 as shown in Figure 11.4. It also may extend the service life of the asset. Something similar occurs if we make some improvements to a building. The value is increased and this added value can be depreciated.

If we can depreciate real property, can we depreciate the house in which we live? Depreciation represents a cost of doing business. Since in most cases we do not "do business" in our own home, our home is not a depreciable asset. You can, however, think of some instances in which a person conducts some business at home. Special depreciation rules apply to that situation.

11.5   DECLINING BALANCE

One of the accelerated methods previously (prior to 1981) used is the declining balance method. When applied to new equipment with a useful life of at least 3 years, the effective rate at which the balance is reduced may be twice the straight-line rate. For this reason, the expression double-declining balance (DDB) is used when this IRS option is applied to new assets. For assets that are not purchased new but are secondhand, the optional rate is 150% of the straight-line rate. In this method, it is the rate that is important since it remains constant throughout the calculations. Formally stated, in the declining-balance method, the amount of depreciation claimed in the previous year is subtracted from the book value (base value) at the beginning of the previous year before computing the next year's depreciation. That is, a constant rate is applied to a balance which is declined each year by the amount claimed in the previous year. For new equipment the rate is calculated by dividing 200% by the number of service life years (SLY) (i.e., 200/SLY). For used equipment the rate is 150% divided by the service life years.

To illustrate, consider the $16,000 piece of equipment used in discussing the straight-line method. We will assume the piece is purchased new at this price. Since the service life of the unit is 5 years, the constant rate to be applied will be 200%/5 = 40%. The calculations for this example are summarized in Table 11.2.

A repetitive process of calculation can be detected. The constant rate of 40% (column 2) is applied to the book value at the end of the previous year (column 3) to obtain the depreciation (column 4). The reduced value of the property is column 3 minus column 4, as shown in column 5. The "*Book Value End of This Year*" for year N is the "*Book Value End of Previous* Year" for year N + 1. It follows that the value in column 3 for year 2 will be the same as the value in column 5 for year l.

Another interesting fact is noted. The amount of depreciation taken over the 5-year service life is less than the depreciable amount. The book value at the end of 5 years is

Table 11.2   Double-Declining-Balance Method

SLY	Rate applied to balance (%)	Book value end of previous year ($)	Depreciation for this year ($)	Book value end of this year ($)
1	40	16,000.00	6,400.00	9,600.00
2	40	9,600.00	3,840.00	5,760.00
3	40	5,760.00	2,304.00	3,456.00
4	40	3,456.00	1,382.40	2,073.60
5	40	2,073.60	829.44	1,244.16
			TOTAL $14,755.84	

$1244.16 and the salvage is $1000. Therefore, $244.16 has not been recovered. Typically, the method is changed to the straight-line approach in the fourth or fifth year to ensure closure on the salvage value. This underlines the fact that the only role played by the depreciable value in the declining-balance method is to set an upper limit on the amount of depreciation that can be recovered. That is, an asset may not be depreciated below a reasonable salvage value. A common mistake is to apply the rate to the *depreciable value* in the first year. The rate is always applied to the *total* remaining book value, which in this example during the first year is $16,000.

If the piece of equipment had been purchased used, for $16,000, the procedure would be the same but the rate would be reduced. In this case, the rate would be 150%/5 or 30%. The 150% calculations are summarized in Table 11.3. In this situation, since $1689.12 in unclaimed depreciation would remain at the end of year 5, the method could be changed to the straight-line approach in the fourth or fifth year with some advantage. A comparison of the double-declining-balance methods and the straight-line method is shown in Figure 11.5.

The proportionately higher rate of recovery in the early service life years is revealed by this figure. More depreciation is available in the first year using the double-declining-balance method ($6400) than in the first 2 years using the straight-line method ($6000). Equipment rental firms that intend to sell the equipment after the first 2 years of ownership are in a good position to capitalize on this feature of the accelerated methods. Of course, if they sell at a price well above the book value, they must consider the impact of the capital gains tax.

11.6   PRODUCTION METHOD

It was stated earlier that the contractor tries to claim depreciation on a given unit of equipment at the same time the equipment is generating profit in order to reduce the tax that might otherwise be payable. The production method allows this since the depreciation is taken based on the number of hours the unit was in production or use for a given year. The asset's

Table 11.3   150 Declining-Balance Method

SLY	Rate (%)	Book value end of previous year ($)	Depreciation for this year ($)	Book value end of this year ($)
1	30	16,000.00	4,800.00	11,200.00
2	30	11,200.00	3,360.00	7,840.00
3	30	7,840.00	2,352.00	5,488.00
4	30	5,488.00	1,646.40	3,841.60
5	30	3,841.60	1,152.48	2,689.12
			TOTAL $13,310.88	

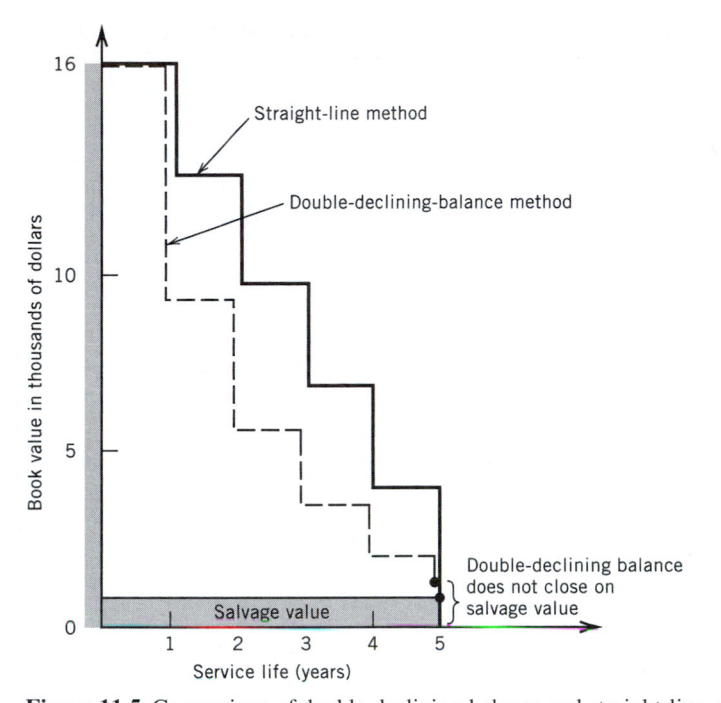

Figure 11.5 Comparison of double-declining-balance and straight-line methods.

cost is prorated and recovered on a per-unit-of-output basis. If the $16,000 equipment unit we have been discussing has a 10,000-hour operation time, the $15,000 depreciable amount is prorated over 10,000 hours of productive service life. This method is popular with smaller contractors since it is easy to calculate and ensures that the depreciation available from the asset will be recovered at the same time the unit is generating profit. A reasonable estimate of the total operating hours for a piece of equipment may be obtained by referring to the odometer on the unit together with the logbook or job cards.

In some cases, unless this method is used, the units may be depreciated during a period when they are not generating income and, consequently, the full benefit of the depreciation deduction may be lost. The objective of the contractor is to have a depreciation deduction available in years in which it can be more effectively applied to reduce taxable income. It may not be possible to defer depreciation and to take it in years in which it can be applied with more advantage. Therefore, the strategy should be to have it available in the years in which profits are likely to be high. The production method ensures that depreciation deduction is available when the machine is productive and theoretically profitable or income producing.

At the time of construction of the Alaska pipeline, contractors with contracts for the access road to parallel the pipeline purchased large equipment fleets in anticipation of project start-up. Then, environmental groups delayed the project several years during which time the contractors were forced to put their equipment fleets in storage. Since these units were not in use and were not productive and profitable during this delay, the contractors claimed no depreciation. Nevertheless, the production method allowed them to apply the depreciation at the proper time when the job mobilized and the units were put into production.

In some situations, the production method might be less desirable. If we own an entrenching machine (service life of 10,000 hours) but only operate it 500 hours per year. using the production method would stretch the period of recovery out over 20 years. If the machine is sold after 5 years, we would have claimed only one-quarter of the available depreciation. One advantage offsetting this apparent disadvantage is that we might have a smaller adjustment to make by way of capital gain on the sale. In such a case, clearly a method other than the production method would be more appropriate and a more balanced way of dealing with the matter.

11.7 DEPRECIATION BASED ON CURRENT LAW

For equipment placed in service prior to 1981, depreciation was calculated using methods such as the straight-line or declining-balance methods described above. For equipment placed in service during the period 1981 to 1986, contractors were required to depreciate using either the Accelerated Cost Recovery System (ACRS) method or the alternate ACRS system. which is equivalent to a straight line prorating of the cost across the life of the asset (e.g., the straight-line or production method). With the ACRS method, equipment is depreciated according to 3-year, 5-year, 10-year and 15-year property classes. For example, light trucks (less than 6.5 tons) are considered to be 3-year property. Most average-weight construction equipment is considered to be 5-year equipment. Some heavy construction equipment such as dredging barges are depreciated over a 10-year, life.

A set of tables defines accelerated depreciation amounts for equipment placed in service after 1986. In contrast to the ACRS method, these tables are referred to as the Modified ACRS or MACRS. Changes to the ACRS system include:

1. 7- and 20-year property life categories have been added.

2. The amount depreciated is calculated using prescribed depreciation methods for each class of equipment. For example, 3-, 5-, 7-, and 10-year property are depreciated using the 200% declining-balance method with a switch to the straight-line method at a time that maximizes the deduction. In addition, the "half-year convention" is used to calculate the first-year depreciation.

3. Certain assets have been reclassified to different property classes. In particular, cars and light general-purpose trucks have been reclassified as 5-year property. Most medium-weight, off-highway construction equipment is now considered 7-year property, while heavy equipment remains 10-year property.

The alternate MACRS method remains the straight-line method and can be used for depreciation as before. The accelerated MACRS values are given in Table 11.4. To better understand the basis of the values in the table, consider the following situation. A $100,000 piece of equipment is to be depreciated using the accelerated MACRS method. It is assumed that the equipment has a 5-year property life. The MACRS table is based on using the 200% declining-balance (DB) method. The rate of depreciation will be 40% (200% divided by 5 years). However, due to the half-year convention, only half of the 200-DB depreciation is taken in the first year. Therefore, the effective percent is 20 and $20,000 can be depreciated in the first year. The remaining value is $80,000. The second-year depreciation is 40% of $80.000, or $32,000. This amounts to 32% of the original $100,000 basis. The balance is now declined to $80,000 − $32,000, or $48,000. Again for the third-year depreciation the 200-DB method yields 40% of $48,000, or $19,200. The depreciation table for this equipment is as follows:

Year	Depreciation	Book Value
1	$20,000	$80,000
2	$32,000	$48,000
3	$19,200	$28,000
4	$11,520	$17,280
5	$11,520	$ 5,760
6	$ 5,760	$0

Table 11.4   MACRS Table for Accelerated Depreciation

Recovery year	Annual recovery (Percent of original depreciable basis)					
	3-year class (200% d.b.)	5-year class (200% d.b.)	7-year class (200% d.b.)	10-year class (200% d.b.)	15-year class (150% d.b.)	20-year class (150% d.b.)
1	33.00	20.00	14.28	10.00	5.00	3.75
2	45.00	32.00	24.49	18.00	9.50	7.22
3	15.00	19.20	17.49	14.40	8.55	6.68
4	7.00	11.52	12.49	11.52	7.69	6.18
5		11.52	8.93	9.22	6.93	5.71
6		5.76	8.93	7.37	6.23	5.28
7			8.93	6.55	5.90	4.89
8			4.46	6.55	5.90	4.52
9				6.55	5.90	4.46
10				6.55	5.90	4.46
11				3.29	5.90	4.46
12					5.90	4.46
13					5.90	4.46
14					5.90	4.46
15					5.90	4.46
16					3.00	4.46
17						4.46
18						4.46
19						4.46
20						4.46
21						2.25

It will be noted that although this is a 5-year property class equipment, it is depreciated out across a 6-year period. Also, following the third year, a switch from 200-DB to straight-line method is made to close on a residual or salvage value of zero. By dividing the depreciation amounts for each year by the original basis of $100.000, it will be seen that the percentage of depreciation in each year is the same as the values given in Table 11.4. To confirm that this method is used to determine the percentages in Table 11.4, the reader should try to calculate the annual depreciation amounts for a $100,000 piece of equipment with a 10-year service life.

11.8   DEPRECIATION VERSUS AMORTIZATION

Depreciation is a legitimate cost of business that recognizes the loss in value of equipment over time. As such, it is an expense and can be deducted from revenues, resulting in a lowering of taxes (e.g., 34 cents per dollar of depreciation as noted earlier). This yields the contractor a tax savings that can be used to replace the equipment. However, this savings would represent only 34% of the original value of the equipment. To provide for the replacement of the equipment at some point in the future, contractors charge the client an amount that provides a fund to purchase new equipment. This practice of charging clients an amount to be used to purchase replacement equipment is referred to as amortizing the equipment. This is a protocol throughout the industry and allows the contractor to accumulate (i.e., escrow) funds for renewing the equipment fleet over time.

For instance, the contractor will charge clients an annual amount of $20,000 for a $100,000 equipment with a service life of 5 years. This provides $100,000 at the end of 5 years to purchase replacement equipment. Of course, due to inflation and escalating prices, a replacement equipment may cost $120,000. This would indicate that the contractor should recover $24,000 per year to escrow the needed $120,000 for a new machine. Part of this amount will be recovered through depreciation due to reduced taxes: $34,000 will be available through the reduction of taxes. The contractor may consider this in calculating the amount of back charge to the client.

Since the amortization charge leads to larger revenue and the possibility of incurring income, the contractor may end up paying some tax on the amount charged to the client to amortize equipment. There is a complex interaction between depreciation and amortization, and this must be studied in the context of each equipment piece and the tax structure of each company.

11.9   INTEREST, INSURANCE, AND TAX (IIT) COSTS

In addition to the amortization/depreciation component, the ownership costs include a charge for other fixed costs that must be recovered by the equipment owner. Throughout the life of the unit, the owner must pay for insurance, applicable taxes, and either pay interest on the note used to purchase the equipment or lose interest on the money invested in equipment if the unit was paid for in cash. These costs are considered together as what can be called the IIT costs. Recovery of these charges is based on percentages developed from accounting records that indicate the proper levels that must be provided during the year to offset these costs. The percentages for each cost are applied to the average *annual value* of the machine to determine the amount to be recovered each hour or year with respect of these cost items.

The average annual value is defined as:

$$AAV = \frac{C(n + 1)}{2n}$$

where AAV is the average annual value, C is the initial new value of the asset, and n is the number of service life years. This expression assumes that the salvage value is zero. What the formula does is level the declining value of the asset over its service life so that a constant average value on an annual basis is achieved. This is indicated graphically in Figure 11.6.

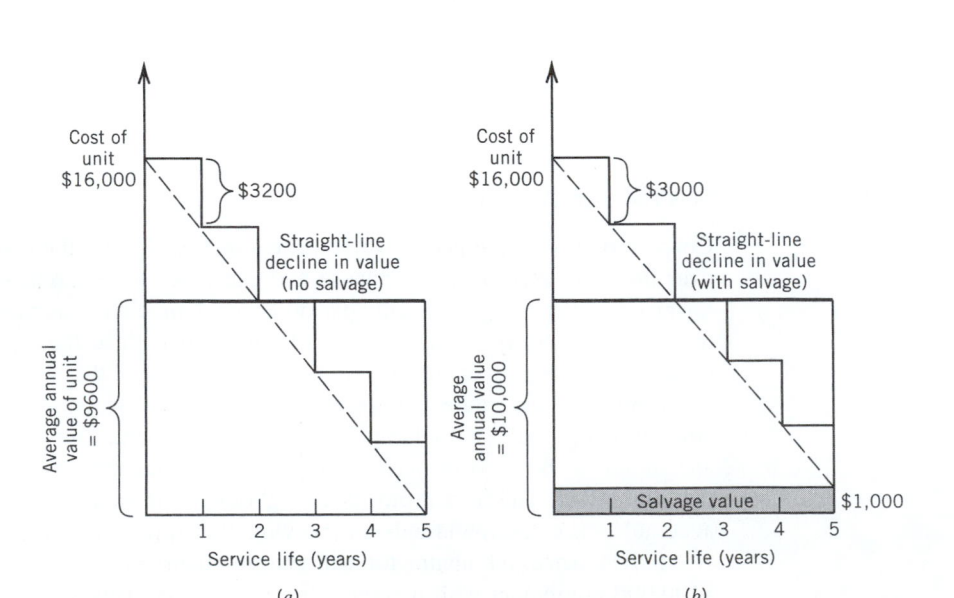

Figure 11.6 Interpretation of average annual value: (*a*) average annual value without salvage value included and (*b*) average annual value considering salvage value.

Applying this formula to a machine with initial capital value of $16,000 and a salvage value of $1000, the average annual value is calculated as:

$$\text{AAV} = \frac{\$16,000 \times (6)}{10} = \$9,600$$

The area under the rectangle in Figure 11.6 representing the average annual value equals the area under the plot representing the straight-line decline in value. Using this fact, the formula can be derived. If we consider the salvage value, the area under the stepped curve is increased by the area of the shaded segment in Figure 11.6b. Therefore, the AAV is increased somewhat. The appropriate expression for AAV including the salvage value is:

$$\text{AAV} = \frac{C(n+1) + S(n-1)}{2n}$$

For the $16,000 piece of equipment considered, this yields

$$\text{AAV} = \frac{\$16,000\,(6) + \$1,000\,(4)}{10} = \$10,000$$

Verification of this expression is left as an exercise for the reader.

Assume that the proper levels of the annual provision to cover IIT costs for the unit are as follows:

$$
\begin{aligned}
\text{Interest} &= 8\% \text{ of AAV} \\
\text{Insurance} &= 3\% \text{ of AAV} \\
\text{Taxes} &= \underline{2\% \text{ of AAV}} \\
\text{Total} &= 13\% \text{ of AAV}
\end{aligned}
$$

The amount to cover these ownership costs must be recovered on an hourly basis by backcharging the owner. Therefore, an estimate of the number of hours the unit will be operational each year must be made. Assume the number of hours of operation for the unit is 2000 hours/year. Then, the IIT cost per hour would be:

$$\text{IIT} = \frac{0.13(\text{AAV})}{2000} = \frac{0.13(9600)}{2000} = \$0.624 \quad \text{or} \quad \$0.62 \text{ per hour}$$

Manufacturers provide charts that simplify this calculation.

The interest component may be a nominal rate or an actual rate or, again, it may reflect some value of the cost of capital to the company. Some contractors also include here a charge for the protective housing or storage of the unit when it is not in use. These adjustments may raise the annual provision to cover IIT costs by from 1 to 5% with the following effect (in this case, each 1% charge may represent a 5 cents per hour increase in the charging rate).

Percentage of AAV(%)	General Provision for IIT Costs ($)	Hourly Rate (Base 2000 hours) ($)
13	1248	0.62
14	1344	0.67
15	1440	0.72
16	1536	0.77
17	1632	0.82
18	1728	0.87

Ultimately it is the competitive situation that sets practical limits to what may be recovered. That is why the tax limitation strategies discussed earlier are of such importance.

Figure 11.7 shows a chart for calculating the hourly cost of IIT. To use the chart the total percent of AAV and the estimated number of annual operating hours are required. Entering the y axis with the percent (use 13% from above) and reading down from the intersection of

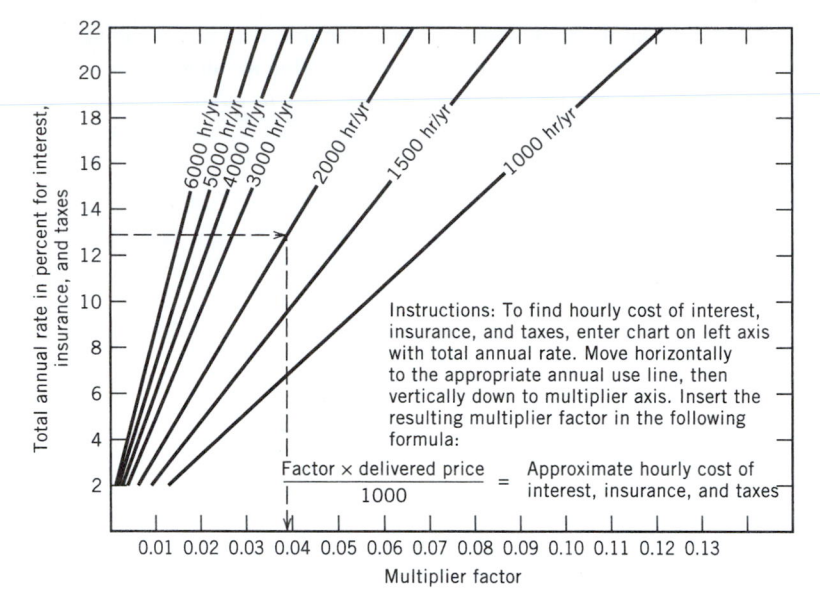

Figure 11.7 Guide for estimating hourly cost of interest, insurance, and taxes (Caterpillar Tractor Co.).

the 13% line with the 2000-hour slant line, the multiplier factor is 0.039. The hourly charge for IIT is calculated as:

$$\text{IIT/hour} = \frac{\text{factor} \times \text{delivery price}}{1000}$$

In the example discussed,

$$\text{IIT/hour} = \frac{0.039 \times 16{,}000}{1000} = \$0.624 \quad \text{or} \quad \$0.62 \text{ per hour}$$

If the amortization/depreciation costs using the straight-line method for the $16,000 unit in the example is $1.50 per hour, the owner must recover $1.50 plus $0.62 or $2.12 per hour for fixed costs.

11.10 OPERATING COSTS

The major components contributing to the operating or variable costs are fuel, oil, grease (FOG), tire replacement (on rubber-wheeled vehicles), and normal repairs. Normally, historical records (purchase vouchers, etc.) are available that help in establishing the rate of use of consumables such as fuel, oil, and tires. Maintenance records indicate the frequency of repair. The function that best represents the repair costs to be anticipated on a unit starts low and increases over the life of the equipment. Since repairs come in discrete amounts, the function has a stepwise appearance (see Fig. 11.8).

The following guidelines for establishing the amount to set aside for repairs are taken from Caterpillar Tractor material.

Guide for Estimating Hourly Repair Reserve. (See Caterpillar Performance Handbook)

To estimate hourly repair costs, select the appropriate multiplier factor from the table below and apply it in the following formula:

$$\frac{\text{Repair factor} \times (\text{delivered price} - \text{tires})}{1000} = \text{estimated hourly repair reserve}$$

	Operating Conditions		
	Excellent	Average	Severe
Track-type tractors	0.07	0.09	0.13
Wheel tractor scrapers	0.07	0.09	0.13
Off-highway trucks	0.06	0.08	0.11
Wheel-type tractors	0.04	0.06	0.09
Track-type loaders	0.07	0.09	0.13
Wheel loaders	0.04	0.06	0.09
Motor graders	0.03	0.05	0.07

The cost of tires on rubber-wheeled vehicles is prorated over a service life expressed in years or hours. Therefore, if a set of tires has an initial cost of $15.000 and a service life of 5000 hours, the hourly cost of tires set aside for replacements is:

$$\text{Hourly cost of tires} = \frac{\$15,000}{5000} = \$3.00$$

11.11 OVERHEAD AND MARKUP

In addition to the direct costs of ownership and operation, general overhead costs must be considered in recovering costs associated with equipment ownership and operation. Overhead charges include items such as the costs of operating the maintenance force and facility including: (1) wages of the mechanics and supervisory personnel, (2) clerical and records support, and (3) rental or amortization of the maintenance facility (i.e., maintenance bays, lifts, machinery, and instruments). The industry practice is to prorate the total charge to each unit in the equipment fleet based on the number of hours it operates as a fraction of the total number of hours logged by the fleet. For instance, if the total number of hours logged by all units in the fleet was 20,000 and a particular unit operated 500 hours, its proportion of the total overhead would be 500/20,000 × 100, or 2.5%. If the total cost of overhead for the year is $100,000, the unit above must recover $2500 in backcharge to the client to cover its portion of the overhead. Overhead rates are updated annually from

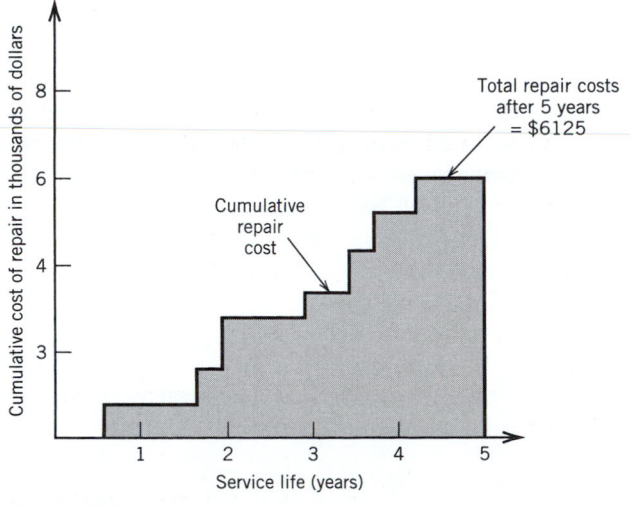

Figure 11.8 Repair cost profile.

operating records to ensure adequate coverage. If overhead costs overrun projections, the coverage will be inadequate and the overrun will reduce profits.

The last component of the total charge associated with a unit of production is the profit expressed as a percentage of total hourly operating costs, which, in turn, may be expressed in cubic yards of material moved or in some other bid-relevant measure. The amount of profit per cubic yard, square foot, or linear foot is a judgment that contractors must make based on their desire to win the contract and the nature of the competition. In a "tight" market where competition is strong, the allowable margin of profit that still allows the bidder to be competitive may be only 1 or 2%. In a "fat" market, where a lot of jobs are available, the demand is greater and the client is ready to pay a higher markup to get the work under way. Competition is bidding higher profit so the amount of profit can be adjusted upward.

Bidding strategy will include attention to the concept of marginal costs, which may permit the acceptance of jobs yielding less than the desired rate of return. In general, bidding, based on margins as low as 1 or 2% is uncomfortably close to what one might call the disaster area; the area of operating *losses*.

REVIEW QUESTIONS AND EXERCISES

11.1   What are the major cost components that must be considered when pricing out a piece of equipment? How can a contractor manipulate amortization for a piece of equipment in order to increase or reduce direct costs charged per unit of production? Why are tires on a rubber-tired vehicle not considered for depreciation?

11.2   You have just bought a new pusher dozer for your equipment fleet. Its cost is $100,000. It has an estimated service life of 4 years. Its salvage value is $12,000.

(a)   Calculate the depreciation for the first and second year using the straight-line and double-declining methods.

(b)   The tax, interest, and insurance components of ownership cost based on average annual value are:

Tax: 2%
Insurance: 2%
Interest: 7%

What cost per hour of operation would you charge to cover interest, tax, and insurance?

11.3   You have just bought a used track-type tractor to add to your production fleet. The initial capitalized value of the tractor is $110,000. The estimated service life remaining on the tractor is 10,000 hours, and the anticipated operating conditions across the remainder of its life are normal. The salvage value of the tractor is $12,000. The tractor was purchased on July 1, 1997.

(a)   What amount of depreciation will you claim for each calendar year between 2007 and 2010?

(b)   What percent of the total depreciable amount is taken in the first year?

(c)   The tax, interest, and insurance components of ownership cost based on average annual value are:

Tax: 3%
Insurance: 2%
Interest: 8%

What cost per hour of operation would you charge to cover interest, tax, and insurance?

(d)   If the total average operating cost for the tractor is $23.50 per hour and the amount of overhead cost prorated to this tractor for the year is $4000, what would be your total hourly cost for the operation of the tractor (during the first year of its service life)?

11.4   Verify the 5- and 7-year property class percentage given in Table 12.4 by applying the 200% d.b. approach to a piece of equipment with a nominal value of $1000. For the 7-year property class, in what year is the switch from 200% d.b. to straight line made based on the percentages given in the table?

Chapter 12

Equipment Productivity

Grader with TopCon 3D-MC

Computer and Total-station

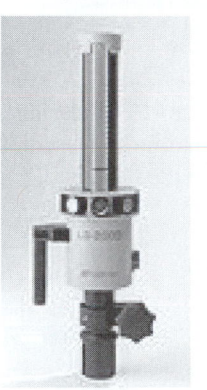

Receiver

Laser Based Machine Control

The Need

Construction equipment using laser control technology can achieve higher levels of productivity. The guiding of road construction equipment in curving contours requires references such as hubs, staking, or elevated string lines. These benchmarks limit productivity, because their installation is slow, subject to human errors, and requires skilled operators to accurately steer the machine using rudimentary control methods. Attempts to guide equipment in curves using radio communication have been tried but this solution is still slow and unreliable.

The Technology

New systems use three modules to control the piece of equipment:

- Survey plans are uploaded in a total station using a computer notebook. The total station converts the digital information into an infrared laser beam.

- A receiver, mounted on the blade of the equipment, intercepts the laser beam emitted by the total station and continuously determines (20 upgrades per second) the blade's current position and grade with respect the theoretical ones defined by the designer plans.

- The interface between the positioning information and the actual steering of the equipment is performed through the use of a control system device, which converts the digital data into machine hydraulic valve pulses.

The main benefit of these systems is the obvious gain of productivity generated by this innovation. According to some research carried out by manufacturers of such guide systems, the laser devices can triple the productivity of equipment on highway projects as well as drastically increase their levels of precision and performance. Laser based systems represent the next generation of equipment controlling devices bringing an alternative to the existing slower and unreliable manual systems.

12.1 PRODUCTIVITY CONCEPTS

Now that a basis for charging each unit of production has been established, the rate of production, or the number of productive units, that can be generated per hour, per day, or other period of time must be considered. Our discussions here will be limited primarily to heavy construction units such as haulers, graders, and dozers. The concepts developed, however, are applicable to all types of construction equipment performing basically repetitive or cyclic operations. The cycle of an equipment piece is the sequence of tasks, which is repeated to produce a unit of output (e.g., a cubic yard, a trip load, etc.).

There are two characteristics of the machine and the cycle that dictate the rate of output. The first of these is the cyclic capacity of the machine or equipment, which establishes the number of units produced per cycle. The second is the cyclic rate or speed of an equipment piece. A truck, for instance, with a capacity of 16 cu yd, can be viewed as producing 16 yd each time it hauls. The question of capacity is a function of the size of the machine, the state of the material that is to be processed, and the unit to be used in measurement. A hauler such as a scraper pan usually has a rated capacity, "struck," versus its "heaped" capacity. The bowl of the scraper can be filled level (struck), yielding one capacity, or can be filled above the top to a heaped capacity. In both cases, the earth hauled tends to take on air voids and bulks, yielding a different weight per unit volume than it had in the ground when excavated (i.e., its in situ location). The material has a different weight-to-volume ratio when it is placed in its construction location (e.g., a road fill, an airport runway) and is compacted to its final density. This leads to three types of measure: (1) bank cubic yards [cu yd (bank)] (in situ volume), (2) loose cubic yards [cu yd (loose)], and (3) compacted cubic yards. Payment in the contract is usually based on the placed earth construction, so that the "pay" unit is the final compacted cubic yard. The relationship between these three measures is shown in Figure 12.1.

The relationship between the bulk or loose volume and the bank volume is defined by the percent swell. In Figure 12.1, the percent swell is 30%. Percent swell is given as:

$$\text{Percent swell} = \left[\left(\frac{1}{\text{load factor}} \right) - 1 \right] \times 100$$

where

$$\text{Load factor} = \frac{\text{pounds per cubic yard} - \text{loose}}{\text{pounds per cubic yard} - \text{bank}}$$

Tables such as Table 12.1 give the load factor for various types of materials indicating their propensity for taking on air voids in the loose state. The higher the load factor, the smaller tendency the material has to "bulk up." Therefore, with a high load factor the loose volume and the in situ volume tend to be closer to one another. Each material has its own characteristic load factor. In the example above, the material has a load factor of 0.77.

$$\text{Percent swell} = \left[\left(\frac{1}{0.77} \right) - 1 \right] \times 100 = 30\%$$

Therefore, we would expect 10 yd of bank material to expand to 13 yd during transport. The shrinkage factor relates the volume of the compacted material to the volume of the bank material. In the example, the shrinkage factor is 10% since the bank cubic yard is reduced by 10% in volume in the compacted state.

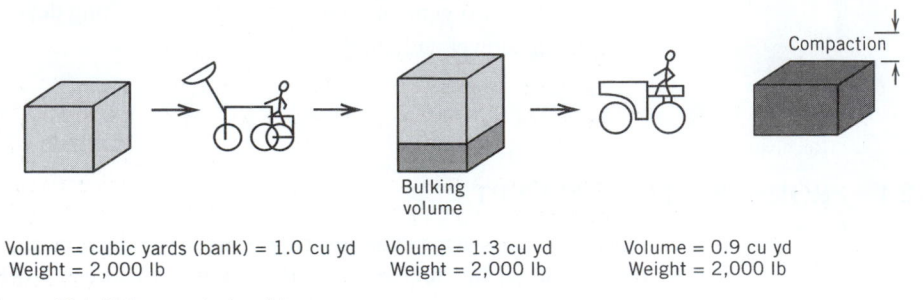

Volume = cubic yards (bank) = 1.0 cu yd   Volume = 1.3 cu yd   Volume = 0.9 cu yd
Weight = 2,000 lb   Weight = 2,000 lb   Weight = 2,000 lb

Figure 12.1 Volume relationships.

Table 12.1 Approximate Material Characteristics[a]

Material	Pounds per Cubic Yard—Bank	Percent of Swell	Load Factor	Pounds per Cubic Yard—Loose
Clay, natural bed	2,960	40	0.72	2,130
Clay and gravel				
Dry	2,960	40	0.72	2,130
Wet	2,620	40	0.72	2,220
Clay, natural bed				
Anthracite	2,700	35	0.74	2,000
Bituminous	2,160	35	0.74	1,600
Earth, loam				
Dry	2,620	25	0.80	2,100
Wet	3,380	25	0.80	2,700
Gravel, $\frac{1}{4}$–2 in.				
Dry	3,180	12	0.89	2,840
Wet	3,790	12	0.89	3,380
Gypsum	4,720	74	0.57	2,700
Iron ore				
Magnetite	5,520	33	0.75	4,680
Pyrite	5,120	33	0.75	4,340
Hematite	4,900	33	0.75	4,150
Limestone	4,400	67	0.60	2,620
Sand				
Dry, loose	2,690	12	0.89	2,400
Wet, packed	3,490	12	0.89	3,120
Sandstone	4,300	54	0.65	2,550
Trap rock	4,420	65	0.61	2,590

[a]The weight and load factor will vary with factors such as grain size, moisture content, and degree of compaction. A test must be made to determine an *exact* material characteristic.

In order to understand the importance of capacity, consider the following situation. A front-end loader has an output of 200 bank cu yd of common earth per hour. It loads a fleet of four trucks (capacity 18 loose cu yd each), which haul the earth to a fill where it is compacted with a shrinkage factor of 10%. Each truck has a total cycle time of 15 min, assuming it does not have to wait in line to be loaded. The earth has a percent swell of 20%. The job requires a volume of 18,000 compacted cu yd. How many hours will be required to excavate and haul the material to the fill? Two types of productive machines are involved: four trucks and a front-end loader. We must see which unit or set of units is most productive. Reference all calculations to the loose cubic yard production per hour. Then the loader productivity (given 20% swell) is

200 cu yd (bank)/hr = 1.2(200) or 240 cu yd (loose)/hr

The truck fleet production is

$$4 \text{ trucks} \times \frac{60 \text{ min/hr}}{15 \text{ min/cycle}} \times 18 \text{ cu yd (loose) truck}$$
$$= 72 \text{ cu yd (loose)} \times 4 \text{ cycle/hr}$$
$$= 288 \text{ cu yd (loose)/hr for 4 trucks}$$

Because the loader production is lower, it constrains the system to a maximum output of 240 cu yd (loose)/hr. We must now determine how many loose cubic yards are represented by 18,000 cu yd (compacted).

$$18,000 \text{ cu yd compacted} = \frac{18,000}{0.9} \quad \text{or} \quad 20,000 \text{ cu yd (bank)}$$
$$20,000 \text{ cu yd (bank)} = 24,000 \text{ cu yd (loose) required}$$

Therefore, the number of hours required is

$$\text{Hours} = \frac{24,000 \text{ cu yd (loose)}}{240 \text{ cu yd (loose)/hr}} = 100$$

This problem illustrates the interplay between volumes and the fact that machines that interact with other machine cycles may be constrained or constraining.

Table 12.2   Typical Rolling Resistance Factors (Caterpillar Tractor Co.)

A hard, smooth, stabilized roadway without penetration under load (concrete or blacktop)	40 lb/ton
A firm, smooth-rolling roadway flexing slightly under load (macadam or gravel-topped road)	65 lb/ton
Snow-packed	50 lb/ton
Loose	90 lb/ton
A rutted dirt roadway, flexing considerably under load; little maintenance, no water (hard clay road, 1 in. or more tire penetration)	100 lb/ton
Rutted dirt roadway, no stabilization, somewhat soft under travel (4–6 in. tire penetration)	150 lb/ton
Soft, muddy, rutted roadway, or in sand	200–400 lb/ton

12.2   CYCLE TIME AND POWER REQUIREMENTS

The second factor affecting the rate of output of a machine or machine combination is the time required to complete a cycle, which determines the cyclic rate. This is a function of the speed of the machine and, in the case of heavy equipment, is governed by (1) the power required, (2) the power available, and (3) the usable portion of the power available that can be developed to propel the equipment unit.

The power required is related to the rolling resistance inherent in the machine due to internal friction and friction developed between the wheels or tracks and the traveled surface. The power required is also a function of the grade resistance inherent in the slope of the traveled way. Rolling resistance in tracked vehicles is considered to be zero, since the track acts as its own roadbed, being laid in place as the unit advances. The friction between track and support idlers is too small to be considered. Rolling resistance for rubber-wheeled vehicles is a function of the road surface and the total weight on the wheels. Tables such as Table 12.2 are available in equipment handbooks giving the rolling resistance in pounds per tons of weight. Figure 12.2 indicates visually the factors influencing rolling resistance and therefore contributing to the required power that must be developed to move the machine.

If tables are not available, a rule of thumb can be used. The rule states that the rolling resistance (RR) is approximately 40 lb/ton plus 30 lb/ton for each inch of penetration of the surface under wheeled traffic. If the estimated deflection is 2 in. and the weight on the wheels of a hauler is 70 tons, we can calculate the approximate rolling resistance as:

$$RR = [40 + 2(30)] \text{ lb/ton} \times 70 \text{ tons} = 7,000 \text{ lb}$$

The second factor involved in establishing the power required is the grade resistance. In some cases, the haul road across which a hauler must operate will be level and, therefore, the slope of the road will not be a consideration. In most cases, however, slopes (both uphill

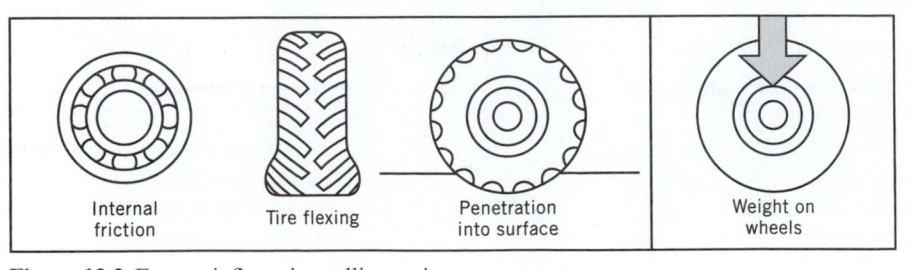

Figure 12.2 Factors influencing rolling resistance.

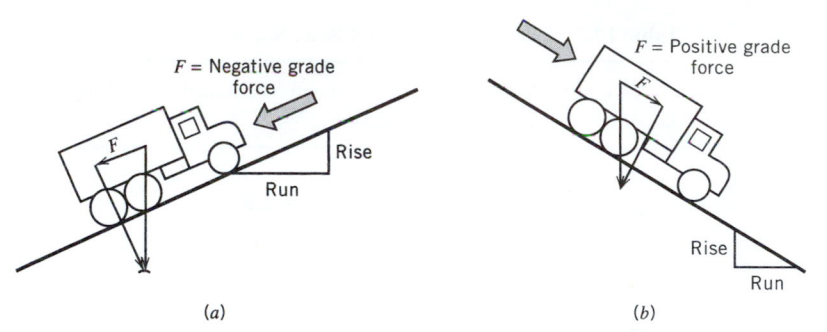

Figure 12.3 Grade resistance: (*a*) negative (resisting) force and (*b*) positive (aiding) force.

and downhill) will be encountered and lead to higher or reduced power requirements based on whether gravity is aiding or resisting movement (see Fig. 12.3).

The percent grade is calculated by the ratio of rise over run, as depicted in Figure 12.3. If, for instance, a slope rises 6 ft in 100 ft of horizontal distance, the percent grade is 6. Similarly, a slope that increases 1.5 ft in 25 ft also has a percent grade of 6. Percent grade is used to calculate the grade resistance (GR) using the following relationship:

$$GR = \text{percent grade} \times 20 \text{ lb/ton/\% grade} \times \text{weight on wheels (tons)}$$

If the 70-ton piece of equipment referred to previously is ascending a 6% grade, the grade resistance is

$$GR = 6\% \text{ grade} \times 20 \text{ lb/ton/\% grade} \times 70 \text{ tons} = 8{,}400 \text{ lb}$$

Assuming the rolling resistance calculated above holds for the road surface of the slope and assuming the equipment is wheeled, the total power required to climb the slope will be

$$\text{Power required} = \text{RR} + \text{GR} = 7{,}000 \text{ lb} + 8{,}400 \text{ lb} = 15{,}400 \text{ lb}$$

If the slope is downward, an aiding force is developed, and the total power required becomes

$$\text{Power required} = \text{RR} - \text{GR} = 7{,}000 \text{ lb} - 8{,}400 \text{ lb} = -1{,}400 \text{ lb}$$

The sign of the grade resistance becomes negative since it is now aiding and helping to overcome the rolling resistance. Since a negative rolling resistance has no meaning, the power required on a downward 6% grade is zero. In fact, the 1,400 lb represents a downhill thrust that will accelerate the machine, and lead to a braking requirement.

Traveled ways or haul roads normally consist of a combination of uphill, downhill, and level sections. Therefore, the power requirement varies and must be calculated for each section. Knowing the power required for each haul road section, a gear range that will

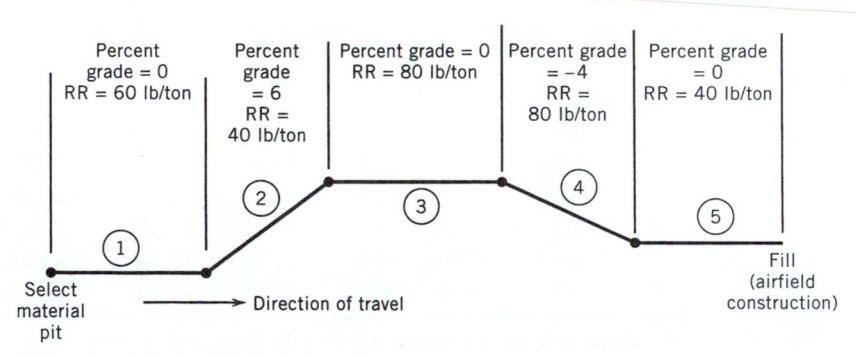

Figure 12.4 Typical haul road profile.

Table 12.3   Calculations for Haul Road Sections[a]

Section	Percent Grade (%)	Grade Resistance (lb)	Rolling Resistance (lb)	Power Required (lb)
1	0	0	4,200	4,200
2	6	8,400	2,800	11,200
3	0	0	5,600	5,600
4	−4	−5,600	5,600	0
5	0	0	2,800	2,800

[a] All calculations assume travel from pit to fill.

provide the required power can be selected. The gear range allows a speed to be developed and, given the speed, we can develop the time required to transit each section and the total time required for a cycle.

Consider the haul road profile shown in Figure 12.4 with rolling resistance and percent grade values as shown. The calculation of power required for each section of the road based on a 70-ton machine is shown in Table 12.3. Given the power requirements, the next section indicates how a gear range is selected. As noted above, this allows determination of the speed across each section and the time required.

12.3   POWER AVAILABLE

The power available is controlled by the engine size of the equipment and the drive train, which allows transfer of power to the driving wheels or power take-off point. The amount of power transferred is a function of the gear being used. Most automobile drivers realize that lower gears transfer more power to overcome hills and rough surfaces. Lower gears sacrifice speed in order to provide more power. Higher gears deliver less power, but allow higher speeds. Manufacturers publish figures regarding the power available in each gear for individual equipment pieces in equipment handbooks that are updated annually. This information can be presented in a tabular format such as that shown in Table 12.4 or in graphical format such as the nomograph shown in Figure 12.5.

For tracked vehicles, the power available is quoted in drawbar pull. This is the force that can be delivered at the pulling point (i.e., pulling hitch) in a given gear for a given tractor type. Power available for a wheeled vehicle is stated in pounds of rimpull. This is the force that can be developed by the wheel at its point of contact with the road surface. Manufacturers also provide information regarding rated power and maximum power. Rated power is the

Table 12.4   Speed and Drawbar Pull (270 hp) (Track-Type Tractor)

Gear	Forward mph	Forward km/h	Reverse mph	Reverse km/h	At Rated rpm lb	At Rated rpm kg	Maximum at Lug lb	Maximum at Lug kg
1	1.6	(2–6)	1.6	(2.6)	52,410	(23,790)	63,860	(28,990)
2	2.1	(3–4)	2.1	(3.4)	39,130	(17,760)	47,930	(21,760)
3	2.9	(4.7)	2.9	(4.7)	26,870	(12,200)	33,210	(15,080)
4	3.7	(6.0)	3.8	(6.1)	19,490	(8,850)	24,360	(11,060)
5	4.9	(7.9)	4.9	(7.9)	13,840	(6,280)	17,580	(7,980)
6	6.7	(10.8)	6.8	(10.9)	8,660	(3,930)	11,360	(5,160)

The Drawbar Pull Forward[a] spans the last four columns (At Rated rpm and Maximum at Lug).

[a] Usable pull will depend on traction and weight of equipped tractor.

level of power that is developed in a given gear under normal load and over extended work periods. It is the base or reference level of power that is available for continuous operation. The maximum power is just what it indicates. It is the peak power that can be developed in a gear for a short period of time to meet extraordinary power requirements. For instance, if a bulldozer is used to pull a truck out of a ditch, a quick surge of power would be used to dislodge the truck. This short-term peak power could be developed in a gear using the maximum power available.

Most calculations are carried out using rated power. If, for example, the power required for a particular haul road section is 25,000 lb based on the procedures described in Section 12.2, the proper gear for the 270-hp track-type tractor is third gear. This is determined by entering Table 12.4 and comparing power required with rated power. Consider the example shown in the shaded area below.

Nomographs are designed to allow quick determination of required gear ranges as well as the maximum speed attainable in each gear. The nomograph shown in Figure 12.5 is for a 35-ton, off-highway truck. To illustrate the use of this figure, consider the following problem. On a particular road construction job, the operator has to choose between two available routes linking the select material pit with a road site fill. One route is 4.6 miles (one-way) on a firm, smooth road with a RR = 50 lb/ton. The other route is 2.8 miles (one-way) on a rutted dirt road with RR = 90 lb/ton. The haul road profile in both cases is level so that grade resistance is not a factor. Using the nomograph of Figure 12.5, we are to determine the pounds pull to overcome rolling resistance for a loaded 35-ton, off-highway truck. The same chart allows determination of the maximum speed.

In order to use the chart, consider the information in the chart regarding gross weight. The weight in pounds ranges from 0 to 280,000 (140 tons). The weights of the truck empty and with a 70,000-lb load (i.e., 35-ton capacity) are indicated by vertical dashed lines intersecting the gross weight axis (top of chart) at approximately 56,000 lb (empty) and 126,000 lb (loaded). For this problem, the loaded line is relevant since the truck hauls loads from pit to fill.

Next consider the slant lines sloping from lower left to upper right. These lines indicate the total resistance (i.e., RR + GR) in increments from 0 to 25% grade. In this problem, there is no grade resistance. In dealing with rolling resistance, it is common to convert it to an equivalent percent grade. Then, the total resistance can be stated in percent grade by

The sum of the rolling resistance and grade resistance that a particular wheel-type tractor and scraper must overcome on a specific job has been estimated to be 10,000 lb. If the "pounds pull-speed" combinations listed below are for this particular machine, what is the maximum reasonable speed of the unit?

Third gear would be selected since the rated rimpull is 12,190 lb. (If the total power required had been in excess of 12,190 lb. we would select second gear because you recall that rated pounds pull should always be used for gear selection. The reserve rimpull of the maximum rating is always available-at reduced speed-to pull the unit out of small holes or bad spots.)

Gear	Speed	Rated	Maximum
		Pounds Rimpull	
1	2.6	38,670	49,100
2	5.0	20,000	25,390
3	8.1	12,190	15,465
4	13.8	7,185	9,115
5	22.6	4,375	5,550

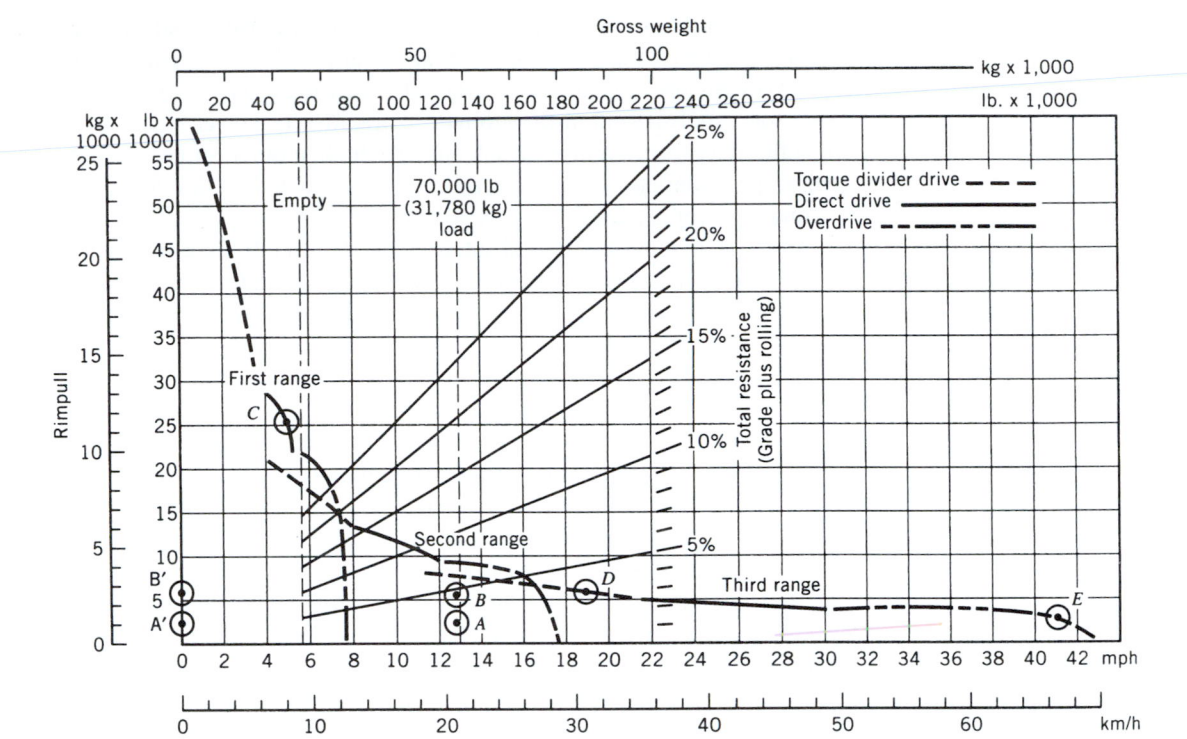

Figure 12.5 Gear requirements chart-35-ton, off-highway truck (Caterpillar Tractor Co.).

adding the equivalent percent grade for rolling resistance to the slope percent grade. To convert rolling resistance to equivalent percent grade, the following expression is used:

$$\text{Equivalent precent grade} = \frac{\text{RR}}{20 \text{ lb/ton/\% grade}}$$

For the rolling resistance values given in the problem, the equivalent percent grades become as follows:

Route	Distance	RR	Equivalent Percent Grade
1	4.6 miles	50 1b/ton	2.5
2	2.8 miles	90 1b/ton	4.5

In order to determine the required pounds pull, the intersection of the slant line representing the equivalent percent grade with the load vertical line is located. This intersection for route 1 is designated point *A* in Figure 12.5. The corresponding point for route two is labeled *B*.

The pounds required value is found using these points by reading horizontally across to the *y* axis, which gives the rimpull in pounds. For route 1, the approximate power requirement is 2,500 lb. The requirement for route 2 is 5,500 lb. Points *A* and *B* are also used to determine the maximum speed along each route.

Consider the curves descending from the upper left-hand corner of the chart to the lower right side. As labeled, these curves indicate the deliverable power available in first, second, and third ranges as well as the speed that can be developed. At 25,000 lb of rimpull, for example, on the *y* axis, reading horizontally to the right the only range delivering this much power is first range (see point *C*). Reading vertically down to the x axis, the speed that can be achieved at this power level is approximately 5 mi/hr.

Proceeding in a similar manner, it can be determined that two ranges, second and third, will provide the power necessary for route 2 (i.e., 5,500 lb). Reading horizontally to the right from point B, the maximum speed is developed in third range at point D. Referencing this point to the x axis, the maximum speed on route 2 is found to be approximately 19 mi/hr. Route 1 requires considerably less power. Again reading to the right, this time from point A, the third range provides a maximum speed of approximately 41 mi/hr (see point E).

Now, having established the maximum speeds along each route and knowing the distances involved, it should be simple to determine the travel times required. Knowledge of the speeds, however, is not sufficient to determine the travel times since the requirements to accelerate and decelerate lower the effective speed between pit and fill. Knowing the mass of the truck and the horsepower of the engine, the classic equation, force = mass × acceleration (F = ma), would allow determination of the time required to accelerate to and decelerate from maximum speed. This is not necessary, however, since the equipment handbooks provide time charts that allow direct readout of the travel time for a route and piece of equipment, given the equivalent percent grade and the distance. These charts for loaded and empty 35-ton trucks are given in Figure 12.6. Inspection of the chart for the

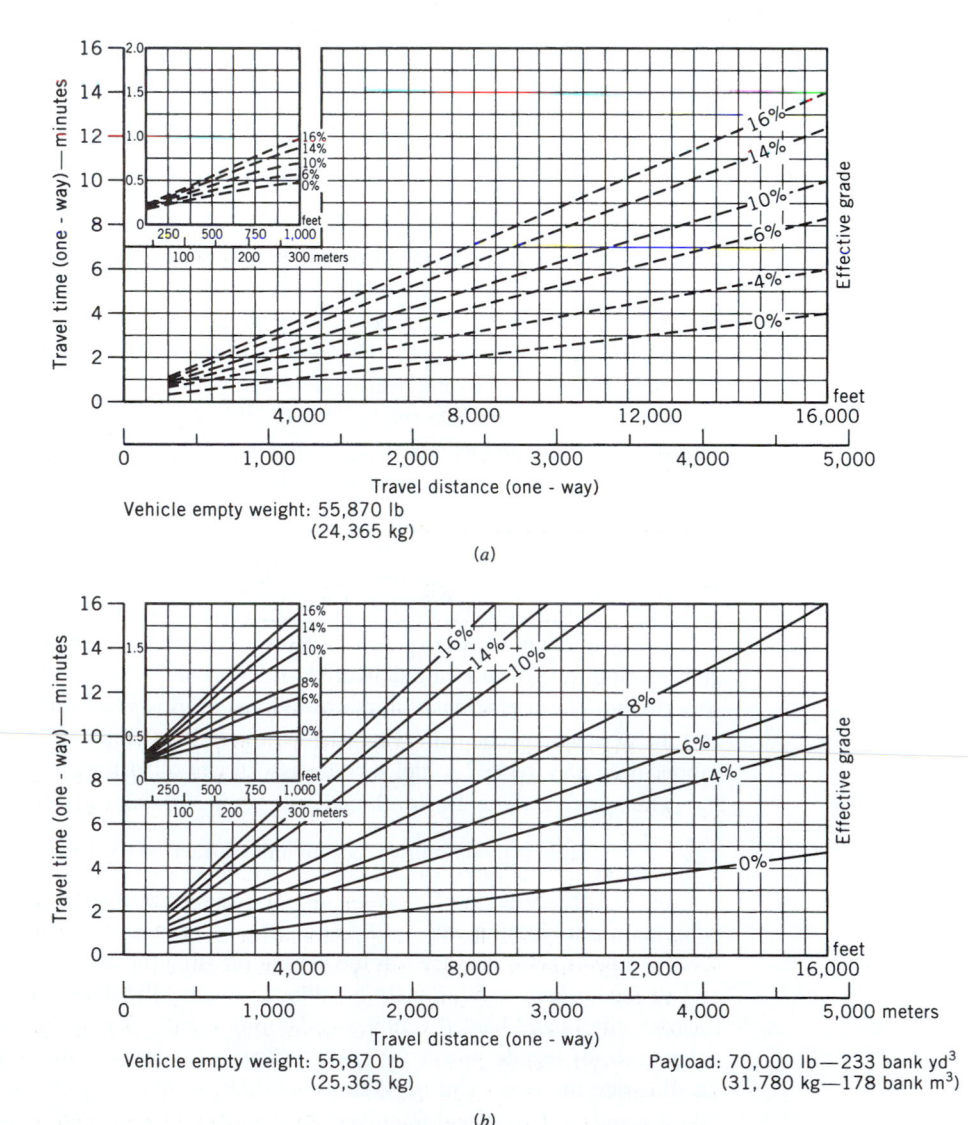

Figure 12.6 Travel time: (a) empty and (b) loaded (Caterpillar Tractor Co.).

loaded truck indicates that the distance to be traveled in feet is shown along the x axis. The equivalent percent grade is again shown as slant lines sloped from lower left to upper right. Converting the mileages given to feet yields the values 24,344 ft for route 1 and 14,784 ft for route 2. Entering the chart, with an equivalent percent grade of 4.5 for route 2, the travel time can be read (on the y axis), as 9.5 min.

A problem develops in reading the travel time for route 1 since the maximum distance shown on the chart is 16,000. One way of reconciling this problem is to break the 24,344 into two segments: (1) 16,000 ft as shown and (2) the 8,344 ft remaining. The assumption is made that the 8,344 ft is traveled at the maximum speed determined previously to be 41 mi/hr. At this speed, the travel time for this segment is

$$T_2 = \frac{8{,}344 \text{ ft } (60 \text{ min/hr})}{(41 \text{ mi/hr})(5{,}280 \text{ ft/mi})} = 2.31 \text{ min}$$

The time for the remaining 16.000 ft is read from the chart as 7.2 min. It is assumed that acceleration and deceleration effects are included in this time. Therefore, the required time for route 1 is also 9.5 min ($T_1 + T_2 = 7.2 + 2.31$). Therefore, the decision as to which route to use is based on wear and tear on the machines, driver skill, and other considerations. This problem illustrates the development of time, given information affecting power required and power available. Using the same procedure, the travel time empty returning to the pit from the fill can be determined and total cycle time can be determined.

12.4 USABLE POWER

To this point, it has been assumed that all of the available power is usable and can be developed. Environmental conditions play a major role in determining whether the power available can be utilized under operating conditions. The two primary constraints in using the available power are the road surface traction characteristics (for wheeled vehicles) and the altitude at which operations are conducted. Most people have watched the tires of a powerful car spin on a wet or slippery pavement. Although the engine and gears are delivering a certain horsepower, the traction available is not sufficient to develop this power into the ground as a driving force. Combustion engines operating at high altitudes experience a reduction in oxygen available within the engine cylinders. This also leads to reduced power.

Consider first the problem of traction. The factors that influence the usable power that can be developed through the tires of wheeled vehicles are the coefficient of traction of the surface being traveled and the weight of the vehicle on the driving wheels.

The coefficient of traction is a measure of the ability of a particular surface to receive and develop the power being delivered to the driving wheels and has been determined by experiment. The coefficient of traction obviously varies based on the surface being traversed and the delivery mechanism (i.e., wheels, track, etc.). Table 12.5 gives typical values for rubber-tired and tracked vehicles on an assortment of surface materials.

The power that can be developed on a given surface is given by the expression:

Usable pounds pull = (coefficient of traction) × (weight on drivers)

In the consideration of rolling resistance and grade resistance, the entire weight of the vehicle or combination was used. In calculating the usable power, *only the weight on the driving wheels* is used, since it is the weight pressing the driving mechanism (e.g., wheels) and surface together. Equipment handbooks specify the distribution of load to all wheels for both empty and loaded vehicles and combinations. The weight to be considered in the calculation of usable power for several types of combinations is shown in Figure 12.7. To illustrate the constraint imposed by usable power, consider the following situation. A 30-yd-capacity, two-wheel tractor-scraper is operating in sand and carrying 26-ton loads.

Table 12.5   Coefficients of Traction

Materials	Rubber Tires	Tracks
Concrete	.90	.45
Clay loam, dry	.55	.90
Clay loam, wet	.45	.70
Rutted clay loam	.40	.70
Dry sand	.20	.30
Wet sand	.40	.50
Quarry pit	.65	.55
Gravel road (loose, not hard)	.36	.50
Packed snow	.20	.25
Ice	.12	.12
Firm earth	.55	.90
Loose earth	.45	.60
Coal, stockpiled	.45	.60

The job superintendent is concerned about the high rolling resistance of the sand (RR = 400 lb/ton) and the low traction available in sand. The question is: Will the tractors have a problem with 26-ton loads under these conditions? The weight distribution characteristics of the 30-yd tractor-scraper are as follows:

	Empty Weight (lb)	Percent	Loaded Weight (lb)	Percent
Drive wheels	50,800	67	76,900	52
Scraper wheels	25,000	33	70,900	48
Total weight	75,800	100	147,800	100

The difference between the total weight empty and loaded is 72,000 lb, or 36 tons. The loaded weight with 26-ton loads would be 127,800 lb. Assuming the same weight distribution given above for fully loaded vehicles, the wheel loads would be as follows:

	Percent	Weight in Pounds
Drive wheels	52	66,456
Scraper wheels	48	61,344
Total	100	127,800

The resisting force (assuming a level haul site) would be

$$\text{Pounds required} = 400\,\text{lb/ton} \times 63.9\,\text{tons} = 25,560\,\text{lb}$$

In Determining Weight on Drivers

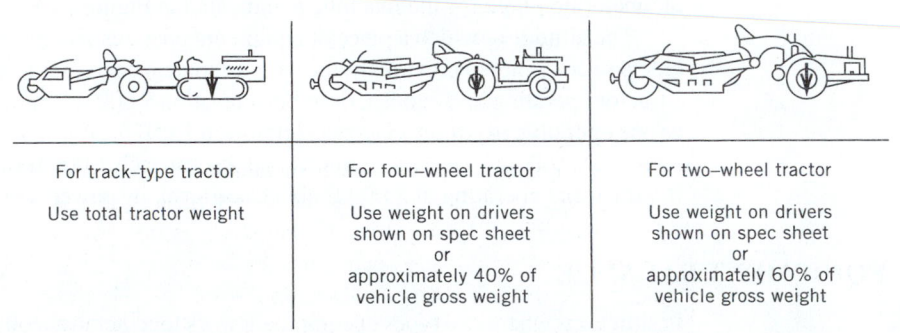

For track–type tractor	For four–wheel tractor	For two–wheel tractor
Use total tractor weight	Use weight on drivers shown on spec sheet or approximately 40% of vehicle gross weight	Use weight on drivers shown on spec sheet or approximately 60% of vehicle gross weight

Figure 12.7 Determination of driver weights.

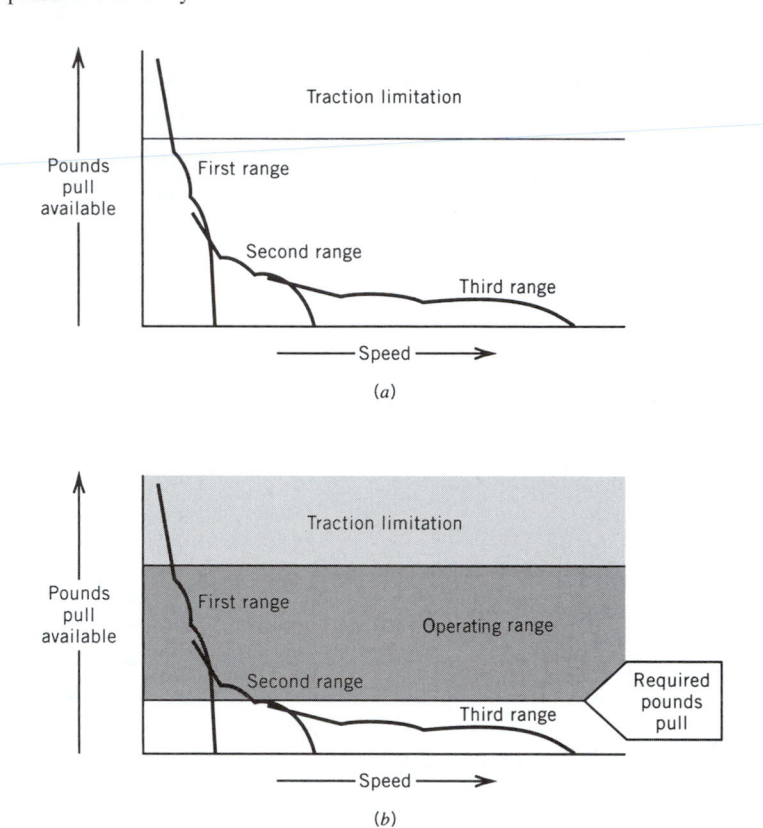

Figure 12.8 Impact of usable power constraints.

The deliverable or usable power is

$$\text{Usable power} = 0.20 \times 66{,}456\,\text{lb} = 13{,}291.20\,\text{lb}$$

Quite obviously, there will be a problem with traction since the required power is almost twice the power that can be developed. The "underfoot" condition must be improved. A temporary surface (e.g., wood or steel planking) could be installed to improve traction. One simple solution would be to simply wet the sand. This yields an increased usable power:

$$\text{Usable power} = 0.40 \times 66{,}456\,\text{lb} = 26{,}582.41\,\text{lb}$$
$$> 25{,}560\,\text{lb}$$

The impact of usable power constraints can be shown graphically (see Fig. 12.8). Now, if the total resistance of the unit (rolling resistance plus grade resistance) is 10,000 lb, then an operating range for the machine is indicated in Figure 12.8b.

The altitude at which a piece of equipment operates also imposes a constraint on the usable power. As noted previously, the oxygen content decreases as elevation increases, so that a tractor operating in Bogotá, Colombia (elevation 8,600 ft), cannot develop the same power as one operating in Atlanta, Georgia (elevation 1,050 ft). A good rule of thumb to correct this effect is as follows: Decrease pounds pull 3% for each 1,000 ft (above 3,000 ft). Therefore, if a tractor is operating at 5,000 ft above sea level, its power will be decreased by 6%.

12.5   EQUIPMENT BALANCE

In situations where two types of equipment work together to accomplish a task, it is important that a balance in the productivity of the units be achieved. This is desirable so that one unit is not continually idle waiting for the other unit to "catch up." Consider the problem of

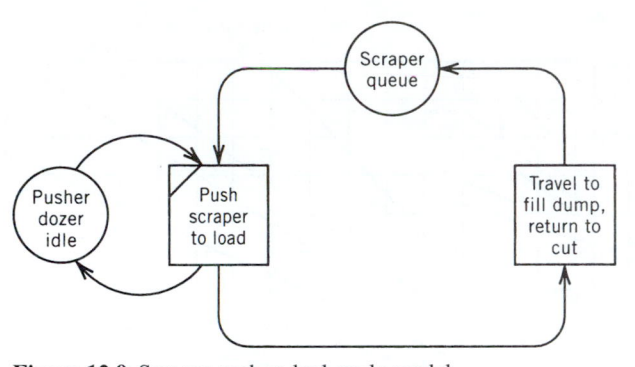

Figure 12.9 Scraper-pusher dual-cycle model.

balancing productivity within the context of a pusher dozer loading a tractor scraper. A simple model of this process is shown in Figure 12.9. The circles represent delay or waiting states, while the squares designate active work activities with associated times that can be estimated. The haul unit is a 30 cu-yd scraper, and it is loaded in the cut area with the aid of a 385-hp pusher dozer. The system consists of two interacting cycles.

Assume that in this case the 30-cu-yd tractor scraper is carrying rated capacity and operating on a 3,000-ft level haul where the rolling resistance (RR) developed by the road surface is 40 lb/ton. Using the standard formula, this converts to

$$\text{Effective grade} = \frac{\text{RR}}{20 \text{ lb/ton/\% grade}} = \frac{40 \text{ lb/ton}}{20 \text{ lb/ton/\% grade}} = 2\% \text{ grade}$$

By consulting the charts given in Figure 12.10, the following travel times can be established:

1. Time loaded to fill: 1.4 min

2. Time empty to return: 1.2 min

Assume further that the dump time for the scraper is 0.5 min and the push time using a track-type pusher tractor is 1.23 min, developed as follows:

$$
\begin{aligned}
\text{Load time} &= 0.70 \\
\text{Boost time} &= 0.15 \\
\text{Transfer time} &= 0.10 \\
\text{Return time} &= \underline{0.28} \\
\text{Total} &= 1.23 \text{ min}
\end{aligned}
$$

Using these deterministic times for the two types of flow units in this system (i.e., the pusher and the scrapers), the scraper and pusher cycle times can be developed, as shown in Figure 12.11.

$$
\begin{aligned}
\text{Pusher cycle} &= 1.23 \text{ min} \\
\text{Scraper cycle} &= 0.95 + 1.2 + 1.4 + 0.5 = 4.05 \text{ min}
\end{aligned}
$$

These figures can be used to develop the maximum hourly production for the pusher unit and for each scraper unit as follows.

Maximum System Productivity (Assuming a 60-min Working Hour)

1. Per scraper

$$\text{Prod (scraper)} = \frac{60 \text{ min/hr}}{4.05 \text{ min}} \times 30 \text{ cu yd (loose)}$$

$$= 444.4 \text{ cu yd (loose)/hr}$$

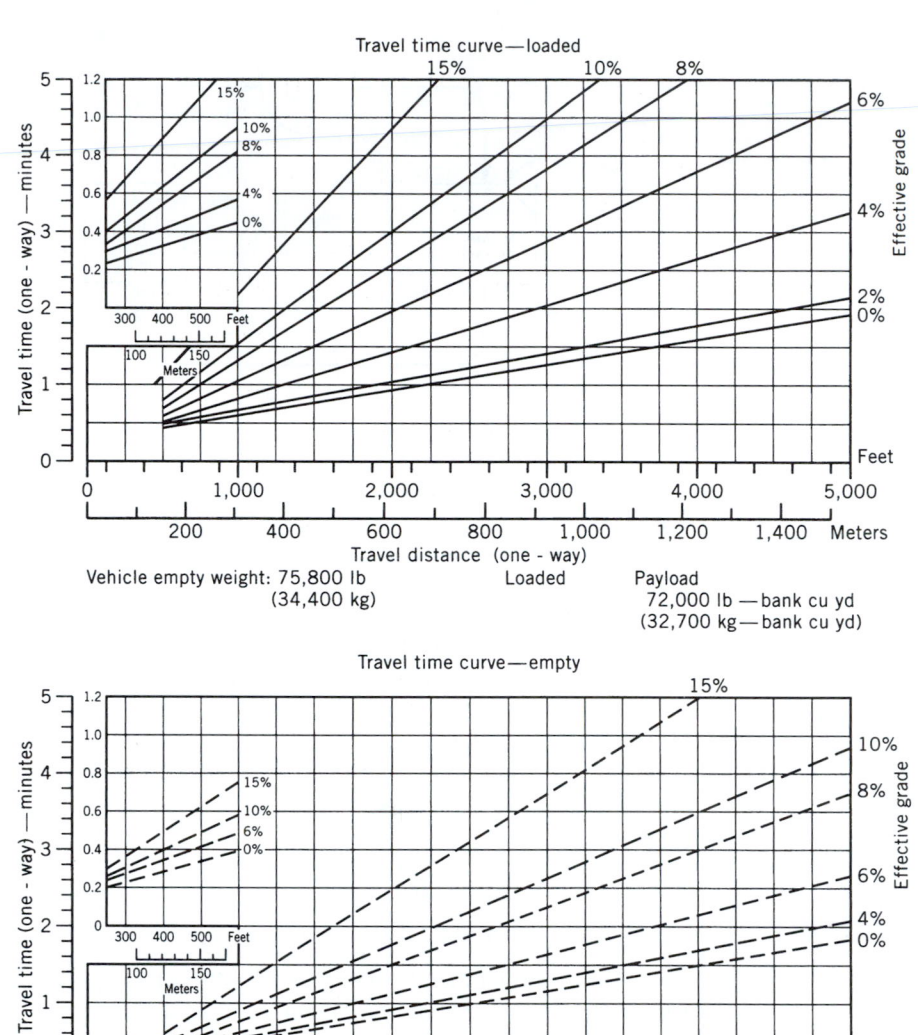

Figure 12.10 Travel time nomographs (Caterpillar Tractor Co.).

2. Based on single pusher

$$\text{Prod (pusher)} = \frac{60}{1.23} \times 30 \text{ cu yd (loose)}$$

$$= 1{,}463.4 \text{ cu yd (loose)/hr}$$

Using these productivities based on a 60-min working hour, it can be seen that the pusher is much more productive than a single scraper and would be idle most of the time if matched to only one scraper. By using a graphical plot, the number of scrapers that are needed to keep the pusher busy at all times can be determined.

The linear plot of Figure 12.12 shows the increasing productivity of the system as the number of scrapers is increased. The productivity of the single pusher constrains the total productivity of the system to 1,463.4 cu yd. This is shown by the dotted horizontal line

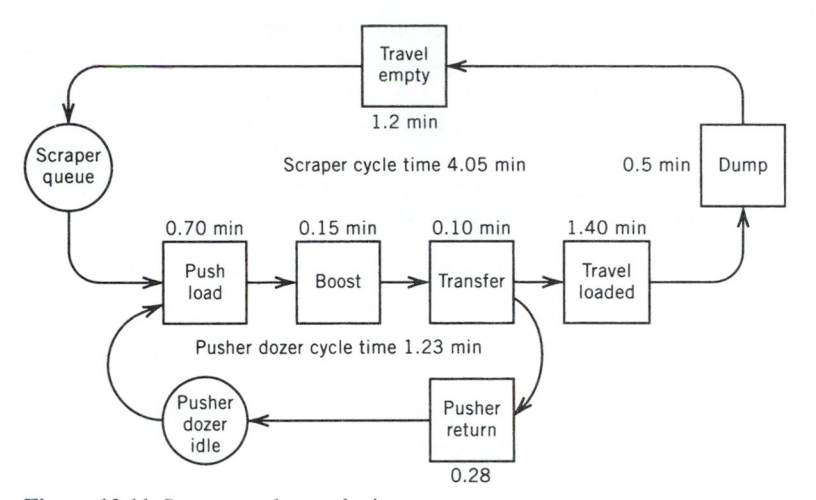

Figure 12.11 Scraper-pusher cycle times.

parallel to the *x* axis of the plot. The point at which the horizontal line and the linear plot of scraper productivity intersect is called the balance point. The balance point is the point at which the number of haul units (i.e., scrapers) is sufficient to keep the pusher unit busy 100% of the time. To the left of the balance point, there is an imbalance in system productivity between the two interacting cycles; this leaves the pusher idle. This idleness results in lost productivity. The amount of lost productivity is indicated by the difference between the horizontal line and the scraper productivity line. For example, with two scrapers operating in the system, the ordinate *AB* of Figure 12.12 indicates that 574.6 cu yd, or a little less than

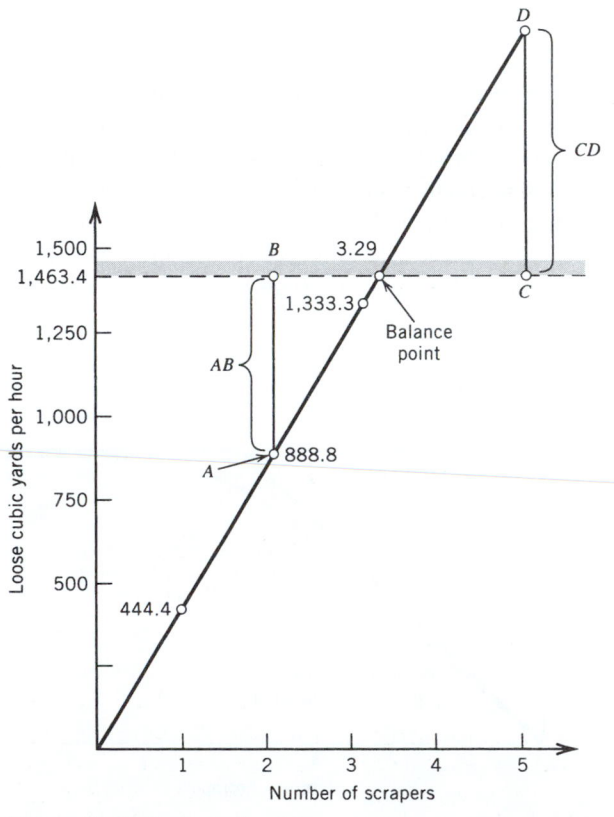

Figure 12.12 Productivity plot.

half of the pusher productivity, is lost because of the mismatch between pusher and scraper productivities. As scrapers are added, this mismatch is reduced until, with four scrapers in the system, the pusher is fully utilized. Now the mismatch results in a slight loss of productivity caused by idleness of the scrapers. This results because, in certain instances, a scraper will have to wait to be loaded until the pusher is free from loading a preceding unit. If five scraper units operate in the system, the ordinate *CD* indicates that the loss in the productive capacity of the scraper because of delay in being push loaded is

$$\text{Productive loss} = 5(444.4) - 1{,}463.4 = 758.6 \text{ cu yd}$$

This results because the greater number of scrapers causes delays in the scraper queue of Figures 12.9 and 12.11 for longer periods of time. The imbalance or mismatch between units in dual-cycle systems resulting from deterministic times associated with unit activities is called *interference*. It is due only to the time imbalance between the interacting cycles. It does not consider idleness or loss of productivity because of random variations in the system activity durations. In most cases, only a deterministic analysis of system productivity is undertaken because it is sufficiently accurate for the purpose of the analyst.

12.6   RANDOM WORK TASK DURATIONS

The influence of mismatches in equipment fleets and crew mixes on system productivity was discussed in the last section in terms of deterministic work task durations and cycle times. In systems where the randomness of cycle times is considered, system productivity is reduced further. The influence of random durations on the movement of resources causes various units to become bunched together and thus to arrive at and overload work tasks. Resulting delays impact the productivity of cycles by increasing the time that resource units spend in idle states pending release to productive work tasks.

Consider the scraper-pusher problem and assume that the effect of random variation in cycle activity duration is to be included in the analysis.

In simple cases such as the two-cycle system model of Figure 12.9, mathematical techniques based on *queuing theory* can be used to develop solutions for situations where the random arrival of scrapers to the dozer can be postulated. In order to make the system amenable to mathematical solution, however, it is necessary to make certain assumptions about the characteristics of the system that are not typical of field construction operations.

Figure 12.13 indicates the influence of random durations on the scraper fleet production. The curved line of Figure 12.13 slightly below the linear plot of production based on

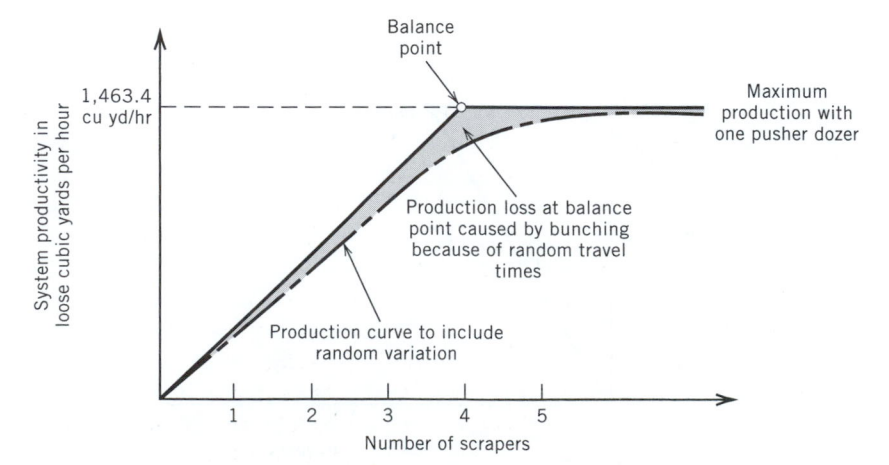

Figure 12.13 Productivity curve to include effect of random cycle times.

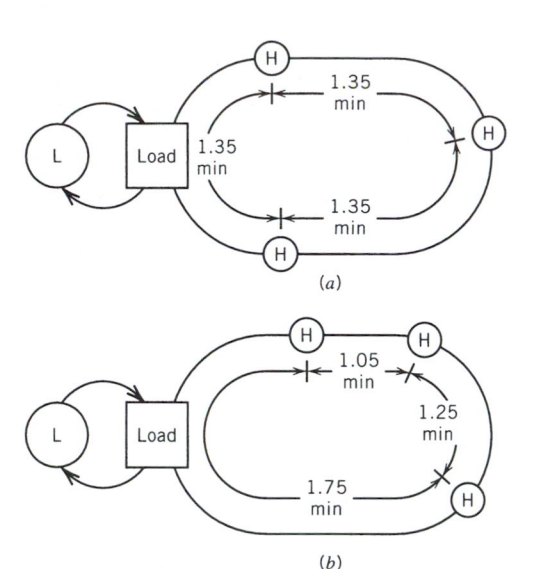

Figure 12.14 Comparison of haul unit cycles.

deterministic work task times shows the reduction in production caused by the addition of random variation of cycle activity times. This randomness leads to bunching of the haulers on their cycle. With deterministic work task times, the haul units are assumed to be equidistant in time from one another within their cycle

In deterministic calculations, all three of the haul units shown in Figure 12.14a are assumed to be exactly 1.35 min apart. In this system, there are three units, and the hauler cycle time is taken as a deterministic value of 4.05 min. In systems that include the effect of random variation of cycle times, "bunching" eventually occurs between the units on the haul cycle. That is, the units do not stay equidistant from one another but are continuously varying the distances between one another. Therefore, as shown in Figure 12.14b, a situation often occurs in which the units on the haul are unequally spaced apart in time from one another. This bunching effect leads to increased idleness and reduced productivity. It is intuitively clear that the three units that are bunched as shown in Figure 12.14b will be delayed for a longer period at the scraper queue since the first unit will arrive to load only 1.05 min instead of 1.35 min in advance of the second unit. The bunching causes units to "get into each other's way." The reduction in productivity caused by bunching is shown as the shaded area in Figure 12.13 and occurs in addition to the reduction in productivity caused by mismatched equipment capacities.

This bunching effect is most detrimental to the production of dual-cycle systems such as the scraper-pusher process at the balance point. Several studies have been conducted to determine the magnitude of the productivity reduction at the balance point because of bunching. Simulation studies conducted by the Caterpillar Tractor Company indicate that the impact of random time variation is the standard deviation of the cycle time distribution divided by the average cycle time. Figure 12.15 illustrates this relationship graphically.

As shown in the figure, the loss in deterministic productivity at the balance point is approximately 10% due to the bunching; this results in a system with a cycle coefficient of variation equal to 0.10. The probability distribution used in this analysis was lognormal. Other distributions would yield slightly different results. The loss in productivity in equipment-heavy operations such as earthmoving is well documented and recognized in the field, mainly because of the capital-intensive nature of the operation and the use of scrapers in both single-unit operations and fleet operations. To some extent, field policies have emerged to counteract this effect by occasionally breaking the queue discipline of

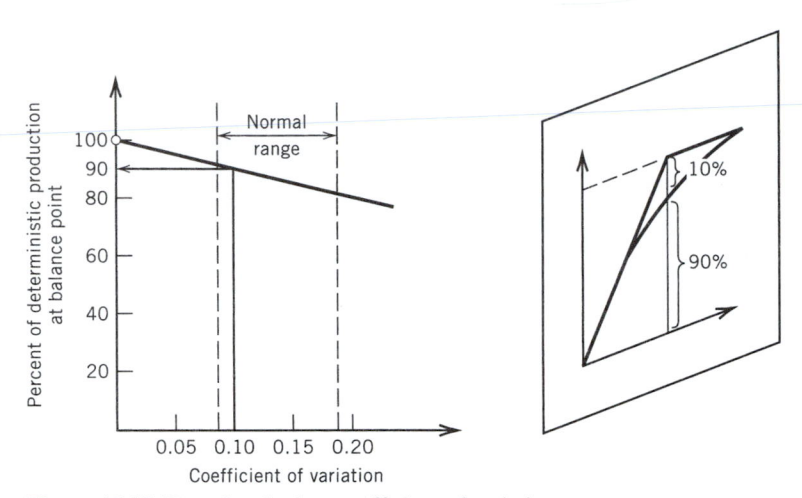

Figure 12.15 Plot of cycle time coefficient of variation.

the scrapers so that they self-load when bunching effects become severe. The resulting increased load and boost time for the scraper add little to the system productivity, but it does break down the bunching of the scrapers.

REVIEW QUESTIONS AND EXERCISES

12.1   A customer estimates that he is getting 30 cu yd (loose) of gypsum in his scraper. Determine the percent overload if the load estimate is correct. The maximum load capacity of the scraper is 84,000 lb.

12.2   Stripping overburden in the Illinois coal belt, the Dusty Coal Company uses 270-hp, track-type tractors (with direct-drive transmissions) and drawn scrapers. The overburden is a very soft loam that weighs 2,800 lb/yd (loose). Estimated rolling resistance factor for the haul road is 300 lb/ton. If the scraper weighs 35,000 lb (empty) and carries 25 loose cubic yards per trip, what is the rolling resistance of the loaded unit? What operating gear and speed do you estimate for the loaded machines on level ground? (See Table 12.4.)

12.3   The ABC Company is planning to start a new operation hauling sand to a ready-mix concrete plant. The equipment superintendent estimates that the company-owned 30-yd wheel tractor scrapers can obtain 26-ton loads. He is concerned about the high rolling resistance of the units in the sand (RR factor 250 lb/ton) and the low tractive ability of the tractors on this job. Will traction be a problem? If so, what do you suggest to help?

12.4   Estimate the cycle time and production of a 30-cu-yd wheel tractor scraper carrying rated capacity, operating on a 4,500-ft level haul. The road flexes under load, has little maintenance, and is rutted. Material is 3000 lb/BCY. The scrapers are push-loaded by one 385-hp, track-type pusher tractor. How many scrapers can be served by this one pusher?

12.5   How many trips would one rubber-tired Herrywampus have to make to backfill a space with a geometrical volume of 5400 cu yd? The maximum capacity of the machine is 30 cu yd (heaped), or 40 tons. The material is to be compacted with a shrinkage of 25% (relative to bank measure) and has a swell fac-

tor of 20% (relative to bank measure). The material weighs 3,000 lb/cu yd (bank). Assume that the machine carries its maximum load on each trip. Check by both weight and volume limitations.

12.6   You own a fleet of 30 cu yd tractor-scrapers and have them hauling between the pit and a road construction job. The haul road is clayey and deflects slightly under the load of the scraper. There is a slight grade (3%) from pit at the fill location. The return road is level. The haul distance to the dump location is 0.5 miles and the return distance is 0.67 miles. Four scrapers are being used.

(a) What is the rimpull required when the scraper is full and on the haul to the fill?

(b) What are the travel times to and from the dump location (see Fig. 12.10)?

(c) The scrapers are push loaded in the pit. The load time is 0.6 min. What is the cycle time of the pusher dozer?

(d) Is the system working at, above, or below the balance point? Explain.

(e) What is the production of the system?

12.7   You are excavating a location for the vault shown below. The top of the walls shown are 1 ft below grade. All slopes of the excavation are $\frac{3}{4}$ to 1 to a toe 1 ft outside the base of the walls. The walls sit on a slab 1 ft in depth. Draw a sketch of the volume to be excavated, break it into components, and calculate the volume. The material from the excavation is to be used in a compacted fill. The front-end loader excavating the vault has an output of 200 bank cubic yards of common earth per hour. It loads a fleet of four trucks (capacity 18 loose cubic yards each) that haul the earth to a fill where it is compacted with a shrinkage factor of 10%. Each truck has a total cycle time of 15 min, assuming it does not have to wait in line to be loaded. The earth has a swell factor of 20%. How many hours will it take to excavate and haul the material to the fill?

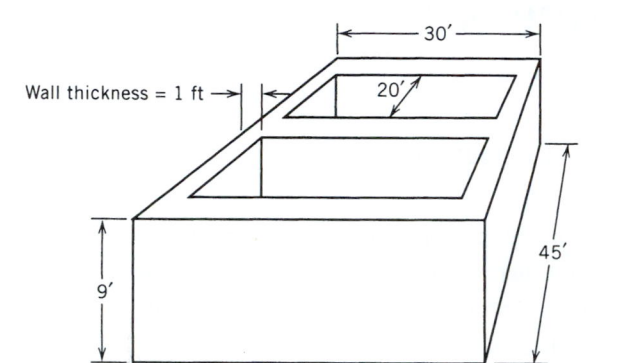

12.8   You have four 35-ton, off-highway trucks hauling from a pit to an airfield job. The haul road is maintained by a patrol grader and has a rolling resistance of 80 lb/ton. The road is essentially level and the distance one-way is 2.1 miles. The gross weight of the truck when loaded is 70 tons.

(a) What is the power required on the haul to the fill location?

(b) What is the maximum speed when hauling to the fill?

(c) What are the travel times to and from the fill (i.e, loaded and empty)?
The trucks are being loaded by a shovel with a 5-yd bucket (assume seven load cycles per truck load). The cycle time for the shovel is 0.5 mins.

(d) What is the total truck cycle time?

(e) What is the production of this system in cubic yards per hour, assuming the trucks carry 35 cu yd per load?

(f) Is the system working at, above, or below its balance point?

(g) If there is probability of major delay on the travel elements to and from the fill of 7% and the mean value of delay is 5 min, what is the new system production?

(h) Is the new system above or below the balance point?

12.9   You are given the following information about a dry-batch paving operation. You are going to use one mixer that has a service rate of 30 services per hour. The dry-batch trucks you use for bringing concrete to the paver have an arrival rate of 7.5 arrivals per hour. Each truck carries 6 cu yd of concrete. You have a total amount of 13,500 cu yd of concrete to pour. You rent a truck at $15 per hour, and the paver at $60 per hour. If the job takes more than 80 hours, you pay a penalty of $140 per hour owing to delays in job completion. On the basis of least cost, determine the number of trucks you should use. Plot the cost versus the number of trucks used.

Chapter 13

Estimating Process

Estimating Using Hand-Held Devices

The Need

Estimators need a way to increase efficiency and lessen the chance for error when collecting estimate information in the field. With Hand-Held Devices, users can have access to the same precision estimating data they use in the office. Cover page information, item numbers and descriptions, assembly numbers and descriptions, WBS codes, variables, and variable help can all be stored on a hand-held device. This ensures that estimators have the information they need to collect all project details necessary to deliver an accurate, complete bid.

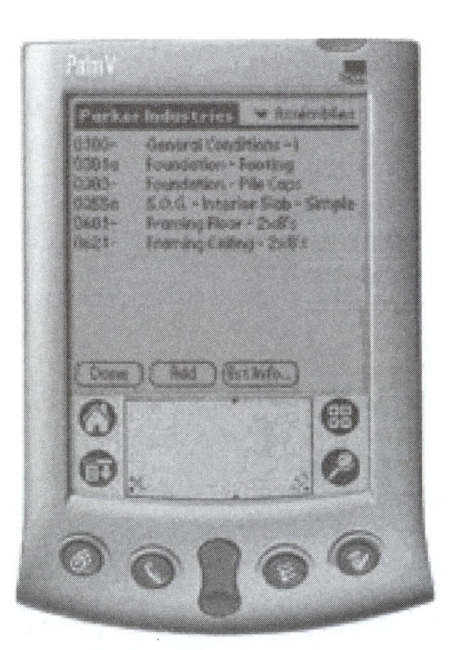

Hand-Held Estimating Device

The Technology

Designed for Hand-Held Devices, estimating programs on personal digital assistants (PDAs) equip estimators with all the tools needed to perform detailed takeoffs remotely. Then, when convenient, data can be transferred to desktop software to instantly generate a detailed estimate or change order. With this type of mobile data acquisition, there's a much better chance of collecting all the necessary dimensional information to create a bid the first time. Specialized programs allow the user to take information from the desktop into the field to use as a checklist for takeoff.

PDA estimating lets users access existing estimate information or create brand new estimates directly in the field. All the project information can be logged in one place, keeping estimators better organized and ready to make additions and changes as needed.

13.1 ESTIMATING CONSTRUCTION COSTS

The key to a good job and successful cost control is the development of a good estimate as the basis for bid submittal. The estimate represents the cost "flight plan" that will be followed by the constructor and that will aid him or her in achieving profit. If the flight plan is unrealistic or contains basic errors, the contractor will lose money on the job. If the estimate is well thought out and correctly reflects the costs that will be encountered in the field, the chances of a profitable job are greatly increased.

Estimating is the process of looking into the future and trying to predict project costs and resource requirements. Studies indicate that one of the major reasons for the failure of construction contracting firms is incorrect and unrealistic estimating and bidding practices. If 20 estimators or contractors were furnished the same set of plans and specifications and told to prepare an estimate of cost and resources, it would be safe to assume there would not be more than 2 estimates prepared on the same basis or from the same units. Therefore, a consistent procedure or set of steps for preparing an estimate is needed to minimize errors and achieve reliable results.

13.2   TYPES OF ESTIMATES

Estimating methods vary in accordance with the level of design detail that is available to the estimator. Prior to the commencement of design, when only conceptual information is available, a comprehensive unit such as a square foot of floor space or a cubic foot of usable space is used to characterize the facility being constructed. The representative unit is multiplied by a price per unit to obtain a gross estimate ($\pm 10\%$ accuracy) of the facility cost. A table of square foot and cubic foot building costs as given in *Building Construction Cost Data* published by the R. S. Means Company is shown in Figure 13.1 Such information is available in standard references and can be used for preliminary cost projections based on minimal design data. This *conceptual estimate* is useful in the schematic or budgetary phase, when design details are not available. The figures developed are of limited use for project control, and their use should be discontinued as soon as design data are available. These estimates are based on documents such as that given in Figure 2.2.

As the level of design detail increases, the designer typically maintains estimates of cost to keep the client informed of the general level of costs to be expected. The production of the plans and specifications usually proceeds in two steps. As noted in Chapter 2, the first step is called *preliminary design* and offers the owner a pause in which to review construction before detail design commences. A common time for this review to take place is at 40% completion of the total design. The preliminary design extends the concept documentation. At this point in the design process, a *preliminary estimate* is prepared by the architect or architect/engineer to reflect expected costs based on more definitive data.

Once the preliminary design has been approved by the owner, final or detail design is accomplished. The detail design phase culminates in the plans and specifications that are given to the constructor for bidding purposes. In addition to these detailed design documents, the architect/engineer produces a final *engineer's estimate* indicating the total job cost minus markup. This estimate should achieve approximately $\pm 3\%$ accuracy since the total design is now available. The owner's estimate is used (1) to ensure that the design produced is within the owner's financial resources to construct (i.e., that the architect/engineer has not designed a gold-plated project) and (2) to establish a reference point in evaluating the bids submitted by the competing contractors.

On the basis of the final drawings and specifications the contractor prepares his estimate of the job's cost to include a markup for profit. This is the *bid estimate*. Both the engineer's and bid estimates require a greater level of effort and a considerable number of estimator hours to prepare. A rough rule of thumb states that the preparation of a bid estimate by the contractor will cost one-fourth of one percent of the total bid price. From the contractor's point of view this cost must be recovered as overhead on jobs that are won. Therefore, a prorate based on the number of successful bids versus total bids must be included in each quotation to cover bid costs on unsuccessful bids.

In building construction, these four levels of estimates are the ones most commonly encountered. To recapitulate, the four types of estimates are:

1. Conceptual estimate
2. Preliminary estimate
3. Engineer's estimate
4. Bid estimate

These four levels of precision reflect the fact that as the project proceeds from concept through preliminary design to final design and the bidding phase, the level of detail increases,

171 | S.F., C.F. and % of Total Costs

				UNIT	UNIT COSTS			% OF TOTAL		
**171 000	S.F. & C.F. Costs**				1/4	MEDIAN	3/4	1/4	MEDIAN	3/4
010	0010	APARTMENTS Low Rise (1 to 3 story)	R171 -100	S.F.	40.20	50.65	67.35			
	0020	Total project cost		C.F.	3.61	4.74	5.95			
	0100	Site work		S.F.	3.34	4.81	7.60	6.30%	10.50%	13.90%
	0500	Masonry			.73	1.87	3.18	1.50%	3.90%	6.50%
	1500	Finishes			4.22	5.40	7.15	8.90%	10.70%	12.90%
	1800	Equipment			1.31	1.99	2.96	2.70%	4.10%	6.30%
	2720	Plumbing			3.13	4.03	5.06	6.70%	8.90%	10.10%
	2770	Heating, ventilating, air conditioning			1.99	2.46	3.56	4.20%	5.60%	7.60%
	2900	Electrical			2.32	3.08	4.19	5.20%	6.70%	8.40%
	3100	Total: Mechanical & Electrical			6.95	8.50	10.90	15.90%	18.20%	22%
	9000	Per apartment unit, total cost		Apt.	31,200	46,700	68,900			
	9500	Total: Mechanical & Electrical		"	5,700	8,400	12,100			
020	0010	APARTMENTS Mid Rise (4 to 7 story)	R171 -100	S.F.	52.80	63.90	77.90			
	0020	Total project costs		C.F.	4.14	5.65	7.85			
	0100	Site work		S.F.	2.06	4.03	7.50	5.20%	6.70%	9.10%
	0500	Masonry			3.23	4.52	6.70	5.20%	7.30%	10.50%
	1500	Finishes			6.55	8.30	10.85	10.40%	11.90%	16.90%
	1800	Equipment			1.66	2.38	3.12	2.80%	3.50%	4.40%
	2500	Conveying equipment			1.20	1.48	1.77	2%	2.20%	2.60%
	2720	Plumbing			3.12	4.93	5.40	6.20%	7.40%	8.90%
	2900	Electrical			3.59	4.77	5.85	6.60%	7.20%	8.90%
	3100	Total: Mechanical & Electrical			9.80	12.35	15.90	17.90%	20.10%	22.30%
	9000	Per apartment unit, total cost		Apt.	38,400	58,300	67,400			
	9500	Total: Mechanical & Electrical		"	12,200	13,700	21,500			
030	0010	APARTMENTS High Rise (8 to 24 story)	R171 -100	S.F.	60.60	73	89.20			
	0020	Total project costs		C.F.	4.98	6.90	8.45			
	0100	Site work		S.F.	1.85	3.55	4.96	2.50%	4.80%	6.10%
	0500	Masonry			3.43	6.20	7.85	4.70%	9.60%	10.70%
	1500	Finishes			6.55	8.40	9.60	9.30%	11.70%	13.50%
	1800	Equipment			1.92	2.37	3.16	2.70%	3.30%	4.30%
	2500	Conveying equipment			1.22	2.02	2.88	2.20%	2.70%	3.30%
	2720	Plumbing			4.54	5.25	6.61	6.90%	9.10%	10.60%
	2900	Electrical			4.15	5.20	7.10	6.40%	7.60%	8.80%
	3100	Total: Mechanical & Electrical			12.25	14.85	18.35	18.20%	21.80%	23.90%
	9000	Per apartment unit, total cost		Apt.	55,700	66,500	73,400			
	9500	Total: Mechanical & Electrical		"	13,500	15,500	16,800			
040	0010	AUDITORIUMS	R171 -100	S.F.	61	85.30	110			
	0020	Total project costs		C.F.	4.01	5.60	8			
	2720	Plumbing		S.F.	3.89	5.15	6.70	5.80%	7%	8.60%
	2900	Electrical			4.96	7.05	9	6.70%	8.80%	10.90%
	3100	Total: Mechanical & Electrical			10.10	13.65	23.70	14.40%	18.50%	23.60%
050	0010	AUTOMOTIVE SALES	R171 -100	S.F.	42.45	50.80	76.60			
	0020	Total project costs		C.F.	3.12	3.56	4.63			
	2720	Plumbing		S.F.	2.26	3.68	4.09	4.70%	6.40%	7.80%
	2770	Heating, ventilating, air conditioning			3.26	4.99	5.40	6.30%	10%	10.30%
	2900	Electrical			3.74	5.60	6.45	7.40%	9.90%	12.30%
	3100	Total: Mechanical & Electrical			7.90	11.55	15.10	16.60%	19.10%	27%
060	0010	BANKS	R171 -100	S.F.	91.05	113	143			
	0020	Total project costs		C.F.	6.60	8.85	11.65			
	0100	Site work		S.F.	9.05	16.60	24.65	7%	13.80%	17.50%
	0500	Masonry			4.60	7.65	16.85	2.90%	5.80%	11.30%
	1500	Finishes			7.75	10.55	13.55	5.50%	7.60%	9.90%
	1800	Equipment			3.63	7.55	16.85	3.20%	8.20%	12.50%
	2720	Plumbing			2.89	4.10	6	2.80%	3.90%	4.90%
	2770	Heating, ventilating, air conditioning			5.65	7.30	9.80	4.90%	7.10%	8.50%

Figure 13.1 Costs based on a representative unit (From *Building Construction Cost Data 1996,* copyright Reed Construction Data, Kingston, MA, 781-585-7880 all rights reserved.).

allowing the development of a more accurate estimate. Estimating continues during the construction phase to establish whether the actual costs agree with the bid estimate. This type of "estimating" is what allows the contractor to project profit or loss on a job after it is in progress.

A listing of estimates commonly developed in conjunction with large and complex industrial projects (e.g., power plants, chemical process plants, and the like) is given in

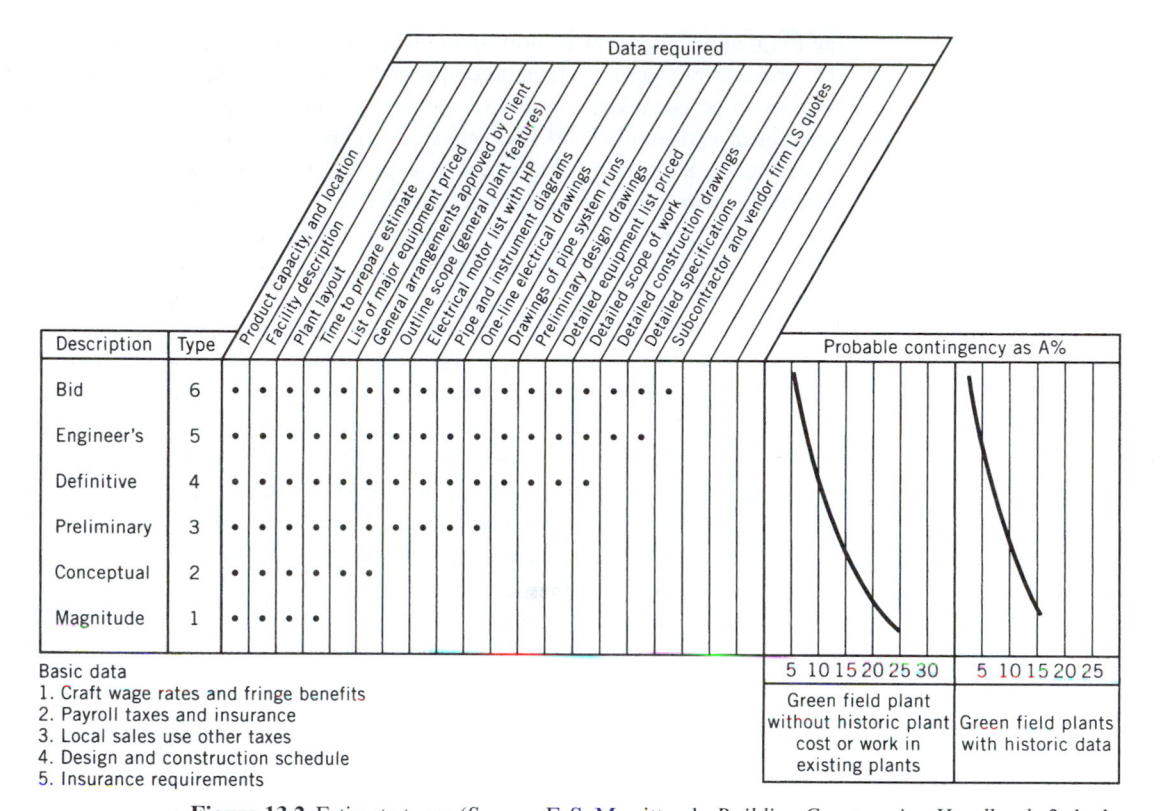

Basic data
1. Craft wage rates and fringe benefits
2. Payroll taxes and insurance
3. Local sales use other taxes
4. Design and construction schedule
5. Insurance requirements

Figure 13.2 Estimate types (*Source:* F. S. Merritt, ed., *Building Construction Handbook*, 3rd ed., New York, McGraw-Hill, 1975).

Figure 13.2. This list includes a magnitude level estimate that is similar in purpose to the conceptual estimate used in building construction. That is, it is used to reflect gross costs for planning and decision purposes before the preliminary and definitive design phases begin. The definitive estimate, as used on complex industrial projects, is a prefinal estimate developed just prior to the production of final drawings and specifications. The definitive estimate can be prepared when all components comprising the project scope definition have been quantitatively determined and priced by using actual anticipated material and labor costs. This estimate is normally prepared when the project scope is defined in terms of firm plot plans, mechanical and process flow diagrams, equipment and material specifications, and engineering and design layouts. The pricing is based on formal vendor quotations for all major items and current predictable market costs for all commodity accounts. The amount of variability inherent in each level of estimate is reflected by the contingency curves shown on the right in Figure 13.2. The variability is, of course, quite high at the magnitude level and decreases to the 3 to 5% range as bid-level documents become available.

13.3 DETAILED ESTIMATE PREPARATION

The preparation of a detailed bid-level estimate requires that the estimator break the project into cost centers or cost subelements. That is, the project is broken down into subcomponents that will generate costs. It is these costs that the estimator must develop on the basis of the characteristic resources required. The word *resource* is used here in the broad sense and applies to the man-hours, materials, subcontracts, equipment-hours, and dollars needed to accomplish the work or meet the requirements associated with the cost center. Typically in construction the cost center relates to some physical subcomponent of the project, such as foundation piles, excavation, steel erection, interior dry wall installation, and the like. Certain non-physical components of the work generate costs, however, and these cost centers must also be considered. Many of the items listed as "indirects" are typical of costs that are not directly connected with physical components or end items in the facility to be constructed. Such items do, however, generate cost that must be recovered. These costs include insurance and bonding premiums, fees for licenses and permits required by contract, expense for special items relating to safety and minority participation programs, and home office overheads projected as allocated to the job. These items are sometimes referred to as *general conditions*, or *general requirements*, although they may or may not be specifically referred to in the contract documents. Accounts relating to these items fall into the categories for conditions of contract and general requirements of the contract. As estimators prepare bids, they have a general framework for cost recovery in mind. In addition, they have a knowledge of the technologies involved in building the project, which allow them to divide projects into individual pieces of work (physical subcomponents, systems, etc.). These work packages consume resources, generating costs that must be recovered from the client. Typically, a chart of cost accounts specific to the company acts as a guide or checklist as the estimator reviews the plans and specifications to highlight what cost centers are present in the contract being estimated.

Although the process of estimating is part art, part science, the estimator generally follows certain steps in developing the estimate:

1. Break the project into cost centers.

2. Estimate the quantities required for cost centers that represent physical end items (e.g., cubic yards of earth, lineal feet of pipe, etc.). For physical systems this procedure is commonly called *quantity takeoff*. For those cost centers that relate to nonphysical items, determine an appropriate parameter for cost calculation (e.g., the level of builder's risk insurance required by the contract or the amounts of the required bonds).

3. Price out the quantities determined in step 2 using historical data, vendor quotations, supplier catalogs, and other pricing information. This pricing may be based on a price per unit (unit cost) basis or a lump-sum (one job) basis. Price development for physical work items may require an analysis of the production rates to be achieved based on resource analysis. If this analysis is used, the estimator must:

    a. Assume work team composition to include number of workers (skilled and unskilled) and equipment required.
    b. On the basis of team composition, estimate an hourly production rate based on the technology being used.
    c. Make an estimate of the efficiency to be achieved on this job, considering site conditions and other factors.
    d. Calculate the effective unit price.

4. Calculate the total price for each cost center by multiplying the required quantity by the unit price. This multiplication is commonly called an *extension*, and this process is called *running the extensions*.

The estimator usually summarizes the values for each cost center on a summary sheet, such as that shown in Figure 13.3.

13.4   DEFINITION OF COST CENTERS

The subdivisions into which the project is divided for detailed cost estimation purposes are variously referred to as:

Jefferson Starship Contractors, Inc.
ESTIMATE SUMMARY
Estimate No.　6692　By:　DWH　　Date: 1 August 2xxx
Owner:　　NASA　　　Project:　Admin Building

Code	Description	MH	Labor	Material	Sub	Owner	Total
01	Site improvements						
02	Demolition						
03	Earthwork						
04	Concrete						
05	Structural steel	1,653	18,768	15,133			33,901
06	Piling						
07	Brick & masonry						
08	Buildings						
09	Major equipment	2,248	26,059	1,794			27,853
10	Piping	2,953	34,518	57,417	1,500	34,541	127,976
11	Instrumentation				33,000		33,000
12	Electrical				126,542		126,542
13	Painting				14,034		14,034
14	Insulation				4,230		4,230
15	Fireproofing			530	1,110		1,640
16	Chemical cleaning						
17	Testing						
18	Const. equipment					35,666	35,666
19	Misc. directs	1,008	10,608	2,050		2,000	14,658
20	Field extra work						
Sub Total Direct Cost		7,862	89,953	76,924	180,416	72,207	419,500
21	Con. tools/sup.			7,361			7,361
22	Field payroll/burden					16,580	16,580
23	Start-up asst.						
24	Ins. & taxes					5,268	5,268
25	Field sprvsn.	480	7,200			2,038	9,238
26	Home off. exp.					2,454	2,454
27	Field emp. ben.					10,395	10,395
Sub Total Indirect Cost		480	7,200	7,361		36,735	51,296
Adjustment Sheets							
Total Field Cost		8,342	97,153	84,285	180,416	108,942	470,796
28	Escalation						
29	Overhead & profit		8,342	5,057	9,021	10,190	32,610
30	Contingency						18,076
31	Total Project Cost						521,482

Figure 13.3 Typical estimate summary.

1. Estimating accounts

2. Line items

3. Cost accounts

4. Work packages

The estimating account is typically defined so as to provide target values for the cost accounts that will be used to collect as-built costs while the job is in progress. Therefore, the end item that is the focus of cost development in the estimating account is linked to a parallel cost account for actual cost information collecting during construction. The cost account expenditures developed from field data are compared with the estimated cost as reflected by the estimating account to determine whether costs are exceeding, underrunning, or coming in on the original estimate values. Therefore, the use of the term *cost account* is not strictly correct during the preparation of bid since this account is not active until the job is in progress and actual cost data are available.

As described in Chapter 6, the term *work package* has become current over the past 20 years and is commonly used to indicate a subdivision of the project that is used both for cost control and scheduling (i.e., time control). When both cost and time control systems are combined into an integrated project management system, work packages are controlled to determine cost versus estimate and time versus schedule.

The subdividing of the project into work packages results in the definition of a work breakdown structure (WBS).

> *A work package is a well-defined scope of work that usually terminates in a deliverable product. Each package may vary in size, but must be a measurable and controllable unit of work to be performed. It also must be identifiable in a numerical accounting system in order to permit capture of both budgeted and actual performance information. A work package is a cost center.*[1]

The breakdown of a project into estimating accounts or work packages (depending on the sophistication of the system) is aided by a comprehensive chart of cost accounts or listing of typical work packages that can be used as a checklist. This checklist or template can be matched to the project being estimated to determine what types of work are present. That is, accounts in the general chart are compared to the project being estimated to determine which ones apply.

13.5 QUANTITY TAKEOFF

The development of the quantities of work to be placed in appropriate units (e.g., square feet, cubic yards, etc.) is referred to as the quantity takeoff (QTO) or quantity surveying.[2] The procedures employed by the estimator to calculate these quantities should incorporate steps to minimize errors. Five of the most common errors experienced during quantity takeoff are:

1. Arithmetic: Errors in addition, subtraction, and multiplication
2. Transposition: Mistakes in copying or transferring figures, dimensions, or quantities
3. Errors of omission: Overlooking items called for or required to accomplish the work
4. Poor reference: Scaling drawings rather than using the dimensions indicated
5. Unrealistic waste or loss factors

The first step in the quantity takeoff procedure is to identify the materials required by each estimating account or work package. Once the types of materials are identified, relevant dimensions are recorded on a spreadsheet so that quantity calculations in the required unit of measure can be made. Calculation of quantities by estimating account or work package has several advantages, not the least of which is the fact that it allows the estimating process to be performed by several estimators, each with a well-defined area of responsibility. No matter how competent an estimator may be in his own field, it is not reasonable to expect him to have an intimate knowledge of all phases of construction. This method enables one estimator to check another estimator's work. It also facilitates computations required to develop the financial "picture" of the job and processing of progress payment requests.

---

[1] James N. Neil, *Construction Cost Estimating for Project Control*, Prentice-Hall, Englewood Cliffs, NJ, 1982, p. 73.

[2] This term is commonly used in the United Kingdom and the Commonwealth countries.

Prepare: Activity list
              Activity material list (estimate)–include work sheets
              Material recap sheets

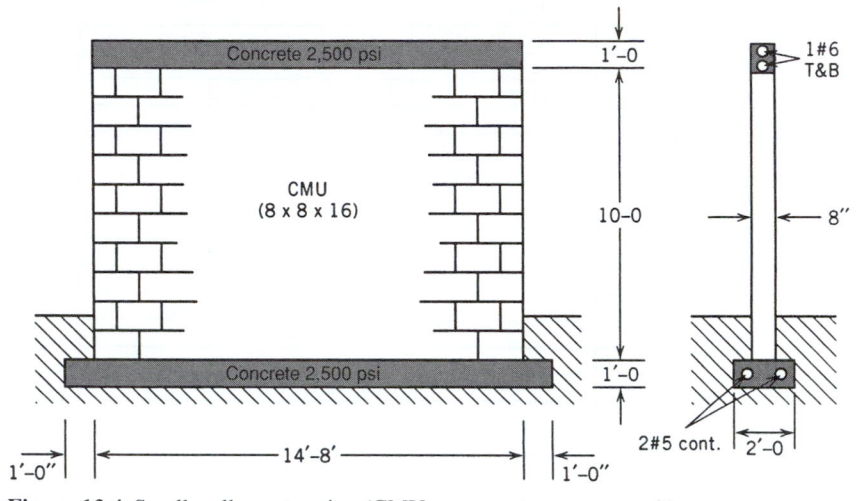

Figure 13.4 Small wall construction (CMU = concrete masonry unit).

When changes occur, only those activities affected must be recalculated. Other procedures require a completely new takeoff.

Before the calculations for the quantity takeoff are performed, detailed working drawings are sometimes required to clarify the contract drawings and specifications or the chosen construction method (e.g., forming techniques). Such a drawing for a small wall is given in Figure 13.4. During construction these details are of tremendous value to the person in the field who is trying to perform the work within the cost guidelines provided. From these drawings and details a checklist should be developed to indicate all of the materials required for each work package. After this checklist has been made, it should be checked against a standard checklist to identify errors of omission.

The actual calculations should be performed on a standard spread sheets to allow for independent check and self-checks. As the calculations for those items shown on the plans progress, each item taken off should be highlighted by a color marker so that those items remaining to be considered are obvious at a glance. Arithmetic should be performed on a calculator or computer that produces hard copy output. This output should be used by the estimator to identify errors. All supporting documentation should be attached to the estimate to aid in checking by other sources or at a later date. The quantities calculated should be exact. Waste and loss factors will be applied later. A materials takeoff sheet for the small wall (Fig. 13.4) is given in Figure 13.5.

A summary, or "recap sheet," should be made. This recap sheet should consist of a listing, by material type, of all the materials required for the entire work item or package. The listing should include total quantities as well as subquantities identified by activity code. The listing should also include appropriate waste and loss factor calculations. An example of a recap sheet is given in Figure 13.6. This example is simple and is included only to demonstrate the nature of quantity development. In practice, most companies will use computerized data bases and spread sheet programs to prepare final estimates. The basic principles of estimating must be well understood, however, to avoid omissions which could prove disastrous at the time of bid submittal.

13.6   METHODS OF DETAILED COST DETERMINATION

After quantities have been determined for accounts that are relevant to the project at hand, the method by which costs will be assigned can be selected. The two methods of cost

Project _____

Activity code	Activity description	Material description	Quantity	Unit	Cost code
1	Layout	Stakes 2 x 4 x 24 8 ea.	10.3	BF	0100
3	Place rebar	#5 st. 2 PCS 16 – 2	32.3	LF	0320
		Tie wire	1	Roll	0320
4	Cost and cure	footing			
		Concrete	1.23	CY	0330
		Curing compound	.25	Gal	0337
5	Erect CMU wall				
		CMU 8 x 8 x 16 stretcher	143	Ea	0412
		CMU 8 x 8 x 16 corner	14	Ea	0412
		CMU 8 x 8 x 16 corner	16	Ea	0412
		Scaffolding 4' x 4' x 6'	2	Sec.	0100
		Mortar	.27	CY	0412
7	Form bond beam				
		2 x 4 (4 – 15' – 0")	43.5	BF	0310
		2 x 2	12.7	BF	0310
		1 x 2	2.0	BF	0310
		3/4" ext ply	60.3	SF	0310
		Snapties 8"	24	Ea	0310
		Nails 8d	1.5	Lb	0310
		Nails 6d	.4	Lb	0310
		Form oil	.07	Gal	0310
8	Place bond beam rebar				
		#6 rebar (str.)	28.67	LF	0320
9	Cost and cure	Bond beam			
		Concrete	.35	CY	0330
		Curing compound	.05	Gal	0337
10	Strip forms and rub bond beam				
		Grout	1	CF	0339.2

Figure 13.5 Activity material list.

determination most frequently used are (1) unit pricing and (2) resource enumeration. If the work as defined by a given estimating account is fairly standard, the cost can be calculated by simply taking *dollar per unit* cost from company records and applying this cost with a qualitative correction factor to the quantity of work to be performed. For instance, if the project calls for 100 lineal feet of pipe and historical data in the company indicate that the pipe can be placed for $65 a lineal foot to include labor and materials, the direct cost calculation for the work would yield a value of $6,500. This value can then be adapted for special site conditions.

Unit pricing values are available in many standard estimating references. The standard references normally give a nationally averaged price per unit. A multiplier is used to adjust the national price to a particular area. These references are updated on an annual basis to keep them current. Among the largest and best known of these services are:

1. R. S. Means Company, *Building Construction Cost Data*
2. F. R. Walker's *The Building Estimator's Reference Book*
3. *The Richardson General Construction Estimating Standards*

Project _____ Wall _____

Description	Activity code	Sub–quantity	Waste	Total quantity	Unit		Cost code
2 x 4 Lumber	Total	53.8	10%	60.0	BF		
	1	10.3					0100
	7	43.5					0310
2 x 4 Lumber	7	12.7	10%	14.0	BF		0310
1 x 2 Lumber	7	2.0	10%	2.25	BF		0310
3/4" Exterior plywood	7	60.3	10%	66	SF		0310
Curing compound	Total	.30		1	Gal		0337
	4	.25					
	9	.05					
Snap ties 8"	7	24.	5%	25	Ea		0310
Nails 8d	7	1.5		3	LB		0310
Nails 6d	7	.4		1	LB		0310
Form oil	7	.07		.25	Gal		0310

Figure 13.6 Construction support materials recap sheet.

These references contain listings of cost line items similar to the cost account line items a contractor would maintain.

The development of direct costs to include overhead and profit for a particular line item using the R. S. Means system is shown in Figure 13.7.

The line items specified in the R. S. Means Construction Cost Data are defined by using the Uniform Construction Index numerical designators. The system assumes a given crew composition and production rate for each line item. In the case illustrated a standard crew designated C-1 can construct 190 SFCA (square foot contact area) of plywood column form per shift (daily output). This underlines the fact that unit pricing data must make some assumption regarding the resource group (i.e., crew, equipment fleet, etc.) and the production rate being used. That is, although unit pricing data are presented in dollars-per-unit format, the cost of the resource group and the rate of production achieved must be considered. The dollars-per-unit value is calculated as follows:

$$\frac{\text{Cost of resources per unit time}}{\text{Production rate of resources}} = \frac{\$/hr}{\text{unit}/hr} = \$/\text{unit}$$

The unit cost is the ratio of resource costs to production rate. The crew composition and assumed cost for the crew are shown in the middle of Figure 13.7.

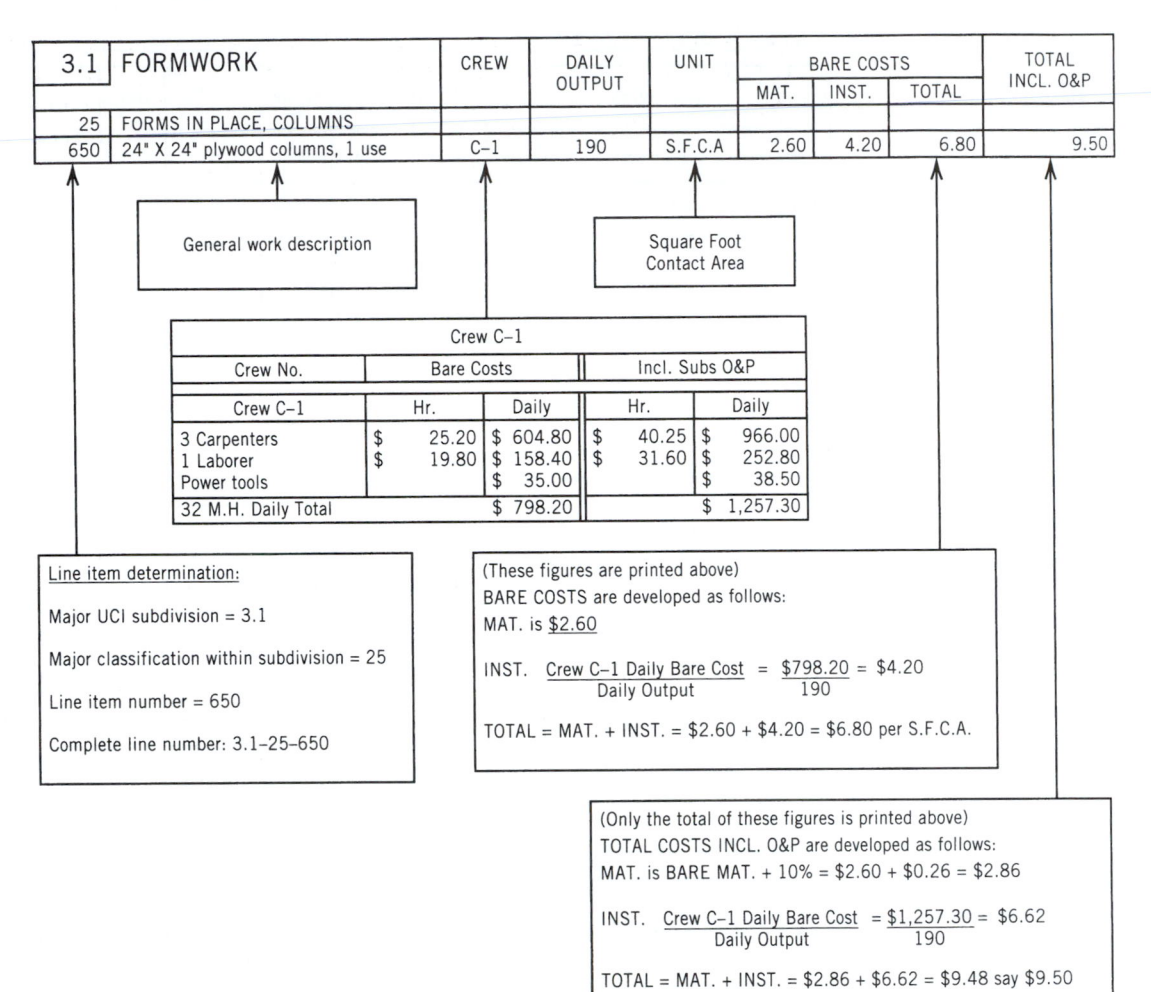

Figure 13.7 Line item cost development using R. S. Means data. "From Means Building Construction Cost Data 1981. Copyright Reed Construction Data, Kingston, MA 781-585-7880; All rights reserved."

In the R. S. Means system two costs are specified for each line item. The bare cost is the direct cost for labor and materials. The total cost includes the cost of burdens, taxes, and subcontractor overhead and profit (inclusive O&P). In Figure 13.7, the bare cost of the C-1 crew is calculated as $798.20 per shift. Therefore, the bare unit installation cost is

$$\frac{\$798.20/\text{shift}}{190 \text{ units/shift}} = \$4.20/\text{SFCA}$$

Combining this installation cost with the material cost per unit of $2.60 yields a bare unit cost for materials and installation of $6.80.

The overhead and profit (O&P) charges associated with labor (as considered in the Means system) are:

1. Fringe benefits (included in bare costs)
2. Workmen's compensation
3. Average fixed overhead
4. Subcontractor overhead
5. Subcontractor profit

In order to adjust the bare costs of installation to include subcontractor's O&P, the appropriate craft values for the members of the craft are located and applied. For the carpenters the total correction is 59.7%, or $15.05 per hour. Therefore, the carpenter rate to include O&P is $40.25 per hour. Similarly, the laborer rate is adjusted to $31.60 per hour to include O&P. A markup of 10% is applied to the power tools, yielding a daily rate of $38.50. The new installation rate to include O&P is

$$\frac{\$1,257.30}{190} = \$6.62/\text{SFCA}$$

The 10% markup is applied to the material cost, resulting in a $2.86 charge per SFCA. The combined cost of materials and installation is $9.48 per SFCA, which is rounded to $9.50.

13.7   PROBLEMS WITH UNIT-COST METHOD

The data that the contractor has available from company records are presented as dollars per unit, and in most cases no records of the crew composition, cost, and production rates is maintained. In fact, the dollars-per-unit value is typically an average of the values obtained on recent jobs. Since on each job the crew composition, costs, and production rates achieved are probably unique to the individual job, the figure represents an aggregate cost per unit. The actual number of man-hours used and the productivity achieved are masked by the dollar-per-unit figure. Unless the resource (i.e., man-hour, etc.) information is kept separately, it has been lost. Therefore, the unit price available from averaging values on previous jobs has to be treated with some caution. Since every job is unique, some of the estimator's intuition must be applied to ensure that the value is adapted to the conditions of the job being estimated. If the conditions of jobs vary very little, however, the application of the unit pricing approach is both practical and efficient.

Clearly, the numerator (cost of resources per unit time) of the unit-cost ratio will vary significantly over time as the costs of labor and machines vary. The costs of all components of the construction process have risen sharply over the past 20 years. This is shown dramatically in the *Engineering News Record* construction and building cost indexes shown in Chapter 2. In order to factor out the inflationary escalation inherent in resource costs, some contractors maintain the ratio of man-hours or resource-hours per hour to production. This establishes a company data-base tied to resource hours required rather than dollars per unit. Therefore, the contractor can retrieve a man-hour or resource-hour (RH) per unit value for each line item. The value is calculated as:

$$\frac{\text{Resource-hours per hour}}{\text{Units per hour}} = \text{RH/unit}$$

The cost per unit can then be calculated by multiplying the resource hours per unit value by the average hourly cost per resource. If it takes 25 resource-hours per unit and the average cost of a resource-hour is $20.00, the unit cost will be $500.00 per unit. This method recognizes that the number of resource-hours required per unit is much more stable over the years than the cost per unit. Therefore, data on resource-hours per unit collected over several years will not be affected by inflationary trends and escalation in the cost of goods and services.

Use of the unit-pricing approach assumes that historical data have been maintained for commonly encountered cost accounts. Data are collected and linked to a reference unit such as a cubic yard or square foot. The costs of materials and installation are aggregated and then presented as a cost per unit. Companies typically accumulate such data either manually or on the computer as a by-product of the job cost system. On a typical job 80 to 90% of the work to be accomplished can be estimated by calculating the number of reference units and multiplying this number by the unit price. Typically the estimator will intuitively adjust this price to reflect special characteristics of the job, such as access restrictions, difficult management environment, and the like. One approach to the quantification of these site and job unique factors is proposed by Louis Dallavia. Although the Dallavia method (*Estimating General Construction Costs*, 1957) is dated, it does reflect, in an approximate way, the factors that are considered by an estimator in adjusting general unit prices to a given project. The system defines a *percent efficiency factor* based on a production range index for each of eight job characteristics. The method of calculating the percent efficiency factor and the table production range indices are shown in Figure 13.8.

13.8   RESOURCE ENUMERATION

Although the unit-pricing approach is sufficiently accurate to estimate the common accounts encountered on a given project, almost every project has unique or special features for which unit-pricing data may not be available. Unusual architectural items that are unique to the structure and require special forming or erection procedures are typical of such work. In such cases, the price must be developed by breaking the special work item into its subfeatures and assigning a typical resource group to each subfeature. The productivity to be achieved by the resource group must be estimated by using either historical data or engineering intuition. The breakdown of the cost center into its subelements would occur much in the same fashion in which the wall of Section 13.5 was subdivided for quantity development purposes. The steps involved in applying the resource enumeration approach are shown in Figure 13.9.

An example of resource enumeration applied to a concrete-placing operation is shown in Figure 13.10. In this example a concrete placement crew consisting of a carpenter foreman, two cement masons, a pumping engineer (for operation of the concrete pump), and seven laborers for placing, screening, and vibrating the concrete has been selected. A concrete pump (i.e., an equipment resource) has also been included in the crew. Its hourly cost has been determined using methods described above. The total hourly rate for the crew is found to be $370.00. The average assumed rate of production for the crew is 12 cu yd/hr. This results in an average labor cost per cubic yard of concrete of $30.83. The line items requiring concrete are listed with the quantities developed from the plans and specifications. Consider the first item that pertains to foundation concrete. The basic quantity is adjusted for material waste. The cost per unit is adjusted to $34.25 based on an efficiency factor for placement of foundation concrete estimated as 90%.

The resource enumeration approach has the advantage over unit pricing in that it allows the estimator to stylize the resource set or crew to be used to the work in question. The rates of pay applied to the resource group reflect the most recent pay and charge rates, and therefore incorporate inflationary or deflationary trends into the calculated price. The basic equation for unit pricing is

$$\frac{\text{Resource cost per unit time}}{\text{Production rate}} = \frac{\$/\text{hr}}{\text{unit/hr}} = \$/\text{unit}$$

In the unit-pricing approach the resource costs and the production rates are the aggregate values of resources and rates accumulated on a number of jobs over the period of historical data collection. With the resource enumeration approach, the estimator specifies a particular crew or resource group at a particular charge rate and a particular production level for the specific work element being estimated. This should yield a much more precise cost-per-unit definition. The disadvantage with such a detailed level of cost definition is the fact that it is time-consuming. Therefore, resource enumeration would be used only on (1) items for which no unit cost data are available (2) "big-ticket" items, which constitute a large percentage of the overall cost of the job and for which such a precise cost analysis may lead to cost

Production Range Index			
	Production Efficiency (%)		
	25 35 45 55 65 75 85 95 100		
Production Elements	Low	Average	High
1. General Economy	Prosperous	Normal	Hard times
Local business trend	Stimulated	Normal	Depressed
Construction volume	High	Normal	Low
Unemployment	Low	Normal	High
2. Amount of Work	Limited	Average	Extensive
Design areas	Unfavorable	Average	Favorable
Manual operations	Limited	Average	Extensive
Mechanized operations	Limited	Average	Extensive
3. Labor	Poor	Average	Good
Training	Poor	Average	Good
Pay	Low	Average	Good
Supply	Scarce	Average	Surplus
4. Supervision	Poor	Average	Good
Training	Poor	Average	Good
Pay	Low	Average	Good
Supply	Scarce	Average	Surplus
5. Job Conditions	Poor	Average	Good
Management	Poor	Average	Good
Site and materials	Unfavorable	Average	Favorable
Workmanship required	First rate	Regular	Passable
Length of operations	Short	Average	Long
6. Weather	Bad	Fair	Good
Precipitation	Much	Some	Occasional
Cold	Bitter	Moderate	Occasional
Heat	Oppressive	Moderate	Occasional
7. Equipment	Poor	Normal	Good
Applicability	Poor	Normal	Good
Condition	Poor	Fair	Good
Maintenance, repairs	Slow	Average	Quick
8. Delay	Numerous	Some	Minimum
Job flexibility	Poor	Average	Good
Delivery	Slow	Normal	Prompt
Expediting	Poor	Average	Good

Example: After studying a project on which he is bidding, a contractor makes the following evaluations of the production elements involved:

Production Element	*% Efficiency*
1. Present economy	75
2. Amount of work	90
3. Labor	70
4. Supervision	80
5. Job conditions	95
6. Weather	85
7. Methods and equipment	55
8. Delays	75
Total	625

As the total of the eight elements is 625, the average value will be 625/8, or 78%.

Figure 13.8 Dallavia method.

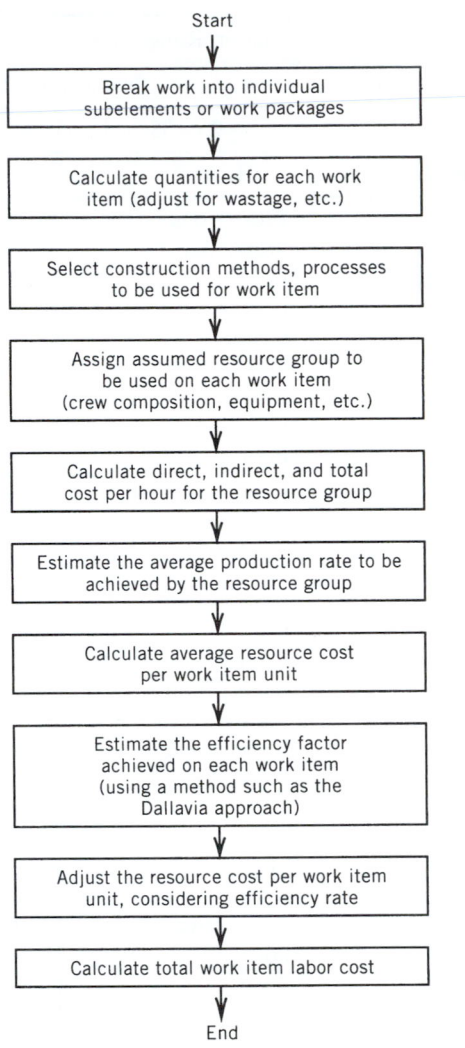

Start

Break work into individual
subelements or work packages

Calculate quantities for each work
item (adjust for wastage, etc.)

Select construction methods, processes
to be used for work item

Assign assumed resource group to
be used on each work item
(crew composition, equipment, etc.)

Calculate direct, indirect, and total
cost per hour for the resource group

Estimate the average production rate to be
achieved by the resource group

Calculate average resource cost
per work item unit

Estimate the efficiency factor
achieved on each work item
(using a method such as the
Dallavia approach)

Adjust the resource cost per work item
unit, considering efficiency rate

Calculate total work item labor cost

End

Figure 13.9 Resource enumeration method of estimating.

savings that may provide the winning margin at bid time, or (3) extremely complex work items on complicated and unique projects for which the use of the unit-pricing approach is deemed inadequate.

13.9 WORK PACKAGE OR ASSEMBLY-BASED ESTIMATING

In this approach to estimate development, a work package or assembly that is commonly encountered in construction is viewed as an estimating group, and appropriate dimensional and cost-related parameters are defined for the package. The wall of Figure 13.4 could be considered an assembly. In this case, the height, width, and depth of the footer, block portion, and cap beam would be specified each time the assembly is encountered. Pricing information for the defined wall would be retrieved from a pricing catalog. Since the reference subelement in this approach is the work package, an extensive listing of assemblies or packages into which the work can be subdivided is maintained.

A concrete footer assembly is shown in Figure 13.11. The relevant data required for takeoff are the dimensional values shown as items A through K. Data regarding the

Concrete Placing Crew			
Quantity	Member	Rate	Total/Hour
1	Carpenter foreman	$40.00	$ 40.00
2	Cement masons	$36.00	$ 72.00
1	Pumping engineer	$38.00	$ 38.00
7	Laborers	$28.00	$196.00
1	Concrete pump	$24.00	$ 24.00
		Crew hourly rate	$370.00

Production rate of crew under normal circumstances (efficiency factor 1) = 12 cu yd/hr.
Average labor cost/cubic yard = $370/12 = $30.83.

Area	Quantity	Percent Waste	Efficiency Factor	Labor Cost/ Cubic Yard	Activity Cost
1. Foundation	53.2	15	0.9	$34.25	$ 1,822
2. Wall to elevation 244.67	52.9	12	0.8	38.54	2,039
3. Slab 10 in.	1.3	30	0.3	102.77	134
4. Beams elevated 244.67	10.5	15	0.7	44.04	462
5. Beams elevated 245.17	9.1	15	0.7	40.44	401
6. Slab elevation 244.67	8.7	10	0.7	40.44	383
7. Interior wall to 244.67	5.5	15	0.4	77.07	424
8. Slab elevation 254.17	6.3	10	0.75	41.11	259
9. Walls 244.67 −254.17	57.2	10	0.8	38.54	2,205
10. Walls 254.17 −267	42.0	10	0.8	38.54	1,619
11. Floors elevated 267	8.9	10	0.9	34.25	305
12. Manhole walls	27.3	10	0.85	36.27	990
13. Roof	14.0	15	0.7	44.04	617
14. Headwall	8.5	10	0.8	38.59	328
Total direct labor cost for concrete					$11,988 say $12,000

Figure 13.10 Labor resource enumeration.

methodology of placement and the relevant specifications are indicated by items (1) to (9) in the figure. Such work-package-based systems can be considered a structured extension of the resource enumeration approach and can be calculated manually. In general, most of these system-based (i.e., assembly-based) systems are computerized and are based on presenting the estimator with individual assemblies. The estimator is interrogated by the computer and provides the dimensional and methodology information in a question-and-answer format. This procedure is shown schematically in Figure 13.12. The estimator goes through the construction systems sequentially, selecting those that are relevant and providing the required data. These data are integrated with information from a pricing catalog. The pricing catalog allows for price, resource, and productivity adjustment. The manual or

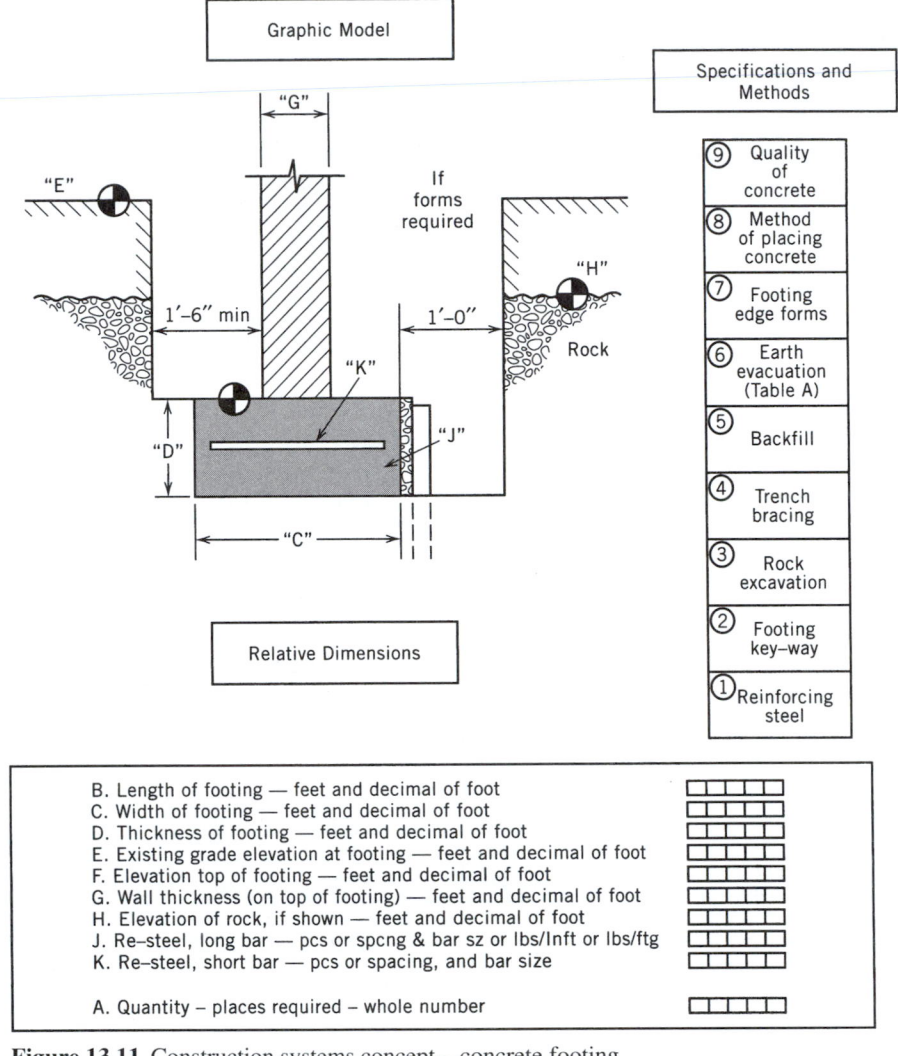

Figure 13.11 Construction systems concept—concrete footing.

computer program integration of these data produces the estimate reports required for bid preparation.

If a manual approach is used to estimate each work package, a work package takeoff sheet is helpful in organizing the collection of data. Such a work package or assembly collection sheet can be organized as shown in Figure 13.13. This form illustrates the development of an estimate for slab on grade in a building project. Material, labor, and equipment resources required for the package are shown on the left side of the sheet, together with target prices for each resource. It is interesting to note that the equipment resources are normally charged on a period basis since partial-day allocation of equipment is not a common practice. The right side of the sheet considers the productivity rate to be used, special notes or characteristics relating to the package, and a total cost summary for the package. This sheet is quite versatile and can be used for earthwork, masonry, and virtually any assembly encountered in the construction of a project. A similar sheet for earthwork is shown in Figure 13.14.

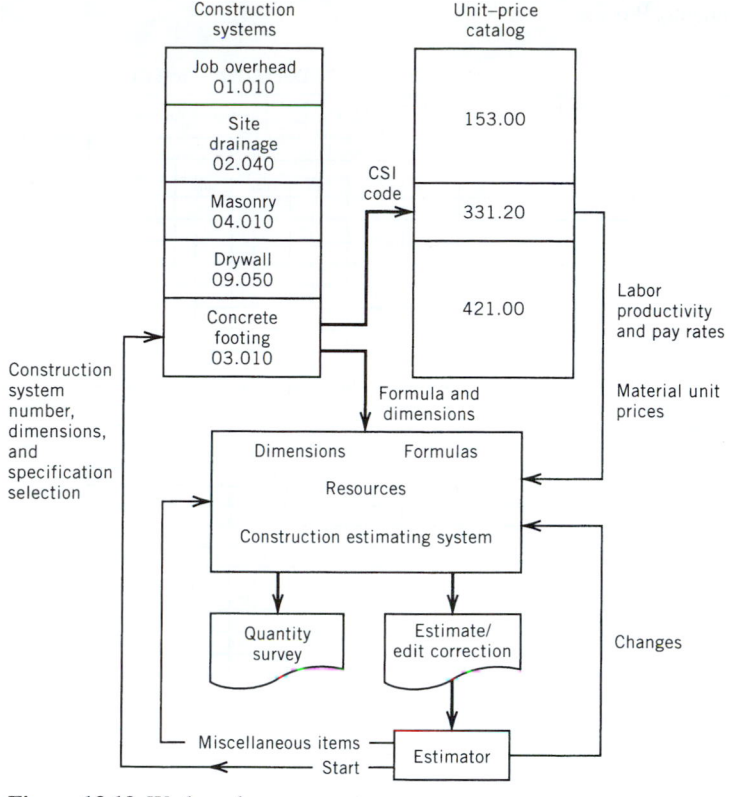

Figure 13.12 Work package concept.

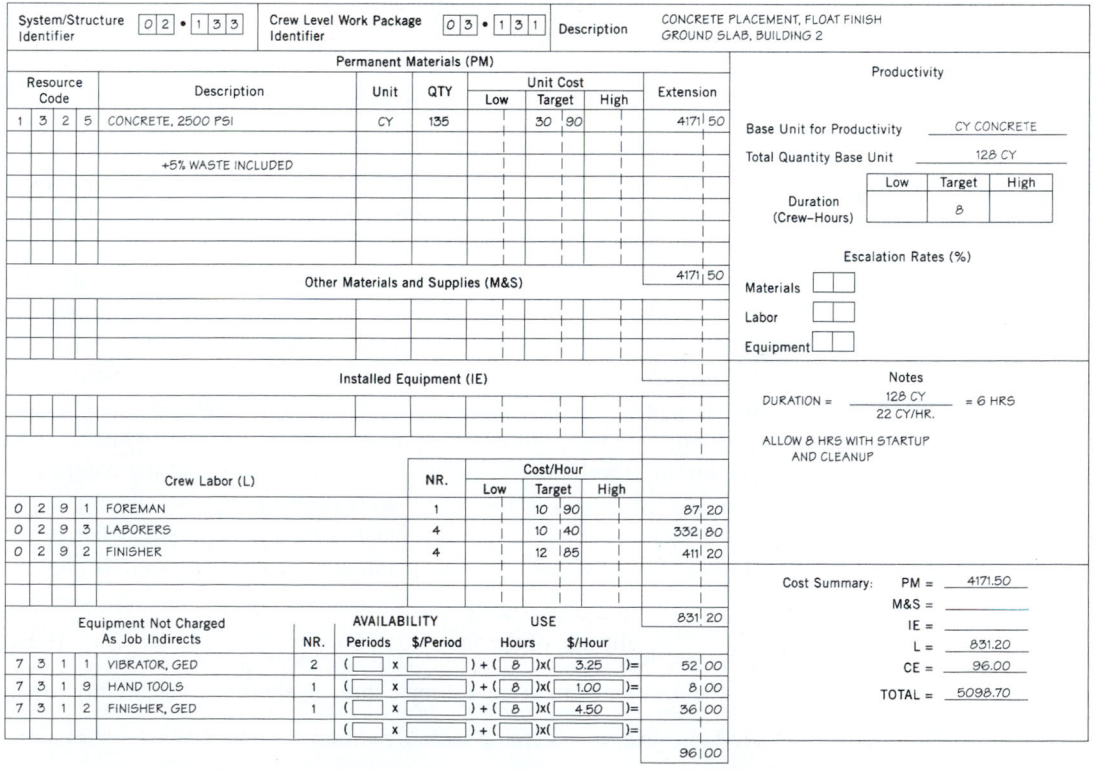

WORK PACKAGE COLLECTION SHEET

System/Structure Identifier: `0 2 · 1 3 3`   Crew Level Work Package Identifier: `0 3 · 1 3 1`   Description: CONCRETE PLACEMENT, FLOAT FINISH GROUND SLAB, BUILDING 2

Permanent Materials (PM)

Resource Code	Description	Unit	QTY	Unit Cost Low	Unit Cost Target	Unit Cost High	Extension
1 3 2 5	CONCRETE, 2500 PSI	CY	135		30 90		4171 50
	+5% WASTE INCLUDED						

Other Materials and Supplies (M&S) — 4171 50

Installed Equipment (IE)

Crew Labor (L)	NR.	Cost/Hour Low	Cost/Hour Target	Cost/Hour High	
0 2 9 1 FOREMAN	1		10 90		87 20
0 2 9 3 LABORERS	4		10 40		332 80
0 2 9 2 FINISHER	4		12 85		411 20

831 20

Equipment Not Charged As Job Indirects	NR.	AVAILABILITY Periods	$/Period	USE Hours	$/Hour	
7 3 1 1 VIBRATOR, GED	2	(☐ x ☐)	+ (8)x(3.25)=			52 00
7 3 1 9 HAND TOOLS	1	(☐ x ☐)	+ (8)x(1.00)=			8 00
7 3 1 2 FINISHER, GED	1	(☐ x ☐)	+ (8)x(4.50)=			36 00
		(☐ x ☐)	+ (☐)x(☐)=			

96 00

Productivity

Base Unit for Productivity: CY CONCRETE

Total Quantity Base Unit: 128 CY

	Low	Target	High
Duration (Crew-Hours)		8	

Escalation Rates (%)

Materials: ☐
Labor: ☐
Equipment: ☐

Notes

DURATION = $\dfrac{128\ CY}{22\ CY/HR.}$ = 6 HRS

ALLOW 8 HRS WITH STARTUP AND CLEANUP

Cost Summary:
PM = 4171.50
M&S = _____
IE = _____
L = 831.20
CE = 96.00
TOTAL = 5098.70

Figure 13.13 Work package collection sheet—concrete slab (*Source:* J. M. Neil, *Construction Cost Estimating and Cost Control*, Englewood Cliffs, NJ: Prentice-Hall, 1982, p. 231).

WORK PACKAGE COLLECTION SHEET

System/Structure Identifier: 0 1 • 3 0 0   Crew Level Work Package Identifier: 0 2 • 1 1 1   Description: EXCAVATION, AREA 3, AND DISPOSAL W/O COMPACTION

Permanent Materials (PM)

Resource Code	Description	Unit	QTY	Unit Cost Low	Target	High	Extension

Productivity

Base Unit for Productivity: BANK CY

Total Quantity Base Unit: 24,000 BCY

	Low	Target	High
Duration (Crew–Hours)		32	

Escalation Rates (%)

Materials: ☐☐
Labor: ☐☐
Equipment: ☐☐

Other Materials and Supplies (M&S)

Installed Equipment (IE)

Notes

LF. = 0.8
CYCLE = 6.3 MIN; ASSUME 45 MIN HOUR

$$BCY/HR = \left(\frac{45 \text{ MIN}}{6.3 \text{ MIN}}\right)(22CY)(0.8LF)(6 \text{ SCRAPERS})$$

$$= 754 \text{ BCY/HR}$$

$$DURATION = \frac{24,000 \text{ BCY}}{754 \text{ BCY}} = 32 \text{ HOURS}$$

ADD 1 DAY FOR BAD WEATHER
∴ PROJECT WILL LAST ONE WORK WEEK

Crew Labor (L)

Resource Code	Description	NR.	Cost/Hour Low	Target	High	Extension
0 8 1 1	FOREMAN	1		13 90		444 80
0 8 1 2	EQUIP OPER, MEDIUM	10		13 40		4288 00
0 8 1 3	SPOTTER	2		10 40		665 60
						5398 40

Cost Summary:
PM = _____
M&S = _____
IE = _____
L = 5,398.40
CE = 22,775.20
TOTAL = 28,173.60

Equipment Not Charged As Job Indirects

Resource Code	Description	NR.	AVAILABILITY Periods	$/Period	USE Hours	$/Hour	Extension
7 3 2 2	SCRAPER, ELEV, 22 CY	6	(1 x 1850–) + (32)x(24.00)=				15,708 00
7 2 0 7	DOZER, D7	3	(1 x 1600–) + (32)x(11.25)=				5,808 00
7 4 1 2	GRADER, 12'	1	(1 x 1000–) + (32)x(8.10)=				1,259 20
			(x) + ( )x( )=				
							22,775 20

Figure 13.14 Work package collection sheet—excavation (*Source:* J. M. Neil, *Construction Cost Estimating and Cost Control*, Englewood Cliffs, NJ: Prentice-Hall, 1982, p. 221).

13.10   SUMMARY

The estimate is the basis for the contractor's bid and, as such, has a significant effect on whether or not a given project is profitable. In building construction the four levels of estimate preparation are (1) conceptual, (2) preliminary, (3) engineer's, and (4) bid.

The first three of these estimates are typically prepared by the architect/engineer and reflect the increasing refinement of the design. Large and complex projects include a magnitude and definitive estimate in addition to those noted above. The bid estimate is a detailed estimate prepared by the contractor. The steps involved in preparing a detailed estimate are shown graphically in Figure 13.15. The project to be estimated is subdivided for cost analysis purposes into estimating accounts or work packages. Quantities for each package or account are developed. These quantities are priced, and the extensions are calculated and checked for errors. At this stage, professional judgment and engineering intuition are utilized to adjust the bid to reflect special or unique factors peculiar to the particular job. Profit margin is also applied at this point, and the bid is revised as required and finalized. Steps 3 through 6 are quantitative in nature and involve the application of formulas and arithmetical concepts. Steps 1 and 2 require professional expertise. Steps 7 and 8 require a good deal of experience and engineering judgment.

In the development of the estimate three methods are commonly used. These are:

1. Unit-pricing or catalog lookup method

2. Resource enumeration

3. Work package/assembly method

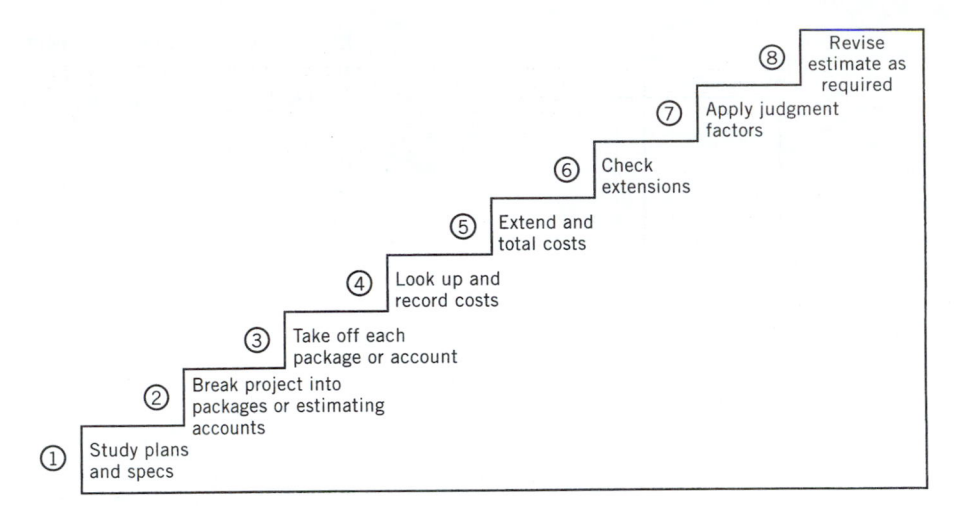

Figure 13.15 Steps in the estimating process.

The work package method can be thought of as an extension of the resource enumeration method. Different methods may be used on different parts of the job. On portions of the job that are very cost sensitive and constitute a large portion of the overall project, methods 2 and 3 may be appropriate. On parts of the work that are standard and straightforward, the unit-pricing approach is normally acceptable. Selection of methods is a tradeoff between the need for accuracy and the cost of obtaining that accuracy. The keys to a successful estimate are (1) the ability to assess the required level of accuracy and (2) the ability to achieve the required level of accuracy at minimal cost.

REVIEW QUESTIONS AND EXERCISES

13.1 Explain the difference between unit-cost estimating methods and resource enumeration methods. When would you use unit cost? When would you use resource enumeration?

13.2 What is meant by contractor O&P? Give three components that are considered in the O&P.

13.3 What is the difference between labor cost and labor productivity? Use a sketch to illustrate.

13.4 The partition wall shown below is to be constructed of 8 × 16 × 6 block. Estimate the cost of the wall to include labor, materials, and contractor O&P using the R. S. Means building cost data or other appropriate estimating reference. The job is located in Cincinnati, Ohio.

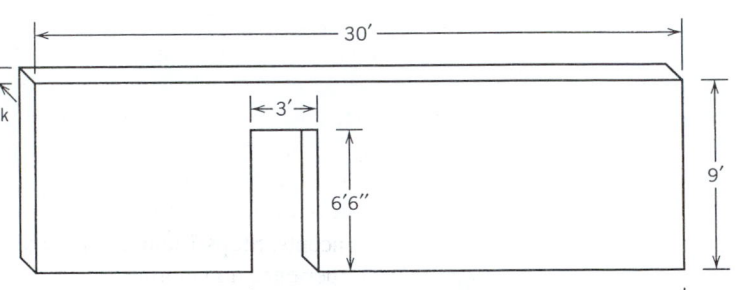

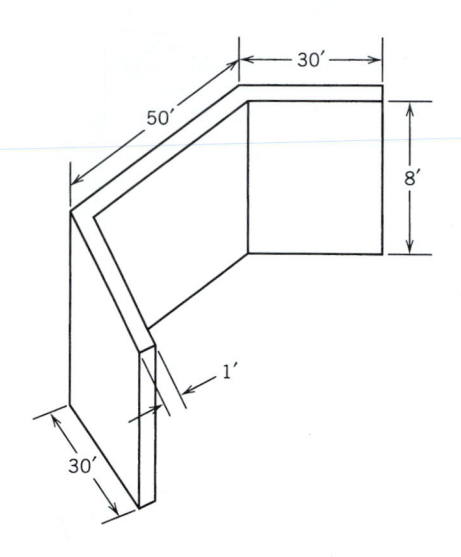

13.5 Given the bridge abutment shown here, determine the number of man-hours required to form the structure using data from R. S. Means or other appropriate reference.

Chapter 14

Construction Labor

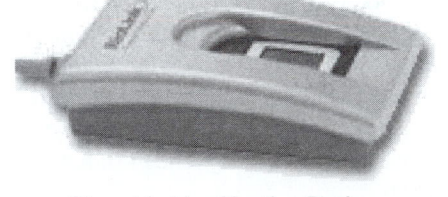

Biometric Identification Devices

Biometric Identification Recognition

The Need

The access control system plays a major role in site security. These days, only a few engineers and managers are needed to wholly control a 20-story building's physical security objects such as elevators, door locks, lights, security system control, etc. through computerized automatic systems. Controling and screening personnel entering a facility is almost impossible for a few security people to physically control. Even though visitors, office workers, and residents are given keys or id-tags after being screened, there is still a possibility that unauthorized people could access important property with fake id's or keys, etc. Recently, biometric technology has arisen as a new, effective, and secure method for identification recognition of personnel.

The Technology

Biometrics are computerized methods of recognizing people based on physical or behavioral characteristics. The main biometric technologies include face recognition, fingerprint, hand geometry, iris, palm prints, signature, and voice. Biometric technologies can work in two modes —authentication (one-to-one matching) and identification (one-to-many matching). However, only three biometrics are capable of the latter—face, finger and iris. This technology has great potential for controlling access to construction work sites as well as other facilities.

14.1 LABOR RESOURCE

The man-power component of the four Ms of construction is by far the most variable and unpredictable. It is, therefore, the element that demands the largest commitment of time and effort from the management team. Manpower or labor has four major aspects that are of interest to management. To properly understand the management and control of labor as a resource, the manager must be aware of the interplay among the following elements:

1. Labor organization
2. Labor law
3. Labor cost
4. Labor productivity

The cost and productivity components were central to the discussion of equipment management in Chapters 11 and 12. Labor includes the added human factor. This element can only be understood in the context of the prevailing legal and organizational climate that is characteristic of the construction industry.

14.2   SHORT HISTORY OF LABOR ORGANIZATIONS

The history of labor organizations begins in the early nineteenth century, and their growth parallels the increasing industrialization of modern society. Initially tradesmen possessing some skill or craft began organizing into groups variously called guilds, brotherhoods, or mechanics societies. Their objectives were to provide members, widows, and children with sickness and death benefits. In addition, these organizations were interested in the development of trade proficiency standards and the definition of skill levels such as apprentice and journeyman. They were often "secret" brotherhoods because such organizations were considered unlawful and illegal conspiracies posing a danger to society.

From the 1840s until the era of the New Deal[1] in the 1930s, the history of labor organizations is the saga of confrontation between management and workers, with the pendulum of power on the management side. With the coming of the New Deal and the need to rejuvenate the economy during the Depression, labor organizations won striking gains that virtually reversed the power relationship between managers and workers. The American Federation of Labor (AFL) was organized by Samuel Gompers in 1886. This was the first successful effort to organize skilled and craft workers such as cabinetmakers, leather tanners, and blacksmiths. Since its inception, the AFL has been identified with skilled craft workers as opposed to industrial "assembly line" type of workers. The Building and Construction Trades Department of the AFL, which is the umbrella organization representing all construction craft unions, was organized in 1908.

The semiskilled and unskilled factory workers in "sweat shop" plants and mills were largely unorganized at the time Gompers started the AFL. Many organizations were founded and ultimately failed in an attempt to organize the industrial worker. These organizations, with euphonious-sounding names, such as Industrial Workers of the World and the Knights of Labor, had strong political overtones and sought sweeping social reforms for all workers. This was particularly attractive to immigrant workers arriving from the socially repressive and politically stagnant atmosphere in Europe. Such organizations attracted political firebrands and anarchists preaching social change and upheaval at any cost. Confrontation with the police was common, and violent riots often led to maiming and killing. The most famous such riot occurred in the Haymarket in Chicago in 1886.

Gompers was seriously interested in protecting the rights of skilled workers and had little interest in the political and social oratory of the unskilled labor organizations. Therefore, separate labor movements representing skilled craft and semiskilled factory workers developed and did not combine until the 1930s. This led to different national and local organizational structures and bargaining procedures that are still utilized and strongly influence the labor picture even today.

---

[1] The administration of President Franklin D. Roosevelt.

In the 1930s, industrial (i.e., factory semiskilled) workers began to organize effectively with the support of legislation evolving during the post-Depression period. The AFL, realizing such organizations might threaten its own dominance, recognized these organizations by bringing them into the AFL camp with the special designation of Federal Locals. Although nominally members, the industrial workers were generally treated as second-class citizens by the older and more established craft unions. This led to friction and rivalry that culminated in the formation of the Committee for Industrial Organizations (CIO). This committee was established in 1935 unilaterally by the industrial locals without permission from the governing body of the AFL. The act was labeled treasonous, and the AFL board ordered the Committee to disband or be expelled. The AFL suspended the industrial unions in 1936. In response, these unions organized as the Congress of Industrial Organizations (CIO), with John L. Lewis of the United Mine Workers as the first CIO president. Following this rift between the industrial and craft union movements, the need to cooperate and work together was apparent. However, philosophical and personal differences prevented this until 1955, at which time the two organizations combined to form the AFL–CIO. This organization remains the major labor entity in the United States today.

14.3 EARLY LABOR LEGISLATION

The courts and legislative bodies of the land have alternately operated to retard or accelerate the progress of labor organizations. The chronology of major items of legislation and the significant events in the labor movement are shown in Table 14.1. At the outset, the law was generally interpreted in order to check organization of labor and, therefore, management was successful in controlling the situation. The most classic illustration of this is the application of the Sherman Antitrust Law to enjoin workers from organizing. The Sherman Antitrust Act had originally been enacted in 1890 to suppress the formation of large corporate trusts and cartels, which dominated the market and acted to fix prices and restrain free trade. The oil and steel interests formed separate cartels in the late nineteenth century to manipulate the market. More recently, Microsoft Corporation has been reviewed by the Justice Department to prevent similar dominance in the computer market. To break up such potential market control, the Antitrust Law provides the government with the power to enjoin corporations from combining to control prices and restrict trade. In 1908 the Supreme Court ruled that the Antitrust Law could be applied to prevent labor from organizing. The argument ran roughly as follows: "If laborers are allowed to organize, they can act as a unit to fix wage prices and restrict free negotiations of wages. This is a restraint of trade and freedom within the labor market." Based on this interpretation, local courts were empowered to issue injunctions to stop labor from organizing. If a factory owner found his workers attempting to organize, he could simply go to the courts and request an injunction forbidding such activity.

In 1914, Congress acted to offset the effect of the Sherman Antitrust Act by passing the Clayton Act. This act authorized employees to organize to negotiate with a particular employer. However, in most cases the employer could demonstrate that the organizing activity was directed by parties outside the employer's shop. This implied that the action was not a local one and, therefore, was subject to action under the Sherman antitrust legislation. Therefore, the injunction remained a powerful management tool in resisting unionization.

14.4 NORRIS–LAGUARDIA ACT

The passing of the Norris–LaGuardia Act heralded the first major movement of the power pendulum away from management and toward labor. This act, sometimes referred to as the Anti-Injunction Act, accomplished what the Clayton Act had failed to do. It specifically stated that the courts could not intercede on the part of management so as to obstruct the

Table 14.1   Chronology of Labor Law and Organization

Labor Law		Labor Movement	
1890	Sherman Antitrust Act	1886	AFL founded by Samuel Gompers; Knights of Labor organized factory workers
1908	Supreme Court supported application to union activity	1905	Industrial Workers of the World
1914	Clayton Act	1908	Building and Construction Trades Department of AFL founded
	Ineffective—individual basis—as court rules		
1931	Davis–Bacon Act	1930s	Take in industrial workers as federal locals
	On federal contracts wages and fringes paid at prevailing rate	1935	Committee for Industrial Organization-AFL ordered disbanding
1932	Norris–LaGuardia (Anti-Injunction Act)	1936	Federal locals (CIO) thrown out
1935	Wagner Act (National Labor Relations Act)	1938	Congress of Industrial Organizations
1938	Fair Labor Standards Act	1940s	Wartime strikes accused of not supporting war effort
	Minimum wages, maximum hours defined		Criminal activities alleged
1943	Smith–Connolly Act (War-Labor Disputes Act)	1955	AFL and CIO reconcile differences and recombine as AFL–CIO
	Reaction to labor in wartime; ineffective		
1946	Hobbs Act—"Anti-Racketeering law"		
	Protects employer from paying kickbacks to labor		
1947	Taft–Hartley (Labor Management Relations Act)		
1959	Landrum–Griffin Act (Labor Management Reporting and Disclosures)		
1964	Title IV Civil Rights Act		

formation of labor organizations. It effectively overrode the Supreme Court interpretation that the Sherman Antitrust Act could be applied to labor organizations. It curtailed the power of the courts to issue injunctions and protected the rights of workers to strike and picket peaceably. It also outlawed the use of "yellow-dog" contracts on the part of management. It was a common practice to have an employee sign a contract upon being hired in which he agreed not to join or become active in any union organization. Such yellow-dog contracts were declared illegal by the Norris–LaGuardia Act. This piece of legislation as interpreted by the Supreme Court during the period of the New Deal effectively freed labor from the constraints of the Sherman Antitrust Act.

14.5   DAVIS–BACON ACT

In 1931, a very far-reaching piece of legislation was passed that even today has a significant impact on the cost of federally funded projects throughout the United States. The Davis–Bacon act provides that wages and fringe benefits on all federal and federally funded projects shall be paid at the "prevailing" rate in the area. The level of prevailing rates is established by the secretary of labor, and a listing of these rates is published with the contract documents so that all contractors will be aware of the standards. To ensure that these rates are paid, the government requires submittal by all contractors of a certified payroll each month to the federal agency providing the funding. These rates are reviewed to determine whether any violations of the Davis–Bacon pay scale have occurred. This act is so far-reaching in its effect because much of public construction at the state and local level may be funded in part by federal grants. A large municipal mass-transit system or wastewater treatment plant, for instance, may be funded in part by a federal agency. In such cases, the prevailing rates must be paid. Since the Department of Labor generally accepts the most recently negotiated *union contract rates* as the prevailing ones, this allows union contractors to bid without fear of being underbid by nonunion contractors paying lower wage rates.

14.6 NATIONAL LABOR RELATIONS ACT

The National Labor Relations Act, also referred to as the Wagner Act, is a landmark piece of legislation that established a total framework within which labor management relations are to be conducted. Its central purposes are to protect union-organizing activity and encourage collective bargaining. Employers are required to bargain in good faith with the properly chosen representatives of the employees. Among other things, it establishes the procedures by which labor can organize and elect representatives. Discrimination against an employee for labor-organizing activities or participation in a union is forbidden by this act.

Employer unfair labor practices defining precisely what actions are not acceptable in management dealings with labor are specified. These practices are summarized in Table 14.2. Comparable unfair practices on the part of labor in dealing with management were not defined. It was assumed that labor was the abused party and would act equitably in its dealings with management. This trust had to be specifically spelled out later in the Taft–Hartley Act.

The act also established a "watch dog" organization to ensure its provisions were properly administered. This organization is the National Labor Relations Board (NLRB). The NLRB acts as the clearinghouse for all grievances and issues leading to complaints by labor against management and vice-versa. It is the highest tribunal below the Supreme Court for settling labor disputes and rules on most issues affecting labor–management relationships.

The act also established the concept of a *closed shop*. For years, labor organizations had fought for the right to force all members of a particular work activity (shop) to be members of a union. If the majority voted for union membership, then in order to work in the shop, a new employee had to belong to the union. This is in contrast to the *open shop* in which employees are not organized and do not belong to a union. The Wagner Act endorsed the concept of the closed shop and made it legal. This concept was later revoked by the Taft–Hartley Act and replaced by the *union shop*. The closed shop was attacked as illegal since it infringed upon a person's "right to work" and freedom of choice regarding union membership. The union shop will be discussed later in the section on the Taft–Hartley Act.

14.7 FAIR LABOR STANDARDS ACT

The Fair Labor Standards Act is commonly referred to as the minimum wage law. It was originally passed in 1938 and establishes the minimum wages and maximum hours for all workers. The minimum wage level is periodically changed to be consistent with changing wage rates. The law defines the 40-hour workweek and time over this amount as overtime. It is, generally, an outgrowth of the child labor abuses that occurred in the nineteenth century. It also forbids discrimination by establishing the concept of "equal pay for equal work." Recent arguments against increasing the minimum wage have hinged on the ideas that certain menial and domestic tasks that could provide unskilled workers with employment have become so expensive that it no longer is reasonable to perform them. The clearing of

Table 14.2   Employer Unfair Labor Practices

Under the National Labor Relations Act, as amended, an employer commits an unfair labor practice if he:

1. Interferes with, restrains, or coerces employees in the exercise of rights protected by the act, such as their right of self-organization for the purposes of collective bargaining or other mutual assistance {Section 8(a)(1)}.

2. Dominates or interferes with any labor organization in either its formation or its administration or contributes financial or other support to it {Section 8(a)(2)}. Thus "company" unions dominated by the employer are prohibited, and employers may not unlawfully assist any union financially or otherwise.

3. Discriminates against an employee in order to encourage or discourage union membership {Section 8(a)(3)}. It is illegal for an employer to discharge or demote an employee or to single him out in any other discriminatory manner simply because he is or is not a member of a union. In this regard, however, it is not unlawful for employers and unions to enter into compulsory union-membership agreements permitted by the National Labor Relations Act. This is subject to applicable state laws prohibiting compulsory unionism.

4. Discharges or otherwise discriminates against an employee because he has filed charges or given testimony under the act {Section 8(a)(4)}. This provision protects the employee from retaliation if he seeks help in enforcing his rights under the act.

5. Refuses to bargain in good faith about wages, hours, and other conditions of employment with the properly chosen representatives of his employees {Section 8(a)(5)}. Matters concerning rates of pay, wages, hours, and other conditions of employment are called mandatory subjects, about which the employer and the union must bargain in good faith, although the law does not require either party to agree to a proposal or to make concessions.

6. Enters into a hot-cargo agreement with a union {Section 8(e)}. Under a hot-cargo agreement, the employer promises not to do business with or not to handle, use, transport, sell, or otherwise deal in the products of another person or employer. Only in the garment industry and the construction industry (to a limited extent) are such agreements now lawful. This unfair labor practice can be committed only by an employer and a labor organization acting together.

Source: From Clough and Sears, *Construction Contracting,* 6th ed., John Wiley & Sons, New York, 1994.

refuse and cutting of grass along roadways were done in former times by hand labor at low wages. Increasing minimum wages make this too expensive.

14.8   UNION GROWTH

Under the provisions of the union legislation of the 1930s, the labor unions began to flourish. As is often the case during periods of transition, where inflexible barriers previously existed, a vacuum in favor of labor developed. The hard line of management was broken, and labor rushed in to organize and exploit the new situation. Along with the benefits accruing to the worker from these events, the inevitable abuses of the unstructured and unrestricted growth soon became apparent. In 1938, the unbridled actions of the unions and their leaders started to swing public opinion against them. Some unions flaunted their newfound power by introducing *restrictive labor practices* and *wartime strikes,* which shut down plants producing critical military supplies. Criminal activities within the unions were widespread and virtually unchecked. In 1943, Congress responded to this changing public perception of unions by passing the War Labor Disputes Act (Smith Connolly Act). This reflected public displeasure with the high-handed tactics and unpatriotic stance of the labor unions. It was designed to limit strikes in critical wartime industries and expedite settlement of disputes. It was largely ineffective but did reflect increasing public support of legislation that would control the prerogatives of labor unions. By 1947, thirty-seven states had enacted some form of labor control bill.

The inroads made by criminal elements active in union activities were recognized by the Hobbs (Anti-Racketeering) Act of 1946. This legislation was enacted to protect employers from threats, force, or violence by union officials extorting payments for "services rendered." Payments requested included commissions for various types of aid and assistance, gifts for controlling labor trouble, and equipment rentals forced on the employers at exorbitant costs. These laws and the continuing difficulties developing from abuse of power on the part of unions set the stage for the enactment of the Taft–Hartley Act of 1947.

14.9   LABOR MANAGEMENT RELATIONS ACT

The Labor Management Relations (Taft–Hartley) Act together with the Wagner Act form the two cornerstones of American labor relations legislation. The Taft–Hartley Act amended the Wagner Act and reversed the swing of the power pendulum once more, still leaving labor in a very strong position, but pushing the pendulum more toward center. It is the first post-Depression law to place effective constraints on the activities of labor. It restructured the makeup and operation of the NLRB, attempting to give management a stronger voice and to balance representation of labor and management. Section 7 of the bill defines the rights of workers to participate in or refrain from union activities. Section 8 provides the counterpoint to the Employer Unfair Practices Section of the Wagner Act. It defines *Union Unfair Labor Practices*, which specify tactics on the part of the labor that are illegal (see Table 14.3). The law also established the Federal Mediation and Conciliation Service, which acts as a third party in trying to expedite a meeting of the minds between unions and management involved in a dispute. This service has been very visible in meeting with representatives of players unions and sports team owners in order to work out the terms of player contracts.

Under the Taft–Hartley legislation, the president is empowered to enjoin workers on strike (or preparing to strike) to work for an 80-day cooling-off period during which time negotiators attempt to reach agreement on contractual or other disputes. This strike moratorium may be invoked in industries where a strike endangers the health of the national economy. The President has utilized his powers under this provision of the law on numerous occasions since 1947.

Section 14(b) is significant in that it redefines the legality of closed-shop operations and defines the *union shop*. A totally closed shop is one in which the worker must be a union member before he or she is considered for employment. As already stated this is declared illegal by the Taft–Hartley Act. The union shop *is* legal. A union shop is one in which a nonmember can be hired. The worker is given a grace period (usually 30 days in manufacturing shops and a shorter period in the construction industry), during which time he or she must become a union member. If the new employee does not become a member, the union can request that the candidate employee be released. Under the closed-shop concept, it was much easier for the unions to block a worker from gaining employment. This could be used to discriminate against a potential employee. The union shop gives the worker a chance to join the union (see Fig. 14.1). If the worker requests membership and the union refuses after 30 days,[2] management can ask the union to show cause why the employee has not been admitted to membership.

---

[2] The period shown in Figure 14.1 is 7 days. This is typical in the construction industry and recognizes the more transient nature of construction work.

Table 14.3   Union Unfair Labor Practices

Under the National Labor Relations Act, as amended, it is an unfair labor practice for a labor organization or its agents:

1. **a.** To restrain or coerce employees in the exercise of their rights guaranteed in Section 7 of the Taft–Hartley Act {Section 8(b)(1)(A)}. In essence Section 7 gives an employee the right to join a union or to assist in the promotion of a labor organization or to refrain from such activities. This section further provides that it is not intended to impair the right of a union to prescribe its own rules concerning membership.
    b. To restrain or coerce an employer in his selection of a representative for collective bargaining purposes {Section 8(b)(1)(B)}.

2. To cause an employer to discriminate against an employee in regard to wages, hours, or other conditions of employment for the purpose of encouraging or discouraging membership in a labor organization {Section 8(b)(2)}. This section includes employer discrimination against an employee whose membership in the union has been denied or terminated for cause other than failure to pay customary dues or initiation fees. Contracts or informal arrangements with a union under which an employer gives preferential treatment to union members are violations of this section. It is not unlawful, however, for an employer and a union to enter an agreement whereby the employer agrees to hire new employees exclusively through a union hiring hall so long as there is no discrimination against nonunion members. Union security agreements that require employees to become members of the union after they are hired are also permitted by this section.

3. To refuse to bargain in good faith with an employer about wages, hours, and other conditions of employment if the union is the representative of his employees {Section 8(b)(3)}. This section imposes on labor organizations the same duty to bargain in good faith that is imposed on employers.

4. To engage in, or to induce or encourage others to engage in, strike or boycott activities, or to threaten or coerce any person, if in either case an object thereof is:

    a. To force or require any employer or self-employed person to join any labor or employer organization, or to enter into a hot-cargo agreement that is prohibited by Section 8(e) {Section 8(b)(4)(A)}.
    b. To force or require any person to cease using or dealing in the products of any other producer or to cease doing business with any other person {Section 8(b)(4)(B)}. This is a prohibition against secondary boycotts, a subject discussed further in Section 14.18 of this text. This section of the National Labor Relations Act further provides that, when not otherwise unlawful, a primary strike or primary picketing is a permissible union activity.
    c. To force or require any employer to recognize or bargain with a particular labor organization as the representative of his employees that has not been certified as the representative of such employees {Section 8(b)(4)(C)}.
    d. To force or require any employer to assign certain work to the employees of a particular labor organization or craft rather than to employees in another labor organization or craft, unless the employer is failing to conform with an order or certification of the NLRB {Section 8(b)(4)(D)}. This provision is directed against jurisdictional disputes, a topic discussed in Section 14.12 of this text.

5. To require of employees covered by a valid union shop membership fees that the NLRB finds to be excessive or discriminatory {Section 8(b)(5)}.

6. To cause or attempt to cause an employer to pay or agree to pay for services that are not performed or not to be performed {Section 8(b)(6)}. This section forbids practices commonly known as featherbedding.

7. To picket or threaten to picket any employer to force him to recognize or bargain with a union:

    a. When the employees of the employer are already lawfully represented by another union {Section 8(b)(7)(A)}.
    b. When a valid election has been held within the past 12 months {Section 8(b)(7)(B)}.
    c. When no petition for a NLRB election has been filed within a reasonable period of time, not to exceed 30 days from the commencement of such picketing {Section 8(b)(7)(C)}.

Source: From Clough and Sears, *Construction Contracting,* 6th ed., John Wiley & Sons, New York, 1994.

All present employees who are members of the Union on the effective date of this agreement shall be required to remain members in good standing of the Union as a condition of their employment.

All present employees who are not members of the Union shall, from and after the 7th day following the date of execution of this agreement, be required to become and remain members in good standing of the Union as a condition of their employment.

All employees who are hired thereafter shall be required to become and remain members in good standing of the Union as a condition of their employment from and after the 7th day of their employment or the effective date of this Agreement, whichever is later, as long as Union membership is offered on the same terms as other members.

Any employee who fails to become a member of the Union or fails to maintain his membership therein in accordance with provisions of the paragraphs of this Section, shall forfeit his rights of employment and the employer shall within two (2) working days of being notified by the Union in writing as to the failure of an employee to join the Union or maintain his membership therein, discharge such employee. For this purpose, the requirements of membership and maintaining membership shall be consistent with State and Federal Laws. The Employer shall not be deemed in default unless he fails to act within the required period after receipt of registered written notice.

(Excerpted from Agreement Between Central Illinois Builders and The United Brotherhood of Carpenters and Joiners of America Local Union No. 44, Champaign-Urbana, Illinois.)

Figure 14.1 Contract typical member clause.

The law also recognizes the concept of "agency" shop. In such facilities a worker can refuse to join the union. The employee, therefore, has no vote in union affairs. The worker must, however, pay union dues since he or she theoretically benefits from the actions of the union and the union acts as his or her "agent." If the union, for instance, negotiates a favorable pay increase, all employees benefit and all are required to financially support the labor representation (i.e., the union negotiators).

Because of the way in which the law regarding closed shop under the Wagner Act was implemented, many workers felt that their constitutional right to work was being abrogated. That is, unless they were already union members, they were not free to work in certain firms. They had no choice. They were forced to either join the union or go elsewhere. The Taft–Hartley Act allows the individual states to enact right-to-work laws that essentially forbid the establishment of totally union shops. States in the South and the Southwest where unions are relatively weak have implemented this feature at the state level. Clough and Sears explain this as follows.

Section 14(b) of the Taft–Hartley Act provides that the individual states have the right to forbid negotiated labor agreements that require union membership as a condition of employment. In other words, any state or territory of the United States may, if it chooses, pass a law making a union-shop labor agreement illegal. This is called the "right-to-work" section of the act, and such state laws are termed right-to-work statutes. At the present writing, 21 states have such laws in force.[3] It is interesting to note that most of these state right-to-work laws go beyond the mere issues of compulsory unionism inherent in the union shop. Most of them outlaw the agency shop, under which nonunion workers must pay as a condition of continued employment the same initiation fees, dues, and assessments as union

---

[3] Alabama, Arizona, Arkansas, Florida, Georgia, Idaho, Iowa, Kansas, Louisiana, Mississippi, Nebraska, Nevada, North Carolina, North Dakota, South Carolina, South Dakota, Tennessee, Texas, Utah, Virginia, and Wyoming now have right-to-work legislation in effect.

employees, but are not required to join the union. Some of the laws explicitly forbid unions to strike over the issue of employment of nonunion workers.[4]

The fact that a right-to-work provision has been implemented can be detected by reading the language of the labor agreements within a given state. In states in which no right-to-work law is in effect, a clause is included indicating that a worker must join the union within a specified period. Such a clause taken from an Illinois labor contract is shown in Figure 14.1. This clause would be illegal in Georgia.

14.10   OTHER LABOR LEGISLATION

The Labor Management Reporting and Disclosure (Landrum–Griffin) Act was passed in 1959 to correct some of the deficiencies of previous legislation. Among its major objectives were (1) the protection of the individual union member, (2) improved control and oversight of union elections, and (3) an increased government role in auditing the records of unions. Misappropriation of union funds by unscrupulous officials and apparent election fraud were the central impetus in enacting this law. Under this law, all unions must periodically file reports with the Department of Labor regarding their organization finances and other activities. The act provides that employers cannot make payments directly to union officials. They can, however, pay dues and fringe benefits to qualified funds of the union for things such as health and welfare, vacation, apprenticeship programs, and the like. Records regarding these funds are subject to review by government auditors.

Title IV of the Civil Rights Act (enacted in 1964) establishes the concept of equal employment opportunity. This legislation was expanded by the Civil Right Act of 1991. It forbids discrimination on the basis of race, color, religion, sex, or national origin. It is administered by the Equal Employment Opportunity Commission (EEOC) and applies to discrimination in hiring, discharge, conditions of employment, and classification. Its application in the construction industry has led to considerable controversy. Individual workers can file an *unfair labor practice charge* against a union because of alleged discrimination. Unions found guilty face *cease-and-desist* orders as well as possible recision of their mandate to act as the authorized employee representative.

Executive Order 11246 issued by President Johnson in 1965 further amplified the government position on equal opportunity. It establishes affirmative action requirements on all federal government or federally funded construction work. It is administered by the Office of Federal Contract Compliance (OFCC). This office is instrumental in establishing the level of minority participation in government work. It has spawned a number of plans for including minority contractors in federally funded projects. Executive Order 11375 (1968) extends Order 11246 to include sex discrimination. Contractors working on federally funded work are required to submit affirmative action reports to the OFCC. If the plan is found to be deficient, the OFCC can suspend or terminate the contract for noncompliance.

14.11   VERTICAL VERSUS HORIZONTAL LABOR ORGANIZATION STRUCTURE

The traditional craft unions are normally referred to as horizontally structured unions. This is because of the strong power base that is located in the union local. Contract negotiations are conducted at the local level and all major decisions are concentrated at the local level. Construction unions are craft unions with a strong local organization. The local normally is run on a day-to-day basis by the *business agent*. Representatives at the individual job sites are called job stewards. The local elects officers and a board of directors on a periodic basis. The local president and business agent may be the same individuals. The bylaws of the local define the organizational structure and particulars of union structure. At the time

---

[4] R. H. Clough and G. A. Sears, *Construction Contracting,* 6th ed., John Wiley & Sons, New York, 1994, p. 372.

of contract negotiations, representatives from the local meet with representatives of the local union contractors to begin discussions. The Associated General Contractors (AGC) in the local area often act as the contractor's bargaining unit. This horizontal structure leads to a proliferation of contracts and a complex bargaining calendar for the contractors' association. If a contractors' group generally deals with 12 craft unions in the local area and renegotiates contracts on an annual or biennial basis, it is obvious that the process of meeting and bargaining can become complicated. Contracts are signed for each union operating in a given area. The national headquarters organizations for construction craft unions normally coordinate areas of national interest to the union, such as congressional lobbying, communication of information regarding recently negotiated contracts, national conventions, printing of newsletters and magazines, seminars, workshops, and other general activities. The real power in most issues, however, is concentrated at the local level. The horizontal organization then is similar to a confederation, with strength at the bottom and coordination at the top.

Vertically structured unions tend to concentrate more of the power at the national level. Significantly, labor contracts are negotiated at the national level. This means a contract is signed at the national level covering work throughout the country. This is considerably more efficient than the hundreds of locally negotiated contracts that are typical of horizontally structured unions. The industrial unions of the CIO have traditional organization in a vertical structure, while the construction unions of the AFL maintain the strong local horizontal structure. The construction elements within industrial unions usually follow the example of the parent union. The construction workers of the United Mine Workers (UMW) are an example of this. They sign a single contract with the mine owners covering all of the crafts from operating engineers to electricians. A list of scales covering all specialties (i.e., craft disciplines) is contained in the national contract. Since the members of the union are mine construction workers first and carpenters, operators, or electricians second, the jealousy regarding so-called craft lines and jurisdiction is less pronounced. It is not uncommon to see an equipment operator in a vertically structured union get down from a tractor and do some small carpentry. This would be impossible in a horizontally organized craft union situation because the carpenters would immediately start a jurisdictional dispute.

14.12   JURISDICTIONAL DISPUTES

In addition to the fragmentation of contracts by craft and local area, one of the major difficulties inherent in the horizontal craft structured union is the problem of craft jurisdiction. Job jurisdiction disputes arise when more than one union claims jurisdiction over a given item of work. This is true primarily because many unions regard a certain type of work as a proprietary right and jealously guard against any encroachment of their traditional sphere by other unions. As technology advances and new products are introduced, the question of which craft most appropriately should perform the work involved inevitably arises. A classical example in building construction is provided by the introduction of metal window and door frames. Traditionally, the installation of windows and doors had been considered a carpentry activity. However, the introduction of metal frames led to disputes between the carpenters and the metal workers as to which union had jurisdiction in the installation of these items. Such disputes can become very heated and lead to a walkout by one craft or the other. This may shut down the job. The contractor is sometimes simply an innocent bystander in such instances. If these disputes are not settled quickly, the repercussions for client and contractor can be very serious, as indicated by the following excerpt from the *Engineering News Record*:

> The nozzle-dispute on the $1-billion Albany, N.Y., mall project has caused hundreds of stoppages on that job, which employs over 2,000 persons. The argument revolves around whether the teamster driving a fuel truck or the operating engineer running a machine shall

hold the nozzle during the fueling operation. Both unions claim the job. Because holding the nozzle involves a certain amount of work, the question is why either union should want it, since regardless of which man does the job, the other still gets paid. The answer undoubtedly is that the union that gets jurisdiction will eventually be able to claim the need for a helper. This particular dispute has been reported as plaguing contractors in many states, including West Virginia, Oklahoma, Missouri, California and Washington.[5]

Although this is a rather extreme example, it is indicative of the jealousies that can arise between crafts.

Concern on the part of unions for jurisdiction is understandable since rulings that erode their area of work ultimately can lead to the craft slowly dwindling into a state of reduced work responsibilities and, eventually, into extinction. Therefore, the craft unions jealously protect their craft integrity. The following clause from a contract indicates how comprehensive the definition of craft responsibility can become.

Scope of Work

This Agreement shall cover all employees employed by the Employer engaged in work coming under all classifications listed under the trade autonomy of the United Brotherhood of Carpenters and Joiners of America.

The trade autonomy of the United Brotherhood of Carpenters and Joiners of America consists of the milling, fashioning, joining, assembling, erection, fastening or dismantling of all material of wood, plastic, metal, fiber, cork and composition, and all other substitute materials and the handling, cleaning, erecting, installing and dismantling of machinery, equipment and all materials used by members of the United Brotherhood.

Our claim of jurisdiction, therefore, extends over the following divisions and sub-divisions of the trade: Carpenters and Joiners; Millwrights; Pile Drivers; Bridge, Dock and Wharf Carpenters; Divers; Underpinners; Timbermen and Core Drillers; Shipwrights, Boat Builders, Ship Carpenters, Joiners and Caulkers; Cabinet Makers, Bench Hands, Stair Builders, Millmen; Wood and Resilient Floor Layers, and Finishers; Carpenter Layers; Shinglers; Siders; Insulators; Acoustic and Dry Wall Applicators; Shorers and House Movers; Loggers, Lumber and Sawmill Workers; Furniture Workers, Reed and Rattan Workers; Shingle Weavers; Casket and Coffin Makers; Box Makers, Railroad Carpenters and Car Builders, regardless of material used; and all those engaged in the operation of woodworking or other machinery required in the fashioning, milling or manufacturing of products used in the trade, or engaged as helpers to any of the above divisions or subdivisions' burning, welding, rigging and the use of any instrument or tool for layout work incidental to the trade. When the term "carpenter and joiner" is used, it shall mean all the subdivisions of the trade. The above occupational scope shall be subject to all agreements between International Representatives.[6]

Jurisdictional disputes present less problem in vertically structured unions since craft integrity is not a matter that determines the strength of the union. All major automobiles are assembled by members of the United Automobile Workers (UAW). The UAW is a typical vertically structured union. Technological changes do not mean the work could be shifted to another union. Therefore, UAW workers can be installing windows today and can be moved to installation of electrical wiring next month. Craft integrity does not have to be jealously protected.

European construction workers are organized into vertically structured unions. National agreements in countries such as Germany cover all workers and are signed periodically defining wage scales and general labor management procedures. Each worker has a primary

---

[5] "Law Productivity: The Real Sin of High Wages," *Engineering News Record,* February 24, 1972.

[6] Excerpted from Agreement Between The United Brotherhood of Carpenters and Joiners of American Local No. 44 Champaign-Urbana, Illinois, and the Central Illinois Builders Chapter of Associated General Contractors of America.

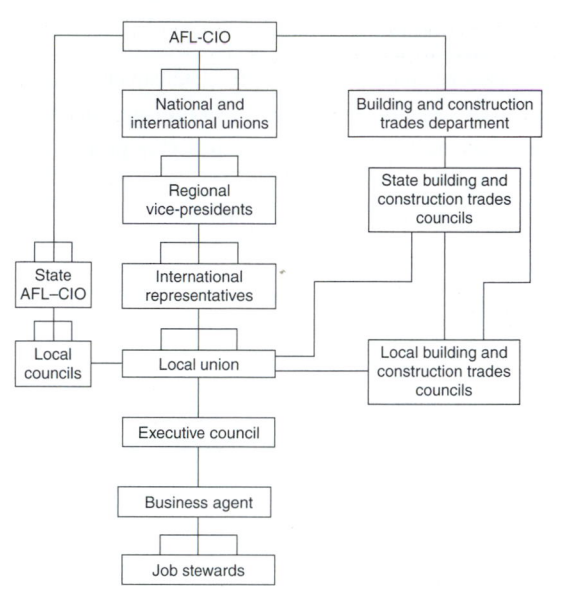

Figure 14.2 Structure typical of an affiliate of the Building and
Construction Trades Department, AFL–CIO.

specialty and is paid at the rate established in the national agreement. Since craft jurisdiction
is not a major issue, it is not unusual to see a worker who is operating a backhoe get down
and work as part of a crew installing shoring. Similar mobility back and forth across craft
lines is common in the United Mine Workers since it is also vertically structured.

14.13   UNION STRUCTURE

The largest labor organization in the United States is the AFL–CIO. The building and
construction trade unions are craft unions and as such are affiliates of the Building and
Construction Trades Department of the AFL–CIO. The structure of affiliates from local
to national level is shown schematically in Figure 14.2. A list of the construction unions
that are within the AFL–CIO is given in Table 14.4. Most construction-related unions are
presently affiliated with the AFL–CIO.

There are two ways a national union may join the AFL–CIO. The first is for an already
established union to apply for a charter. The other is for the federation to create a new union
from a related group of locals that are not members of any national union but are directly
associated with the AFL–CIO.

The top governing body of the AFL–CIO is the biennial convention. Between conventions, the executive council runs the affairs of the federation. The members are the president,
secretary-treasurer, and several vice-presidents elected by the majority at the convention
(usually from among the presidents of the national unions). The president has the authority to rule on any matters concerning the constitution or a convention decision between
meetings of the council.

The AFL–CIO maintains trade departments at the level directly below the executive
council. The mission of these major sections is to further unionization in the appropriate
industry or trade. They also aid in the settlement of jurisdictional disputes between the
members in their department. Disputes with a union in another department are appealed to
the executive council. Departments also represent their members before Congress and other
government agencies. The Building and Construction Trades Department is responsible for
all construction craft unions.

Table 14.4 AFL–CIO Construction Unions

1. International Association of Bridge, Structural, Ornamental, and Reinforcing Iron Workers
2. International Association of Heating and Frost Insulators and Asbestos Workers
3. International Brotherhood of Boilermakers, Iron Ship Builders, Blacksmiths, Forgers, and Helpers
4. International Brotherhood of Electrical Workers
5. International Brotherhood of Painters, and Allied Trades
6. International Union of Bricklayers and Allied Craft Workers
7. International Union of Elevator Constructors
8. International Union of Operating Engineers
9. Laborers International Union of North America
10. Operative Plasterers and Cement Masons' International Association
11. Sheet Metal Workers' International Association
12. United Association of Journeymen and Apprentices of the Plumbing and Pipe Fitting Industry of the United States and Canada
13. United Brotherhood of Carpenters and Joiners of America
14. United Union of Roofers, Water proofers and Allied Workers

14.14 NATIONAL UNIONS

National unions are defined as those unions having collective bargaining agreements with different employers in more than one state and federal employee unions with exclusive bargaining rights. Because of their assumed role of collective bargaining in many areas, the national unions have become increasingly powerful. In construction unions, however, the locals still play the most important role in collective bargaining and, therefore, power still resides at the local level.

Each union has exclusive jurisdiction to function as the workers' representative in its trade or branch of industry. The jurisdiction of most unions is at least partially set forth in their charter and constitution. As the unions' outlook and purposes have changed or as their members' jobs have altered, many unions have changed their jurisdiction as well.

The daily conduct of union business is in the hands of the national president, whose influence is a big factor in deciding what issues the union executive board will discuss and vote on. What the president decides will have an effect on the general public as well as on the union. The president's more important powers are to decide on constitutional matters, issue or revoke local charters, hire or fire union employees, and sanction strikes. Most actions involving the powers of the president can be appealed to the board or to the convention.

The organizer or representative of the union provides contact between the locals and the national headquarters and attempts to gain new members for the union and to set up new locals. The organizer is the union advisor to all of the locals within his area and must explain national policies to them. At the same time, he informs the national level of local problems.

14.15 STATE FEDERATIONS AND CITY CENTRALS

State federations are concerned mostly with lobbying for needed legislation and public relations on the state level. They are composed of locals whose national union is a member of the AFL-CIO. Conventions are held annually where programs of interest to all of the state's workers are concerned.

City centrals are concerned more with economics, serving as a clearinghouse for locals and aiding in dealings with employers. They have become increasingly involved in general community affairs and activities that may indirectly benefit their members.

Joint boards and trade councils are composed of locals involved in similar trades or industries. Their principal duty is to ensure that workers present a unified front in collective bargaining and obtain uniform working conditions in their area. A joint board or council is usually required for unions with more than three locals in the same region. The joint board is made up of all locals of the same national union, while the trades council is composed of locals of different national unions in related trades in the same industry.

The prototype for local trades councils is the Building and Construction Trades Council, which has its higher-level counterpart in the Building and Construction Trades Department of the AFL–CIO. Its problems are not limited to labor–management relations; it is often involved in settling ticklish jurisdictional disputes. The Building and Construction Trades Council provides craft unions with an important advantage characteristic to the industrial unions: the ability to present a united front in dealings with management. Some councils negotiate city-wide agreements with employers or see that the agreements of their member locals all expire on the same date. They have a great deal of influence with the locals but may not make them act against national union policy.

14.16   UNION LOCALS

The locals are the smallest division of the national union. They provide a mechanism through which the national union can communicate with its members at the local level. Locals provide for contact with other workers in the same trade and are a means by which better working conditions are obtained, grievances are settled, and educational and political programs are implemented. They may be organized on an occupational or craft basis or on a plant or multiplant basis. In the building industry, it is common to have locals for each craft in large cities. The local officials who preside over the committees and the general meeting are the president, vice-president, treasurer, and various secretaries. They are usually unpaid or paid only a small amount and continue to work at their trade. They perform their union duties in their spare time. In small locals, a financial secretary will take care of the local books and records; but in large locals a trained bookkeeper is employed for this purpose.

The most important local official is the business agent, a full-time employee of the local. He exercises a great deal of leadership over the local and its affairs through the advice he provides to the membership and elected officials. He is usually trained and experienced in labor relations and possesses a large amount of knowledge of conditions on which other members are poorly informed.

The business agent's duties cover the entire range of the local's activities. He helps settle grievances with employers, negotiates agreements, points out violations of trade agreements, and operates the union hiring hall. He is also an organizer, trying to get unorganized workers into the union. Only locals with a large membership can afford a full-time agent, and over one-half of the locals employing agents are in the building trades where there is a greater need due to the transient nature of the work. For the locals who do not have enough money to employ their own business agent, an agent is usually maintained by the city central or state federation.

The shop steward is not a union official but is the representative who comes in closest contact with the members. He must see that union conditions are maintained on the job and handle grievances against the employer. The steward is a worker on the job site elected by his peers.

14.17 UNION HIRING HALLS

One of the salient features of construction labor is its transient nature. Construction workers are constantly moving from job site to job site and company to company. It is not uncommon for a construction worker to be employed by five or six different contractors in the same year. The union hiring hall provides a referral service that links available labor with contractor's requests. Following each jobs, a worker registers with the union hall and is referred to a new job site as positions become available. The procedures governing operation of the union hiring hall constitute an important part of the agreement between the union and the contractor. Articles of the labor contract specify precisely how the hiring hall is to operate. Although there are small variations from craft to craft and region to region, similar procedures are commonly used for referring workers through the union hall.

14.18 SECONDARY BOYCOTTS

The legality of boycotts to influence labor disputes has been an issue of primary importance throughout the history of labor–management relations. A boycott is an action by one party to exert some economic or social pressure on a second party with the intent of influencing the second party regarding some issue. A *secondary* boycott is one in which party A who has a dispute with party B attempts to bring pressure on B by boycotting party C who deals with B and who can bring strong indirect pressure on B to agree to some issue. This is shown schematically in Figure 14.3. If the electrical workers in a plant fabricating small appliances go to the factory and form a picket line to get an agreement, there is a primary boycott in progress. If, however, the workers send some of their members into the town and put pickets up at stores selling appliances from the plant, a secondary boycott is established. The store owners are a third party (C) being pressured to influence the factory to settle with the workers. The Taft–Hartley Act declared the use of a secondary boycott to be illegal.

In the construction industry, such secondary boycotts occur on sites with both union and nonunion workers when a union attempts to force a nonunion subcontractor to sign a union contract. In such cases, the union will put up a picket line at the entrance to the work site, in effect, to picket or boycott the nonunion subcontractor. Tradition among labor unions, however, demands that no union worker can cross another union's picket line. Therefore, the actual effect of the union picket line will be to prevent all union workers from entering the site. This may cause the shutdown of the entire site pending resolution of the nonunion subcontractor's presence on the site. In this situation, the general or prime contractor is a

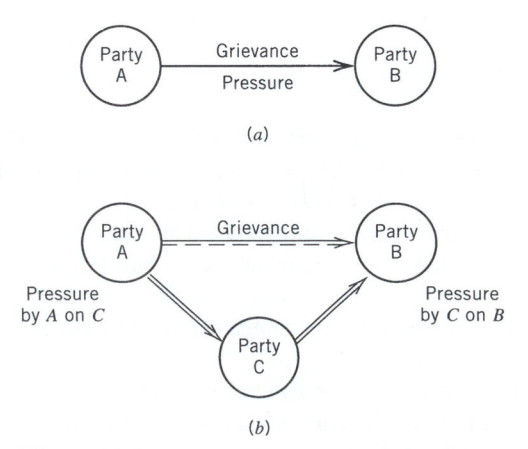

Figure 14.3 Types of boycotts: (*a*) primary boycott and (*b*) secondary boycott.

third party being pressured by the union to influence the nonunion subcontractor. This is called *common situs* picketing. In 1951, the U.S. Supreme Court ruled this practice to be a secondary boycott and, therefore, illegal under the Taft–Hartley Act. The high court made this ruling in the case of the Denver Building and Construction Trades Council.

Following this decision, the doctrine of "separate gates" was developed to deal with secondary boycott problems. Under this policy, the prime contractor establishes a separate or alternate gate for the nonunion subcontractor with whom the union has a dispute. The union is then directed to place its picket line at this gate rather than the main project gate. If it fails to comply, it can be enjoined from boycotting. Other union personnel entering the site can enter at the main gate without crossing the picket line of another union.

Certain interpretations of the secondary boycott have essentially provided exceptions in the construction industry. Unions normally have attempted to refrain from handling goods or products from nonunion shops. Such materials are called "hot cargo," and unions have bargained for hot-cargo contract clauses that, in effect, prevent a contractor from handling such materials from nonunion fabricators. This is a secondary boycott in the sense that the contractor becomes an innocent third party in the dispute between the union and the fabricator or product supplier. The Landrum–Griffin Act provides that such hot-cargo or subcontractor clauses that ban use of these materials or contact with these open shop units are illegal. An exemption is made, however, for the construction industry. As noted by Clough and Sears:

> Subcontractor agreements typically require the general contractor to award work only to those subcontractors who are signatory to a specific union labor contract or who are under agreement with the appropriate union.[7]

The Supreme Court also ruled in 1967 that prefabrication clauses that ban the use of certain prefabricated materials are exempted from the secondary boycott legislation if such prefabricated items threaten the craft integrity and eliminate work that would normally be done on site. Union carpenters, for instance, might refuse to install prefabricated door units since the doors and the frames are preassembled in a factory off site. This eliminates assembly work that could be done on site and endangers the union scope of work. Use of such prefabricated items could lead to the decay of the craft's jurisdiction and integrity. Therefore, the use of such clauses in labor agreements is not considered to be an unlawful practice in these instances.

14.19   OPEN-SHOP AND DOUBLE-BREASTED OPERATIONS

In recent years escalating union wage settlements have led to an upsurge in the number of open-shop contractors successfully bidding on large contracts. Restrictive work rules and high wages have made it difficult for union contractors to be competitive in some market areas. In an open-shop firm, there is no union agreement and workers are paid and advanced on a merit basis. The largest group of open-, or merit, shop contractors is represented by the Associated Builders and Contractors (ABC). Traditionally, open-shop contractors have bid successfully in the housing and small building market where the required skill level is not high. Union contractors have dominated the more sophisticated building and heavy construction markets based on their ability to attract skilled labor with higher wages and benefits.

Large open-shop contractors have been willing to meet or exceed the union wage rates in order to avoid the costly work delays associated with jurisdictional disputes and restrictive work rules. In some cases, the unions have responded by signing project agreements that relax certain work rules for the duration of a given job.

---

[7] R.H. Clough and G. A. Sears, *Construction Contracting,* 6th ed., John Wiley & Sons, New York, 1994, p. 376.

In order to be able to bid in both open-shop and union formats, some firms have organized as *double-breasted* contractors. Large firms will have one subsidiary that operates with no union contracts. A separately managed company will be signed to all union contracts. In this way, the parent firm can bid both in union shop markets[8] and in markets in which the lower-priced open shop encourages more cost-competitive bidding.

14.20   LABOR AGREEMENTS

Just as the contractor enters into a contract with the client, with vendors supplying materials (i.e., purchase orders), and with subcontractors working under his direction, if union labor is utilized, the contractor also enters into contracts, or labor agreements, with each of the craft unions with whom he deals. These contracts usually cover a 1- or 2-year period and include clauses governing the reconciliation of disputes, work rules, wage scales, and fringe benefits. The wages are normally defined in step increases throughout the period of the contract. These step increases are normally contained in the addendum to the labor contract.

The opening sections of the agreement typically provide methods for reconciling disputes that can arise between the contractors and the union during the life of the contract. To handle disputes, articles in the contract set up a joint conference committee to reconcile disputes and provide for arbitration procedures for disputes that cannot be settled by the committee. Typical contracts also include provisions governing:

1. Maintenance of membership
2. Fringe benefits
3. Work rules
4. Apprentice program operation
5. Wages (addendum)
6. Hours
7. Worker control and union representation
8. Operation of the union hiring hall
9. Union area
10. Subcontractor clauses (see Section 14.18)
11. Special provisions

Fringe benefits are economic concessions gained by unions covering vacation pay, health and welfare, differentials in pay due to shift, contributions by the contractor to apprenticeship programs, and so-called industrial advancement funds. These are paid by the contractor in addition to the base wage and garnish the salary of the worker. The building and construction trades councils for each union area normally print summaries of contract wage and fringe benefit provisions that assist in the preparation of payroll. Such a summary is shown in Figure 14.4.

Workrules are an important item of negotiation and have a significant effect on the productivity of workers and the cost of installed construction. A typical work rule might require that all electrical materials on site will be handled by union electricians. Another might require that all trucks moving electrical materials on site be driven by union electricians. Such provisions can lead to expensive tradesmen doing work that could be done by less-expensive crafts or laborers. Therefore, work rules become major topics of discussion during the period of contract negotiation.

---

[8] The owner may specify that union labor is to be used, or Davis–Bacon rulings may dictate that union rates will prevail.

CRAFT AND BUSINESS REPRESENTATIVE	WAGE RATE PER HOUR	FOREMAN	OVERTIME RATE	W—WELFARE P—PENSION A—APPRENTICE V—VACATION	TRAVEL PAY SUBSISTENCE	AUTOMATIC WAGE INCREASE	AUTOMATIC FRINGE INCREASES	EXPIRATION DATE
Asbestos Workers Local No. 18 Robert J. Scott, BR 946 North Highland Indianapolis, Indiana 46202	$21.57		Double	W—$1.95 P—$1.95 A—$0.38	$45 per day			5-31-98
Boilermakers Local No. 60 George Williams, BR 400 North Jefferson Peoria, Illinois 61603	$22.30	$1.50—F $3.00—GF	Double	V—$3.85 Deduct W—$5.21 P—$1.95 A—$0.10	$40 per day	$3.00 9-1-97		8-31-98
Carpenters Local No. 44 Gene Stirewalt, BR 212 W. Hill St. Champaign, Illinois 61820	$19.70	12%	Double	W—$0.52 P—$0.90 A—$0.20 IAF—$0.10		$1.00 10-15-98	$0.35 4-15-98	4-15-99
Cement Finishers Local No. 143 Francis E. Ducey, BR 212 ½ South First St. Champaign, Illinois 61820	$16.89	$1.50 15% GF	Double	W—$0.55		$0.90 1-24-98		7-24-98
Electricians Local No. 601 Jack Hensler, BR 212 South First St. Champaign, Illinois 61820	$22.25	10% 20%—GF	Double	W—$0.70 A—0.2%		$1.00—11-1-97 $0.50—5-1-98 $0.50—11-1-98		4-30-99

Figure 14.4 Labor organizations and wage rates. (Figures provided are for demonstration only and do not purport to be accurate.)

14.21   LABOR COSTS

The large number of contributions and burdens associated with the wage of a worker makes the determination of a worker's cost to the contractor a complex calculation. The contractor must know how much cost to put in the bid to cover the salary and associated contributions for all of the workers. Assuming that the number of carpenters, ironworkers, operating engineers, and other craft workers required is known and the hours for each can be estimated, the average hourly cost of each craft can be multiplied by the required craft hours to arrive at the total labor cost. The hourly average cost of a worker to *the contractor* consists of the following components:

1. Direct wages
2. Fringe benefits
3. Social security contributions (FICA)
4. Unemployment insurance
5. Workmen's compensation insurance
6. Public liability and property damage insurance
7. Subsistence pay
8. Shift pay differentials

The direct wages and fringe benefits can be determined by referring to a summary of wage rates such as the one shown in Figure 14.4.

All workers must pay social security on a portion of their salary. For every dollar the worker pays, the employer must pay a matching dollar. The worker pays a fixed percent on every dollar earned up to a cutoff level. After the annual income has exceeded the cutoff level, the worker (and the worker's employer) need pay no more. The FICA contribution in 2004 was required on the first $87,900 of annual income at the rate of 7.65 percent. Therefore, a person making $87,900 or more in annual income would contribute $6724.35 and the person's employer or employers, the contractor, would contribute a like amount.

Unemployment insurance contributions are required of all employers. Each state sets a percent rate that must be paid by the employer. The premiums are escrowed on a monthly or quarterly basis and sent periodically to the state unemployment agency. The amount to be paid is based on certified payrolls submitted by the employer at the time of paying this contribution. The fund established by these contributions is used to pay benefits to workers who are temporariliy out of work through no fault of their own.

The states also require employers to maintain Workmen's Compensation Insurance for all workers in their employ. This insurance reimburses the worker for injuries incurred in the course of employment. Labor agreements also specifically state this requirement. This recognizes the employer's responsibility to provide a safe working environment and the employer's obligation to provide support to disabled workers. Without this insurance, workers injured in the course of their work activity could become financially dependent on the state. The rates paid for workmen's compensation are a function of the risk associated with the work activity. The contribution for a pressman in a printing plant is different from that of a worker erecting steel on a high-rise building. A typical listing of construction specialties and the corresponding rates is given in Table 14.5. Similar summaries of workmen's compensation rates are printed in the Quarterly Cost Roundup issues of the *Engineering News Record*. The rates are quoted in dollars of premium per $100 of payroll. The rate for an ironworker, for example, is $29.18 (or 29.2%) per hundred dollars of payroll paid to ironworkers and structural steel erectors. The premium paid for public liability and property damage (PL and PD) insurance is also tied to the craft risk level and given in Table 14.5.

Table 14.5 Building Craft Wage and Insurance Rates[a]

Locals	Wages	Pension	Health and Welfare	Vacation	Apprentice Training	Misc	Workmen's Compensation[b]	Public Liability[c]	Property Damage[c]
Asbestos Workers	$20.30	$1.20	$1.10		$0.20		$12.18	$2.00	$1.10
Boilermakers	$20.50	$1.50	$2.10		$0.04		$12.92	$0.74	$0.72
Bricklayers	$18.70	$1.00	$1.10	$1.30			$ 7.10	$0.76	$0.54
Carpenters	$18.90	$1.00	$1.00			$0.16 prom.	$11.34	$0.80	$0.52
Cement Masons	$17.80	$ .90	$ .80		$0.04	$0.40 bldg.	$ 5.02	$0.82	$0.58
Electricians	$20.90	1.1%	0.9%	0.8%	0.05%		$ 4.38	$0.34	$0.42
Operating Engineers	$18.70	$1.50	$1.00		$0.14	$0.20 admin.	$11.22	$1.86	$2.00
Iron Workers	$19.20	$1.14	$1.30	$1.00	$0.14		$29.18	$3.00	$1.88
Laborers	$12.50	$0.66	$0.40			$0.10 educ.	$ 7.50	$0.38	$0.40
Painters	$18.90	$1.30	$1.30			$0.40 bldg.	$ 7.18	$0.26	$0.88
Plasterers	$18.34	$1.10	$ .80		$500/yr	$0.20 prom.	$ 6.96	$0.78	$0.54
Plumbers	$21.50	$1.00	$1.30		$0.22	$0.12 prom.	$ 5.60	$0.58	$1.18
Sheet Metal	$20.40	$1.40	$1.00		$0.08	$0.04 natl. $0.18 ind.	$ 7.14	$0.41	$0.40

Unemployment 5.0%
Social Security 7.65%

[a] These rates are only indicative of wage and insurance rates. They are not representative of current data.
[b] Rates are applied per $100 of pay.
[c] Public liability. Maximum coverage under these rates—$5000/person, $10,000 per accident

For higher coverage—$10,000/$20,000: 1.26 × basic rate
$25,000/50,000: 1.47 × basic rate
$50,000/100,000: 1.59 × basic rate
$300,000/300,000: 1.78 × basic rate

Property damage: Maximum coverage under these rates—$5000/person, $25,000 per accident
For higher coverage—$25,000/100,000: 1.23 × basic rate
$50,000/100,000: 1.30 × basic rate

When a construction project is underway, accidents occurring as a result of the work can injure persons in the area or cause damage to property in the vicinity. If a bag of cement falls from an upper story of a project and injures persons on the sidewalk below, these persons will normally seek a settlement to cover their injuries. The public liability (PL) arising out of this situation is the responsibility of the owner of the project. Owners, however, normally pass the requirement to insure against such liability to the contractor in the form of a clause in the *general conditions* of the construction contract. The general conditions direct the contractor to have sufficient insurance to cover such public liability claims. Similarly, if the bag of cement falls and breaks the windshield of a car parked near the construction site, the owner of the car will seek to be reimbursed for the damage. This is a property damage situation that the owner of the construction project becomes liable to pay. Property damage (PD) insurance carried by the contractor (for the owner) covers this kind of liability. Insurance carriers normally quote rates for PL and PD insurance on the same basis as for workmen's compensation insurance. Therefore, to provide PL and PD insurance, the contractor must pay $3.00 for PL and $1.88 for each one hundred dollars of steel erector salary paid on the job. These rates vary over time and geographical area and can be reduced by maintaining a safe record of operation. The total amount of premium is based on a certified payroll submitted to the insurance carrier.

Subsistence is paid to workers who must work outside of the normal area of the local. As a result, they incur additional cost because of their remoteness from home and the need to commute long distances or perhaps live away from home. If an elevator constructor in Chicago must work in Indianapolis for 2 weeks, he will be outside of the normal area of his local and will receive subsistence pay to defray his additional expenses.

Shift differentials are paid to workers in recognition that it may be less convenient to work during one part of the day than during another. Typical provisions in a sheet metal worker's contract are given in Figure 14.5. In this example, the differential results in an add-on to the basic wage rate. Shift differential can also be specified by indicating that a worker will be paid for more hours than he works. A typical provision from a California ironworkers contract provides the following standards for shift work: (a) If two shifts are

---

A shift differential premium of twenty (20) cents per hour will be paid for all time worked on the afternoon or second shift, and a shift differential of thirty (30) cents per hour will be paid for all time worked on the night or third shift as follows:

(1) *First Shift.* The day, or first, shift will include all Employees who commence work between 6 A.M. and 2 P.M. and who quit work at or before 6 P.M. of the same calendar day. No shift differential shall be paid for time worked on the day, or first, shift.

(2) *Second Shift.* The afternoon, or second, shift shall include all Employees who commence work at or after 2 P.M. and who quit work at or before 12 midnight of the same calendar day. A shift differential premium of twenty (20) cents per hour shall be paid for all time worked on the afternoon, or second, shift.

(3) *Third Shift.* The night, or third, shift shall include all Employees who commence work at or after 10 P.M. and who quit work at or before 8 A.M. of the next following calendar day. A shift differential premium of thirty (30) cents per hour shall be paid for all time worked on the night, or third, shift.

(4) *Cross Shift.* Where an Employee starts work during one shift, as above defined, and quits work during another shift, as above defined, said Employee shall not be paid any shift differential premium for time worked, if any, between the hours of 7 A.M. and 3 P.M.; but shall be paid a shift differential of twenty (20) cents per hour for all time worked, if any, between the hours of 3 P.M. and 11 P.M. and a shift differential premium of thirty (30) cents per hour for all time worked, if any, between the hours of 11 P.M. and 7 A.M.

Figure 14.5 Shift work provision.

Compute the average hourly cost to a contractor of an ironworker involved in structural steel erection in a subsistence area. The ironworker works on the second shift of a three-shift job and works six 10-hour days per week. The workers work 7 hours and are paid 8 hours under the shift pay agreement. Additional PL and PD insurance for $50,000/$100,000 coverage is desired. Use 6.2% FICA and 5.0% for unemployment insurance.

	Hours Worked	Straight Time-Hours (ST)	Premium Time (PT)
Monday–Friday	$5 \times 7 = 35$	$5 \times 8 = 40$	
	$5 \times 3 = 15$	$5 \times 3 = 15$	$1 \times 5 \times 3 = 15$
Saturday	$1 \times 7 = 7$	$1 \times 8 = 8$	$1 \times 1 \times 8 = 8$
	$1 \times 3 = 3$	$1 \times 3 = 3$	$1 \times 1 \times 3 = 3$
	60	66	26

Base Rate = $19.20
ST 66 hours @ $19.20 = $1267.20
PT 26 hours @ $19.20 = $ 499.20

Gross Pay $1766.40

Fringes:		
	Health and Welfare	$1.30 \times 66 = \$ 85.80$
	Pension	$1.14 \times 66 = \$ 75.24$
	Vacation	$1.00 \times 66 = \$ 66.00$ (deferred wage)
	Apprenticeship training	$0.14 \times 66 = \$ 9.24$
		$3.58 \times 66 = \$236.28$

WC = $29.18

PL $1.59 \times 3.00 = \$4.77$
PD $1.30 \times 1.88 = \$2.44$
  Total = $36.39 per $100.00 of Payroll

WC, PL, and PD $= \$36.39 \times \dfrac{1267.20}{100} = \$461.13$

FICA $= 0.0765 \times (\$1766.40 + \$66) = \$140.18$
Unemployment $= 0.05 \times (\$1766.40 + \$66) = \$91.62$
Subsistence = 6 days $\times$ $20.00/day = $120.00

Total Cost = Base + Fringes + WC, PL, PD + UNEMPL + FICA + SUBS = $2815.61

Average Hourly Cost (to contractor) $= \dfrac{\$2815.61}{60} = \$46.93$

Figure 14.6 Sample wage calculation.

in effect, each shift works 7.5 hours for 8 hours of pay and (b) if three shifts are in effect, each shift works 7 hours for 8 hours of pay. This means that if a three-shift project is being worked the ironworker will receive overtime for all time worked over 7 hours. In addition, he will be paid 8 hours pay for 7 hours work. Calculation of shift pay will be demonstrated in the following section.

14.22 AVERAGE HOURLY COST CALCULATION

A typical summary[9] of data regarding trade contracts in given areas is presented in Table 14.5. A worksheet showing the calculation of an ironworker's hourly cost to a contractor is shown in Figure 14.6. It is assumed that the ironworker is working in a subsistence area

[9] Although representative, data in this table are not current. Such information is dynamic and changes continuously.

on the second shift of a three-shift job during June and will be paid 8 hours for 7 hours of work (i.e., shift differential).

The ironworker works 10-hour shifts each day for 6 days, or 60 hours for the week. It is important to differentiate between those hours that are straight-time hours and those that are premium hours. Insurance premiums and fringe benefit contributions are based on straight-time hours. Social security and unemployment insurance contributions are calculated using the total income figure. The Hours Worked column breaks the weekday and Saturday hours into straight-time and premium-time components. Since the worker receives a shift differential, the first 7 hours are considered straight time and the other 3 hours are paid at overtime rate. The straight-time hours corresponding to the hours worked are shown in the second column. Eight hours are paid for the first 7 hours worked. The overtime is double time. The single-time portion, or first half of the double time, is credited to straight time. The second half of the double time is credited to the premium-time column. Based on the column totals the worker works 60 hours and will be paid 66 straight-time hours and 26 premium hours.

By consulting Table 14.5, it can be determined that the base wage rate for ironworkers is $19.20 per hour. This yields a straight-time wage of $1267.20 (66 hours) and premium pay of $499.20 (26 hours). Total gross pay is $1766.40.

Fringes are based on straight-time hours, and the rates are given in the contract wage summary. The fringes paid by the contractor to union funds amount to $3.58 per hour. The vacation portion of the fringe is considered to be a deferred income item and, therefore, is subject to FICA. It is also used in the calculation of unemployment insurance contribution.

The amounts to be paid to the insurance carrier for workmen's compensation (WC), PL, and PD can be taken from Table 14.5. The contract calls for increased PL and PD rates. The bodily injury (PL) portion and the property damage coverage are to be increased to cover $50,000 per person/$100,000 per occurrence. This introduces a multiplier of 1.59 for the PL rate and 1.30 for the PD rate (see footnotes at the bottom of Table 14.5). The total rate per $100 of payroll for WC, PL, and PD is $36.39. This is applied to the straight-time pay of $1267.20 and gives a premium to be escrowed of $461.13.

Both FICA and unemployment insurance are based on the total gross pay plus the deferred vacation fringe. Subsistence is $20.00 per day not including travel pay and time to travel to the site (not included in this calculation). By summing all of these cost components, the contractor's total cost becomes:

Gross pay	$1766.40
Fringes	$ 236.28
WC, PL, PD	$ 461.13
FICA	$ 140.18
Unemployment	$ 91.62
Subsistence	$ 120.00
Total	$2815.61

Hourly rate = $2815.61/60 = $46.93 or approx. $47.00

This is considerably different from the base wage rate of $19.20 per hour. A contractor relying on the wage figure only to come up with an estimated price will grossly underbid the project and "lose his shirt."

It is particularly important to verify that the WC, PL, PD rate being used for a worker is the correct one. Particularly hazardous situations result in rates as high as $44 per $100 of payroll (e.g., tunneling). However, if a worker is simply installing miscellaneous metals, he should not be carried as a structural steel erector. The difference in the rates between the two specialties can be significant. It should also be noted that the rates given in Table 14.5 are for a particular geographical area and are the so-called manual rates. The manual rate is the one used for a firm for which no safety or experience records are available. These rates can be substantially reduced for firms that evidence over years of operation that they have an extremely safe record. This provides a powerful incentive for contractors to be safe. If the WC, PL, PD rate can be reduced by 30%, the contractor gains a significant edge in bidding against the competition.

The calculation of the hourly average wage indicates the complexity of payroll preparation. A contractor may deal with anywhere from 5 to 14 different crafts, and each craft union has its own wage rate and fringe benefit structure. Union contracts normally require that the payroll must be prepared on a weekly basis, further complicating the situation. In addition, all federal, state, and insurance agencies to which contributions or premiums must be paid require certified payrolls for verification purposes. Because of this, most contractors with a work force of any size use the computer for payroll preparation. Data are collected by field personnel using time cards. These time records are submitted to clerical personnel who prepare them for submittal to the computer. Most firms have in-house computers for this purpose. Some firms may utilize service bureaus to provide this payroll preparation function. Charges for this service run in the vicinity of $\frac{1}{2}$ to 1% of the total payroll amount.

REVIEW QUESTIONS AND EXERCISES

14.1   What is meant by the following terms:

(a) Yellow-dog contract

(b) Agency shop

(c) Subcontractor clause

14.2   What is a secondary boycott? Name two types of secondary boycotts. Does the legislation forbidding secondary boycotts apply to construction unions? Explain.

14.3   What is a jurisdictional dispute? Why does this kind of dispute present no problem in District 50 locals?

14.4   What are the basic differences between the AFL as a labor union and the CIO type of union?

14.5   What will be the impact on double-breasted operations and the right-to-work provision of the Taft-Hartley legislation if labor is able to revoke existing practices regarding common situs picketing?

14.6   Answer the following questions true (T) or false (F):

(a) _____ Some state laws authorize use of closed shops.

(b) _____ A union can legally strike a job site in order to *enforce* the provision of a subcontractor clause.

(c) _____ The Teamsters union is the largest member of the AFL–CIO.

(d) _____ Open-shop operations have caused construction labor unions to rethink their position vis à vis union contractors.

(e) _____ The right-to-work clause of the Taft-Hartley law allows the individual states to determine whether union shops are legal.

(f) _____ The unit-price contract is an incentive-type negotiated contract.

(g) _____ The local AFL craft unions have very little authority and are directed mainly by the national headquarters of AFL–CIO.

(h) _____ The Sherman Antitrust Law was originally designed to prevent the formation of large corporations or cartels that could dominate the market.

(i) _____ The business agent is the representative of the union charged with enforcing the work rules of the labor agreement.

(j) _____ A submittal must be verified for accuracy in accordance with contract plans and specifications.

(k) _____ The Sherman Antitrust Act, enacted in 1890, was used to suppress the formation of large trusts and cartels, which dominated the market and acted to fix prices and restrain free trade.

(l) _____ Yellow-dog contracts were used by employers to encourage their employees to join and become active in union organizations.

(m) _____ Under the Taft–Hartley legislation, the president of the United States is empowered to enjoin workers on strike (or preparing to strike) to work for a 90-day cooling-off period during which time negotiators attempt to reach agreement on contractual or other disputes.

(n) _____ The National Labor Relations Act was enacted to protect union-organizing activity and encourage collective bargaining.

(o) _____ In an open-shop working environment, workers are paid based on which union hall they belong to.

(p) _____ The calculation of fringe benefits is based on gross pay, whereas FICA is based on straight-time hours.

(q) _____ If a general contractor does not feel like paying worker's compensation fees, then the contractor does not have to. Each state has appropriated funds that will cover this option.

14.7 Compute the average hourly cost of a carpenter to a contractor. Assume the work is in a subsistence area and the daily subsistence rate is $19.50. The carpenter works the second shift on a two-shift project where a project labor contract establishes a "work 7 pay 8 hour" pay basis for straight time. He works 6 days, 10 hours a day. In addition to time and a half for overtime Monday through Friday, the contract calls for double time for all work on weekends. Use 6.2% FICA and 5.0% for unemployment insurance. Assume all data relating to the WC, PL, PD, fringes, and wage are as given in Table 14.5.

14.8 Identify the local labor unions that operate in your region. List the relevant business agents and the locations of the hiring halls.

14.9 List the labor unions that you consider would be involved in a project similar to the gas station project of Appendix I.

14.10 Visit a local contractor and a local hiring hall and determine the procedure to be followed in the hiring of labor.

Chapter 15

Cost Control

Digital Hardhat System

The Need

The cost and time required to travel between construction sites limits the ability of personnel to quickly respond to problems at remote sites and to communicate issues between all necessary decision makers. Also, it is difficult to organize and transmit multimedia project information (digital pictures, video, electronic documents, and audio recordings) so that others can access current project information in an intuitive and timely manner. The Digital Hardhat (DHH) technology enables dispersed users to capture and communicate multimedia field data to collaboratively solve problems, and collect and share information. The DHH is a pen-based personal computer with special Multimedia Facility Reporting System software that allows the field representative to save multimedia information into a project-specific database, which is then accessible to others through the World Wide Web.

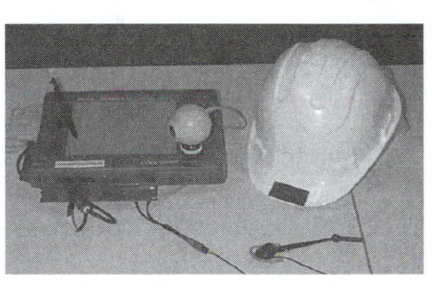

The mobile unit and a hardhat

The Technology

The Digital Hardhat (DHH) is a pen-based personal computer (PC) running a Windows operating system, which is used to collect multimedia information such as text, sound, video, and images. This pen-based computer can also be used to communicate between the construction site and other locations using various connection methods including a wireless network connection, which enables personnel to roam around the site and video teleconference live with others to solve problems collaboratively. In addition, special software called Multimedia Facility Reporting (MFR) System allows the field representative to save multimedia information into a project-specific database accessible through the internet. The project information collected through the system will help

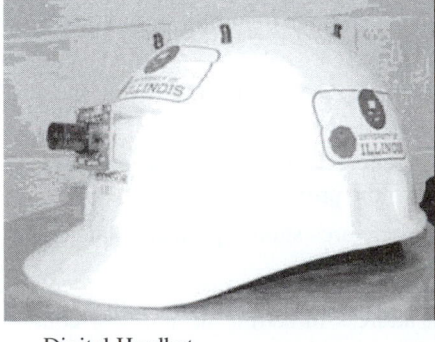

Digital Hardhat

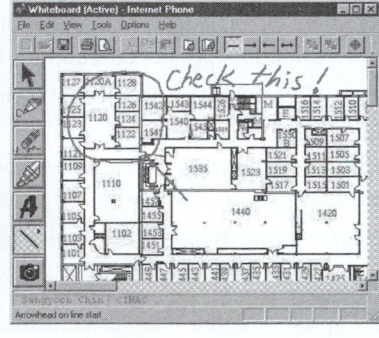

Using a whiteboard

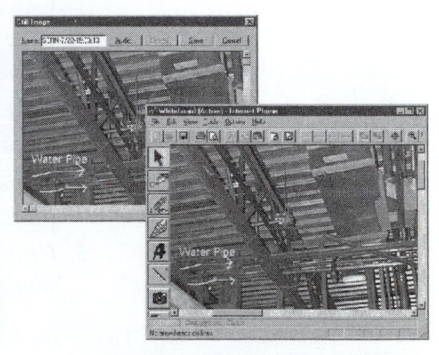

Multimedia Information on MFR

document site conditions, progress, and problems in an organized manner so the information can be retrieved easily as needed by any project participant. In the application of this system, immediate reductions in travel cost will be the most obvious benefit; however, costs associated with more quickly resolving issues, reducing construction claims, and fewer time delays will be the ultimate benefit of this technology.

15.1 COST CONTROL AS A MANAGEMENT TOOL

The early detection of actual or potential cost overruns in field construction activities is vital to management. It provides the opportunity to initiate remedial action and increases the chance of eliminating such overruns or minimizing their impact. Since cost overruns increase project costs and diminish profits, it is easy to see why both project management and upper-level management must become sensitive to the costs of all project activities.

An important byproduct of an effective cost reporting system is the information that it can generate for management on the general cost performance of field construction activities. This information can be brought to bear on problems of great interest to project management. The determination of current project status, effectiveness of work progress, and preparation of progress payment requests require data generated by both project planning and cost control reporting systems. Project cost control data are important not only to project management in decision-making processes but also to the company's estimating and planning departments because these data provide feedback information essential for effective estimates and bids on new projects. Thus a project control system should both serve current project management efforts and provide the field performance database for estimating future projects.

15.2 PROJECT COST CONTROL SYSTEMS

The design, implementation, and maintenance of a project cost control system can be considered a multistep process. The five steps, shown schematically in Figure 15.1, form the basis for establishing and maintaining a cost control system. The following questions regarding each step in the implementation of the cost control system must be addressed.

1. *Chart of Cost Accounts.* What will be the basis adopted for developing estimated project expenditures, and how will this basis be related to the firm's general accounts and accounting functions? What will be the level of detail adopted in defining the project cost accounts, and how will they interface with other financial accounts?

2. *Project Cost Plan.* How will the cost accounts be utilized to allow comparisons between the project estimate and cost plan with actual costs as recorded in the field? How will the project budget estimate be related to the construction plan and schedule in the formation of a project cost control framework?

3. *Cost Data Collection.* How will cost data be collected and integrated into the cost reporting system?

4. *Project Cost Reporting.* What project cost reports are relevant and required by project management in its cost management of the project?

5. *Cost Engineering.* What cost engineering procedures should project management implement in its efforts to minimize costs?

These are basic questions that management must address in setting up the cost control system. The structure of cost accounts will be discussed in this chapter.

Step 1 Chart of cost accounts
Step 2 Project cost plan
Step 3 Cost data collection
Step 4 Project cost reporting
Step 5 Cost engineering

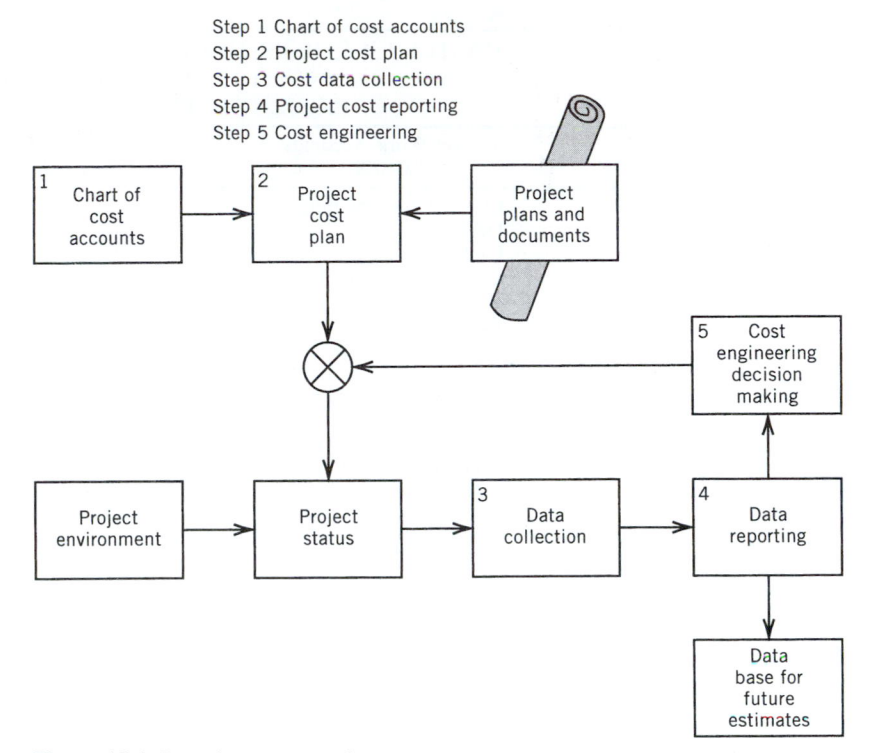

Figure 15.1 Steps in cost control.

15.3 COST ACCOUNTS

The first step in establishing a cost control system for a construction job is the definition of project-level cost centers. The primary function of the cost account section of a chart of accounts is to divide the total project into significant control units, each consisting of a given type of work that can be measured in the field (see Fig. 15.2). Once job cost accounts are established, each account is then assigned an identifying code known as a cost code. Once segregated by associated cost centers, all the elements of expense (direct labor, indirect labor, materials, supplies, equipment costs, etc.) constituting work units can be properly recorded by cost code.

The design, structure, and development of a cost coding system and its associated set of expense accounts have a significant impact on the cost management of a company or project. The job cost accounting system is essentially an accounting information system. Therefore, management is free to establish its own chart of accounts in any way that helps it in reaching specific financial and cost control objectives, whether these objectives are related to general company performance, to the control of a specific project, or to specific contract requirements.

15.4 COST CODING SYSTEMS

A variety of cost coding systems exist in practice, and standard charts of accounts are published by organizations such as the American Road Builders Association, Associated General Contractors, and the Construction Specifications Institute. In many industries, cost codes have a company-wide accounting focus emphasizing expense generation based on a departmental breakdown of the firm. In some construction firms, cost systems have a structured sequence corresponding to the order of appearance of the various trades or types of

MASTER LIST OF PROJECT COST ACCOUNTS Subaccounts of General Ledger Account 80.000 PROJECT EXPENSE				
Project Work Accounts 100–699			Project Overhead Accounts 700–999	
100		Clearing and grubbing	700		Project administration
101		Demolition		.01	Project manager
102		Underpinning		.02	Office engineer
103		Earth excavation	701		Construction supervision
104		Rock excavation		.01	Superintendent
105		Backfill		.02	Carpenter foreman
115		Wood structural piles		.03	Concrete foreman
116		Steel structural piles	702		Project office
117		Concrete structural piles		.01	Move in and move out
121		Steel sheet piling		.02	Furniture
240		Concrete, poured		.03	Supplies
	.01	Footings	703		Timekeeping and
	.05	Grade beams		.01	security
	.07	Slab on grade		.02	Timekeeper
	.08	Beams		.03	Watchmen
	.10	Slab on forms	705		Guards
	.11	Columns		.01	Utilities and services
	.12	Walls		.02	Water
	.16	Stairs		.03	Gas
	.20	Expansion joint		.04	Electricity
	.40	Screeds	710		Telephone
	.50	Float finish	711		Storage facilities
	.51	Trowel finish	712		Temporary fences
	.60	Rubbing	715		Temporary bulkheads
	.90	Curing	717		Storage area rental
245		Precast concrete	720		Job sign
260		Concrete forms	721		Drinking water
	.01	Footings	722		Sanitary facilities
	.05	Grade beams	725		First-aid facilities
	.07	Slab on grade	726		Temporary lighting
	.08	Beams	730		Temporary stairs
	.10	Slab	740		Load tests
	.11	Columns	750		Small tools
	.12	Walls	755		Permits and fees
270		Reinforcing steel	756		Concrete tests
	.01	Footings	760		Compaction tests
	.12	Walls	761		Photographs
280		Structural steel	765		Surveys
350		Masonry	770		Cutting and patching
	.01	8-in. block	780		Winter operation
	.02	12-in. block	785		Drayage
	.06	Common brick	790		Parking
	.20	Face brick			Protection of adjoining property
	.60	Glazed tile	795		Drawings
400		Carpentry	796		Engineering
440		Millwork	800		Worker transportation
500		Miscellaneous metals	805		Worker housing
	.01	Metal door frames	810		Worker feeding
	.20	Window sash	880		General clean-up
	.50	Toilet partitions	950		Equipment
560		Finish hardware		.01	Move in
620		Paving		.02	Set up
680		Allowances		.03	Dismantling
685		Fencing		.04	Move out

Figure 15.2 List of typical project expense (cost) accounts.

Table 15.1　Classification of Accounts: Major Divisions
in Uniform Construction Index

Cost Centers			
0	Conditions of the contract	9	Finishes
1	General requirements	10	Specialties
2	Site work	11	Equipment
3	Concrete	12	Furnishings
4	Masonry	13	Special construction
5	Metals	14	Conveying system
6	Carpentry	15	Mechanical
7	Moisture prevention	16	Electrical
8	Doors, windows, and glass		

construction processes typical of the company's construction activity. In most construction companies, detailed project cost accounts such as those shown in Figure 15.2 are used. This method recognizes the fact that construction work is project oriented and that to achieve the cost management goal of maximizing profit, projects must be accounted for individually. One project may be a winner while another is losing money. Such situations may be masked in the accounting system unless job cost accounts are maintained on a project-by-project basis. Therefore, both billings (revenue) and cost (work in progress) accounts are typically maintained for each project. The actual account descriptions or designations vary in accordance with the type of construction and the technologies and placement processes peculiar to that construction. Building contractors, for instance, are very interested in accounts that describe the cost aspects of forming and casting structural concrete as used in building frames. Heavy construction contractors, on the other hand, are interested in earthwork-related accounts such as grading, ditching, clearing and grubbing, and machine excavation. Standard cost accounts published by the American Road Builders Association emphasize these accounts, while the Uniform Construction Index (UCI), published by the Construction Specifications Institute, emphasizes building-oriented accounts. A breakdown of the major classifications within the UCI cost account system is shown in Table 15.1. A portion of the second level of detail for classifications 0 to 3 is shown in Figure 15.3.

15.5　PROJECT COST CODE STRUCTURE

The UCI Master Format code as used by the R. S. Means *Building Construction* Cost Data identifies three levels of detail. At the highest level the major work classification as given in Table 15.1 is defined. Also at this level major subdivisions within the work category are established. For instance, 30-level accounts pertain to concrete while 031 accounts are accounts specifically dealing with concrete forming. In a similar manner, 032 accounts are reserved for cost activity associated with concrete reinforcement.

At the next level down, a designation of the physical component or subelement of the construction is established. This is done by adding three digits to the work classification two-digit code. For instance, the three-digit code for footings is 158. Therefore, the code 031158 indicates an account dealing with concrete forming costs for footings.

At the third and lowest level, digits specifying a more precise definition of the physical subelement are used. For instance an account code of 0311585000 can indicate that this account records costs for forming concrete footings of a particular type (see Fig. 15.4). At this level the refinement of definition is very great, and the account can be made very sensitive to the peculiarities of the construction technology to be used. Further refinement could differentiate between forming different types of footings with different types of material.

0 Conditions of the Contract		0270.	Site Improvements
0000-0099.	unassigned	0271.	Fences
		0272.	Playing fields
1 General Requirements		0273.	Fountains
0.100.	Alternates of Project Scope	0274.	Irrigation systems
		0275.	Yard improvements
0.101-0109.	unassigned	0276-0279.	unassigned
0110.	Schedules and Reports	0280.	Lawns and Planting
0111-0119.	unassigned	0281.	Soil Preparation
0120.	Samples and Shop Drawings	0282.	Lawns
		0283.	Ground covers and other plants
0121-0129.	unassigned	0284.	Trees and shrubs
0130.	Temporary Facilities	0285-0289.	unassigned
0131-0139.	unassigned	0290.	Railroad Work
0140.	Cleaning Up	0291-0294.	unassigned
0141-0149.	unassigned	0295.	Marine Work
0150.	Project closeout	0296.	Boat Facilities
0151-0159.	unassigned	0297.	Protective Marine Structures
0160.	Allowances	0298.	Dredging
0161-0169.	unassigned	0299.	unassigned
2 Site Work		*3 Concrete*	
0200.	Alternates	0300.	Alternates
0210-0209.	unassigned	0301-0309.	unassigned
0120.	Clearing of Site	0310.	Concrete Formwork
0211.	Declination	0311-0319.	unassigned
0212.	Structures moving	0320.	Concrete Reinforcement
0213.	Clearing and grubbing	0321-0329.	unassigned
0214-0219.	unassigned	0330.	Cast-in-Place Concrete
0220.	Earthwork	0331.	Heavyweight aggregate concrete
0221.	Site grading	0332.	Lightweight aggregate concrete
0222.	Excavating and backfilling		
0223.	Dewatering	0333.	Post-tensioned concrete
0224.	Subdrainage	0334.	Nailable concrete
0225.	Soil poisoning	0335.	Specially finished concrete
0226.	Soil compaction control		
0227.	Soil stabilization	0336.	Specially placed concrete
0228-0229.	unassigned	0337-0339.	unassigned
0230.	Piling	0340.	Precast Concrete
0231-0234.	unassigned	0341.	Precast concrete panel
0235.	Caissons	0342.	Precast structural concrete
0236-0239.	unassigned		
0240.	Shoring and bracing	0343.	Precast prestressed concrete
0241.	Sheeting		
0242.	Underpinning	0344-0349.	unassigned
0243-0249.	unassigned	0350.	Clementitious Decks
0250.	Site drainage	0351.	Poured gypsum deck
0251-0254.	unassigned	0352.	Insulating concrete roof decks
0255.	Site utilities		
0256-0259.	unassigned	0353.	Cementitious unit decking
0260.	Roads and Walks		
0261.	Paving	0354-0399.	unassigned
0262.	Curbs and gutters		
0263.	Walks		
0264.	Road and parking Appurtenances		
0265-0269.	unassigned		

Figure 15.3 Detailed codes for classification within Uniform Construction Index.

| 031 | Concrete Formwork | | | | | | | | | |

031	Struct C.I.P. Formwork	CREW	DAILY OUTPUT	LABOR-HOURS	UNIT	1996 BARE COSTS				TOTAL INCL. O&P
						MAT.	LABOR	EQUIP.	TOTAL	
158	FORMS IN PLACE, FOOTINGS Continuous wall, 1 use C–1									
5000	Spread footings, 1 use		305	.105	SFCA	1.51	2.50	.09	4.10	5.75

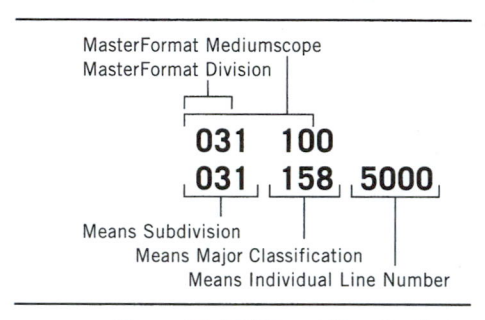

MasterFormat Mediumscope
MasterFormat Division

031 100
031 158 5000

Means Subdivision
Means Major Classification
Means Individual Line Number

Figure 15.4 UCI cost (line item) structure in the master format code.

At this level, the cost engineer and construction manager have a great deal of flexibility in reflecting unique aspects of the placement technology that lead to cost fluctuations and thus must be considered in defining cost centers.

Large and complex projects in industrial and energy-related construction may require cost codes that reflect additional information, such as the project designation, the year in which the project was started, and the type of project. Long and complex codes in excess of 10 digits can result. An example of such a code is shown in Figure 15.5. This code, consisting of 13 digits, specifically defines the following items:

1. Year in which project was started (2004)
2. Project control number (15)
3. Project type (5 for power station)
4. Area code (16 for boiler house)
5. Functional division (2, indicating foundation area)
6. General work classification (0210, indicating site clearing)
7. Distribution code (6, indicating construction equipment)

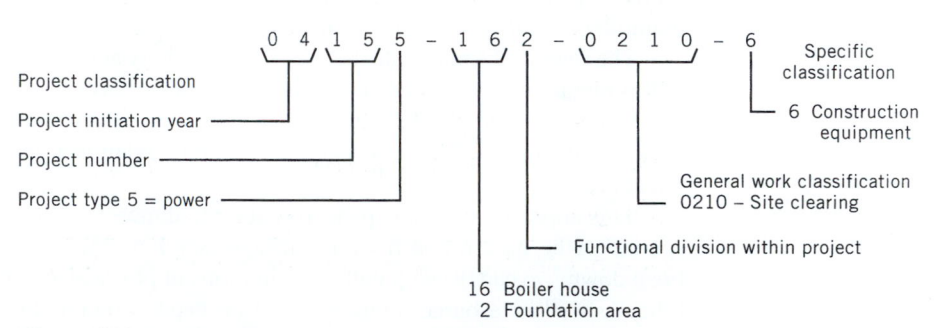

Figure 15.5 Classification of accounts: typical data structure for a computerized cost code.

The distribution code establishes what type of resource is being costed to the work process (i.e., clearing), the physical subelement (i.e., foundations) in what area of which project. Typical distribution codes might he as follows:

1. Labor
2. Permanent materials
3. Temporary materials
4. Installed equipment
5. Expendables
6. Construction equipment
7. Supply
8. Subcontract
9. Indirect

Clearly, a high concentration of information can be achieved by proper design of the cost code. Such codes are also ideally suited for data retrieval, sorting, and assembly of reports on the basis of selected parameters (e.g., all construction equipment costs for concrete forming on project 10 started in a given year). The desire to cram too much information into cost codes, however, can make them so large and unwieldy that they are confusing to upper-level management.

15.6 COST ACCOUNTS FOR INTEGRATED PROJECT MANAGEMENT

In large and complex projects, it is advantageous to break the project into common building blocks for control both of cost and time. The concept of a common unit within the project that integrates both scheduling and cost control has led to the development of the work breakdown approach. The basic common denominator in this scheme is the work package, which is a subelement of the project on which both the cost and time data are collected for project status reporting. The collection of time and cost data based on work packages has led to the term *integrated project management*. That is, the status reporting function has been integrated at the level of the work package. The set of work packages in a project constitutes its work breakdown structure (WBS).

The work breakdown structure and work packages for control of a project can be defined by developing a matrix similar to the one shown in Figure 15.6. The columns of this matrix are defined by breaking the down project into physical subcomponents. Thus we have a hierarchy of levels that begins with the project as a whole and, at the lowest level, subdivides the project into physical end items such as foundations and areas. As shown in Figure 15.6, the project is subdivided into systems. The individual systems are further divided into disciplines (e.g., civil, mechanical, electrical). The lowest level of the hierarchy indicates physical end items (foundation 1, etc.). Work packages at this lowest level of the hierarchy are called control accounts.

The rows of the matrix are defined by technology and responsibility. At the lowest level of this hierarchy, the responsibilities are shown in terms of tasks, such as concrete, framing, and earthwork. These tasks imply various craft specialties and technologies. Typical work packages then are defined as concrete tasks on foundation 1 and earthwork on foundations 1 and 2.

This approach can be expanded to a three-dimensional matrix by considering the resources to be used on each work package (see Fig. 15.7). Using this three-dimensional breakdown, we can develop definition in terms of physical subelement, task, and responsibility, as well as resource commitment. A cost code structure to reflect this matrix structure is given in Figure 15.8. This 15-digit code defines units for collecting information in terms

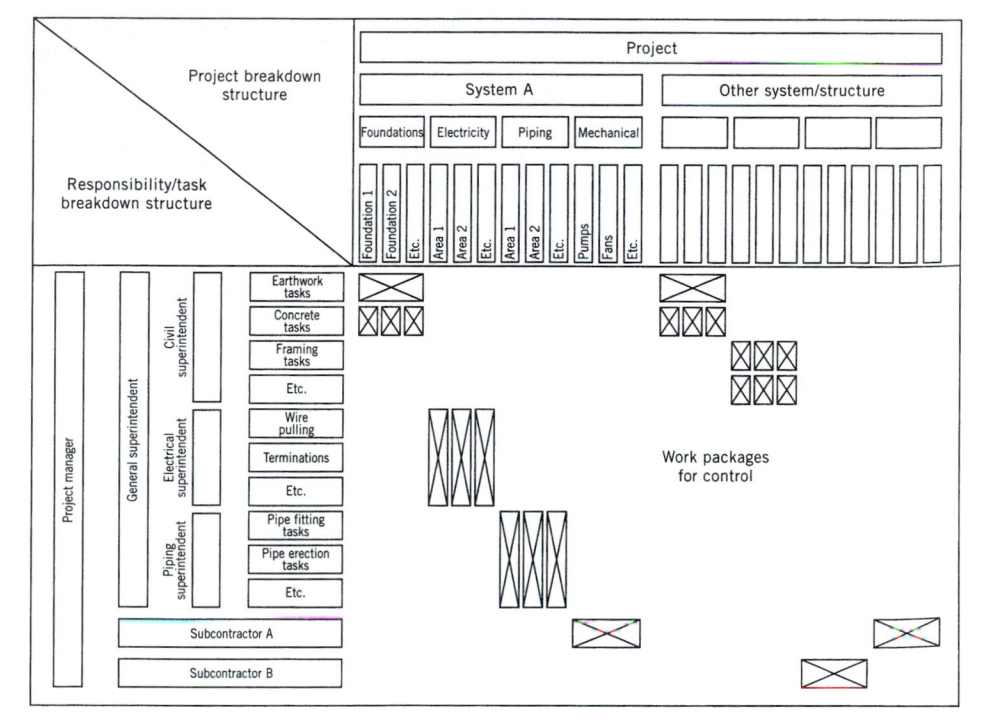

Figure 15.6 Project control matrix.

of work package and resource type. Resource usage in terms of monetary units, quantities, man-hours, and equipment-hours for a foundation in the boiler building would be collected under work package code 121002. If this work relates to placement and vibration of concrete by using a direct chute, the code is expanded to include the alphanumeric code DF441. The resource code for the concrete is 2121. Therefore, the complete code for concrete in the boiler building foundations placed by using a chute would be 121002-DF441-2121.

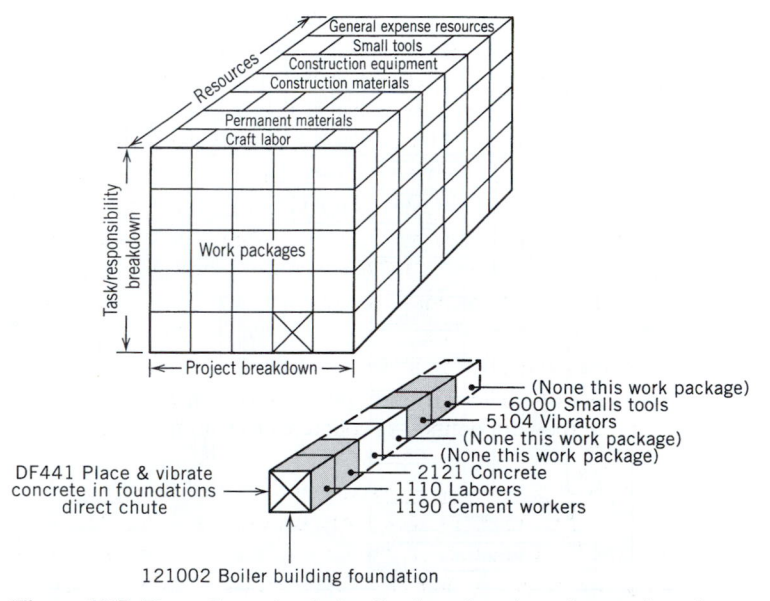

Figure 15.7 Three-dimensional visualization of work-package-oriented cost accounts.

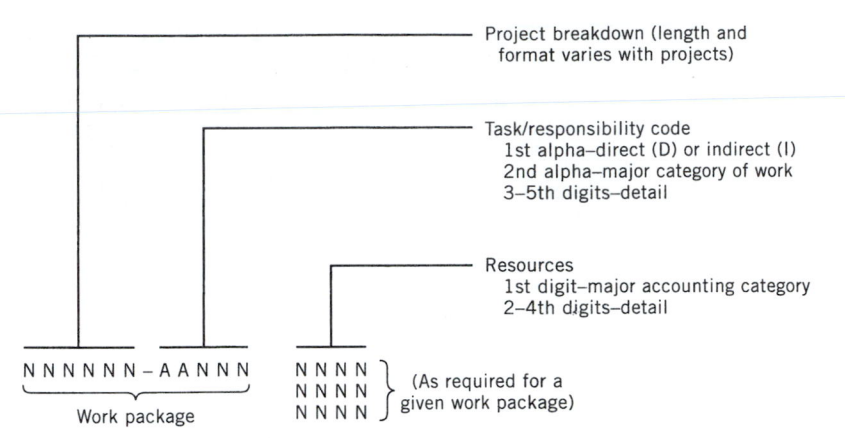

Figure 15.8 Basic cost code structure.

This code allows collection of cost data at a very fine level. Scheduling of this work is also referenced to the work package code as shown in Figure 15.9. The schedule activities are shown in this figure as subtasks related to the work package.

15.7   EARNED VALUE METHOD

One widely accepted way of calculating progress on complex projects using a work or account based breakdown system is the "earned value" approach. This system of determining project progress addresses both schedule status (e.g., on schedule, behind schedule, etc.) and cost status (e.g., over budget, etc.). This method of tracking cost and schedule was originally implemented by the Department of Defense in the late 1970s to help better

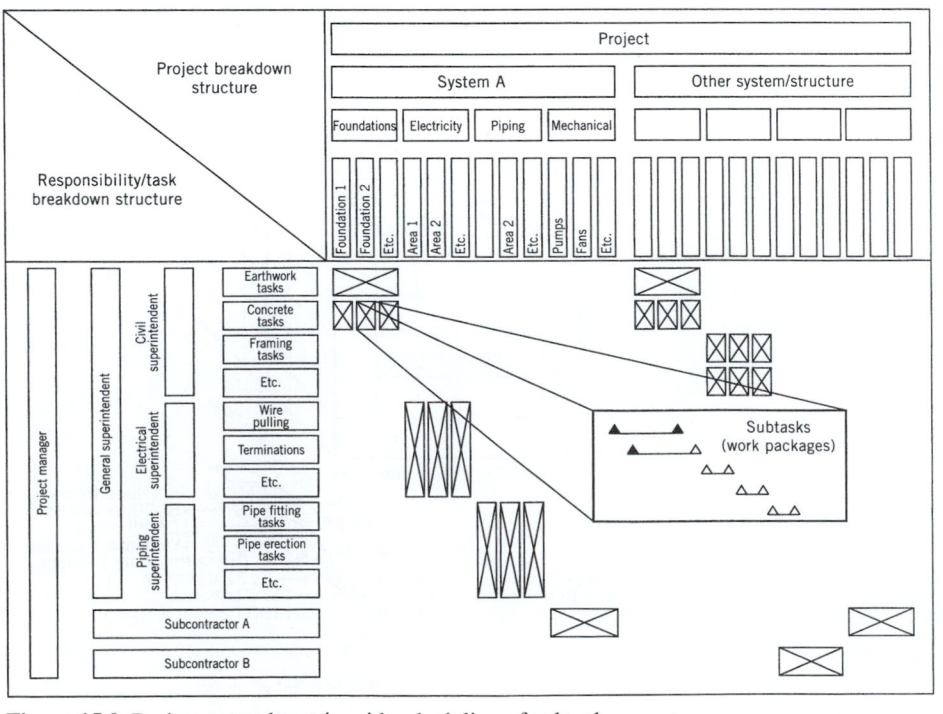

Figure 15.9 Project control matrix with scheduling of subtasks.

control complex projects. The system was called the Cost and Schedule Control Systems Criteria or C/SCSC. This method of monitoring contracts proved to be so effective that other government agencies (e.g., Department of Energy, etc.) adopted C/SCSC as a means of maintaining oversight on complex projects such as nuclear and conventional power plants. Private owners such as power companies implemented similar systems since reporting to various government authorities encouraged or required the use of C/SCSC and earned value concepts. Ultimately, owners of complex industrial projects began to use the system as well.

The idea of earned value is based upon a rigorous development of percent complete of the budgeted costs associated with individual work packages or line items. Each work package has an initial budget or estimate which is defined as the Budgeted Cost at Completion or BCAC. As work proceeds on an individual work package or account, assessment of the percent complete is made at various study dates. The initial schedule establishes an expected level of work completion as of the study date. The level of expected production is often shown as an S-Curve plotting the cost or units of production (e.g., units produced, work hours expended, etc.) against time. This cost/production curve is referred to as the baseline. At any given time (study date), the units of cost/production indicated by the baseline are called the Budgeted Cost of Work Scheduled (BCWS).

The tracking system requires that field reports provide information about the Actual Cost of Work Performed (ACWP) and the Actual Quantity of Work Performed (AQWP). The "earned value" is the Budgeted Cost of Work Performed (BCWP). The relative values for a given work package or account at a given point in time (see Fig. 15.10) provide information about the status in terms of cost and schedule variance. The six parameters which form the foundation of the "earned value" concept are:

BCWS: Budgeted Cost of Work Scheduled = Value of the baseline at a given time

ACWP: Actual Cost of Work Performed − Measured in the field

BCWP: Budgeted Cost of Work Performed = [% Complete] × BCAC

BCAC: Budgeted Cost At Completion = Contracted Total Cost for the Work Package

AQWP: Actual Quantity of Work Performed − Measured in the field

BQAC: Budgeted Quantity at Completion − Value of the Quantity Baseline as Projected at a given Point.

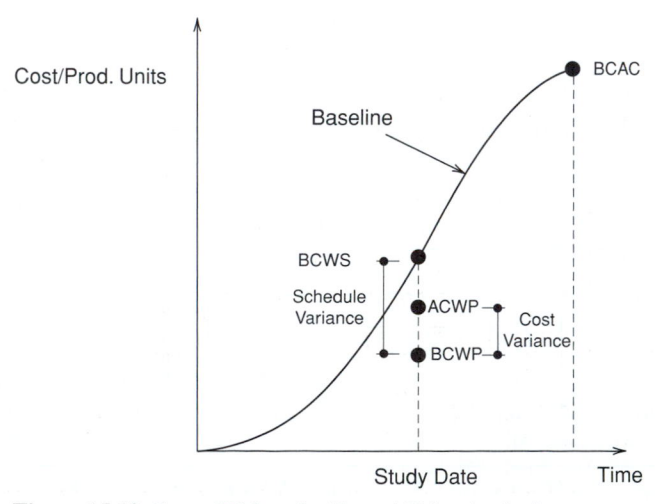

Figure 15.10 Control Values for Earned Value Analysis.

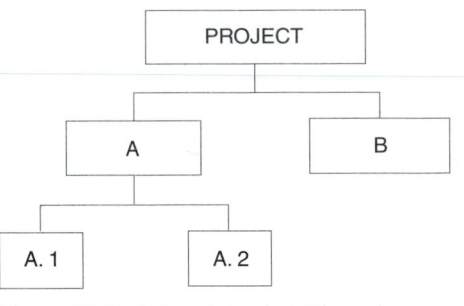

Figure 15.11 A Simple Project Hierarchy.

In order to put these terms into context, consider the small project shown in Figure 15.11. The project consists of two control accounts—"A" and "B". "A" consists of two sub-accounts, A.1 and A.2. The study date (e.g., September 1, etc.) information for these work packages is given in Table 15.2. In this example, the budget is expressed in worker hours so the baseline for control is in worker hours. The estimated total number of worker hours for this scope of work is 215 (the sum of the estimated worker hours for A.1, A.2, and B). The BCWP or earned value for a given work package is given as:

$$BCWP_i = PC_i \times BCAC_i$$

where i is the work package or account label, and PC is the percent complete as of the study date.

The percent complete (PC) for each package is based on the ratio of the Actual Quantities (AQWP) divided by the Budgeted Quantity at Completion (BQAC) based on the latest quantity assessment. If we know the original quantity estimate is 100 units but updated information indicates that a total of 120 units will be required to complete the work, completion of 50 units would not indicate 50 percent complete. The correct PC would be 50/120 (e.g., AQWP/BQAC).

Based on the information in Table 15.2, the PC for each work package in the small project would be:

$$PC (A.1) = 35/105 = 0.333$$
$$PC(A.2) = 60/77 = 0.780$$
$$PC(B) = 100/125 = 0.800$$

Then

$$BCWP \ (Project) = .333 \ (100) + .78 \ (50) + .8 \ (65) = 33.3 + 39 + 52 = 124.3$$

Table 15.2 Study Date Data for Simple Project

	BCAC	ACWP	BQAC	AQWP	PC (%)	BCWP	ECAC
A							
A.1	100	40	105	35	33.3	33.3	120
A.2	50	35	77	60	78.0	39.0	45
B	65	50	125	100	80.0	52.0	62.5
TOTAL	215	125	—	—	57.8	124.3	227.5

Project PC (PPC) = Total BCWP ÷ Total BCAC = 124.3 ÷ 215 = 57.8%
$ECAC_i$ = Estimated Cost at Completion for Work Package i = $ACWP_i \div PC_i$

Therefore, the Project Percent Complete (PPC) for the small project is:

$$PPC \{124.3/215\} \times 100 = 57.8 \text{ percent}$$

This simple example illustrates several points:

1. The PC for a given package is based on the ratios of the AQWP/BQAC.
2. The PPC is calculated by relating the total BCWP (i.e., earned value) to the total BCAC for the project scope of work.
3. The total work earned is compared to the work required. The values of units to be earned are based on the originally budgeted units in an account/work package and the percent earned is based on the latest projected quantity of units at completion.

Worker hours are used to here to demonstrate the development of the PPC. However, other cost or control units may be used according to the needs of management.

It is very important to know that schedule and cost objectives are being achieved. Schedule and cost performance can be characterized by cost and schedule variances as well as cost performance and schedule performance indices. These values in C/SCSC are defined as follows:

$$CV, \text{ Cost Variance} = BCWP - ACWP$$
$$SV, \text{ Schedule Variance} = BCWP - BCWS$$
$$CPI, \text{ Cost Performance Index} = BCWP/ACWP$$
$$SPI, \text{ Schedule Performance Index} = BCWP/BCWS$$

Figures 15.12 a, b, and c plot the values of BCWP, ACWP, and BCWS for the small project data given in Table 15.2. At any given study date, management will want to know what are the cost and schedule variance for each work packages. The variances can be calculated as follows:

$$CV \text{ (A.1)} = BCWP \text{ (A.1)} - ACWP \text{ (A.1)} = 33.3 - 40 = -6.7$$
$$CV \text{ (A.2)} = BCWP \text{ (A.2)} - ACWP \text{ (A.2)} = 39 - 35 = +4$$
$$CV \text{ (B)} = BCWP \text{ (B)} - ACWP \text{ (B)} = 52 - 50 = +2$$

Since the CV values for A.2 and B are positive, those accounts are within budget (i.e., the budgeted cost earned is greater than the actual cost). In other words, less is being paid in the field than was originally budgeted. The negative variance for A.1 indicates it is overrunning budget. That is, actual cost is greater than the cost budgeted.

This is confirmed by the values of the CPI for each package.

$$CPI \text{ (A.1)} = 33/40 < 1.0 \text{ A value less that } 1.0 \text{ indicates cost overrun of budget.}$$
$$CPI \text{ (A.2)} = 39/35 > 1.0$$
$$CPI \text{ (B)} = 52/50 > 1.0 \text{ Values greater than } 1.0 \text{ indicate actual cost less than budgeted cost}$$

The schedule variances for each package are as follows:

$$SV \text{ (A.1)} = BCWP \text{ (A.1)} - BCWS \text{ (A.1)} = 33.3 - 50 = -16.7$$
$$SV \text{ (A.2)} = BCWP \text{ (A.2)} - BCWS \text{ (A.2)} = 39 - 32 = +7$$
$$SV \text{ (B)} = BCWP \text{ (B)} - BCWS \text{ (B)} = 52 - 45 = +7$$

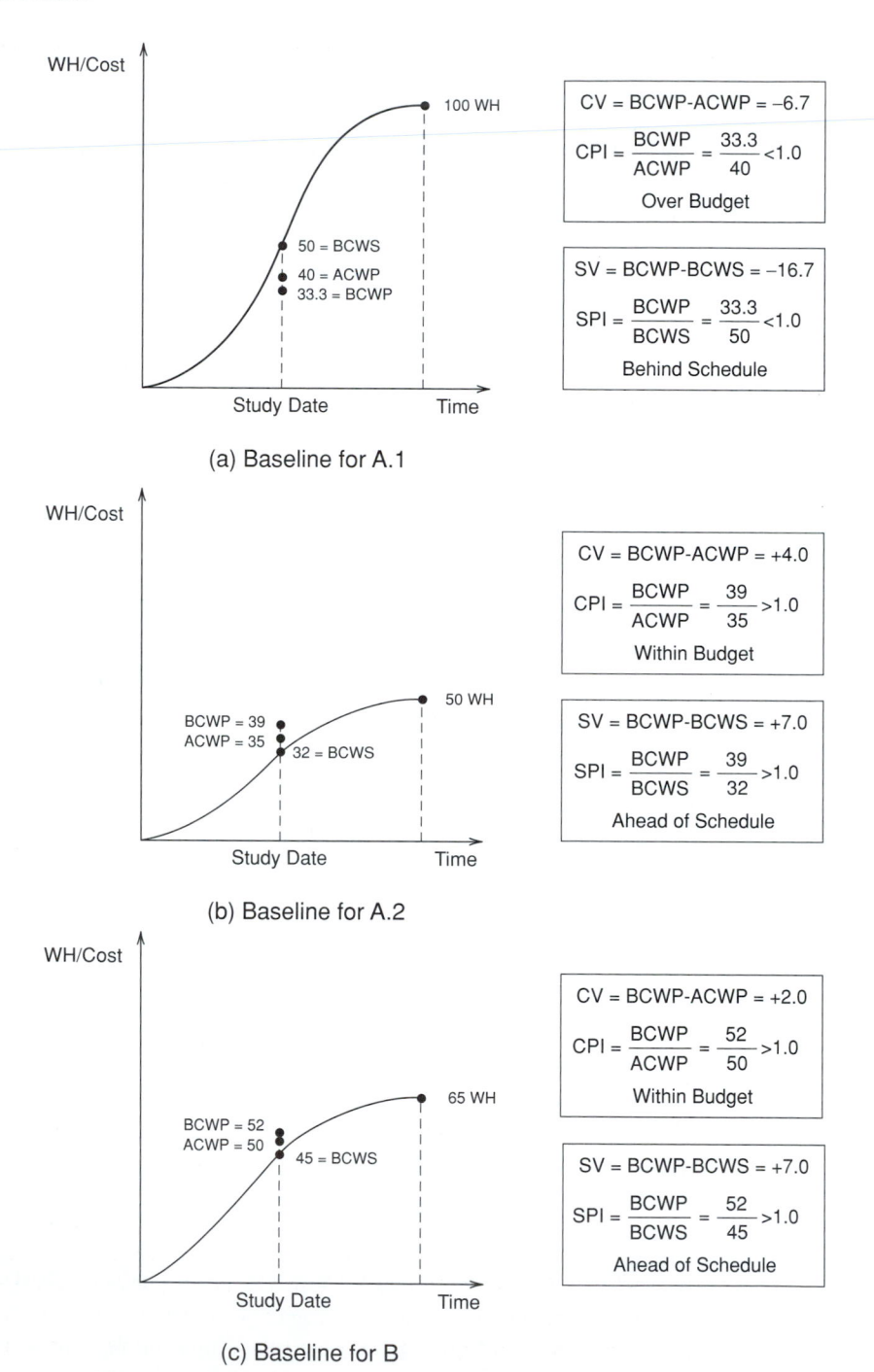

Figure 15.12 States of Control Account for Single Project.

The positive values for A.2 and B indicate that these items are ahead of schedule. The negative value for A.1 indicates a scheduling problem. The calculation of the SPI values will confirm this assessment. Overall, it can be stated that A.2 and B are ahead of schedule and below cost while A.1 is behind schedule and over cost.

Six scenarios for permutations of ACWP, BCWP, and BCWS are possible as established by Singh (Singh, 1991). The various combinations are shown in Figure 15.13 and Table 15.3.

The reader is encouraged to verify the information in Table 15.3.

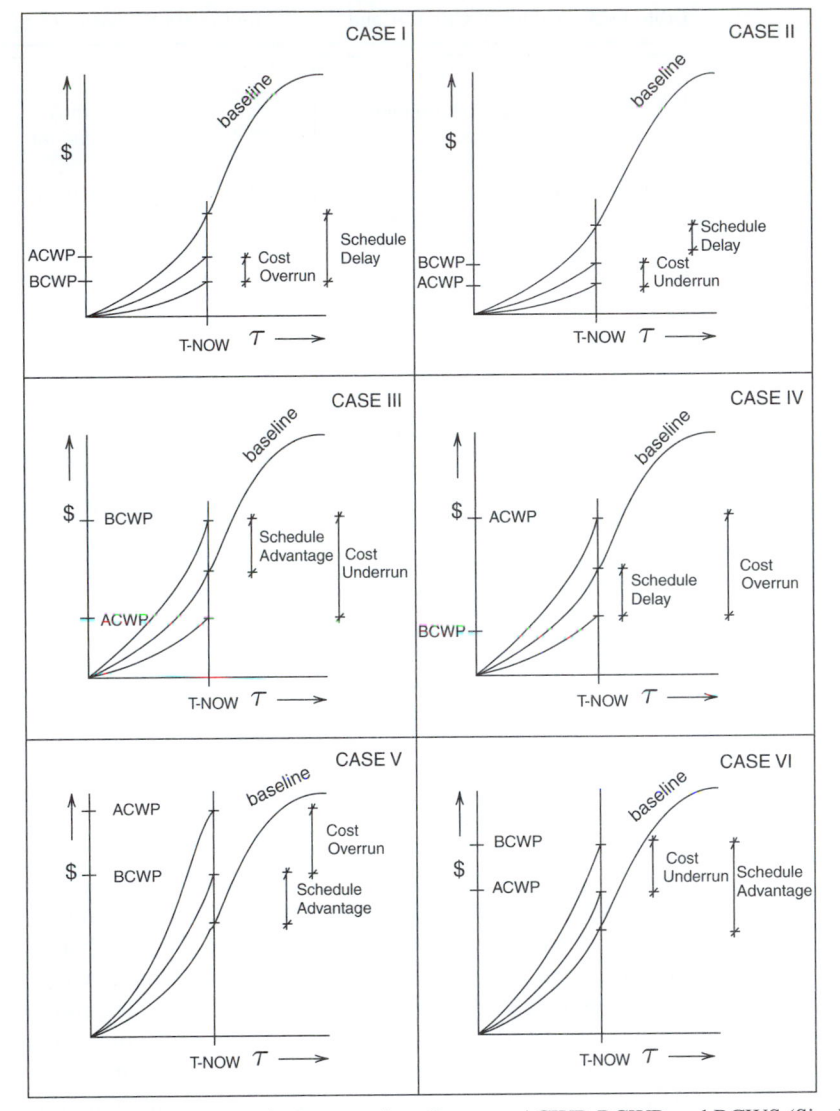

Figure 15.13 Scenarios for Permutations Between ACWP, BCWP, and BCWS (Singh, 1991).

The "earned" value approach requires a comprehensive knowledge of work packaging, budgeting, and scheduling. It is a data intensive procedure and requires the acquisition of current data on the ACWP and AQWP for each work package or account. It is a powerful tool, however, when management is confronted with complex projects consisting of hundreds of control accounts. In large projects consisting of thousands of activities and control accounts, it is a necessity. Without it, projects can quickly spiral out of control. A more detailed presentation of this topic is beyond the scope of this chapter. The interested reader should refer to current government publications which describe the Earned Value Management System (EVMS) and the inherent procedures associated with its implementation.

15.8   LABOR COST DATA COLLECTION

The purpose of the payroll system is to (1) determine the amount of and disburse wages to the labor force, (2) provide for payroll deductions, (3) maintain records for tax and other purposes, and (4) to provide information regarding labor expenses. The source document

Table 15.3   Values of CPI, CV, and SPI, SV for the Six Scenarios (Singh, 1991)

CPI < 1, CV < 0 → overrunning cost	CPI > 1, CV > 0 → within budget
SPI < 1, SV < 0 → behind schedule	SPI < 1, SV < 0 → behind schedule
CASE I	CASE II
CPI > 1, CV > 0 → within budget	CPI < 1, CV < 0 → overrunning cost
SPI > 1, SV > 0 → ahead of schedule	SPI < 1, SV < 0 → behind schedule
CASE III	CASE IV
CPI < 1, CV < 0 → overrunning cost	CPI > 1, CV > 0 → within budget
SPI > 1, SV > 0 → ahead of schedule	SPI > 1, SV > 0 → ahead of schedule
CASE V	CASE VI

used to collect data for payroll is a daily or weekly time card for each hourly employee similar to that shown in Figure 15.14. This card is usually prepared by foremen, checked by the superintendent or field office engineer, and transmitted via the project manager to the head office payroll section for processing. The makeup of the cards is such that the foreman or timekeeper has positions next to the name of each employee for the allocation of the time worked on appropriate cost subaccounts. The foreman in the distribution made in Figure 15.14 has charged 4 hours of A. Apple's time to an earth excavation account and 4 hours to rock excavation. Apple is a code 15 craft, indicating that he is an operating engineer (equipment operator). As noted, this distribution of time allows the generation of management information aligning work effort with cost center. If no allocation is made, these management data are lost.

The flow of data from the field through preparation and generation of checks to cost accounts and earnings accumulation records is shown in Figure 15.15.

This data structure establishes the flow of raw data or information from the field to management. Raw data enter the system as field entries and are processed to service both payroll and cost accounting functions. Temporary files are generated to calculate and produce checks and check register information. Simultaneously, information is derived from the field entries to update project cost accounts. These quantity data are not required by the financial accounting system and can be thought of as management data only.

From the time card, the worker's ID (badge number), pay rate, and hours in each cost account are fed to processing routines that cross check them against the worker data (permanent) file and use them to calculate gross earnings, deductions, and net earnings. Summations of gross earnings, deductions, and net earnings are carried to service the legal reporting requirements placed on the contractor by insurance carriers (Public Liability and Property Damage, workmen's compensation), the unions, and government agencies (e.g., Social Security and Unemployment).

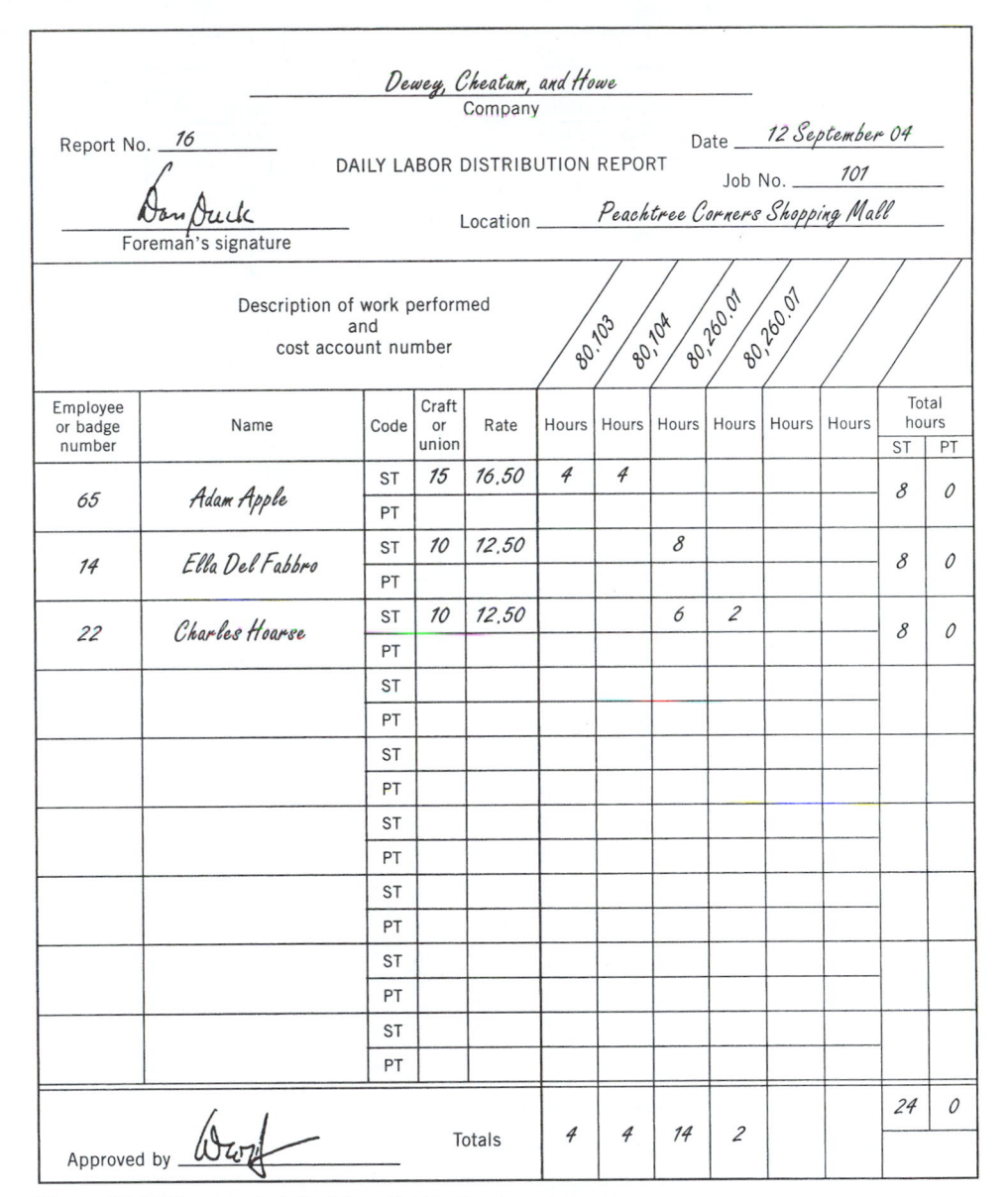

Figure 15.14 Foreman's daily labor distribution report.

15.9   CHARGES FOR INDIRECT AND OVERHEAD EXPENSE

Contractor-incurred expenses associated with the construction of a given facility relate to:

1. Direct cost consumed in the realization of a physical subelement of the project (e.g., labor and material costs involved in pouring a slab).

2. Production support costs incurred by the project-related support resources or required by the contractor (e.g., superintendent's salary, site office costs, various project related insurances) costs associated with the operation and management of the company as a viable business entity (e.g., home office overhead, such as the costs associated with preparation of payroll in the home office, preparation of the estimate, marketing, salaries of company officers).

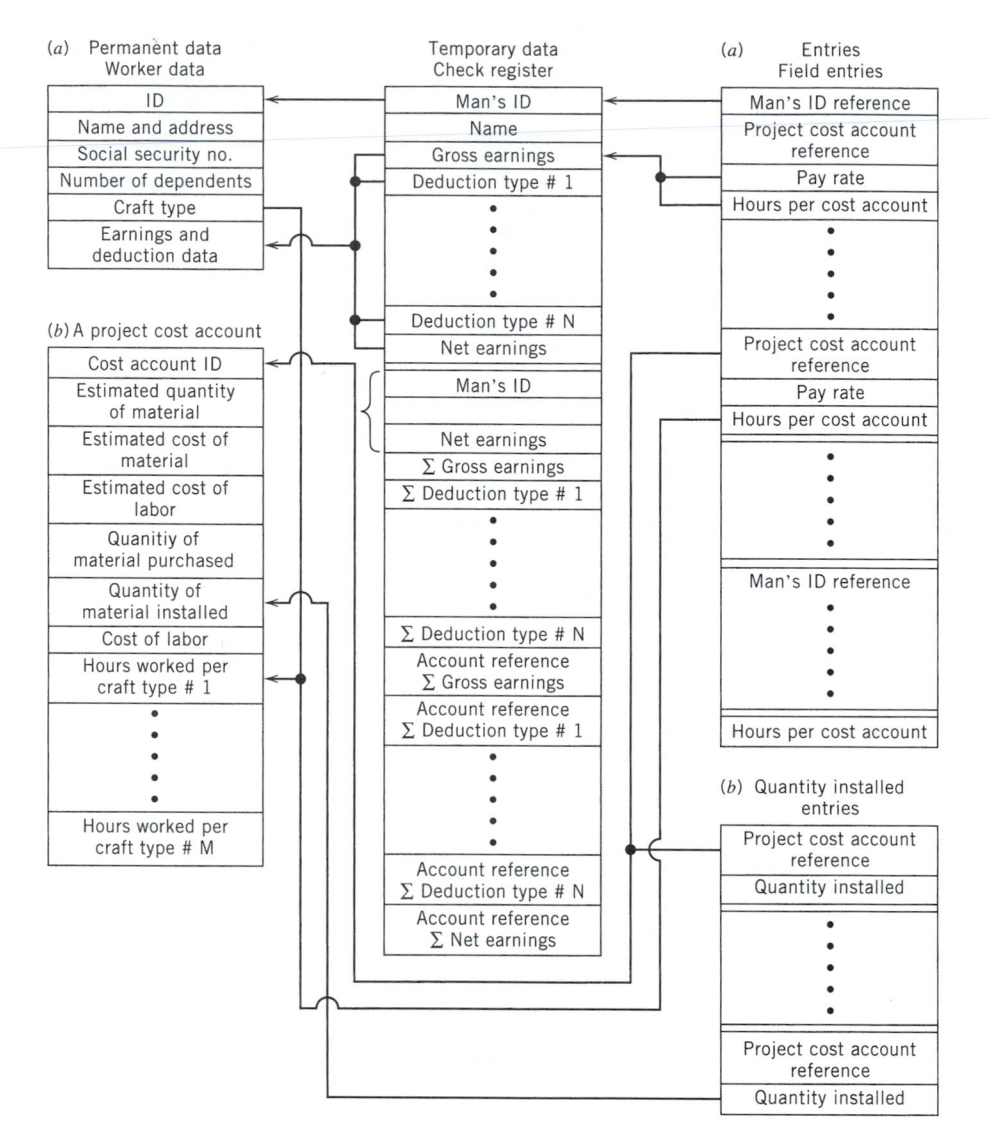

Figure 15.15 Payroll data structure.

The production support costs are typically referred to as project indirect costs. The home office charges are normally referred to as home office overhead. All of these costs must be recovered before income to the firm is generated. The home office overhead, or general and administrative (G&A) expense, can be treated as a period cost and charged separately from the project (direct costing). On the other hand, they may be prorated to the job and charged to the job cost overhead accounts and the work-in-progress expense ledger accounts (absorption costing).

15.10 PROJECT INDIRECT COSTS

Job-related indirect costs such as those listed in the labor cost report of Figure 15.16 (e.g., haul trash) are typically incurred as part of the on-site related cost associated with realizing the project. As such, they are charged to appropriate accounts within the job cost system. The level and amount of these costs should be projected during the estimating phase and included in the bid as individual estimate line items. Although it is recommended that job indirect

CENTURY CENTER BLDG #5
ATLANTA, GA

LABOR COST REPORT
HALCON CONSTRUCTORS, INC.
ATLANTA DIVISION

WEEK 57
WEEK ENDING 10/11

PAGE 1
PROJECT NUMBER 13-5265

Cost Code	Cost Code Information			Quantity		Unit Price		Cost		Projected Cost	
	Description	Units	% Comp	Estimated	Actual	Estimated	Actual	Estimated	Actual	To Date Over/Under	To Complete Over/Under
111	**This Week**		2	1	1	200.000	248.000		248	48	
	Haul Trash	Wk	68	50	34		92.765	10,000	3,154	3,646-	1,716-
112	**This Week**		1	1	1	217.391	543.000		543	326	
	Daily Clean	Wk	68	69	47	217.391	311.596	15,000	14,645	4,428	2,072
115	**This Week**		1	1	1		13.000		13	1,747	No Budg.
130	Safety	Wk	81	69	56		206.304	15,000	11,553	204-	144-
131	Protect Trees	Ls	100					500	31	621-	
132	Shoring	Ls	100	18	18	1,388.889	1,345.389	25,000	24,217	783-	Comp.
307	**This Week**		1	1	5		19.600		98	4	
	Hand Exc	Cy	97	725	705	18.793	19.569	13,625	13,796	547	15
310	Dewater	Ls	100					2,000	2,060	60	Comp.
312	Bkfl Hand	Cy	83	6,000	5,000	1.500	1.291	9,000	6,453	1,047-	209-
316	Fine Gr	Sf	99	15,000	14,830	.167	.135	2,500	1,999	501-	Comp.

Figure 15.16 Labor cost report (some typical line items).

be precisely defined during estimate development, many contractors prefer to handle these charges by adding a flat rate amount to cover them. Under this approach, the contractor calculates the direct costs (as defined above) and multiplies these charges by a percentage factor to cover both project indirects and home office fixed overhead. To illustrate, assume that the direct costs for a given project are determined to be $200,000. If the contractor applies a fixed factor of 20% to cover field indirects and home office overhead, the required flat charge would be $40,000. If he adds 10% for profit, his total bid amount would be $264,000.

The estimate summary shown in Figure 13.3 establishes line items for indirect charges and calculates them on an item-by-item basis (rather than applying a flat rate). Typical items of job-related indirect cost that should be estimated for recovery in the bid are those listed in Figure 15.2 as project overhead accounts (700–999). This is the recommended procedure since it is felt that sufficient information is available to the contractor at the time of bid to allow relatively precise definition of these job-related indirect costs. The R. S. Means method of developing overhead and profit (illustrated in Fig. 13.7) represents a percentage rate approach that incorporates a charge into the estimate to cover overhead on a line item-by-item basis. This is essentially a variation of the flat rate application described above.

15.11   FIXED OVERHEAD

Whereas the project indirect charges are unique to the job and should be estimated on a job-by-job basis, home office overhead is a more or less fixed expense that maintains a constant level not directly tied to individual projects. In this case, the application of a percentage rate to prorate or allocate home office expense to each project is accepted practice, since it is not reasonable to try to estimate the precise allocation of home office to a given project. Rather, a percentage prorate or allocation factor is used to incorporate support of home office charges into the bid.

The calculation of this home office overhead allocation factor is based on:

1. The general and administrative (G&A) (home office) expenses incurred in the past year

2. The estimated sales (contract) volume for the coming year

3. The estimated gross margin (i.e., markup) for the coming year. This procedure is illustrated in the following example (Adrian, 1998).

Step 1: Estimate of Annual Overhead (G&A Expense)

Last year's G&A	$270,000
10% inflation	27,000
Firm growth	23,000
Estimated G&A	$320,000

Step 2: Estimate of $ of Cost Basis for Allocation

Estimated volume	$4,000,000
Gross margin	20% = $800,000
Labor and material	$3,200,000

Step 3: Calculate Overhead Percent

$$\frac{\text{Overhead costs estimated (G\&A)}}{\text{Labor and material estimate}} = \frac{320,000}{3,200,000} = 10\%$$

Step 4: Cost to Apply to a Specific Project

Estimated labor and material costs	$500,000
Overhead to apply (@ 10 percent)	50,000
	$550,000

In the example, the anticipated volume for the coming year is $4,000,000. The G&A expense for home office operation in the previous year was $270,000. This value is adjusted for inflation effects and expected expansion of home office operations. The assumption is that the overhead allocation factor will be applied to the direct labor and materials costs. These direct costs are calculated by factoring out the 20% gross margin. Gross margin, in this case, refers to the amount of overhead and profit anticipated.

Direct costs amount to $3,200,000. The $320,000 in G&A costs to be recovered indicate a 10% prorate to be applied against the $3,200,000 of direct costs. This means that an overhead amount of $50,000 would be added to a contract bid based on $500,000 of direct cost to provide for G&A cost recovery. The profit would be added to the $550,000 base recovery amount.

15.12   CONSIDERATIONS IN ESTABLISHING FIXED OVERHEAD

In considering costs from a business point of view, it is common to categorize them either as variable costs or fixed costs. Variable costs are costs directly associated with the production process. In construction they are the direct costs for labor, machines, and materials as well as the field indirect costs (i.e., production support costs). These costs are considered variable since they vary as a function of the volume of work underway. Fixed costs are incurred at a more or less constant rate independent of the volume of work in progress. In order to be in business, a certain minimum of staff in the home office, space for home office operations, telephones, supplies, and the like must be maintained, and costs for these items are incurred. These central administrative costs are generally constant over a given range of sales/construction volume. If volume expands drastically, home office support may have to be expanded also. For purposes of analysis, however, these costs are considered fixed or constant over the year. Fixed costs are essentially the general and administrative costs referred to above.

As described in Section 15.11, the level of G&A (fixed) costs can be estimated by referring to the actual costs incurred during the previous year's operation. The method of projecting fixed overhead as a percentage of the estimated total direct costs projected for the coming year is widely used. Since the fixed overhead incurred in the previous year is typically available as a percentage of the previous year's total sales volume, a simple conversion must be made to reflect it as a percentage of the total direct cost. The formula for this conversion is

$$P_c = \frac{P_s}{(100 - P_s)}$$

where P_c = percentage applied to the project's total direct cost for the coming year

P_s = percentage of total volume in the reference year incurred as fixed or G&A expense

If, for instance, $800,000 is incurred as home office G&A expense in a reference year in which the total volume billed was $4,000,000, the P value would be 20% ($800,000/$4,000,000 × 100). The calculated percentage to be added to direct costs estimates for the coming year to cover G&A fixed overhead would be

$$P_c = \frac{20}{100 - 20} = 25\%$$

If the direct cost estimate (e.g., labor, materials, equipment, and field indirects) for a job is $1,000,000, $250,000 would be added to cover fixed overhead. Profit would be added to the total of field direct and indirects plus fixed overhead. The field (variable) costs plus the fixed overhead (G&A) charge plus profit yield the bid price. In this example, if profit is included at 10%, the total bid would be $1,375.000. It is obvious that coverage of the field overhead is dependent on generating enough billings to offset both fixed and variable costs.

Certain companies prefer to include a charge for fixed overhead that is more responsive to the source of overhead support. The assumption here is that home office support for management of certain resources is greater or smaller, and this effect should be included in charging for overhead. For instance, the cost of preparing payroll and support for labor in the field may be considerably higher than the support needed in administering materials procurement and subcontracts. Therefore, a 25% rate for fixed overhead is applied to labor and equipment direct cost, while a 15% rate on materials and subcontract costs is used. If differing fixed overhead rates are used on various subcomponents of the field (variable) costs in the bid, the fixed overhead charge will reflect the mix of resources used. This is shown in Table 15.4 in which a fixed rate of 20% on the total direct costs for three jobs is compared to the use of a 25% rate on labor and equipment and a 15% rate on materials and subcontracts.

It can be seen that the fixed overhead amounts using the 25/15% approach are smaller on jobs 101 and 102 than the flat 20% rate. This reflects the fact that the amount of labor and equipment direct cost on these projects is smaller than the materials and subcontract costs. The assumption is that support requirements on labor and equipment will also be proportionately smaller. On job 102, for instance, it appears that most of the job is sub-contracted with only $200,000 of labor and equipment in house. Therefore, the support costs for labor and equipment will be minimal, and the bulk of the support cost will relate to management of materials procurement and subcontract administration. This leads to a significant difference in fixed overhead charge when the 20% flat rate is used, as opposed to the 25/15% modified rates (i.e., $440,000 vs. $350,000).

On job 103, the fixed overhead charge is the same with either of the rate structures, since the amount of labor and equipment cost is the same as the amount of the materials and subcontract cost.

Table 15.4   Comparison of Fixed Overhead Rate Structures

			20% on Total Direct	25% on Labor and Equipment; 15% on Material and Subcontracts
Job 101	Labor and equipment	$ 800,000	$160,000	$200,000
	Materials and subcontracts	$1,200,000	240,000	180,000
			$400,000	$380,000
Job 102	Labor and equipment	200,000	$ 40,000	$ 50,000
	Materials and subcontracts	2,000,000	400,000	300,000
			$440,000	$350,000
Job 103	Labor and equipment	700,000	$140,000	$175,000
	Materials and subcontracts	700,000	140,000	105,000
			$280,000	$280,000

It should be obvious that in tight bidding situations use of the stylized rate system, which attempts to better link overhead costs to the types of support required, might give the bidder an edge in reducing his bid. Of course, in the example given (i.e., the 25/15% rate vs. 20%) the 20% flat rate would yield a lower overall charge for fixed overhead on labor- and equipment-intensive jobs. The main point is that the charge for fixed overhead should be reflective of the support required. Because the multiple rate structure tends to reflect this better, some firms now arrive at fixed overhead charges by using this approach rather than the flat rate applied to total direct cost.

REVIEW QUESTIONS AND EXERCISES

15.1 As a construction project manager, what general categories of information would you want to have on a cost control report to properly evaluate what you think is a developing overrun on an operation, "place foundation concrete," that is now under way and has at least 5 weeks to go before it is completed?

15.2 What are the major functions of a project coding system?

15.3 List advantages and disadvantages of the UCI coding system.

15.4 Assume you are the cost engineer on a new $12 million commercial building project. Starting with your company's standard cost code, explain how you would develop a project cost code for this job. Be sure the differences in purpose and content between these two types of cost codes are clear in your explanation. Specify any additional information that may be needed to draw up the project cost code.

15.5 Develop a cost code system that gives information regarding:

   a. When project started
   b. Project number
   c. Physical area on project where cost accrued
   d. Division in Uniform Construction Index
   e. Subdivision
   f. Resource classification (labor, equipment)

15.6 The following planned figures for a trenching job are available:

Quantity	Resources (hours)	Cost
Excavation—	Machines 1000	$100,000
second hauling	Labor 5000	$100,000
100,000 cu yd	Trucks 2000	$ 62,500

At a particular time during the construction, the site manager realizes that the actual excavation will be in the range of 110,000 cu yd. Based on the new quantity, he figures that he will have 30,000 cu yd left.
From the main office, the following job information is available:

Resources		Cost
Machines	895 hours	85,000
Labor	6011 man-hours	79,000
Trucks	1684 hours	50,140

What would concern you as manager of this job?

15.7 Categorize the following costs as (a) direct, (b) project indirect, or (c) fixed overhead:

Labor

Materials

Main office rental

Tools and minor equipment

Field office

Performance bond

Sales tax

Main office utilities

Salaries of managers, clerical personnel, and estimators

15.8 The following data are available on Del Fabbro International, Inc. The fixed (home office) overhead for the past year was $365,200. Total volume was $5,400,000. It is assumed that G&A costs will account for $1,080,000 of this volume. Del Fabbro uses a profit markup of 10%. The estimating department has indicated that the direct and field indirects for a renovation job will be $800,000. What bid price should be submitted to ensure proper coverage of fixed overhead? Assume a 5% inflation factor and a 12% growth factor in the calculation.

15.9 Calculate the cost and scheduling variances for each of the work packages shown. What is the percent complete for the entire package?

	WORK HOURS			QUANTITIES		
	EST	ACT	FORECAST	EST	ACT	FORECAST
A	15000	8940	15500	1000	600	1100
B	2000	1246	1960	200	93	195
C	500	356	510	665	540	680

15.10 Draw a Hierarchical diagram of the work packages given, using the WBS code values. Calculate the BCWP and *percent complete* for all codes and work packages to include A.00 and B.00. Finally, compute the total percent complete of the project.

CODE	DESCRIPTION	WORK HOURS			QUANTITIES		
		EST	ACT	FORECAST	EST	ACT	FORECAST
A. 00	E/W Duct	440					
A.10	Partitions	230	150	225	25	14	25
A.20	Hangers	210	130	220	3	2.2	3.8
B.00	N/W Duct	645					
B.10	Partitions	370	75	390	50	12	48
B.20	Hangers	275	85	260	16	4.5	16

15.11 Given the following diagrams of progress on individual work packages of a project answer the following questions:

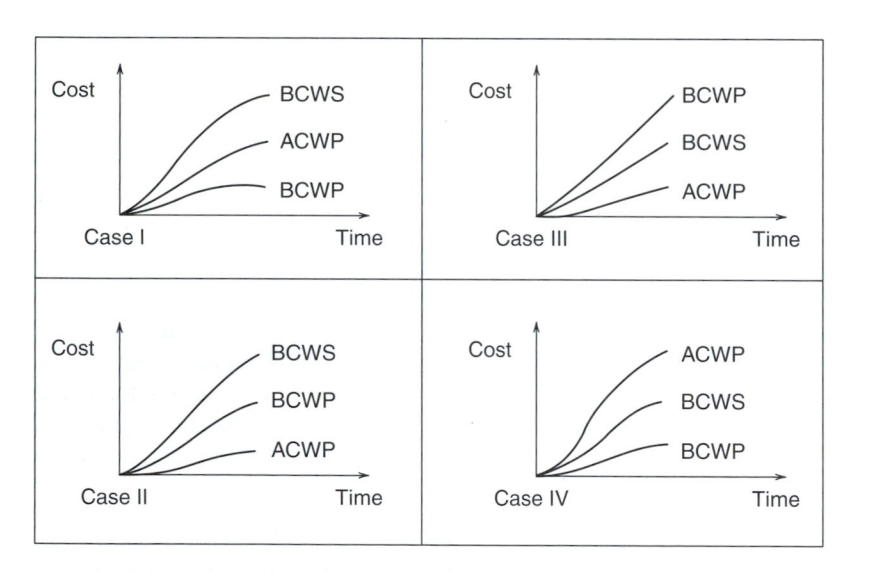

(a) For Case 1, is the project ahead or behind schedule?

(b) For Case 2, is the project over or under cost?

(c) For Case 3, is the Cost Performance Index greater than 1?

(d) For Case 1, is the SPI greater than 1? Explain by calculation.

(e) For Case 4, is the project on schedule and budget or not? Explain.

Chapter 16

Material Management

Fiber Reinforced Polymer Rebar

Dry Dock #4 Pearl Harbor, Hawaii.

The Need

Reinforced concrete is a very common building material for the construction of facilities and structures. As a complement to concrete's very limited tensile strength, steel rebar has been an effective and cost-efficient reinforcement. However, insufficient concrete cover, poor design or workmanship, and the presence of large amounts of aggressive agents in the concrete as well as environmental factors all can lead to cracking of the concrete and corrosion of the steel rebar. For instance, in the United States, almost 40% of bridges are structurally deficient or functionally obsolete largely due to cracking and corrosion.

The Technology

Composite materials made of fibers embedded in a polymeric resin, also known as fiber-reinforced polymers (FRPs), have become an alternative to steel reinforcement for concrete structures. Aramid fiber-reinforced polymer (AFRP), carbon fiber-reinforced polymer (CFRP), and glass fiber-reinforced polymer (GFRP) rods are commercially available products for use in the construction industry. They have been proposed for use in lieu of steel reinforcement or steel prestressing tendons in nonprestressed or prestressed concrete structures. The problems of steel corrosion are avoided with the use of FRPs because FRP materials are nonmetallic and noncorrosive. In addition, FRP materials exhibit several properties including high tensile strength, which make them suitable for use as structural reinforcement. Fiberglass rebar may be a suitable alternative to steel reinforcing in architectural concrete, in concrete exposed to de-icing or marine salts, and in concrete used near electromagnetic equipment.

Caissons and port facilities.

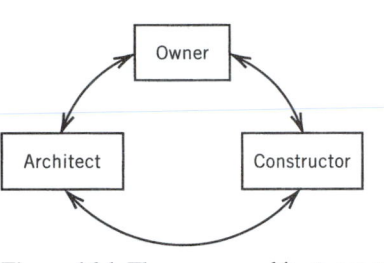

Figure 16.1 The owner–architect–construction relationship.

16.1   MATERIAL MANAGEMENT PROCESS

In the traditional contractual relationship, the owner contracts with a general contractor or construction manager to build his facilities and with an architect to perform the design. The general contractor, through this contract with the owner, is obligated to perform the work in accordance with the architect's instructions, specifications, and drawings. Thus, the architect is the owner's agent during the design and construction of a project. The lines of communication between the three parties are established as shown in Figure 16.1.

The materials that comprise facilities in building construction are subject to review by the architect or design professional. The contractor usually delegates responsibility for some of the categories of work involved in the project to subcontractors and suppliers. This delegation is accomplished through subcontracts and purchase orders. As a result of this delegation, a distinct life cycle evolves for the materials that make up the project. The four main phases of this cycle are depicted in Figure 16.2.

16.2   THE ORDER

When the contract for construction is awarded, the contractor immediately begins awarding subcontracts and purchase orders for the various parts of the work. How much of the work is subcontracted depends on the individual contractor. Some contractors subcontract virtually all of the work in an effort to reduce the risk of cost overruns and to have every cost item assured through stipulated-sum subcontract quotations. Others perform almost all the work with their own field forces.

The subcontract agreement defines the specialized portion of the work to be performed and binds the contractor and subcontractor to certain obligations. The subcontractor, through the agreement, must provide all materials and perform all work described in the agreement. The Associated General Contractors (AGC) of America publish the *Standard Subcontract Agreement* for use by their members.

A sample of this agreement can be found in Appendix G. Most contractors either adopt a standard agreement, such as that provided by the AGC, or implement their own agreement. In most cases, a well-defined and well-prepared subcontract is used for subcontracting work.

All provisions of the agreement between the owner and contractor are made part of the subcontract agreement by reference. The most important referenced document in the subcontract agreement is the General Conditions. Procedures for the submittal of shop drawings and samples of certain materials are established in the General Conditions. The General Conditions provide that "Where a Shop Drawing or Sample is required by the

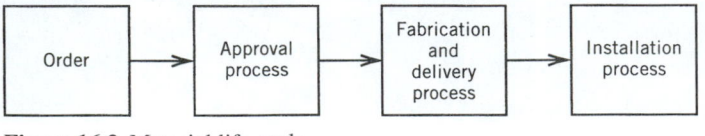

Figure 16.2 Material life cycle.

Special Purchase Order

HCB HENRY C BECK COMPANY

VENDOR:

MAIL INVOICE TO:
HENRY C. BECK COMPANY
1210 S. Old Dixie Highway
Jupiter, Florida 33458

DATE:

CHG. TO JOB #21330 _____

SHIP TO: 1210 S. Old Dixie Highway / Jupiter, Florida 33458

QUANTITY	ARTICLE	U.P.	AMOUNT	COST CODE

STATE AND LOCAL SALES TAXES MUST BE SET OUT SEPARATELY ON INVOICE

Invoice in Triplicate
To Above Address
No Later Than 25th of Month —— Vendor's Acceptance (when required)   SUPT. OR PROJECT MGR.
Show S.P.O. Number On Invoice   WHITE (ORIGINAL) - VENDOR'S COPY   PINK   - SUPERINTENDENT'S COPY
                                 CANARY   –JOB OFFICE COPY   GOLDENROD -PROJECT MANAGER'S COPY
                                        (MAIL TO DALLAS WITH INVOICE)

Figure 16.3 Field purchase order (courtesy of Henry C. Beck Co.).

Contract Documents. . . , any related Work performed prior to ENGINEER's review and approval of the pertinent submittal will be at the sole expense and responsibility of the CONTRACTOR."

The purchase order is a purchase contract between the contractor and the supplier. This document describes the materials to be supplied, their quantities, and the amount of the purchase order.

Purchase orders vary in complexity and can be as simple as a mail order house (e.g., Sears) order form, or as complex as the construction contract itself. When complex and specially fabricated items are to be included in the construction, very detailed specifications and drawings become part of the purchase order. Some typical purchase order forms are shown in Figures 16.3 and 16.4. Figure 16.3 shows a form for field-purchased items procured from locally available sources. These items are usually purchased on a cash-and-carry basis. The purchase order in this case is used primarily to document the purchase for record-keeping and cost accounting purposes (rather than as a contractual document). A more formal purchase order used in a contractual sense is shown in Figure 16.4. It is used in the purchase of more complex items from sources that are remote to the site.

Regardless of the complexity of the transaction, certain basic elements are present in any purchase order. Five items can be identified as follows:

1. Quantity or number of items required.
2. Item description. This may be a standard description and stock number from a catalog or a complex set of drawings and specifications.
3. Unit price.
4. Special instructions.
5. Signatures of agents empowered to enter into a contractual agreement.

Letter or transmittal form accompanying this order when mailed to Vendor should show the number of shop drawings and/or samples to be furnished and the address to which they must be sent; also the address to which Vendor is to mail correspondence relating to this order.

PURCHASE ORDER

HCB HENRY C BECK COMPANY   No.

VENDOR
ADDRESS
_____ 19 _____

JOB:
Job Mailing Address:

Please ship the following to HENRY C. BECK COMPANY, at

SHIP VIA:

It is agreed that shipment will be made on or before _____ or right is reserved to cancel order.

IMPORTANT NOTE: It is IMPERATIVE in the interest of prompt payment that all invoices be rendered in the original with two (2) copies. Mail together with two (2) copies of bills of lading and/or other papers to JOB at address above.

ITEM NO.	QUANTITY	DESCRIPTION	UNIT	AMOUNT

SALES or USE TAX (is) (is not) included in amounts shown above.   HENRY C. BECK COMPANY

F. O. B.

TERMS: By_____

See above IMPORTANT NOTE for invoicing instructions. They MUST be complied with.

Accepted:_____

Show above order number on invoices, and on the outside of each package containing Shipment.   By_____

Figure 16.4 Formal purchase order (courtesy of Henry C. Beck Co.).

For simple purchase orders, the buyer normally prepares the order. If the vendor is dissatisfied with some element of the order, he may prepare his own purchase order document as a counterproposal.

The special instructions normally establish any special conditions surrounding the sale. In particular, they provide for shipping and invoicing procedures. An invoice is a billing document that states the billed price of shipped goods. When included with the shipped goods, it also constitutes an inventory of the contents of the shipment. One item of importance in the order is the basis of the price quotation and responsibility for shipment. Price quotations normally establish an FOB location at which point the vendor will make the goods available to the purchaser. FOB means Free On Board and defines the fact that the vendor will be responsible for presenting the goods free on board at some mutually

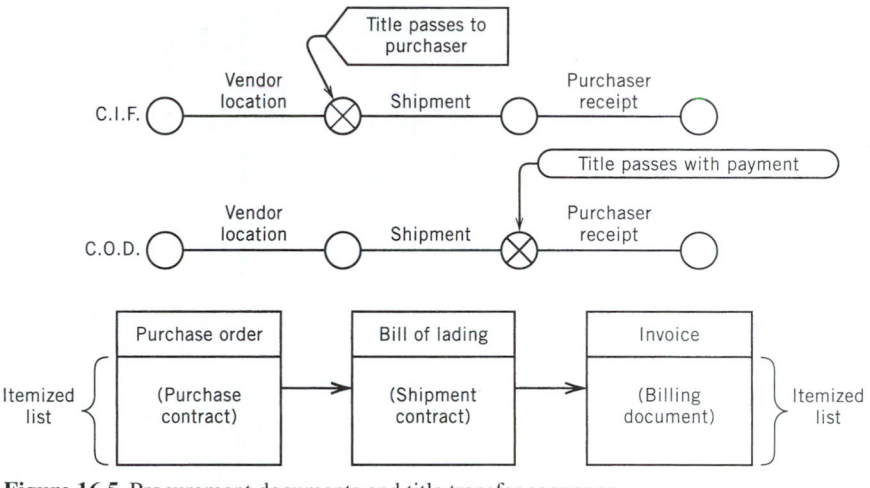

Figure 16.5 Procurement documents and title transfer sequence.

agreed-upon point such as the vendor's sales location, factory, or the purchaser's yard or job site. This is important because if the FOB location is other than the vendor's location, the vendor is indicating that the price includes shipment. The vendor may quote the price as cost, insurance, and freight (CIF). This indicates that the quoted price includes item cost plus the shipment cost to include freight and insurance expenses to the FOB location.

In the event the vendor ships the goods, it is of interest to establish at what point in time title of ownership passes from the vendor to the purchaser. This is established by the *bill of lading*. The bill of lading is a contractual agreement between a common carrier and a shipper to move a specified item or group of goods from point A to point B at a contracted price. If ownership passes to the purchaser at the vendor's location, the contract for shipment is made out between the purchaser and the common carrier. In cases in which the vendor has quoted a CIF price, he acts as the agent of the purchaser in retaining a carrier and establishing the agreement on behalf of the purchaser. The bill of lading is written to pass title of ownership at the time of pickup of the goods by the common carrier at the vendor's location. In such cases, if the common carrier has an accident and damages the goods during transfer, the purchaser must seek satisfaction for the damage since he is the owner.

If goods are to be paid cash on delivery (i.e., COD), the title of ownership passes at the time of payment. In such cases, the bill of lading is between vendor and common carrier. If damage should occur during shipment, recovery of loss falls to the vendor as owner.

The sequence of events in CIF and COD transactions is shown in Figure 16.5. This figure also indicates the relationship between order, bill of lading, and invoice. A typical bill of lading memorandum and invoice are shown in Figures 16.6 and 16.7.

The invoice normally states the payment procedures and establishes trade discounts that are available to the purchaser if payment is made in a timely fashion. Trade discounts are incentives offered by the vendor for early payment. If the purchaser pays within a specified period, he must pay the stated price minus a discount. Failure to pay within the discount period means that the full price is due and payable. Terminology relating to trade discounts is as follows:

1. ROG/AOG: The discount period begins upon receipt of goods (ROG) or arrival of goods (AOG).

2. 2/10 NET 30 ROG: This expression appearing on the invoice means 2% can be deducted from the invoiced amount if the contractor pays within 10 days of AOG/ROG. Full payment is due within 30 days of AOG/ROG.

This Shipping Order

Carbon, and retained by the Agent.

Shipper's No. _____

_____
(Name of Carrier)

Carrier's No. _____

RECEIVED, subject to the classifications and tariffs in effect on the date of the issue of this Bill of Lading.

at   **HALLANDALE, FLA.**   **Date** _____ **20xx**   **From**   **MEADOW STEEL PRODUCTS, INC.**

The property described below, in apparent good order, except as noted (contents and condition of contents of packages unknown), marked consigned, and destined as indicated below, which said carrier (the word carrier being understood throughout this contract as meaning any person or corporation in possession of the property under the contract) agrees to carry to its usual place of delivery of said destination, if on its own route, otherwise to deliver to another carrier on route to said destination. It is mutually agreed, as to each carrier of all or any of said property over all or any portion of said route to destination, and as to each party at any time interested in all or any of said property, that every service to be performed hereunder shall be subject to all the terms and conditions of the Uniform Domestic Straight Bill of Lading set forth (1) in Official Southern, Western and Illinois Freight Classification in effect on the date thereof, if this is a rail or rail-water shipment, or (2) in the applicable motor carrier classification or tariff if this is a motor carrier shipment.

Shipper hereby certifies that he is familiar with all the terms and conditions of the said bill of lading, including those on the back thereof, set forth in the classification or tariff which governs the transportation of this shipment, and the said terms and conditions are hereby agreed to by the shipper and accepted for himself and his assigns.

Consigned to _____

(Mail or street address of consignee—For purposes of notification only.)

Destination _____ State _____ Zip _____ County _____

Delivery
Address* _____
(*To be filled in only when shipper desires and governing tariffs provide for delivery thereof.)

Route _____

Delivering Carrier _____ Car or Vehicle Initials _____ No. _____

Figure 16.6 Typical bill of lading (courtesy of Augusta Meadow Steel Products, Inc.).

No. Packages	Kind of Package, Description of Articles, Special Marks, and Exceptions	*WEIGHT (Subject to Correction)	Class or Rate	Check Column
	REINFORCING STEEL ACCESSORIES			
			50	

*If the shipment moves between two ports by a carrier by water, the law requires that the bill of lading shall state whether it is carrier's or shipper's weight.

NOTE: Where the rate is dependent on value, shippers are required to state specifically in writing the agreed or declared value of the property.

Agreed or declared value of the property is hereby specifically stated by the shipper to be not exceeding
_____ per _____

+ The fibre boxes used for this shipment conform to the specifications set forth in the box maker's certificate thereon, and all other requirements of the Consolidated Freight Classification.

Subject to Section 7 of Conditions of applicable bill of lading, if this shipment is to be delivered to the consignee without recourse on the consignor the consignee shall sign the following statement:

The carrier shall not make delivery of this shipment without payment of freight and all other lawful charges.

_____
(Signature of Consignor)

If charges are to be prepaid, write or stamp here: "To Be Prepaid."

TO BE PREPAID

Received $ _____ to apply in prepayment of the charges on the property described hereon.

Agent or Cashier

Per _____
(The signature here acknowledges only the amount prepaid.)

+ Shipper's imprint in lieu of stamp; not a part of Bill of Lading approved by the Interstate Commerce Commission.

$ _____
Charges Advanced:

Agent must detach and retain this Shipping Order and must sign the Original Bill of Lading.

MEADOW STEEL PRODUCTS, INC. Shipper, Per _____

Permanent post-office address of shipper **1804 SO. 31st AVE., HALLANDALE, FLA. 33009**

Figure 16.6 (*Continued*).

281

Bibb Steel & Supply Company

INCORPORATED

INVOICE

FABRICATED STRUCTURAL STEEL

AREA CODE 912-788-7373
POST OFFICE BOX 3007
4105 BROADWAY
MACON. GA. 31205

SOLD TO	Bellamy Brothers Contracting Co., Inc.	DATE	December 29, 1978
ADDRESS	P. O. Box 218 Ellenwood, Georgia 30049	CUSTOMER'S ORDER NO.	
SHIP TO	Same @ Fulton-Clayton Cty's, Georgia Proj. # ACI-85-1 (154) 72 & PR-8500-2 (121)		Bridge #1
VIA	O. T.	TERMS: ~~XXXXXXXXXXXXX~~ Net 10th Prox.	

1% PER MO. INTEREST CHARGED AFTER MATURITY.

✓	QUANTITY	DESCRIPTION	UNIT PRICE	GROSS	DISCOUNT	NET AMOUNT
		Revised lateral bracing connection for bridge #1 as designed				
		by Bibb Steel and approved by Georgia D.O.T.				$ 11 672 00
		3% Georgia Sales Tax				350 16
		1% MARTA Tax				116 72
						$ 12 138 88
		RECEIVED				
		DATE 1-12-79				
		BELLAMY BROS., INC.				
		1654 SULLIVAN RD.				
		COLLEGE PARK, GA 30339				
		Completes Contract-Bridge #1				

INVOICE NO.
No 801

RECEIVED ABOVE IN GOOD CONDITION

RECEIVED BY

THANKS

Figure 16.7 Typical invoice (courtesy of Bibb Steel & Supply Company).

3. 2/10 PROX NET 30: A 2% cash discount is available if invoice is paid not later than the 10th of the month following ROG. Payment is due in full by the end of the following month.

4. 2/10 E.O.M.: The discount (2%) is available to the 11th of the month following ROG. Payment in full is due thereafter.

Trade discounts received are treated as earned income in financial statements.

The special conditions of the purchase order may include a "hold harmless" clause. Such clauses protect one of the parties to the purchase order from liability arising out of damages resulting from the conditions of the purchase order. A transit concrete mix company, for instance, may have the contractor submit his orders on their forms holding the vendor harmless for damages arising out of delivery of the concrete to the site. Thus, if the transit mix truck should back across a gas main on the site, rupturing it during normal delivery, liability for repair costs will accrue to the contractor since the concrete vendor is "held harmless." The converse could, of course, occur if the contractor uses his own purchase order form, which holds him harmless in such an event. These situations are not covered by normal liability insurance since such "contractually accruing" liability is considered to be outside the realm of normal liability. If the language of the order is prepared by the contractor, the hold harmless clause will operate to protect him. If the vendor's language is used, the special conditions will hold him harmless in these damages situations.

For the contractor's protection, reference is made in complex purchase orders (requiring special fabrication) to the contractor specifications and other documents that define the materials to be supplied. Specifications detail the required *shop drawings, product data*, and *samples* that must be submitted for approval prior to fabrication and delivery. The provisions of the purchase order and the subcontract agreement require the subcontractor and supplier to obtain approval for their materials.

16.3 APPROVAL PROCESS

The contract drawings prepared by the architect are generally not specific enough to facilitate accurate fabrication of the materials involved. Therefore, to produce the necessary materials for a project, subcontractors and suppliers must provide details that further amplify the contract drawings. These details can be classified into three groups: (1) shop drawings, (2) product data, and (3) samples.

Shop drawings are defined in the General Conditions as "All drawings, diagrams, illustrations, schedules, and other data or information which are specifically prepared or assembled by or for CONTRACTOR and submitted by CONTRACTOR to illustrate some portion of the Work." The detailing, production, and supplying of shop drawings are the sole responsibility of the contractor or the contracted agent. However, the design professional is responsible for verification that the supplied shop drawings correctly interpret the contract documents. Dimensions, quantities, and coordination with other trades are the responsibility of the contractor. Approved shop drawings become the critical working drawings of a project and are considered a part of the contract documents. Typically, shop drawings are submitted for materials such as reinforcing steel, formwork, precast concrete, structural steel, millwork, casework, metal doors, and curtain walls.

Product data may be submitted to illustrate the performance characteristics of the material items described by the shop drawings or may be submitted as verification that a standard product meets the contract specifications. Product data are illustrations, standard schedules, performance charts, instructions, brochures, diagrams, and other information furnished by the contractor to illustrate a material, product, or system for some portion of the work. Mill test reports, concrete mix designs, masonry fire rating tests, curtain wall wind test reports, and mechanical equipment performance tests are examples of product data.

Product data are particularly important when a subcontractor or supplier is submitting data on a product that is a variance from the contract specifications. The architect carefully analyzes the submitted data prior to rendering an approval of the substitution. Also, the product data are used extensively to coordinate the materials used by the mechanical and electrical subcontractors. The contractor must communicate the product data between these major subcontractors to ensure proper performance of their portion of the work.

Samples usually involve the finishes of a project and are physical examples of materials to be supplied. The architect may require samples of plastic laminate finishes for doors and counters, flooring, wall coverings, paint, stucco, precast concrete, ceilings, and other items. These are used by the architect in developing the overall building finish scheme.

The approval process involving shop drawings, product data, and samples has several substages that are critical to the material life cycle. These are: (1) submission by the subcontractor or supplier, (2) review of the submittal by the contractor, (3) review by the architect or design professional, and (4) return of submittal to the subcontractor or supplier.

At the time of awarding subcontracts and purchase orders, the contractor usually establishes the quantity, size, and other requirements for all submittals. In most cases, several blue line prints (usually six) are required when shop drawings are submitted for approval. The product data quantities required may range from three to six copies. The copies of a submittal may vary depending on the number of other subcontractors or vendors that must receive approved copies to coordinate their work. In all cases, careful planning of the quantity of submittals will expedite the other substages by eliminating the handling of unnecessary copies of submittals.

Timing of submittals is of utmost importance in the effective processing of material submittals. Subcontracts and purchase orders often contain language such as "all submittals must be made immediately" or "fifteen (15) days after execution of this agreement, all submittals must be made." In most cases, contractors do not preplan in detail the required submittal data from a subcontractor or supplier. The result is a landslide of submittals, most of which are not necessary, in the early stages of the project. Thus, field office personnel waste time sorting and determining the most critical submittals. A well-planned approach to scheduling submittals will ensure timely processing and better control of required submittals.

Once a submittal is received by the contractor, the process of checking for conformance with the intent of the contract documents is performed. A submittal, whether it is a shop drawing, product data, or sample, is governed by the contract drawings and specifications. The contractor's field or main office personnel in charge of submittals may make notations and comments to the designer or his engineers to clarify portions of the submittal or to correct the submittal. The submittal represents specific details of the project and is of primary importance in coordination, as well as depicting exactly what a supplier or subcontractor is providing. The contractor is required by the general conditions to *clearly* note to the architect or design professional any variation from the contract documents.

The amount of time involved in the contractor's review of submittals may vary from 1 to 5 days, depending on the nature of the submittal and its correctness. Reinforcing steel and structural steel shop drawings typically require the greatest amount of time. Also, schedules such as doors, hardware, and door frames consume a great deal of time because of the minute details that must be checked. However, the time expended in submittal processing by the contractor can most easily be controlled at this substage. It must be remembered that time spent in reviewing, checking, and coordinating submittals is one of the most effective methods of ensuring a highly coordinated and smooth running project.

Once the contractor has completed the review of a submittal, the document is transmitted to the architect for approval. The contractor may indicate on the transmittal the date when approval is needed. Here again, the amount of time required for the architect to review a submittal depends on its complexity and whether or not other engineers (i.e., mechanical, electrical, or structural) must participate in the review. As a general rule, 2–3 weeks is a good estimate for the time required by the architect to complete the review and return the submittal.

The period when a submittal is in the hands of the architect is probably the most critical substage of the approval process for materials. During this critical substage, the contractor's submittal can be "lost in the shuffle" if the architect's activities are not monitored daily. The most common method of monitoring submittals is through the use of a submittal log, which indicates the date, description, and quantity of each submittal. From this log the contractor can develop a listing of critical submittals to monitor on a daily basis. Once the submittal leaves the contractor's control in the field office, its return must be followed constantly or valuable time will be wasted.

The final substage of the approval process for a material item is the return of the submittal to the supplier or subcontractor. The submittal may be in one of the following four states when returned to the architect:

1. Approved.

2. Approved with noted corrections; no return submittal needed.

3. Approved with noted corrections; however, a final submittal is required.

4. Not approved; resubmit.

The first through third designations would release the vendor or subcontractor to commence fabrication and delivery. The fourth stage would require that the approval process be repeated. In some cases the disapproval by the architect is due to a subcontractor or supplier not communicating clearly through the submittal of the information needed. A meeting between all parties may then be arranged to seek a reasonable solution.

When the approval process is completed, the material has been accepted as part of the project. Its details have been carefully reviewed for conformance with the contract documents. Also, through this process, the item has been coordinated with all trades involved in its installation and verified for inclusion into the project. The material is now ready for fabrication and delivery.

16.4 FABRICATION AND DELIVERY PROCESS

As a submittal is returned to the subcontractor or supplier, the needed delivery date to meet the construction schedule is communicated on the transmittal, verbally, or through other correspondence. In any event, delivery requirements are established and agreed on. The supplier or subcontractor may be required to return to the contractor corrected file and field-use drawings, product data, or samples. These are used to distribute to the contractor's field personnel (i.e., superintendent or foreman) and the other subcontractors and suppliers that must utilize these final submittals.

Of the four phases of a material's life cycle the fabrication and delivery process is the most critical. Generally, the largest amount of time is lost and/or gained in this phase. The duration of the fabrication and delivery process depends directly on the nature of the material and the amount of physical transformation involved. For these reasons, the contractor must employ every available method of monitoring materials throughout the fabrication and delivery process.

Contractors generally devote the largest amount of time and effort to controlling and monitoring the fabrication and delivery phase. The term *expediting* is most commonly used to describe monitoring methods in this phase of a material item's life. Methods used to ensure timely fabrication and delivery may range from using checklists developed from the job schedule to actually including this phase as a separate activity on a job schedule. Unfortunately, the fabrication and delivery usually only become activities on the job schedule when the delivery becomes a problem. Extremely critical items requiring extended fabrication times often warrant visits by the contractor to the fabrication facility to ensure the material is actually in fabrication, and proceeding on schedule.

At the completion of fabrication, the delivery of the material is made and the final phase of the life cycle is begun. Materials delivered are checked for compliance with the approved submittal as regards quality, quantity, dimensions, and other requirements. Discrepancies are reported to the subcontractor or supplier. These discrepancies, whether they be shortages or fabrication errors, are subjected to the same monitoring and controlling processes as the entire order. Occasionally they become extremely critical to the project and must be given a great deal of attention until delivery is made.

16.5 INSTALLATION PROCESS

The installation process involves the physical incorporation into the project of a material item. Depending on how effectively materials are scheduled and expedited, materials arriving at the job site may be installed immediately, partially installed and partially stored, or completely stored for later installation. When storage occurs, the installation process becomes directly dependent on the effective storage of materials.

One of the most important aspects of the effective storage of materials is the physical protection of material items. Careful attention must be given to protection from weather hazards such as prevention of water damage or even freezing. Another important aspect is protection against vandalism and theft. Finish hardware, for instance, is generally installed over a considerable time period. A secure hardware room is usually set aside where it is sorted, shelved, and organized to accommodate the finish hardware installation process.

Location of materials stored outside the physical building on the project site or within the building must be carefully planned and organized to facilitate effective installation. In high-rise-building construction material storage, each floor can be disastrous if careful planning is not used. For instance, materials stored concurrently on a floor may include plumbing and electrical rough-in materials, ductwork, window wall framing, glazing materials, drywall studs, and other items. The magnitude of the amount of materials involved warrants meticulous layout of materials. Equally important is the storage of materials to facilitate hoisting with a minimal amount of second handling. Reinforcing steel, for instance, may be organized in a "lay-down" area and then directly hoisted as needed. Adequate lay-down areas must be provided within reach of vertical hoisting equipment.

16.6   MATERIAL TYPES

Building construction materials can be logically grouped into three major categories: (1) bulk materials that require little or no fabrication, (2) manufacturer's standard items that require some fabrication, and (3) items that are fabricated or customized for a particular project. Grouping materials into categories can be of value in determining which materials warrant major contractor control efforts. Obviously, material items that require fabrication have longer life cycles because of submittal requirements and fabrication. These materials require a great deal of control by the contractor.

The bulk material category includes those materials that require very little vendor modification and can be delivered from vendor storage locations to the job site with very little fabrication delay. Table 16.1 lists examples of typical bulk materials in building construction projects. These materials usually require only a 1- to 5-day delivery time, following execution of purchase order or subcontract and approved submittals. Submittal requirements generally include only product and performance data.

Manufacturer's standard material items include materials that are usually stocked in limited quantities and are manufactured for the project after the order is executed and submittals are approved. Table 16.2 illustrates typical materials that are included in this category. Submittal requirements include detailed shop drawings, product and performance data, and samples. Finish materials such as paints, wall coverings, floor coverings, and

Table 16.1   Typical Bulk Materials

Paving materials
Fill materials—crushed stone, soil, sand, etc.
Damproofing membrane
Lumber and related supplies
Form materials—plywood, post shores, etc.
Ready-mix concrete
Wire mesh
Stock reinforcing steel and accessories
Masonry
Stock miscellaneous metals
Soil and waste piping
Water piping
Electrical conduit
Electrical rough-in materials—outlet boxes, switch boxes, etc.
Caulking and sealants

Table 16.2 Typical Standard Material Items

General Materials

    Fencing materials
    Formwork systems—metals and fiberglass pans, column forms, etc.
    Brick paving
    Brick or ceramic veneers
    Standard structural steel members
    Metal decking
    Waterproofing products
    Insulation products
    Built-up roof materials
    Caulking and sealants
    Standard casework and millwork
    Special doors
    Metal-framed windows
    Finish hardware and weather-stripping
    Ceramic and quarry tile
    Flooring materials
    Acoustical ceilings
    Paints and wallcoverings
    Lath and plaster products
    Miscellaneous specialties
    Equipment—food service, bank, medical, incinerators, etc.
    Building furnishings
    Special construction items—radiation protection, vaults, swimming pools, integrated ceilings
    Elevators, escalators, dumbwaiters, etc.

Mechanical and Plumbing Equipment and Materials

    Fire protection equipment
    Water supply equipment
    Valves
    Drains
    Clean-outs
    Plumbing fixtures
    Gas-piping accessories
    Pumps
    Boilers
    Cooling towers
    Control systems
    Air-handling equipment
    Refrigeration units (chillers)

plastic laminates require a fully developed finish design for the project. Development of the finish design can have serious consequences on ordering and delivery of finish materials. Manufacturing and delivery times generally range from 3 to 12 weeks for these materials. These extended manufacturing and delivery times place considerable importance on planning and controlling these materials.

The fabricated category of construction materials must conform to a particular project's unique requirements. The fabricated item, however, is composed of or results from modification of standard components. Table 16.3 illustrates materials that fall into this category. Submittals required include highly detailed shop drawings, product data, and samples.

Table 16.3   Typical Fabricated Materials Items

Electrical Equipment and Materials

Busduct
Special conduit
Switchboards and panels
Transformers
Wire
Trim devices
Lighting fixtures
Underfloor duct
Communications devices
Motors and starters
Motor control centers
Electric heaters
Fire alarm equipment
Lightning protection equipment
Concrete reinforcement
Structural steel
Precast panels and decks
Stone veneers
Miscellaneous and special formed metals
Ornamental metals
Millwork
Custom casework and cabinetwork
Sheet metal work
Sheet metal veneers
Hollow metal doors and frames
Wood and plastic laminate doors
Glass and glazing
Storefront
Window walls and curtain walls

Fabrication and delivery times range from 2 weeks for items such as reinforcing steel and precast concrete to 10–12 weeks for curtainwall systems, doors and frames, and similar items.

REVIEW QUESTIONS AND EXERCISES

16.1   Name four important items of information that should be on a typical purchase order.

16.2   What are four good sources of price information about construction materials?

16.3   What is meant by the following expressions?

    a. CIF

    b. 2/10 E.O.M.

    c. 2/10 net 30

    d. ROG

    e. Bill of lading

16.4   Visit a local architect's office and ascertain how product data are obtained and used.

16.5   Visit a local building contractor and determine how he handles control of submittals from subcontractors to architect/engineer. What system does he use to ensure the job will not be held up due to procurement and approval delays?

16.6   Visit a construction site and determine what procedures are used for verifying receipt arrival and ensuring proper storage of materials at the site.

16.7   Determine what procedures are used for removing waste materials from a local construction (building) site. Is there any scrap value in these materials? Explain.

16.8   Determine the local prices for some bulk materials such as concrete, sand, cement, steel mesh, bricks, and lumber and compare them to the periodically published prices in the *Engineering News Record*.

16.9 Select a particular material item (e.g., concrete) and follow its material handling process from the local source through final installation in the building. What special equipment is needed (if any)?

16.10 What types of special materials handling equipment can be identified on local building sites? Do they take advantage of certain properties of the material being handled (e.g., the fluidity of concrete)?

Chapter 17

Safety

Pipe manipulator mounted on
the excavator.

Safety in Trenches

The Need

Even though heavy construction equipment such as a crane or backhoe
excavators is used to perform the task of pipe laying in the trench, workers are
required to be inside the trench to guide the excavation, pipe laying, and final
alignment. Work place safety has become a major concern in the construction
industry over the past few decades, and trench cave-ins have caused serious
and often fatal injuries to workers in the United States. It has become of
crucial importance to implement the use of new technologies to prevent
accidents in trench excavation and pipe installation.

   Diverse approaches such as shoring, shielding, and sloping have been
applied to protect workers from cave-ins in trenching and pipe laying
operations. However, even when support systems are used, the danger of
cave-ins still exist due to the nature of the soil and unexpected circumstances.
The Construction Automation and Robotics Laboratory (CARL) at North
Carolina State University has developed an alternative which involves
advanced new technology: the prototype robotic excavation and pipe
installation system called Pipeman.

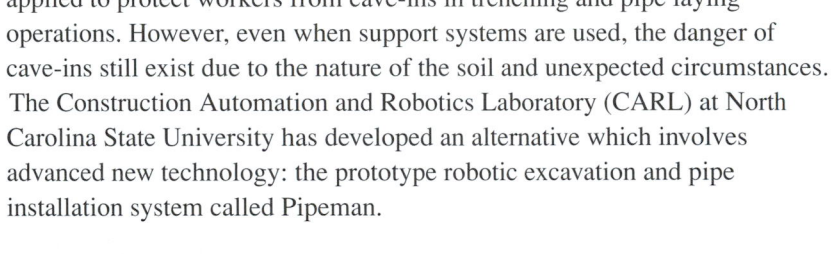

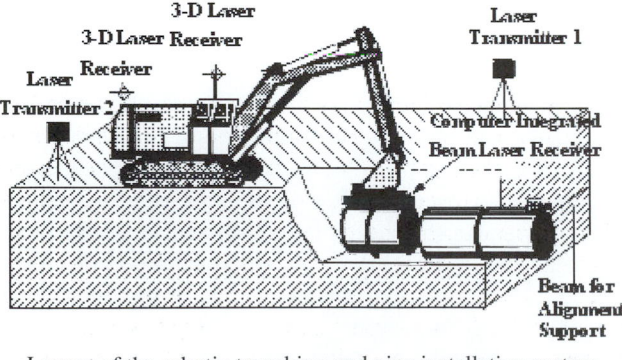

Layout of the robotic trenching and pipe installation system
(Huang & Bernold, 1993).

Overview of pipe manipulator.

The Technology

The basic Pipeman concept consists of a 3-D spatial positioning system (SPS), which is
interfaced with an excavator to provide the location of the excavator and a beam laser. A
pipe manipulator prototype is attached to the bucket of the excavator, which is capable of
handling pipes of various sizes. A beam laser is also used to help the operator align pipes.

Integration of SPS with a CAD system will update the excavator position in real-time and provide an as-built drawing of pipe laying.

The main components of the concept are the man-machine interface, actuation system, laser beam and feedback system. The man-machine interface is used to keep the operator in a safe area and allow him to guide the Pipeman intelligently, while Pipeman works in a hazardous environment.

17.1   NEED FOR SAFE PRACTICE

A disabling injury or fatal accident on the job site has negative impact on operations at many levels. Accidents cost money and affect worker morale. Because of the type of work involved in construction, many dangers exist both for the workers and for the public. For this reason, the subject of safety offers one area of noncontroversial mutual interest between management and the work force. The necessity of safe operations and of protecting and conserving lives by preventing accidents is understood by all.

Although the fatality rate in construction has been reduced within recent years, the improvement in safety record achieved by the construction industry still lags seriously behind that achieved in other hazardous industries. The annual number of fatalities in the construction industry in 2003 exceeded the number of combat deaths during the first 18 months of armed conflict in Iraq (period 2003–2004). Construction is a dangerous business.

It is the contractor's responsibility to see that everything possible is done to provide a safe working environment for the work force and the public in general. The factors that motivate safe practices at the job site are generally identified as follows:

1. Humanitarian concerns
2. Economic costs and benefits
3. Legal and regulatory considerations

Society has taken the position that because of the high health and accident potential intrinsic to the construction industry, the contractor must accept the liabilities associated with this hazardous environment and make an appropriate commitment to safe practice and accident prevention.

17.2   HUMANITARIAN CONCERNS

It is normally accepted that day-to-day living has intrinsic risks that may result in members of the society being subjected to mental and physical hardship. One of the functions of society is to minimize pain and suffering. Particularly at the level of the work site, society has defined the principle that the employer is responsible for providing a safe environment for the work force. This is based on humanitarian concern. If, for instance, a worker loses a leg because of a job-related accident and is confined to a wheelchair, the worker is, in a sense, a casualty of the workplace. Through his desire to be a participating member of society and support members of his family, the worker is injured. Society has traditionally shouldered the responsibility for this limitation on a worker's abilities. Over the past 120 years, the principle of employer liability for death and injury resulting from accidents or health hazards occurring at the workplace has been firmly established in common law. The courts have further charged the employer with the following five responsibilities[1]:

1. To provide a reasonably safe workplace

---

[1] Lee E. Knack, in *Handbook of Construction Management and Organization,* Bonny and Frein (eds.), Van Nostrand Reinhold, New York, 1973, Chapter 25.

2. To provide reasonably safe appliances, tools, and equipment

3. To use reasonable care in selecting employees

4. To enforce reasonable safety rules

5. To provide reasonable instructions regarding the dangers of employment

Mandatory requirements for the employer to make formal provision for injuries and deaths on the job resulted in the enactment of workmen's compensation laws in all fifty states during the first half of the twentieth century.

In 1884, Germany enacted the first workmen's compensation act, followed by Austria in 1887 and England in 1897. The U.S. federal government passed the first American compensation act in 1908 covering government employees. Following several legal battles, the Supreme Court, in 1917, declared that states could enact and enforce compulsory Workmen's Compensation Laws under their power to provide for the public health, safety, and welfare.

17.3   ECONOMIC COSTS AND BENEFITS

Safety costs can be broken into three categories as follows:

1. Direct cost of previous accidents

   a. Insurance premiums and ratings
   b. Mandatory accident prevention methods
   c. Records, safety personnel

2. Direct cost of each accident occurrence

   a. Delay to project
   b. Uninsured damages

3. Indirect cost

   a. Investigation
   b. Loss of skilled workers
   c. Loss of equipment
   d. Lost production

Direct costs from previous accidents come primarily in the form of insurance premiums, which have a significant effect on a contractor's operating expense. Workmen's compensation and liability insurance premiums can be calculated using either manual or merit rating systems. Manual rating is based on the past losses of the industry as a whole. The premium rate for compensation is normally set by the individual state Compensation Rating Bureaus. Many states are guided by or actually have their rates set by the National Council on Compensation Insurance (NCCI). The premium rates are based on factors such as classification of operations, rates of pay, the frequency and severity of accidents in a particular classification, increases in the cost of cases, and the attitudes of various industrial compensation commissions. The rates as set and approved by each state insurance commissioner are known as the manual (standard) rates. These manual rates are published periodically in the *Engineering News Record* (ENR) *Quarterly Cost Roundup* issues. A listing of some of the rates as reported in the R. S. Means *Building Construction Cost Data* is given in Figure 17.1.

The merit rating system bases premiums on a particular company's safety record. High-risk (high-accident-rated) companies are therefore penalized with higher premiums than those paid by companies with low accident rates. In this way, a good safety program can result in substantial financial savings to a company. Higher returns on jobs can be realized, and the ability to bid lower and win more jobs is greatly enhanced.

State	Carpentry — 3 stories or less (5651)	Carpentry — interior cab. work (5437)	Carpentry — general (5403)	Concrete Work — NOC (5213)	Concrete Work — flat (flr., sdwk.) (5221)	Electrical Wiring — inside (5190)	Excavation — earth NOC (6217)	Excavation — rock (6217)	Glaziers (5462)	Insulation Work (5479)	Lathing (5443)	Masonry (5022)	Painting & Decorating (5474)	Pile Driving (6003)	Plastering (5480)	Plumbing (5183)	Roofing (5551)	Sheet Metal Work (HVAC) (5538)	Steel Erection — door & sash (5102)	Steel Erection — inter., ornam. (5102)	Steel Erection — structure (5040)
GA	17.96	9.28	20.73	13.67	10.87	6.80	14.05	14.05	12.74	11.08	16.74	13.30	12.36	31.91	12.09	6.44	28.45	11.02	7.94	7.94	20.82
HI	11.66	9.32	39.81	13.93	9.03	8.89	10.81	10.81	14.80	20.18	9.51	17.35	7.68	28.43	19.16	5.36	40.64	8.32	14.10	14.10	27.94
ID	19.47	7.92	19.01	15.29	8.40	5.46	8.93	8.93	12.65	20.04	8.08	15.49	17.76	23.92	11.20	7.89	29.71	11.67	8.91	8.91	33.93
IL	23.15	14.22	23.28	37.47	14.13	10.95	12.08	12.08	35.98	29.13	14.87	23.41	16.53	52.50	18.54	16.13	39.95	19.17	24.90	24.90	80.46
IN	10.50	3.84	8.29	7.80	4.12	3.25	5.28	5.28	6.34	9.18	5.01	6.14	5.27	14.99	5.93	3.63	15.91	5.30	5.64	5.64	15.31
IA	8.81	6.56	14.86	15.22	5.78	4.94	5.29	5.29	11.73	11.99	6.21	12.64	10.60	19.81	10.94	7.47	23.45	7.66	12.14	12.14	35.92
KS	18.46	7.67	12.78	16.79	10.33	6.78	8.14	8.14	10.19	15.28	9.98	15.27	9.54	26.31	19.37	7.89	33.94	9.19	8.33	8.33	26.52
KY	16.47	15.26	24.34	22.12	9.53	8.95	13.67	13.67	15.96	14.68	10.62	22.00	19.83	30.44	13.94	10.38	32.83	15.92	13.21	13.21	45.21
LA	29.63	22.95	26.01	20.86	14.80	10.48	24.73	24.73	23.88	20.82	14.00	21.27	27.63	54.39	21.42	13.78	49.11	21.35	15.51	15.51	66.41
ME	13.43	10.42	43.59	24.29	11.12	10.02	14.75	14.75	13.35	15.39	14.03	17.26	16.07	31.44	17.95	11.12	32.07	17.70	15.08	15.08	37.84

Figure 17.1 Compensation insurance base rates for construction workers (selected states and crafts).

Once the premiums reach a value of $1000, the contractor is eligible for a merit system rating. That is, the cost of the premium will be individually calculated with the safety record of the company being the critical consideration. Under the merit system, there are two basic methods utilized to incorporate the safety record into the final cost of the premium. These are referred to as the *experience* rating and *retrospective* rating methods.

Most insurance carriers use the experience rating method, which is based on the company's record for the past 3 years not including the most recent preceding year. In this system, an experience modification rate is multiplied by the manual rate to establish the premium for a given firm. Data on losses, the actual project being insured, and other variables are considered in deriving the experience modification rate (EMR). If the company has an experience modification rate of 75%, it will pay only 75% of the manual premium. Good experience ratings (EMRs) can lead to significant savings. Clough and Sears (Wiley, 1994) illustrate this with the following example:

> *Assume that a building contractor does an annual volume of $10 million worth of work. Considering a typical amount of subcontracting and the cost of materials, this general contractor's annual payroll will be of the order of magnitude of $2.5 million. If his present workmen's compensation rate averages about 8%, his annual premium cost will be about $200,000. Now assume that an effective accident prevention program results in an experience modification rate (EMR) of 0.7. This will result in a reduction of the annual premium cost to about $140,000 for this contractor. Annual savings on the order of $60,000 are thereby realized on the cost of this one insurance coverage alone.*

Retrospective rating is somewhat like self-insurance. It is basically the same as experience rating except for one point. It utilizes the loss record of the contractor for the previous year or other defined retrospective period to compute the premium. This can raise or lower the premium cost based on performance during the retrospective period. The starting point or basis for this method is again the manual premium. A percentage (usually 20%) of the standard premium resulting from applying the experience modification factor to the manual rate is used to obtain the basic premium. The retrospective rate is then calculated as

$$\text{Retrospective rate} = (\text{Tax multiplier}) \times \{\text{Basic premium} + [(\text{Incurred loss}) \times (\text{Loss conversion factor})]\}$$

The incurred loss is the amount paid out to settle claims over the retrospective period. The loss conversion factor is a percentage loading used to weight the incurred losses to cover general claims investigation and adjustment expenses. The tax multiplier covers premium taxes that must be paid to the state. If the data for a given company are as follows:

Manual premium	$25,000
Experience modification rate	0.75
(25% credit)	

then

Standard premium = 0.75($25,000) = $18,750
Basic premium @ 20% of standard = 0.20($18,750) = $3750
Loss conversion factor = 1.135 (derived from experience)
Tax multiplier = 1.03 (based on state tax)
Incurred losses = $10,000

Then

Retrospective premium = 1.03[3750 + (1.135 × $10,000)]
= $15,553

This is a nice savings over the standard premium of $18,750 and provides the contractor with a clear incentive to minimize the incurred losses. By so doing, the contractor can expect a large premium rebate at the end of the year.

17.4   UNINSURED ACCIDENT COSTS

In addition to the cost of insurance premiums, additional direct costs for things such as the salary of the safety engineer and his staff as well as costs associated with the implementation of a good safety program can be identified. The precise amount of the costs associated with other safety cost categories is more difficult to assess, and these costs can be thought of as additional uninsured costs resulting from accidents. Typical uninsured costs associated with an accident are shown in Table 17.1.

Although varying slightly from source to source, hidden losses of this variety have been estimated to be as much as nine times the amount spent on comprehensive insurance. In addition to the costs noted in Table 17.1, another cost is that of paying an injured employee to show up for work even if he cannot perform at his best. This is common practice for minor injuries. This is done to avoid recording a lost time accident, which might impact the insurance premium. While it is very common to return injured workers to work, it is

Table 17.1   Uninsured Costs

Injuries	Associated Costs
1. First-aid expenses	1. Difference between actual losses and amount recovered
2. Transportation costs	2. Rental of equipment to replace damaged equipment
3. Cost of investigations	3. Surplus workers for replacement of injured workmen
4. Cost of processing reports	4. Wages or other benefits paid to disabled workers
	5. Overhead costs while production is stopped
	6. Loss of bonus or payment of forfeiture of delays

Wage Losses	Off the Job Accidents
1. Idle time of workers whose work is interrupted	1. Cost of medical services
2. Man-hours spent in cleaning up accident area	2. Time spent on injured workers' welfare
3. Time spent repairing damaged equipment	3. Loss of skill and experience
4. Time lost by workers receiving first aid	4. Training replacement worker
	5. Decreased production of replacement
	6. Benefits paid to injured worker or dependents

Production Losses	Intangibles
1. Product spoiled by accident	1. Lowered employee morale
2. Loss of skill and experience	2. Increased labor conflict
3. Lowered production of worker replacement	3. Unfavorable public relations
4. Idle machine time	

Source: From Lee E. Knack, in *Handbook of Construction Management and Organization*, Bonny and Frein (eds.), Van Nostrand Reinhold, New York, 1973, Chapter 25.

important that they not be returned to work too soon to avoid their being reinjured or injured more seriously.

The following situation illustrates the additional losses resulting from hidden costs. At a large industrial construction site, the survey party chief was on the way to the office to get a set of plans. The survey crew was to lay out four machine foundations that morning. The wooden walkway beneath the party chief collapsed. Due to the confusion resulting from the accident, work activity on the entire site was impacted. The party chief was in the hospital for 5 weeks with a shattered pelvis. Another party chief who was unfamiliar with the site was assigned to the surveying crew. As a result, the four machine foundations were constructed 2 ft farther west than called for in the plans. After this was discovered, six laborers worked for 20 hours removing the reinforced concrete. The survey crew of four spent another 5 hours laying out the foundations. Four carpenters worked 20 more hours preparing new forms. Five more hours were required for the ironworkers to place the steel reinforcement. The total indirect cost was approximately $5000. Although this activity was not on the critical path, if it had been, liquidated damages might have been charged to the contractor. Still, the accident resulted in costs amounting to one week's pay for the employees affected and the cost of material that had to be replaced.

17.5   FEDERAL LEGISLATION AND REGULATION

The federal government implemented a formal program of mandatory safety practices in 1969 with the passage of the Construction Safety Act as an amendment to the Contract Work Hours Standard Act. This legislation requires contractors working on federally funded projects to meet certain requirements to protect the worker against health and accident hazards. In addition, certain reporting and training provisions were established. This program of required procedures has been referred to as a *physical* approach to achieving safety. That is, regulations are prescribed that are designed to minimize the possibility of an unsafe condition arising. A typical physical measure of this type is the requirement to install guard rails around all open floors of a multistory building during construction. Guard rails are needed anytime there is change in elevation of 6 feet and the worker is not protected by a personnel fall arrest system, warning line, or warning attendant (used to watch workers and warn them if they're too close to falling).

Furthermore, physical measures are implemented to minimize injury in the event of an accident. An example of this is the requirement to wear a safety belt when working with high steel, and the installation of safety nets to protect a man who slips and falls. This physical approach is in contrast to the behavioral approach that is designed to make all levels of the work force from top management to the laborer think in a safe way and thus avoid unsafe situations. Research on the behavioral approach is discussed in detail in Levitt and Samelson, *Construction Safety Management*.

Shortly after the passage of the Construction Safety Act, a more comprehensive approach to mandatory safe practices was adopted in the form of the Williams–Steiger Occupational Safety and Health Act (OSHA) passed by Congress in 1970. This act established mandatory safety and health procedures to be followed by all firms operating in interstate commerce.

Under this act, all employers are required to provide "employment and a place of employment which are free from recognized hazards that are causing or are likely to cause death or serious physical harm to his employees." The provisions of the Construction Safety Act were included in the act by reference. The provisions of OSHA fall within the jurisdiction of the secretary of labor. In 1971, the law was implemented by publishing the code of Federal Regulation (CFR) 1926 that specifically refers to the construction industry and CFR 1910 that pertains to General Standards. Many existing standards issued by various standards organizations, including the American National Standards Institute (ANSI), were

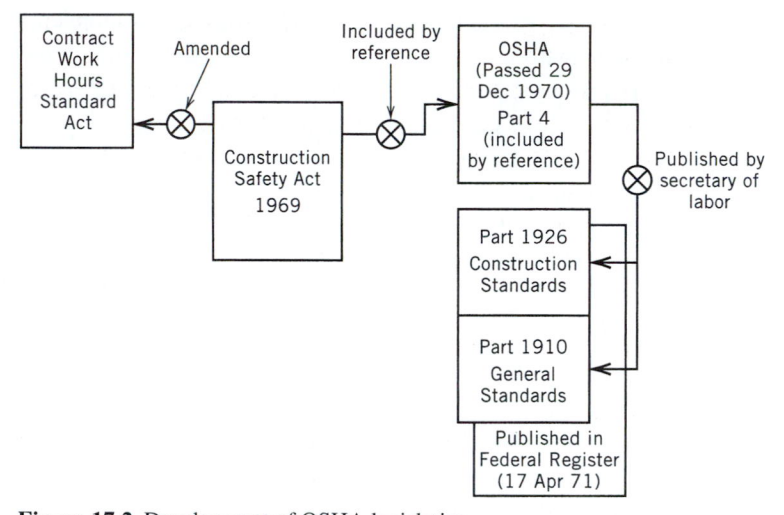

Figure 17.2 Development of OSHA legislation.

included in the basic law by reference. The schematic development of the legislation is shown in Figure 17.2 OSHA has a service for providing update information on standards that is designed to aid in keeping the six volumes of regulations current. The regulations are divided as follows:

Volume I: General Industry Standards

Volume II: Maritime Standards

Volume III: Construction Standards

Volume IV: Other Regulations and Procedures

Volume V: Field Operation Manual

Volume VI: Industrial Hygiene Manual

Under the OSHA legislation, the assistant secretary of labor for occupational safety and health administers and enforces OSHA through the labor department's Occupational Safety and Health Administration with its 10 regional offices around the country. The Occupational Safety and Health Review Commission is designed by OSHA as the review body to which citations for alleged violations and proposed penalties can be appealed. Research on safety topics is under the control of the National Institute for Occupational Safety and Health (NIOSH), which is part of the Department of Health and Human Services.

The OSHA legislation allows individual states to establish programs that operate in place of the federal program. Many contractors prefer a state-operated program since this affords a closer contact with the inspector and the regulatory agency. Appeals and requests for clarification are handled at a state level. This expedites the process of responding to citations and rectification of disputes if differences of opinion arise. The state agency must establish that it is able to administer the law as effectively as the federal government. Presently, 26 states operate approved state plans.

17.6   OSHA REQUIREMENTS

Employers must "make, keep and preserve, and make available to representatives of the Secretaries of Labor, and Health and Human Services" records of recordable occupational injuries and illnesses (Williams–Steiger Act, 1970). Any fatal or serious accidents must be reported to OSHA within 8 hours. Certain records of job-related fatalities, injuries, and illnesses must be maintained by firms having eight or more employees. The two key forms that must be available for review when a compliance officer makes an inspection are:

1. OSHA 300: This is a log that summarizes each reportable case as a single line entry and must be posted for employee inspection (see Fig. 17.3). The instructions that accompany the OSHA recordkeeping forms do include the following instructions. "You must post the Summary (Form 300A) only–not the Log (Form 300)–by February 1 of the year following the year covered by the form and keep it posted until April 30 of that year." OSHA allows the use of industry-generated forms similar to the OSHA forms as long as they have at least the same information as the OSHA forms.

2. First report of injury: Example in Figure 17.4.

These records must be preserved for 5 years.

The employer is also required to post at the work site records of citations and notices of employees' rights. There have been strong drives to change OSHA so that employers with scattered work sites, as in construction, can be allowed to maintain the required records at a central location (e.g., home office).

17.7 HOW THE LAW IS APPLIED

The 10 OSHA regional and area offices employ inspectors whose duties include visits to active projects to determine if the builders are conforming to the regulations. As noted, there are "State Plan States" that have their own laws, rules, and regulations. The OSHA legislation allows this as long as the program is "at least as effective" as the federal OSHA provisions. Indiana, for instance, operates a state system that simply adopted all of the federal OSHA rules, regulations, changes, and so forth. Michigan, on the other hand, writes and publishes its own rules, some of which go above and beyond what federal OSHA rules require.

An inspection can be initiated at random by OSHA or state safety inspectors or by an employee (or his union) who submits a written statement to the labor department that he believes there is a violation that threatens physical harm or imminent danger. All inspections must be on an unannounced basis during the working day. Rulings by the Supreme Court, however, require that an inspection warrant be obtained from the proper authority if access to the site is denied. It is not unusual that the contractor requires a warrant prior to permitting entry to the work site. This essentially changes the "surprise" nature of the inspection and allows the work site supervisor to prepare for the inspection.

The inspection is divided into four parts:

1. An opening conference with the employer.

2. Selection of a representative of the employees and of the employer to accompany the inspector on a tour of the workplace.

3. The walk-around inspection. The inspector is allowed to talk with any employees.

4. The closing conference during which the inspector discusses the conditions and practices he observed that might be safety or health violations. Only "alleged violations" are discussed. Any Safety Orders will follow via U.S. mail 2 weeks to 3 months after the actual inspection. However, correction of any items mentioned by the safety officer in the closing conference for which there is no basis for appeal should be corrected immediately. Fines may be proposed with the citations. The employer is allowed 15 days to appeal a penalty.

The appeals of citations, penalties, or abatement periods are made through a procedure to the Occupational Safety and Health Review Commission or appropriate state board. The

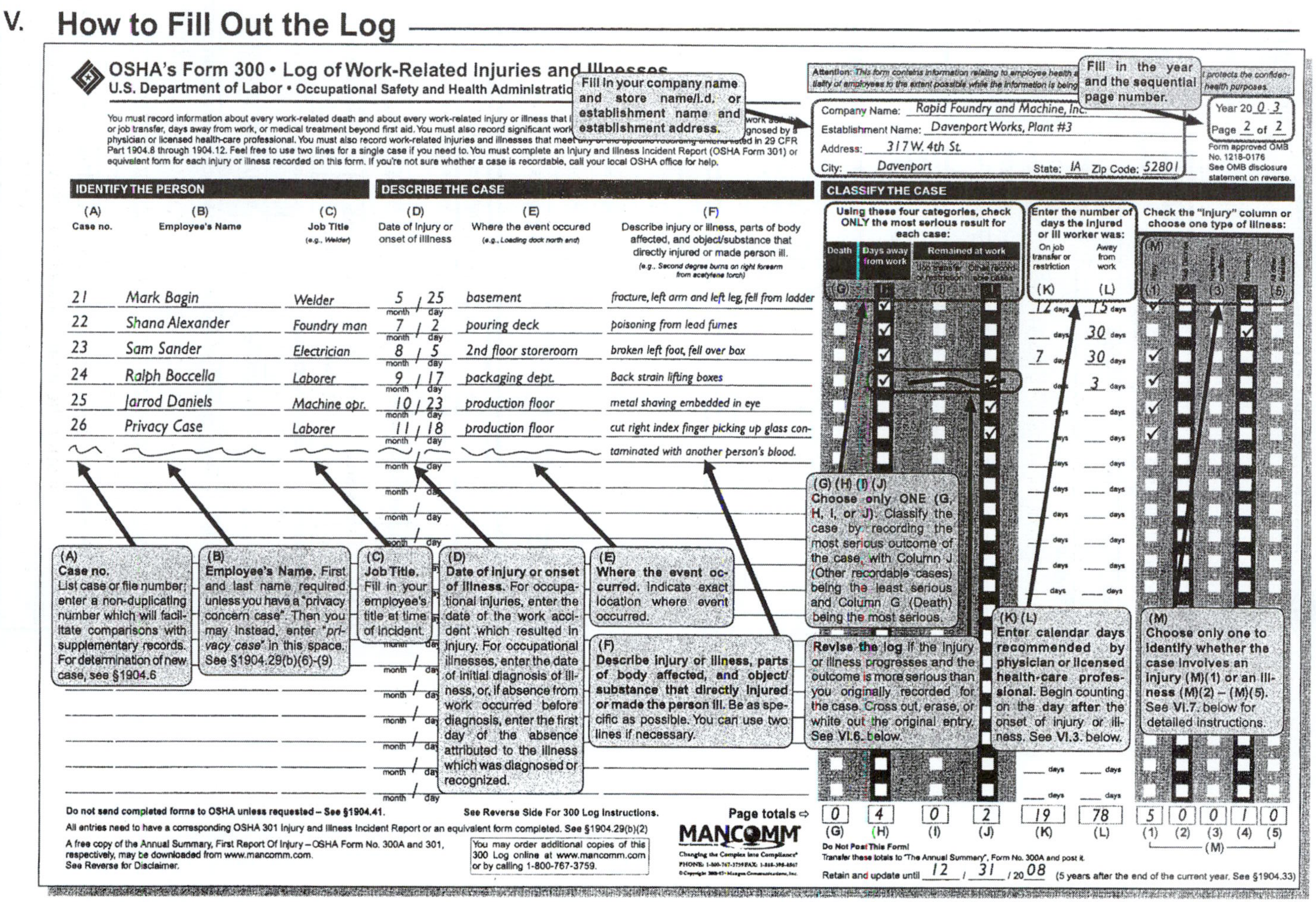

V. How to Fill Out the Log

Figure 17.3 OSHA Form 300.

Indiana Worker's Compensation
First Report of Employee Injury/Illness

Please Return Completed Form to: 402 W. Washington St., Room W196
Indianapolis, IN 46204-2753
(317) 232-3808

FOR WORKER'S COMPENSATION BOARD USE ONLY		
JURISDICTION	JURISDICTION CLAIM NUMBER	PROCESS DATE

PLEASE TYPE or PRINT IN INK

NOTE: Your Social Security Number is being requested by this state agency in order to pursue its statutory responsibilities. Disclosure is voluntary and you will not be penalized for refusal.

EMPLOYEE INFORMATION

SOCIAL SECURITY NUMBER	DATE OF BIRTH	SEX ◯MALE ◯FEMALE ◯UNKNOWN	OCCUPATION/JOB TITLE	NCCI CLASS CODE

LAST NAME	FIRST	MIDDLE	MARITAL STATUS ◯UNMARRIED ◯MARRIED ◯SEPARATED ◯UNKNOWN	DATE HIRED	STATE OF HIRE	EMPLOYEE STATUS

ADDRESS (INCL ZIP)

| HRS/DAY | DAYS/WK | AVG WG/WK | PAID DAY OF INJ ▢ SALARY CONT'D ▢ |

| PHONE | # OF DEPENDENTS | WAGE $ | PER ◯HR ◯DAY ◯WK ◯MO ◯YR ◯OTHER: |

EMPLOYER INFORMATION

EMPLOYER (NAME, ADDRESS, CITY, STATE, ZIP)	EMPLOYER FEDERAL ID#	SIC CODE	INSURED REPORT NUMBER
	LOCATION #	EMPLOYER'S LOCATION ADDRESS (IF DIFFERENT)	
	PHONE #		
	CARRIER/ADMINISTRATOR CLAIM NUMBER	REPORT PURPOSE CODE	

Actual Location of Accident/Exposure (if not on employer's premises):

CARRIER/CLAIMS ADMINISTRATOR INFORMATION

CLAIMS ADMINISTRATOR (NAME, ADDRESS, PHONE NO)	CARRIER FEDERAL ID#	CHECK IF APPROPRIATE ▢ SELF INSURANCE
	▢ INSURANCE CARRIER	POLICY/SELF-INSURED NUMBER
PHONE:	▢ THIRD PARTY ADMIN	POLICY PERIOD FROM TO
AGENT NAME	CODE NUMBER	

OCCURRENCE/TREATMENT INFORMATION

DATE OF INJ/EXP	TIME OF OCCURRENCE __M	DATE EMPLOYER NOTIFIED	TYPE OF INJURY/EXPOSURE	TYPE CODE
LAST WORK DATE	TIME WORKDAY BEGAN	DATE DISABILITY BEGAN	PART OF BODY	PART CODE
RTW DATE	DATE OF DEATH	INJURY/EXPOSURE OCCURRED ON EMPLOYER'S PREMISES? ▢ YES ▢ NO	CONTACT NAME	PHONE NUMBER

DEPARTMENT OR LOCATION WHERE ACCIDENT/EXPOSURE OCCURRED	ALL EQUIPMENT, MATERIALS, OR CHEMICALS INVOLVED IN ACCIDENT
SPECIFIC ACTIVITY ENGAGED IN DURING ACCIDENT/EXPOSURE	WORK PROCESS EMPLOYEE ENGAGED IN DURING ACCIDENT/EXPOSURE

HOW INJURY/EXPOSURE OCCURRED. DESCRIBE THE SEQUENCE OF EVENTS AND INCLUDE ANY RELEVANT OBJECTS OR SUBSTANCES

CAUSE OF INJURY CODE

NAME OF PHYSICIAN/HEALTH CARE PROVIDER	INITIAL TREATMENT ▢ NO MEDICAL TREATMENT ▢ MINOR: BY EMPLOYER ▢ MINOR: CLINIC/HOSP ▢ EMERGENCY CARE ▢ HOSPITALIZED >24 HRS ▢ FUTURE MAJOR MEDICAL/ LOST TIME ANTICIPATED
WITNESSES (NAME, PHONE #)	DATE ADMINISTRATOR NOTIFIED	
DATE PREPARED	PREPARER'S NAME	TITLE	PHONE NUMBER	

An employer's failure to report an occupational injury or illness may result in a $50 fine (IC 22-3-4-13)

STATE FORM 34401 (R8 2/96)

Figure 17.4 Typical first report of injury/illness form.

Table 17.2   OSHA Standards Most Commonly Cited for Violations

Section	Subject	Section	Subject
1926.500	Guardrails, Handrails, Covers	1926.100	Head Protection
.451	Scaffolding	.552	Materials, Hoists, Personnel
.450	Ladders		Hoists, Elevators
.350	Gas Welding and Cutting	.50	Medical Services, First Aid
.401	Grounding and Bending	.501	Stairways
.550	Cranes and Derricks	.300	General Requirements, Hand and
.25	Housekeeping		Power Tools
.152	Flammable and Combustible	.651	Excavation
	Liquids	.51	Sanitation
.400	General Electrical	.28	Personal Protective Equipment
.402	Electrical Equipment Installation	.102	Eye and Face Protection
	and Maintenance		
.150	Fire Protection	.302	Power-operated Hand Tools
.652	Trenching	.351	Arc Welding and Cutting
.601	Motor Vehicles	.105	Safety Nets

commission, after a hearing and review, can affirm, modify, or vacate the citation, proposed penalty, and abatement period.

Table 17.2 is a list of OSHA standards representing job site physical hazards that are commonly cited as violations. In addition to those listed, the 1988 provision requiring firms to inventory and label hazardous materials is the most highly cited violation. This Hazard Communication (HazCom) Standard also requires that employees be trained in the safe use of such materials. Material Safety Data sheets (MSDs) must be maintained for each hazardous substance. The program addresses the need to make employees more aware of the chemical hazards in the work place.

If, during the course of an inspection, a violation is noted, a written citation is given to the employer and the area where the violation occurs will be posted. A reasonable length of time shall be granted the employer for correction of the violation.

These violations, and failure to abate in the given time, incur monetary violations up to $70,000. Serious violations incur a mandatory fine of $7000. Failure to abate within the given time period can result in a fine of $7000 a day for the period the violation persists.

17.8   SAFETY RECORD KEEPING

Documentation under the Williams–Steiger Act is required as follows: "Every employer who is covered under this act must keep occupational injury and illness records for his employees in the establishment in which his employees usually report to work."

The OSHA laws require employers to keep both a log of recordable occupational injuries and illnesses and a supplementary record of each injury or illness. These records must be kept up to date and should be available to government representatives.

These records are also used to compile the annual accident report (OSHA 300), which must be posted in a prominent place in the establishment available to the employees.[2] Also

---

[2] The OSHA 300 (or equivalent) must be kept up to date at all times.

the poster entitled "Safety and Health Protection on the Job" shall be posted in a similar manner.

The only employers excluded from this portion of the act are those who are already reporting this material under the Federal Coal Mine Health and Safety Act or the Federal Metal and Nonmetallic Mine Safety Act.

Recordable occupational illnesses and injuries are those that result from a work accident or from exposure to the work environment and lead to fatalities, lost workdays, transfer to another job (temporary or permanent), termination/limitation of employment, or treatment beyond simple first aid measures. Also, those cases involving loss of consciousness or restriction of work or motion are recordable.

Reporting at the job site level breaks into six reporting levels as follows:

1. First aid log
2. First report of injury log
3. Supervisor's accident investigation report
4. Project accident report
5. OSHA required Injury Report (OSHA 300)
6. Fatality or major accident report

The first aid log is kept on the job and lists every treatment given. The first report of injury log is required by the workmen's compensation laws in most states. It is prepared to record every personal injury that requires off-site medical treatment regardless of whether the employee lost time from work or not. The supervisor's accident investigation report is prepared by the foreman for each recordable accident and places special emphasis on identifying methods by which the accidents could be prevented in the future. A typical project accident report form is shown in Figure 17.5. It is a monthly summary of disabling injuries and lost time sent to the home office. The form shown is a report of information on each disabling injury required by OSHA and kept at the job site. Finally, as noted above, any fatality or accident that hospitalizes three or more employees must be reported to the OSHA area director within 8 hours.

17.9 SAFETY PROGRAM

A good job site safety program should be founded on:

1. Safety indoctrination of all new personnel arriving on the site
2. Continuous inspection for possible safety hazards
3. Regular briefings to increase the safety awareness of personnel at all levels
4. Written programs and documentation specifying all safety activities

If workers or supervisors flagrantly neglect safety rules and regulations, warnings should be considered.

It is good practice to personally brief each employee arriving on site regarding job procedures. A briefing sheet such as the one shown in Figure 17.6 is an effective aid for conducting this type of briefing. This focuses the worker's attention at the outset on the importance of safety and indicates management's interest in this phase of the job. Safety rules and regulations such as those shown in Figure 17.7 should be available in "handout" form and be conspicuously posted around the job site.

General safety meetings conducted by the safety engineer should be held at least once a month with supervisors at the foreman and job steward level. A typical report for such a

```
┌─────────────────────────────────────────────────────────────────┐
│ Job Name Peachtree Shopping Mall Job No. 10-100 Location Atlanta, │
│ Georgia Month April 20xx                                          │
│                                                                   │
│ This report should be completed and mailed to the Safety Branch   │
│ of the Industrial Relations Department in the Atlanta office by    │
│ the fifth day of the month.                                       │
│                                    Project Superintendent _____  │
│                                                                   │
│ This figure may be taken from payroll                             │
│   records. In the case of fractions use                           │
│   the nearest whole number. Do not in-    1. Average number of    │
│   clude subcontractors or others.            employees _____     │
│                                                                   │
│ Figure actual hours worked whether                                │
│   straight time or overtime. Includes     2. Total hours worked   │
│   only those on our payroll.                 by all employees ___ │
│                                                                   │
│ Record only those injuries that cause                             │
│   death, permanent disability (loss of a                          │
│   finger, etc.), or loss of time beyond                           │
│   the day on which the accident oc-                               │
│   curred. No matter what time of day                              │
│   the injury may occur, if the employee   3. Number of:           │
│   returns to his regular job at the start    Temporary disabling  │
│   of his next regular shift, the injury is     injuries _____    │
│   not counted. If he does not return at      Permanent disabling  │
│   that time, it must be counted as a disa-     injuries _____    │
│   bling injury.                              Deaths _____        │
│                                              Total disabling      │
│                                                injuries for this  │
│                                                month _____       │
│                                                                   │
│ For temporary injuries, count the actual                          │
│   calendar days lost, excluding the day                           │
│   of injury. If the injured employee has  4. Number of days lost  │
│   not returned by the end of the month,      as a result of:      │
│   make an estimate of projected num-         Temporary disabling  │
│   ber of lost days. For deaths and per-        injuries _____    │
│   manent injuries, use the number of         Permanent disabling  │
│   days specified in the standard table.        injuries _____    │
│                                              Death _____         │
│                                              Total days lost      │
│                                                attributable to    │
│                                                this month _____  │
└─────────────────────────────────────────────────────────────────┘
```

Figure 17.5 Project accident report.

meeting is shown in Figure 17.8. The objective of these meetings is primarily to heighten the safety awareness of supervisors directly in charge of workers. These foreman-level personnel in turn should hold at least one "tool box" safety meeting each week to transfer this awareness to the work force and discuss safety conditions with their crew. The report format includes a record of those in attendance, the first aid report, and a description of the safety topics discussed. In addition to the general safety meetings, each job should have a designated safety committee that meets regularly. The members of the safety committee should include key supervisory personnel and craftsmen with an alertness to potential danger and a genuine desire to prevent accidents and injuries. One of the purposes of the safety committee should be to make suggestions as to how to improve overall job safety. Therefore, the members appointed should be sensitive to safety and innovative in devising safe methods.

August 2, 20xx

PEACHTREE SHOPPING MALL
Atlanta, Georgia

Welcome to the job! ABC Construction Company is interested in you, and during your employment with us, we will exert every effort to make this job pleasant, with a good working atmosphere. On the other hand, your skills, ability, and performance are most important and essential to the successful completion of the project. To set up and complete a good job, certain rules and regulations must be established. For our mutual benefit, these rules and regulations are as follows:

WORKING RULES AND REGULATIONS

Employment
The Project Manager, or his duly authorized representative, will do all the hiring on the job.

Identification
Employees shall wear a company badge at all times, in full view, above the waist, on an outer garment. Badge numbers will be used in gate clearance, payroll, and timekeeping identification.

Hours of Work
The regular workday will begin as per individual instructions, with a lunch period of one-half hour at a designated time. The workweek shall be five days, Monday through Friday.
All employees will be at their work locations, ready to start work at work time. All employees are expected to remain at work until the authorized quitting time, at which time they may put up their tools and leave their place of work. Loitering in the change rooms and/or other places during working hours, or late starting of work and early quitting of work will be subject to proper disciplinary measures.

Checking In and Out
Employees are to check in and out at starting and quitting time. Infractions of this rule will be treated with appropriate disciplinary measures. Employees authorized to leave the project during regular working hours must check out with the timekeeper.

Issuing, Care, and Use of Tools
Certain company tools will be issued to journeymen and apprentices, or the foreman on a check or receipt system. Tools (while issued) must be properly used and maintained. A toolroom clearance will be required on termination. Loss of or damage to tools will be noted on the employee's record.

A Day's Work
Each employee on the job is expected to perform a full day's work. Your willingness, cooperation, and right attitude will go a long way in accomplishing this objective.

Conduct on the Job
Good conduct on the job is essential to the overall welfare of all employees and the daily progress of the job. Therefore, conduct including, but not limited to, the following violations will be subject to appropriate disciplinary action or discharge.

Figure 17.6 Job briefing sheet.

Theft of company's or employees' property
Recurring tardiness
Leaving company's premises without proper authorization
Possession and/or use of intoxicants and/or narcotics on company's premises
Willful damage to company's materials, tools, and/or equipment
Engaging in horseplay (including shouting to passers-by)
Insubordination
Gambling
Fighting on company premises
Sleeping on the job
Failure to observe established safety rules and regulations

Housekeeping

Good housekeeping is essential to the safe and efficient construction of the job and is the responsibility of each employee. Work areas, stairways, walkways, and change rooms shall be kept clean at all times.

Safety Rules

Established safety rules and regulations will be observed and followed by all employees in the best interest of accident-free operations.

All unsafe working conditions should be reported to your immediate foreman, who in turn reports it to the company safety engineer.

All employees will be required to wear proper clothing above and below the waist. Hard hats must be worn by all employees and visitors while on the construction site.

Pay Period

Wednesday thru Tuesday is the pay period, with pay day on Friday of each week.

Use of First Aid Facilities

First aid facilities are available at the job site and direct contracts have been established with local doctors, hospitals, and emergency crews for accidents of a serious nature. All injuries, regardless of severity, must be reported to the employee's supervisor, field safety supervisor, and/or first aid immediately upon occurrence. Insurance regulations make this requirement mandatory.

Sanitary Facilities

Adequate sanitary facilities are provided on the job site and are to be used by all employees. We request your cooperation in maintaining these facilities in a clean and orderly condition.

Raincoats and Boots

Raincoats and boots are supplied to employees where the conditions of the job being performed require them.

Remaining in Work Areas

Each employee must remain on the job site and at his work location at all times during regular working hours, unless authorized to leave by his supervisor.

Absenteeism

Unauthorized absenteeisms will result in termination of employment. An employee who must be absent or late should call 999-9000 and report to timekeeper.

Your cooperation in observing the rules and regulations for the job will show proper consideration for other employees and will be appreciated by the company.

If you agree to and will abide by the above, please sign and return to our field supervisor, Charles Hoarse.

cc: Employee File

Figure 17.6 (*Continued*)

ABC Contractors and Engineers

760 Spring Street, N.W., Atlanta, Georgia 30308
(404) 999-9000

July 30, 20xx

Re: OCCUPATIONAL SAFETY & HEALTH ACT 1970 (Construction) (OSHA)

Employers, owner, contractors, subcontractors, superintendents, or foremen in charge shall not direct or permit an employee to work under conditions that are not in compliance with the above code.

Where one contractor is selected to execute the work of the project, he shall assure compliance with the requirements of this code from his employees as well as all subcontractors.

Every employee shall observe all provisions of the above codes that directly concern or affect his conduct. He shall use the safety devices provided for his personal protection and he shall not tamper with or render ineffective any safety device or safeguard.

1.	*Overhead Hazards*	All employees shall be provided with HARD HATS and shall use HARD HATS.
2.	*Falling Hazards*	Every hole or opening in floors, roofs, platforms, etc., into or through which a person may fall shall be guarded by a barrier sufficient to PREVENT FALLS.
3.	*Slipping Hazards*	Scaffolds, platforms, or other elevated working surfaces covered with ice, snow, grease, or other substances causing slippery footing shall be removed, turned, sanded, etc., to ensure safe footing.
4.	*Tripping*	Areas where employees must work shall be kept *reasonably free* from accumulations of dirt, debris, scattered tools, materials, and sharp projections.
5.	*Projecting Nails*	Projecting nails in boards, planks, and timbers shall be *removed*, hammered, or *bent over* in a safe way.
6.	*Riding of Hoisting Equipment*	No employee shall ride on or in the load bucket, sling, platform, ball, or hook.
7.	*Lumber & Nail Fastenings*	Lumber used for temporary structures must be sound. Nails shall be driven full length and shall be of the proper size, length, and number. The proper use of double-headed nails is not prohibited.
8.	*Guard Rail or Safety Rail*	Should be 2 × 4 at a height of 35″–37″ plus a midrail of 1 × 4. The hand rail shall be smooth and free from splinters and protruding nails. Other material or construction may be used provided the assembly *assures* equivalent *safety*.
9.	*Toe Boards*	Shall extend 4″ above platform level and shall be installed *where needed* for the safety of those working below.
10.	*Protection Eye Equipment*	Eye protection shall be provided by employers and *shall* be used for cutting, chipping, drilling, cleaning, buffing, grinding, polishing, shaping, or surfacing masonry, concrete, brick, metal, or similar substances. Also for the use and handling of corrosive substances.
11.	*Protective Apparrel*	*Waterproof boots* where required shall have safety insoles unless they are the overshoe type. *Waterproof clothing* shall be supplied to the employee required to work in the rain.

Figure 17.7 Job safety rules and regulations.

12. *Safety Belts & Lines*	Shall be *arranged* so that a free fall of no more than 6″ will be allowed.
13. *Stairways*	Temporary stairways shall not be less than 3 feet in width and shall have treads of no less than 2 inch × 10 inch plank. Must have hand rails. (See #8.)
14. *Smoking*	Prohibited in areas used for gasoline dispensing and fueling operations or other *high hazard fire areas.*
15. *Flammable*	*Flammable liquid shall be kept in safety cans* or approved use and storage containers. Suitable grounding to prevent the buildup of static charges shall be provided on all flammable liquid transfer systems.
16. *Sanitation*	*Toilet facilities* shall be provided and made available in sufficient number to accommodate all employees.
17. *Drinking Water*	A supply of *clean* and *cool* potable water shall be provided in readily accessible locations on all projects.
18. *Salt Tablets*	Shall be made available at *drinking stations* when required.
19. *Excavations*	Material and other superimposed loads shall be placed at least 3 feet back from the edge of any excavation and shall be piled or retained so as to prevent them from falling into the excavation. Sides and slopes of excavation shall be stripped of loose rocks or other material. Slopes shall be at an angle of 45 degrees or less (*1 on 1 slope*).
20. *Structural Steel Erection*	When erection connections are made, 20% of the bolts in each connection must be drawn up wrench tight. At least 2 *bolts* must be used at each end of the member. *No loads* shall be placed on a framework until the permanent bolting is complete. Only employees of the structural steel erector engaged in work directly involved in the steel erection shall be permitted to work under any single-story structural steel framework that is not in true alignment and *permanently bolted.*
21. *Use of Ladders*	Ladders shall be provided to give access to floors, stagings, or platforms. Ladders shall be maintained in a safe condition at all times. Ladders shall be securely *fastened top* and *bottom* as well as braced where required. Ladders leading to floors, roofs, stagings, or platforms shall extend at least 3 feet above the level of such floors, stagings, or platforms.
22. *Scaffolds*	All scaffolding shall be constructed so as to *support* 4 times the anticipated working load, and shall be braced to prevent lateral movement. Planks shall overhang their end supports not less than 6″ or more than 12″. 2″ planking may span up to and including 10′. The minimum *width* of any planked platform shall be 18 inches. *Guard rails* and *toe rails* shall be provided on the open sides and ends of all scaffold platforms more than 8′ high (see #8).
23. *Rigging, Ropes, and Chains*	All rope, chains, sheaves, and blocks shall be of sufficient strength, condition, and size to safely raise, lower, or sustain the imposed load *in any position.* *Wire rope* shall be used with power-driven hoisting machinery. No rope shall be used when visual inspection of the rope shows marked signs of corrosion, misuse, or *damage.* All load hooks shall have *safety clips.* Loads that tend to swing or turn during hoisting shall be controlled by a *tag line* whenever practicable.

Figure 17.7 (*Continued*)

24. *Welding and Cutting*	Oxygen from a cylinder or torch shall never be used for *ventilation.* Shields or goggles must be worn where applicable. *Cradles* shall be used for lifting or lowering cylinders.
25. *Cranes & Derricks*	All cranes and derricks shall be equipped with a properly operating boom angle *indicator* located within the normal view of the operator. Every derrick and crane shall be operated by a designated person. A copy of the *signals in use* shall be posted in a conspicuous place on or near each derrick or crane. Cranes and derricks shall have a *fire extinguisher* attached.
26. *Trucks*	Trucks shall not be backed or dumped in places where men are working nor backed into a hazardous location unless guided by a person so stationed on the side where he can see the truck driver and the space in back of the vehicle.

The above items do not encompass all the construction safety regulations as they pertain to OSHA but are intended as a guide to the ever present hazards and primary causes of accidents in our industry.

Figure 17.7 (*Continued*)

ABC Construction Company
Job 10-100
Peachtree Shopping Mall
Atlanta, Georgia
Sept. 1, 20xx

GENERAL SAFETY MEETING #7

Safety Slogan for the Week:

"Be Alert, Don't Get Hurt."

C. Hoarse—Safety Supervisor
A. Apple—Carpenter Foreman
D. Duck—Surveyor
M. Maus—Laborer
D. Halpin—Field Engineer
R. Woodhead—Tool Room

Subcontractors Present:

Live Wire Electric
Henry Purcell
James Wallace

The First Aid Report for August 15 to August 31 Was Given. There Were:

First Aid	7
Doctor's Cases	0
Lost Time Injuries	0

Figure 17.8 Safety meeting minutes.

SHORTCUTS

All of us, supposedly, at one time or another, have been exposed to possible injury by short cutting when a few extra steps would have meant the safe way. We did so as kids when we jumped the fence instead of using the gate and we do so as men when we cross streets by jaywalking instead of using the intersection. Accident statistics plainly indicate the fact that people disregard the fact that minor safety violations may have very serious results.

In construction work, short cutting can be deadly. All of us know of cases in which this kind of thoughtless act resulted in a serious injury. For instance, an ironworker tried to cross an opening by swinging on reinforcing rods, his hands slipped, and he fell about 20 feet to a concrete floor. If he had bothered to take a few moments to walk around the building, he would still be tying rods.

The safe way is not always the shortest way and choosing the safe way is your *Personal Responsibility*. When you are told to go to work in a particular area, you are expected to take the safe route, not an unsafe short route. We cannot be your guardian angel; that is one thing you will have to do for yourself.

If you are told to go to work in some place that has no safe access, report this fact to your foreman so that necessary means of access can be provided.

Ladders and scaffolds are provided for high work; use them. Even though a high job may take only a few moments, DO NOT CLIMB ON FALSE WORK, or on some improvised platform.

Your first responsibility is to yourself. Remember that ladders, steps, and walkways have been built to save you trouble and to save your neck, too. Use them always.

Gambling a few minutes and a little energy against a possible lifetime of pain and misery is a poor bet.

GENERAL DISCUSSION

Flagmen must control all the back-up operations on this job.

Traffic—Be on the alert for moving vehicles, our area is slippery. *Don't* walk beside moving equipment.

Injuries—Report all injuries to your foreman immediately.

Charlie Hoarse

C. Hoarse, Safety Supervisor

Figure 17.8 (*Continued*)

REVIEW QUESTIONS AND EXERCISES

17.1 What factors should motivate a contractor to have a safe operation and a good safety program?

17.2 What factors influence the rate assigned to a contractor for workmen's compensation insurance?

17.3 What are two major economic benefits of a good construction program?

17.4 Explain organizing for safety.

17.5 What actions could you as the contractor take to instill a sense of safety among your workers?

17.6 Observe several construction sites and ascertain details of their safety program. If possible attend a tool box safety meeting. Then prepare a list of both good and bad examples of safety practice.

17.7 Using OSHA regulations as a guide, determine what are the accepted safety standards for:

a. Guard rails

b. Exposed reinforcing steel

c. Protection of openings

d. Man hoists

17.8 Many construction workers resist the use of safety helmets, goggles, and protective mittens and clothing despite the fact that they are designed to protect them. Give several reasons why this practice persists.

Appendices

[1] Reprinted by permission of the Engineers Joint Contract Documents Committee (EJCDC). Copyright 2002. National Society of Professional Engineers, 1420 King Street, Alexandria, VA 22314; American Consulting Engineers Council, 1015 15th Street N.W., Washington, DC 20005; American Society of Civil Engineers, 1801 Alexander Bell Drive, Reston, VA 20191-4400. To order AGC contract documents, phone 1-800-AGC-1767 or fax your request to 703-837-5405, or visit the AGC website at www.agc.org.

[2] Builders Association of Chicago, Inc.

[3] Reproduced with the permission of the Associated General Contractors of America.

Appendix A

Typical Considerations Affecting the Decision to Bid

TYPICAL CONSIDERATIONS: THE DECISION TO BID (OR NOT)[1]

A. **Goals and Present Capabilities of Your Company (Plans for Growth, Type of Work, Market Conditions)**

1. It is quite reasonable to actually want to stay where you are if you are satisfied with a situation of making a good living and staying active in work.
   - If so, is this job the kind you like doing? Does it have a good profit potential?

2. If you wish to grow larger, how fast do you wish to grow? Do you have the people and capital to do so?
   - Will the project to be bid help you in your growth?
   - Or will you have to bid it low just to keep your present men and equipment working, thus tying them up and postponing growth? (If you prefer type 1 goals, this latter strategy may be fine.)

3. *Type of work.* Which type of work do you presently have the capability and experience to do? What types of work do you want to do in the future? Can you handle this particular project now? Will it give you good experience for the type of work you want to do in the future?

4. Consider the present and future competitive market conditions in this type of work.
   - Is it possible to earn a fair and reasonable profit? Or is the competition heavy?
   - Think of the job as an investment of your time, your talent, and your money. It should earn a good return—in money, in satisfaction, and pride; or provide some other return.

B. **Location of the Work**

1. Is the project located in an area in which you normally like to operate?

2. If not, would too large a portion of your time be consumed traveling to and from this job?

3. Do you have an associate or assistant who you believe can do a good job of supervising the job if you cannot often visit the site yourself?

4. Do you plan to expand your area of operations anyway, and if so, is this job in an area in which you want to expand?

C. **Time and Place for Bid**

1. When is the bid due (day and hour)? Will you have time to prepare an accurate and careful estimate? (For example, if you need 2 weeks to prepare a good bid and only 4 days remain, don't bid the job.)

---

[1] Based on material prepared by Prof. Boyd C. Paulson, Jr., Stanford University.

2. Where is the bid to be submitted? How will you get it there? Do you have to allow 2 or 3 days for the mail?

3. Are there special rules for late delivery? For faxing last-minute changes?

D. How to Obtain Plans and Specifications

1. If you are a prime contractor, you must find out who will provide the plans and specifications.

- Is there a fee? How much?
- Is there a deposit? How much? Is it refundable?
- Is a plans room open and available? Where? What hours?

2. If you are a subcontractor, you want to know which prime contractors have plans and specifications.

- Will they give you a copy of those that apply to your work?
- Do they have a plans room for subcontractors? Where? What hours?
- Can you get your plans and specifications directly from the owner? Fees? Deposits? How much? Refunds?

E. Legal and Other Official Requirements

1. *Licensing.* Some states, counties, cities, and towns require that a contractor have a license to work in their area.

- If required, it is a legal necessity.
- In some cases, unlicensed contractors can be fined without it.
- Unlicensed contractors may not have recourse to the courts, even if wronged.
- Especially note this when working on local government-funded projects.

2. *Prequalification* may be required. If so, documents such as a financial statement, a statement of work in progress and experience, as well as a past litigation and performance history will be required.

3. *Bonding*

- Does project require (a) bid bond? (b) performance bond? (c) payment bond?
- What is your bonding limit?
- Can you qualify for bonds on this project?

F. Scope of Work

1. What is the approximate size of the project (or subcontract):

(a) In dollars—is it within your financial and bonding limits?

(b) In major units of work (e.g., earth-moving equipment, cubic yards of concrete, pounds of steel, etc.) is it within the capacity of your available manpower and equipment resources?

2. What are the major types of work on the project or subcontract?

(a) Are they the kind your company prefers to do?

(b) Are they the kind your company is qualified to do?

3. How much time is available to complete the work?

(a) When does it start; when does it finish?

(b) How much other work do you plan to have going at that time? Can you handle this job as well?

G.   Comparison of Resources

Compare the resources available to you to those that will be needed (order of magnitude only) on the job to be bid.

1. *Men*: Do you have a supervisor or foreman for the job? Can you get the laborers and craftsmen that who be needed?

2. *Equipment*: What major items of equipment (truck, crane, loader, etc.) will be needed? Do you own it already? Will it be available? Can you purchase new equipment? Can you rent or lease the equipment you will need?

3. *Money*: Will loans or credit be needed? How much? Can you get the financing needed?

H.   Summary

All of these items should be considered in making the decision to bid or not bid on a particular job.

- This is an *executive decision*.
- It is a decision *you* as the contractor must make.

Appendix B

Performance and Payment Bonds

CONTRACT PERFORMANCE BOND*

Bond No. 31-0120-42879-96-2

KNOW ALL MEN: That we   Ryan Construction Corp.
P. O. Box 16, Zionsville, IN 46077-0493

(here insert the name and address or legal title of the Contractor) hereinafter called the Principal, and

United States Fidelity and Guaranty Company
135 N. Pennsylvania Street, Indianapolis, IN 46204

hereinafter called the Surety or Sureties, are held and firmly bound unto The Trustees of Indiana University, hereinafter called the Owner, in the sum of:

Eight Hundred Twenty Two Thousand and 00/100 Dollars ($822,000.00)

for payment whereof the Principal and the Surety or Sureties bind themselves, their heirs, executors, administrator, successors and assigns, jointly and severally, firmly, by these presents.

WHEREAS, the Principal has, by means of a written Agreement, dated: June 10, XXXX, entered into a contract with the Owner for

Lilly Clinic Expansion at Adult Outpatient Center
Indiana University Medical Center, Indianapolis, IN
Bid Package No. 3
IUPUI#961-5262-3

a copy of which Agreement is by reference made a part hereof:

NOW THEREFORE, the condition of this Obligation is such that, if the Principal shall faithfully perform the Contract on his part and shall fully indemnify and save harmless the Owner from all cost and damage which he may suffer by reason of failure to do so and shall fully reimburse and repay the Owner all outlay and expense which the Owner may incur in making good any such default, then this Obligation shall be null and void, otherwise it shall remain in full force and effect.

CONTRACT PERFORMANCE BOND
(Page 1 of 2 Pages)

Disclaimer:
*This document is representative. Language in actual bonding documents should be verified by a legal professional.

The said surety for value received hereby stipulates and agrees that no change, extension of time, alteration or addition to the terms of the contract, or to the work to be performed thereunder or the specifications accompanying them, shall in any way affect its obligations on this bond, and it does hereby waive notice of any such change, extension of time, alteration, or addition to the terms of the contract, or to the work or to the specifications.

PROVIDED, however that no suit, action or proceeding by reason of any default whatever shall be brought on this Bond after two years from the date of final payment.

AND PROVIDED, that any alterations which may be made in the terms of the Contract, or in the work to be done under it, or the giving by the Owner of any extension of time for the performance of the Contract, or any other forbearance on the part of either the Owner or the Principal to the other shall not in any way release the Principal and the Surety or Sureties, or either or any of them, their heirs, executors, administrators, successors or assigns from their liability hereunder, notice to the Surety or Sureties of any such alterations, extension or forbearance being hereby waived.

Signed and Sealed this 10th day of June, XXXX

In presence of:

Michael Ryan)	Ryan Construction Corp. (SEAL)
Michael Ryan) as to	By: Daniel Ryan
Corporate Secretary)	Daniel Ryan, President
_____)	United States Fidelity and Guaranty
) as to	Company (SEAL)
_____)	By: U.S. Grant
	U. S. Grant, Attorney-In-Fact
_____)	_____(SEAL)
) as to	
_____)	_____

CONTRACT PERFORMANCE BOND
(Page 2 of 2 Pages)

LABOR AND MATERIAL PAYMENT BOND*

Bond No. 31-0120-42879-96-3

KNOW ALL MEN BY THESE PRESENTS, THAT _____
Ryan Construction Corp., P. O. Box 16, Zionsville, IN 46077-0493

as Principal, hereinafter called Principal, and _____
United States Fidelity and Guaranty Company, 135 N. Pennsylvania St., Indianapolis, IN 46204

as Surety, hereinafter called Surety, are held and firmly bound unto <u>The Trustees of Indiana University</u>

as Obligee, for the use and benefit of claimants as hereinbelow defined, in the amount of
<u>Eight Hundred Twenty Two Thousand and 00/100 Dollars ($822,000)</u>, for the payment whereof Principal and Surety bind themselves, their heirs, executors, administrators, successors and assigns, jointly and severally, firmly by these presents.

WHEREAS, Contractor has by written agreement dated <u>June 10, XXXX</u>, entered into a contract with the Obligee, for

> Lilly Clinic Expansion at Adult Outpatient Center
> Indiana University Medical Center, Indianapolis, IN
> Bid Package No. 3
> IUPUI#961-5262-3

which contract is by reference made a part hereof, and is hereinafter referred to as the contract.

NOW, THEREFORE, THE CONDITION OF THIS OBLIGATION is such that, if principal shall promptly make payment to all claimants as hereinafter defined, for all labor and material used or reasonably required for use in the performance of the Contract, then this obligation shall be void; otherwise, it shall remain in full force and effect, subject, however, to the following conditions:

1. A claimant is defined as one having a direct contract with the Principal or with a Subcontractor of the Principal for labor, material, or both, used or reasonably required for use in the performance of the Contract, labor and material being construed to include that part of water, gas, power, light, heat, oil, gasoline, telephone service or rental of equipment directly applicable to the Contract.

2. The above named Principal and Surety hereby jointly and severally agree with that every claimant as herein defined, who has not been paid in full before the expiration of a period of ninety (90) days after the date on which the last of such claimant's work or labor was done or performed, or materials were furnished by such claimant, may sue on this bond for the use of such claimant, prosecute the suit to final judgment for such sum or sums as may be justly due claimant, and have execution thereon. Obligee shall not be liable for the payment of any costs or expenses of any such suit.

Disclaimer:
*This Document is representative. Language in actual bonding documents should be verified by a legal professional.

3.   No suit action shall be commenced hereunder by any claimant:

   a)   Unless claimant, other than one having a direct contract with the Principal, shall have given written notice to any two of the following: the Principal, Obligee or the Surety above named, within ninety (90) days after such claimant did or performed the last of the work or labor, or furnished the last of the materials for which said claim is made, stating with substantial accuracy the amount claimed and the name of the party to whom the materials were furnished, or for whom the work or labor was done or performed. Such notice shall be served by mailing the same by registered mail or certified mail, postage prepaid, in an envelope addressed to the Principal, Obligee or Surety, at any place where an office is regularly maintained for the transaction of business, or served in any manner in which the legal process may be served in the state in which the aforesaid project is located, save that such service need not be made by a public officer.

   b)   After the expiration of one (1) year following the date on which Principal ceased Work on said Contract, it being understood, however, that if any limitation embodied in this bond is prohibited by any law controlling the construction hereof such limitation shall be deemed to be amended so as to equal to the minimum period of limitation permitted by such law.

   c)   Other than in a state court of competent jurisdiction in and for the county or other political subdivision of the state in which the Project, or any part thereof, is situated, or in the United States District Court for the District in which the Project, or any part thereof, is situated, and not elsewhere.

4.   The amount of this bond shall be reduced by and to the extent of any payment or payments made in good faith hereunder, inclusive of the payment by Surety of Mechanics' Liens which may be filed of record against said improvement, whether or not claim for the amount of such lien be presented under and against this bond.

Signed and sealed this 10th day of June, XXXX

_____
(Witness)

(RYAN CONSTRUCTION CORPORATION
((Principal)   (Seal)
(
(
(BY: _____
       Title: Daniel Ryan, President

_____
(Witness)

(UNITED STATES FIDELITY & GUARANTY
CO.
((Surety)   (Seal)
(
(
(BY: _____
       Title: U.S. Grant,
              Attorney-In-Fact

Appendix C

Standard Form of Agreement Between Owner and Contractor on the Basis of a Stipulated Price

SUGGESTED FORM OF AGREEMENT
BETWEEN OWNER AND CONTRACTOR FOR
CONSTRUCTION CONTRACT (STIPULATED PRICE)

Prepared by

ENGINEERS JOINT CONTRACT DOCUMENTS COMMITTEE

and

Issued and Published Jointly By

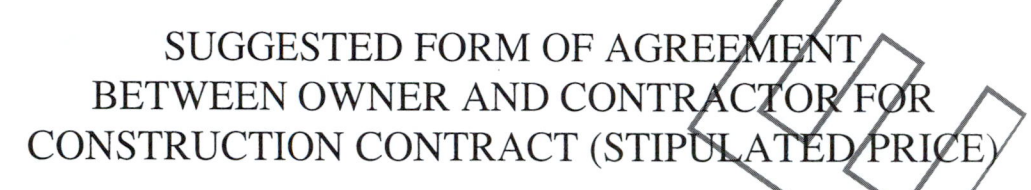

PROFESSIONAL ENGINEERS IN PRIVATE PRACTICE
a practice division of the
NATIONAL SOCIETY OF PROFESSIONAL ENGINEERS

_____

AMERICAN COUNCIL OF ENGINEERING COMPANIES

_____

AMERICAN SOCIETY OF CIVIL ENGINEERS

This document has been approved and endorsed by

The Associated General Contractors of America

Knowledge for Creating
and Sustaining
the Built Environment

Construction Specifications Institute

This Suggested Form of Agreement has been prepared for use with the Standard General Conditions of the Construction Contract (C-700, 2002 Edition). Their provisions are interrelated, and a change in one may necessitate a change in the other. The language contained in the Suggested Instructions to Bidders (C-200, 2002 Edition) is also carefully interrelated with the language of this Agreement. Their usage is discussed in the Commentary on EJCDC Construction Documents. See also Guide to the Preparation of Supplementary (C-800, 2002 Edition).

National Society of Professional Engineers
1420 King Street, Alexandria, VA 22314-2715

American Council of Engineering Companies
1015 15th Street, N.W., Washington, DC 20005

American Society of Civil Engineers
1801 Alexander Bell Drive, Reston, VA 20191-4400

Introduction

This Suggested Form of Agreement between Owner and Contractor for Construction Contract (Stipulated Price) ("Agreement") has been prepared for use with the Guide to the Preparation of Instructions to Bidders ("Instructions")(C-200, 2002 Edition) and with the Standard General Conditions of the Construction Contract ("General Conditions")(C-700, 2002 Edition). Their provisions are interrelated, and a change in one may necessitate a change in the others. For guidance in the preparation of Supplementary Conditions and coordination with Instructions to Bidders, see Guide to the Preparation of Supplementary Conditions ("Supplementary Conditions")(C-800, 2002 Edition). See also Suggested Bid Form ("Bid Form") (C-410, 2002 Edition). The EJCDC has not prepared a suggested form of Advertisement or Invitation to Bid because such documents will vary widely to conform to statutory requirements.

This form and the other Bidding Documents prepared and issued by the EJCDC assume acceptance of the Project Manual concept of the Construction Specifications Institute which provides for an organizational format for location of all bound documentary information for a construction project, namely: Bidding Requirements (which term refers to the Advertisement or Invitation to Bid, the Instructions, and any Bid Form that may be suggested or prescribed, all of which provide information and guidance for all Bidders) and the Contract Documents (defined in Article 1 of the General Conditions), which include the Agreement, bonds and certificates, the General Conditions, the Supplementary Conditions, the Drawings, and the Specifications. The Bidding Requirements are not considered part of the Contract Documents because much of their substance pertains to the relationships prior to the award of the Contract and has little effect or impact thereafter and because many contracts are awarded without going through the bidding process. In some cases, however, the actual Bid may be attached as an exhibit to the Agreement to avoid extensive retyping. (The terms "Bidding Documents" and "Bidding Requirements" are defined in Article 1 of the General Conditions.) The Project Manual concept is explained in the Manual of Practice issued by the Construction Specifications Institute.

Suggested language is presented herein with "Notes to User" to assist in preparing the Agreement. Much of the language should be usable on most projects, but modifications and additional provisions will often be necessary. The suggested language has been coordinated with the other standard forms produced by the EJCDC. When modifying the suggested language or writing additional provisions, the user must check the other documents thoroughly for conflicts and coordination of language usage and make appropriate revisions in all affected documents.

Refer to the discussions in EJCDC's Recommended Competitive Bidding Procedures for Construction Projects ("Bidding Procedures") (No. 1910-9-D, 1987 Edition) (to be reissued in 2002) on the particular paragraphs of which frequent reference is made below.

For brevity, paragraphs of the Instructions to Bidders are referenced with the prefix "I," those of the Bid Form are referenced with the prefix "BF," and those of this Agreement are referenced with the prefix "A."

NOTES:

1. EJCDC publications may be ordered from:

NSPE headquarters
1420 King Street
Alexandria VA 22314-2715
703-684-2800
www.nspe.org

ASCE headquarters
1801 Alexander Bell Drive
Reston, VA 20191-4400
800-548-2723
www.asce.org

ACEC headquarters
1015 15th Street NW
Washington DC 20005
202-347-7474
www.acec.org

EJCDC
SUGGESTED FORM OF AGREEMENT
BETWEEN OWNER AND CONTRACTOR FOR

CONSTRUCTION CONTRACT(STIPULATED PRICE)

THIS AGREEMENT is by and between _____

(Owner) and _____

(Contractor).

Owner and Contractor, in consideration of the mutual covenants set forth herein, agree as follows:

ARTICLE 1 - WORK

1.01   Contractor shall complete all Work as specified or indicated in the Contract Documents. The Work is generally described as follows:

ARTICLE 2 - THE PROJECT

2.01   The Project for which the Work under the Contract Documents may be the whole or only a part is generally described as follows:

ARTICLE 3 - ENGINEER

3.01   The Project has been designed by

(Engineer), who is to act as Owner's representative, assume all duties and responsibilities, and have the rights and authority assigned to Engineer in the Contract Documents in connection with the completion of the Work in accordance with the Contract Documents.

ARTICLE 4 - CONTRACT TIMES

4.01   Time of the Essence

A.   All time limits for Milestones, if any, Substantial Completion, and completion and readiness for final payment as stated in the Contract Documents are of the essence of the Contract.

4.02   Dates for Substantial Completion and Final Payment

A.   The Work will be substantially completed on or before _____, _____, and completed and ready for final payment in accordance with Paragraph 14.07 of the General Conditions on or before _____, _____.

[or]

4.02   Days to Achieve Substantial Completion and Final Payment

A.   The Work will be substantially completed within _____ days after the date when the Contract Times commence to run as provided in Paragraph 2.03 of the General Conditions, and completed and ready for final payment in accordance with Paragraph 14.07 of the General Conditions within _____ days after the date when the Contract Times commence to run.

4.03   Liquidated Damages

A.   Contractor and Owner recognize that time is of the essence of this Agreement and that Owner will suffer financial loss if the Work is not completed within the times specified in Paragraph 4.02 above, plus any extensions thereof allowed in accordance with Article 12 of the General Conditions. The parties also recognize the delays, expense, and difficulties involved in proving in a legal or arbitration proceeding the actual loss suffered by Owner if the Work is not completed on time. Accordingly, instead of requiring any such proof, Owner and Contractor agree that as liquidated damages for delay (but not as a penalty), Contractor shall pay Owner $_____ for each day that expires after the time specified in Paragraph 4.02 for Substantial Completion until the Work is substantially complete. After Substantial Completion, if Contractor shall neglect, refuse, or fail to complete the remaining Work within the Contract Time or any proper extension thereof granted by Owner, Contractor shall pay Owner $_____ for each day that expires after the time specified in Paragraph 4.02 for completion and readiness for final payment until the Work is completed and ready for final payment.

NOTES TO USER
1.   *Where failure to reach a Milestone on time is of such consequence that the assessment of liquidated damages for failure to reach one or more Milestones on time is to be provided, appropriate amending or supplementing language should be inserted here.*

ARTICLE 5 - CONTRACT PRICE

5.01   Owner shall pay Contractor for completion of the Work in accordance with the Contract Documents an amount in current funds equal to the sum of the amounts determined pursuant to Paragraphs 5.01.A, 5.01.B, and 5.01.C below:

A.   For all Work other than Unit Price Work, a Lump Sum of:

_____   ($_____)
(words)   (numerals)

All specific cash allowances are included in the above price and have been computed in accordance with paragraph 11.02 of the General Conditions.

B.   For all Unit Price Work, an amount equal to the sum of the established unit price for each separately identified item of Unit Price Work times the estimated quantity of that item as indicated in this paragraph 5.01.B:

As provided in Paragraph 11.03 of the General Conditions, estimated quantities are not guaranteed, and determinations of actual quantities and classifications are to be made by Engineer as provided in Paragraph 9.07 of the General Conditions. Unit prices have been computed as provided in Paragraph 11.03 of the General Conditions.

UNIT PRICE WORK

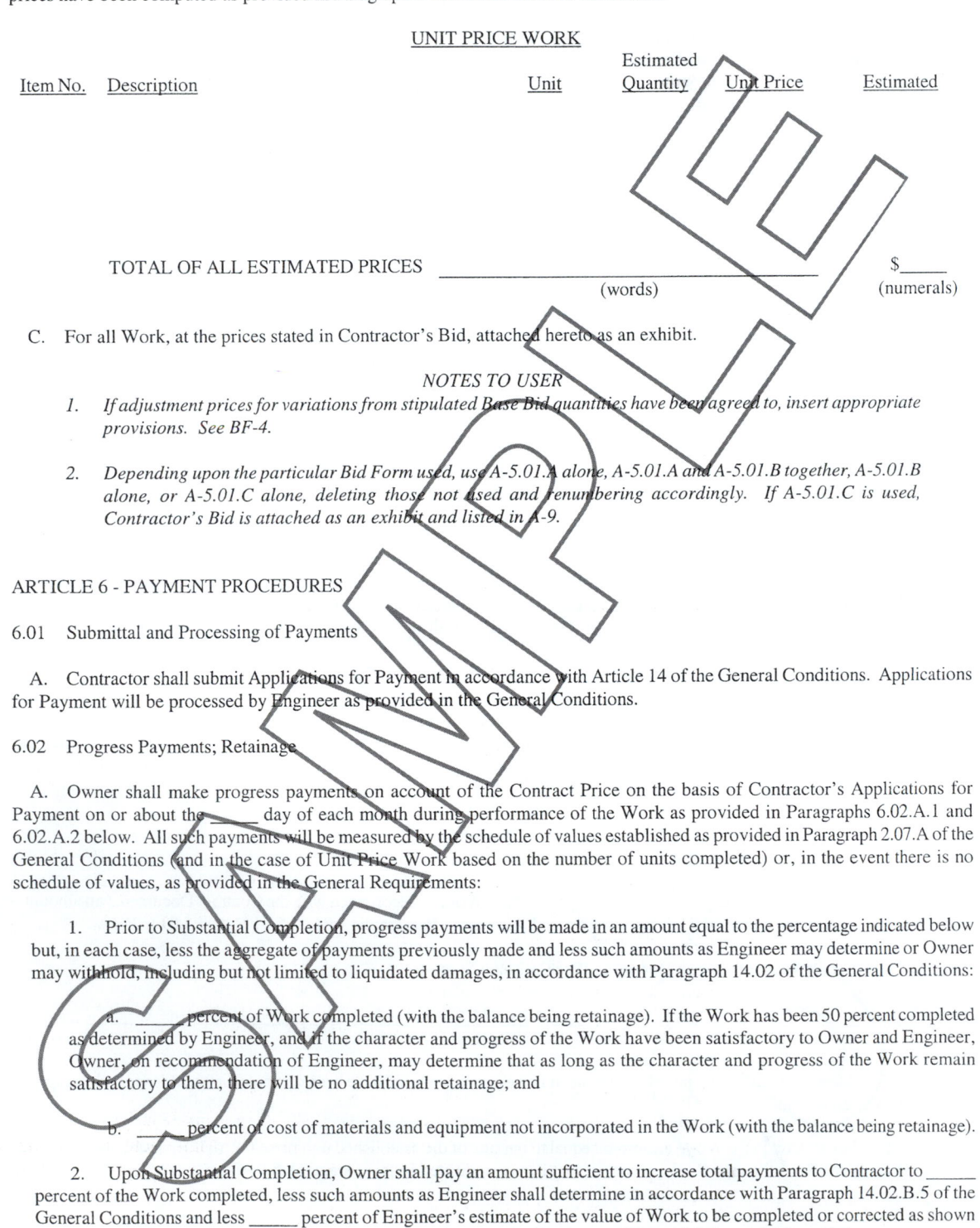

Item No.	Description	Unit	Estimated Quantity	Unit Price	Estimated

TOTAL OF ALL ESTIMATED PRICES _____ $_____
(words)   (numerals)

C. For all Work, at the prices stated in Contractor's Bid, attached hereto as an exhibit.

NOTES TO USER

1. If adjustment prices for variations from stipulated Base Bid quantities have been agreed to, insert appropriate provisions. See BF-4.

2. Depending upon the particular Bid Form used, use A-5.01.A alone, A-5.01.A and A-5.01.B together, A-5.01.B alone, or A-5.01.C alone, deleting those not used and renumbering accordingly. If A-5.01.C is used, Contractor's Bid is attached as an exhibit and listed in A-9.

ARTICLE 6 - PAYMENT PROCEDURES

6.01 Submittal and Processing of Payments

A. Contractor shall submit Applications for Payment in accordance with Article 14 of the General Conditions. Applications for Payment will be processed by Engineer as provided in the General Conditions.

6.02 Progress Payments; Retainage

A. Owner shall make progress payments on account of the Contract Price on the basis of Contractor's Applications for Payment on or about the _____ day of each month during performance of the Work as provided in Paragraphs 6.02.A.1 and 6.02.A.2 below. All such payments will be measured by the schedule of values established as provided in Paragraph 2.07.A of the General Conditions (and in the case of Unit Price Work based on the number of units completed) or, in the event there is no schedule of values, as provided in the General Requirements:

   1. Prior to Substantial Completion, progress payments will be made in an amount equal to the percentage indicated below but, in each case, less the aggregate of payments previously made and less such amounts as Engineer may determine or Owner may withhold, including but not limited to liquidated damages, in accordance with Paragraph 14.02 of the General Conditions:

      a. _____ percent of Work completed (with the balance being retainage). If the Work has been 50 percent completed as determined by Engineer, and if the character and progress of the Work have been satisfactory to Owner and Engineer, Owner, on recommendation of Engineer, may determine that as long as the character and progress of the Work remain satisfactory to them, there will be no additional retainage; and

      b. _____ percent of cost of materials and equipment not incorporated in the Work (with the balance being retainage).

   2. Upon Substantial Completion, Owner shall pay an amount sufficient to increase total payments to Contractor to _____ percent of the Work completed, less such amounts as Engineer shall determine in accordance with Paragraph 14.02.B.5 of the General Conditions and less _____ percent of Engineer's estimate of the value of Work to be completed or corrected as shown on the tentative list of items to be completed or corrected attached to the certificate of Substantial Completion.

6.03   Final Payment

A.   Upon final completion and acceptance of the Work in accordance with Paragraph 14.07 of the General Conditions, Owner shall pay the remainder of the Contract Price as recommended by Engineer as provided in said Paragraph 14.07.

ARTICLE 7 - INTEREST

7.01   All moneys not paid when due as provided in Article 14 of the General Conditions shall bear interest at the rate of _____ percent per annum.

ARTICLE 8 – CONTRACTOR'S REPRESENTATIONS

8.01   In order to induce Owner to enter into this Agreement Contractor makes the following representations:

A.   Contractor has examined and carefully studied the Contract Documents and the other related data identified in the Bidding Documents.

B.   Contractor has visited the Site and become familiar with and is satisfied as to the general, local, and Site conditions that may affect cost, progress, and performance of the Work.

C.   Contractor is familiar with and is satisfied as to all federal, state, and local Laws and Regulations that may affect cost, progress, and performance of the Work.

D.   Contractor has carefully studied all: (1) reports of explorations and tests of subsurface conditions at or contiguous to the Site and all drawings of physical conditions in or relating to existing surface or subsurface structures at or contiguous to the Site (except Underground Facilities) which have been identified in the Supplementary Conditions as provided in Paragraph 4.02 of the General Conditions and (2) reports and drawings of a Hazardous Environmental Condition, if any, at the Site which has been identified in the Supplementary Conditions as provided in Paragraph 4.06 of the General Conditions.

NOTES TO USER
1.   *If the reports and/or drawings referred to in A-8.01.D do not exist, either modify A-8.01.D or delete A-8.01.D and renumber accordingly.*

E.   Contractor has obtained and carefully studied (or assumes responsibility for doing so) all additional or supplementary examinations, investigations, explorations, tests, studies, and data concerning conditions (surface, subsurface, and Underground Facilities) at or contiguous to the Site which may affect cost, progress, or performance of the Work or which relate to any aspect of the means, methods, techniques, sequences, and procedures of construction to be employed by Contractor, including any specific means, methods, techniques, sequences, and procedures of construction expressly required by the Bidding Documents, and safety precautions and programs incident thereto.

NOTES TO USER

   1.   If the reports and/or drawings referred to in A-8.01.D do not exist, delete the phrase "additional or supplementary" in the first sentence of A-8.01.E.

F.   Contractor does not consider that any further examinations, investigations, explorations, tests, studies, or data are necessary for the performance of the Work at the Contract Price, within the Contract Times, and in accordance with the other terms and conditions of the Contract Documents.

G.   Contractor is aware of the general nature of work to be performed by Owner and others at the Site that relates to the Work as indicated in the Contract Documents.

H.   Contractor has correlated the information known to Contractor, information and observations obtained from visits to the Site, reports and drawings identified in the Contract Documents, and all additional examinations, investigations, explorations, tests, studies, and data with the Contract Documents.

I.   Contractor has given Engineer written notice of all conflicts, errors, ambiguities, or discrepancies that Contractor has discovered in the Contract Documents, and the written resolution thereof by Engineer is acceptable to Contractor.

J.   The Contract Documents are generally sufficient to indicate and convey understanding of all terms and conditions for performance and furnishing of the Work.

ARTICLE 9 - CONTRACT DOCUMENTS

9.01   Contents

   A.   The Contract Documents consist of the following:

      1.   This Agreement (pages 1 to _____ inclusive).

      2.   Performance bond (pages _____ to _____, inclusive).

      3.   Payment bond (pages _____ to _____, inclusive).

      4.   Other bonds (pages _____ to _____, inclusive).

         a.   _____ (pages _____ to _____, inclusive).

         b.   _____ (pages _____ to _____, inclusive).

         c.   _____ (pages _____ to _____, inclusive).

      5.   General Conditions (pages _____ to _____, inclusive).

      6.   Supplementary Conditions (pages _____ to _____, inclusive).

      7.   Specifications as listed in the table of contents of the Project Manual.

8.   Drawings consisting of _____ sheets with each sheet bearing the following general title: _____ [or] the Drawings listed on attached sheet index.

9.   Addenda (numbers _____ to _____, inclusive).

10.   Exhibits to this Agreement (enumerated as follows):

    a.   Contractor's Bid (pages _____ to _____, inclusive).

    b.   Documentation submitted by Contractor prior to Notice of Award (pages _____ to _____, inclusive).

    c.   _____.

11.   The following which may be delivered or issued on or after the Effective Date of the Agreement and are not attached hereto:

    a.   Notice to Proceed (pages _____ to _____, inclusive).

    b.   Work Change Directives.

    c.   Change Order(s).

B.   The documents listed in Paragraph 9.01.A are attached to this Agreement (except as expressly noted otherwise above).

C.   There are no Contract Documents other than those listed above in this Article 9.

D.   The Contract Documents may only be amended, modified, or supplemented as provided in Paragraph 3.04 of the General Conditions.

ARTICLE 10 - MISCELLANEOUS

10.01 Terms

A.   Terms used in this Agreement will have the meanings stated in the General Conditions and the Supplementary Conditions.

10.02 Assignment of Contract

A.   No assignment by a party hereto of any rights under or interests in the Contract will be binding on another party hereto without the written consent of the party sought to be bound, and, specifically but without limitation, moneys that may become due and moneys that are due may not be assigned without such consent (except to the extent that the effect of this restriction may be limited by law), and unless specifically stated to the contrary in any written consent to an assignment, no assignment will release or discharge the assignor from any duty or responsibility under the Contract Documents.

10.03 Successors and Assigns

A.   Owner and Contractor each binds itself, its partners, successors, assigns, and legal representatives to the other party hereto, its partners, successors, assigns, and legal representatives in respect to all covenants, agreements, and obligations contained in the Contract Documents.

10.04 Severability

A.   Any provision or part of the Contract Documents held to be void or unenforceable under any Law or Regulation shall be deemed stricken, and all remaining provisions shall continue to be valid and binding upon Owner and Contractor, who agree that the Contract Documents shall be reformed to replace such stricken provision or part thereof with a valid and enforceable provision that comes as close as possible to expressing the intention of the stricken provision.

10.05 Other Provisions

NOTES TO USER

1.   If Owner intends to assign a procurement contract (for goods and services) to the Contractor, see Notes to User at Article 23 of Suggested Instructions to Bidders for Procurement Contracts (EJCDC No. P-200, 2000 Edition) for provisions to be inserted in this Article.

2.   Insert other provisions here if applicable.

IN WITNESS WHEREOF, Owner and Contractor have signed this Agreement in duplicate. One counterpart each has been delivered to Owner and Contractor. All portions of the Contract Documents have been signed or identified by Owner and Contractor or on their behalf.

NOTES TO USER
1.   See I-21 and correlate procedures for format and signing between the two documents.

This Agreement will be effective on _____, _____ (which is the Effective Date of the Agreement).

OWNER: CONTRACTOR:

_____   _____

By: _____   By: _____

Title: _____   Title: _____

          [CORPORATE SEAL]   [CORPORATE SEAL]

Attest: _____   Attest: _____

Title: _____   Title: _____

Address for giving notices: Address for giving notices:

_____   _____

_____   _____

_____   _____

                                                              License No.: _____
(If Owner is a corporation, attach evidence of authority to   (Where applicable)
sign. If Owner is a public body, attach evidence of authority
to sign and resolution or other documents authorizing
execution of Owner-Contractor Agreement.)   Agent for service or process: _____

                                                              _____

                                                              (If Contractor is a corporation or a partnership, attach evidence
                                                              of authority to sign.)

**Engineers Joint Documents Committee
Design and Construction Related Documents
Instructions and License Agreement**

Instructions

Before you use any EJCDC document:
1. Read the License Agreement. You agree to it and are bound by its terms when you use the EJCDC document.

2. Make sure that you have the correct version for your word processing software.

How to Use:
1. While EJCDC has expended considerable effort to make the software translations exact, it can be that a few document controls (e.g., bold, underline) did not carry over.

2. Similarly, your software may change the font specification if the font is not available in your system. It will choose a font that is close in appearance. In this event, the pagination may not match the control set.

3. If you modify the document, you must follow the instructions in the License Agreement about notification.

4. Also note the instruction in the License Agreement about the EJCDC copyright.

License Agreement

You should carefully read the following terms and conditions before using this document. Commencement of use of this document indicates your acceptance of these terms and conditions. If you do not agree to them, you should promptly return the materials to the vendor, and your money will be refunded.

The Engineers Joint Contract Documents Committee ("EJCDC") provides **EJCDC Design and Construction Related Documents** and licenses their use worldwide. You assume sole responsibility for the selection of specific documents or portions thereof to achieve your intended results, and for the installation, use, and results obtained from **EJCDC Design and Construction Related Documents**.

You acknowledge that you understand that the text of the contract documents of **EJCDC Design and Construction Related Documents** has important legal consequences and that consultation with an attorney is recommended with respect to use or modification of the text. You further acknowledge that EJCDC documents are protected by the copyright laws of the United States.

License:
You have a limited nonexclusive license to:

1. Use **EJCDC Design and Construction Related Documents** on any number of machines owned, leased or rented by your company or organization.

2. Use **EJCDC Design and Construction Related Documents** in printed form for bona fide contract documents.

3. Copy **EJCDC Design and Construction Related Documents** into any machine-readable or printed form for backup or modification purposes in support of your use of **EJCDC Design and Construction Related Documents**.

You agree that you will:
1. Reproduce and include EJCDC's copyright notice on any printed or machine-readable copy, modification, or portion merged into another document or program. All proprietary rights in **EJCDC Design and Construction Related Documents** are and shall remain the property of EJCDC.

2. Not represent that any of the contract documents you generate from **EJCDC Design and Construction Related Documents** are EJCDC documents unless (i) the document text is used without alteration or (ii) all additions and changes to, and deletions from, the text are clearly shown.

You may not use, copy, modify, or transfer EJCDC Design and Construction Related Documents, or any copy, modification or merged portion, in whole or in part, except as expressly provided for in this license. Reproduction of EJCDC Design and Construction Related Documents in printed or machine-readable format for resale or educational purposes is expressly prohibited.

If you transfer possession of any copy, modification or merged portion of EJCDC Design and Construction Related Documents to another party, your license is automatically terminated.

Term:
The license is effective until terminated. You may terminate it at any time by destroying **EJCDC Design and Construction Related Documents** altogether with all copies, modifications and merged portions in any form. It will also terminate upon conditions set forth elsewhere in this Agreement or if you fail to comply with any term or condition of this Agreement. You agree upon such termination to destroy **EJCDC Design and Construction Related Documents** along with all copies, modifications and merged portions in any form.

Limited Warranty:
EJCDC warrants the CDs and diskettes on which **EJCDC Design and Construction Related Documents** is furnished to be free from defects in materials and workmanship under normal use for a period of ninety (90) days from the date of

delivery to you as evidenced by a copy of your receipt.

There is no other warranty of any kind, either expressed or implied, including, but not limited to the implied warranties of merchantability and fitness for a particular purpose. Some states do not allow the exclusion of implied warranties, so the above exclusion may not apply to you. This warranty gives you specific legal rights and you may also have other rights which vary from state to state.

EJCDC does not warrant that the functions contained in **EJCDC Design and Construction Related Documents** will meet your requirements or that the operation of **EJCDC Design and Construction Related Documents** will be uninterrupted or error free.

<u>Limitations of Remedies:</u>
EJCDC's entire liability and your exclusive remedy shall be:

1. the replacement of any document not meeting EJCDC's "Limited Warranty" which is returned to EJCDC's selling agent with a copy of your receipt, or

2. if EJCDC's selling agent is unable to deliver a replacement CD or diskette which is free of defects in materials and workmanship, you may terminate this Agreement by returning EJCDC Document and your money will be refunded.

In no event will EJCDC be liable to you for any damages, including any lost profits, lost savings or other incidental or consequential damages arising out of the use or inability to use **EJCDC Design and Construction Related Documents** even if EJCDC has been advised of the possibility of such

damages, or for any claim by any other party.

Some states do not allow the limitation or exclusion of liability for incidental or consequential damages, so the above limitation or exclusion may not apply to you.

<u>General:</u>
You may not sublicense, assign, or transfer this license except as expressly provided in this Agreement. Any attempt otherwise to sublicense, assign, or transfer any of the rights, duties, or obligations hereunder is void.

This Agreement shall be governed by the laws of the State of Virginia. Should you have any questions concerning this Agreement, you may contact EJCDC by writing to:

Arthur Schwartz, Esq.
General Counsel
National Society of Professional Engineers
1420 King Street
Alexandria, VA 22314

Phone: (703) 684-2845
Fax: (703) 836-4875
e-mail: aschwartz@nspe.org

You acknowledge that you have read this agreement, understand it and agree to be bound by its terms and conditions. You further agree that it is the complete and exclusive statement of the agreement between us which supersedes any proposal or prior agreement, oral or written, and any other communications between us relating to the subject matter of this agreement.

Appendix D

Standard Form of Agreement Between Owner and Contractor on the Basis of Cost-Plus

SUGGESTED FORM OF AGREEMENT BETWEEN OWNER AND CONTRACTOR FOR CONSTRUCTION CONTRACT (COST-PLUS)

Prepared by

ENGINEERS JOINT CONTRACT DOCUMENTS COMMITTEE

and

Issued and Published Jointly By

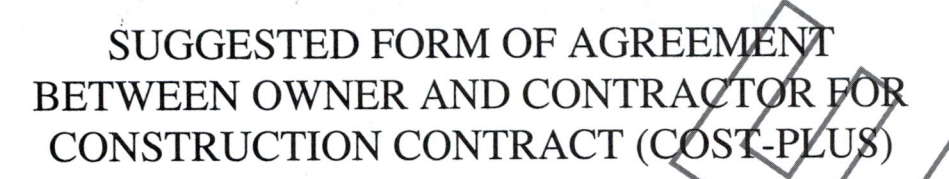

PROFESSIONAL ENGINEERS IN PRIVATE PRACTICE
A practice division of the
NATIONAL SOCIETY OF PROFESSIONAL ENGINEERS

———————————

AMERICAN COUNCIL OF ENGINEERING COMPANIES

———————————

AMERICAN SOCIETY OF CIVIL ENGINEERS

———————————

This document has been approved and endorsed by

The Associated General Contractors of America

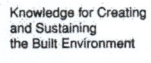
Knowledge for Creating
and Sustaining
the Built Environment

Construction Specifications Institute

This Suggested Form of Agreement has been prepared for use with the Standard General Conditions of the Construction Contract (C-700, 2002 Edition). Their provisions are interrelated, and a change in one may necessitate a change in the other. The language contained in the Suggested Instructions to Bidders (C-200, 2002 Edition) is also carefully interrelated with the language of this Agreement. Their usage is discussed in the Commentary on EJCDC Construction Documents. See also Guide to the Preparation of Supplementary (C-800, 2002 Edition).

National Society of Professional Engineers
1420 King Street, Alexandria, VA 22314-2715

American Council of Engineering Companies
1015 15th Street, N.W., Washington, DC 20005

American Society of Civil Engineers
1801 Alexander Bell Drive, Reston, VA 20191-4400

Introduction

This Suggested Form of Agreement between Owner and Contractor (Cost-Plus) ("Agreement") has been prepared for use with the Guide to the Preparation of Instructions to Bidders ("Instructions")(C-200, 2002 Edition) and with the Standard General Conditions of the Construction Contract ("General Conditions")(C-700, 2002 Edition). Their provisions are interrelated, and a change in one may necessitate a change in the others. For guidance in the preparation of Supplementary Conditions and coordination with Instructions to Bidders, see Guide to the Preparation of Supplementary Conditions ("Supplementary Conditions")(C-800, 2002 Edition). See also Guide to Preparation of the Bid Form ("Bid Form")(C-400, 2002 Edition). The EJCDC has not prepared a suggested form of Advertisement or Invitation to Bid because such documents will vary widely in response to statutory requirements.

This form and the other Bidding Documents prepared and issued by the EJCDC assume acceptance of the Project Manual concept of the Construction Specifications Institute which provides for an organizational format for location of all bound documentary information for a construction project, namely: Bidding Requirements (which term refers to the Advertisement or Invitation to Bid, the Instructions, and any Bid Form that may be suggested or prescribed, all of which provide information and guidance for all Bidders) and the Contract Documents (defined in Article 1 of the General Conditions), which include the Agreement, bonds and certificates, the General Conditions, the Supplementary Conditions, the Drawings, and the Specifications. The Bidding Requirements are not considered part of the Contract Documents because much of their substance pertains to the relationships prior to the award of the Contract and has little effect or impact thereafter and because many contracts are awarded without going through the bidding process. In some cases, however, the actual Bid may be attached as an exhibit to the Agreement to avoid extensive retyping. (The terms "Bidding Documents" and "Bidding Requirements" are defined in Article 1 of the General Conditions.) The Project Manual concept is explained in the Manual of Practice issued by the Construction Specifications Institute.

Suggested language is presented herein with "Notes to User" to assist in preparing the Agreement. Much of the language should be usable on most projects, but modifications and additional provisions will often be necessary. The suggested language has been coordinated with the other standard forms produced by the EJCDC. When modifying the suggested language or writing additional provisions, the user must check the other documents thoroughly for conflicts and coordination of language usage and make appropriate revisions in all affected documents.

Refer to the discussions in EJCDC's Recommended Competitive Bidding Procedures for Construction Projects ("Bidding Procedures") (No. 1910-9-D, 1987 Edition) (to be reissued in 2002) on the particular paragraphs of which frequent reference is made below.

For brevity, paragraphs of the Instructions to Bidders are referred to with the prefix "I," and those of this Agreement with the prefix "A."

NOTES:

1.   EJCDC publications may be ordered from:

NSPE headquarters
1420 King Street
Alexandria VA 22314-2715
703-684-2800
www.nspe.org

ASCE headquarters
1801 Alexander Bell Drive
Reston, VA 20191-4400
800-548-2723
www.asce.org

ACEC headquarters
1015 15th Street NW
Washington DC 20005
202-347-7474
www.acec.org

EJCDC
SUGGESTED FORM OF AGREEMENT
BETWEEN OWNER AND CONTRACTOR FOR CONSTRUCTION
CONTRACT (COST-PLUS)

THIS AGREEMENT is by and between _____ (Owner)

and _____ (Contractor).

Owner and Contractor, in consideration of the mutual covenants set forth herein, agree as follows:

ARTICLE 1 - WORK

1.01   Contractor shall complete all Work as specified or indicated in the Contract Documents. The Work is generally described as follows:

ARTICLE 2 - THE PROJECT

2.01   The Project for which the Work under the Contract Documents may be the whole or only a part is generally described as follows:

ARTICLE 3 - ENGINEER

3.01   The Project has been designed by

(Engineer), who is to act as Owner's representative, assume all duties and responsibilities, and have the rights and authority assigned to Engineer in the Contract Documents in connection with the completion of the Work in accordance with the Contract Documents.

ARTICLE 4 - CONTRACT TIMES

4.01   Time of the Essence

   A.   All time limits for Milestones, if any, Substantial Completion, and completion and readiness for final payment as stated in the Contract Documents are of the essence of the Contract.

4.02   Dates for Substantial Completion and Final Payment

   A.   The Work will be substantially completed on or before _____, _____, and completed and ready for final payment in accordance with Paragraph 14.07 of the General Conditions on or before _____, _____

   [or]

4.02   Days to Achieve Substantial Completion and Final Payment

   A.   The Work will be substantially completed within _____ days after the date when the Contract Times commence to run as provided in Paragraph 2.03 of the General Conditions, and completed and ready for final payment in accordance with Paragraph 14.07 of the General Conditions within _____ days after the date when the Contract Times commence to run.

4.03   Liquidated Damages

   A.   Contractor and Owner recognize that time is of the essence of this Agreement and that Owner will suffer financial loss if the Work is not completed within the times specified in Paragraph 4.02 above, plus any extensions thereof allowed in accordance with Article 12 of the General Conditions. The parties also recognize the delays, expense, and difficulties involved in proving in a legal or arbitration proceeding the actual loss suffered by Owner if the Work is not completed on time. Accordingly, instead of requiring any such proof, Owner and Contractor agree that as liquidated damages for delay (but not as a penalty), Contractor shall pay Owner $_____ for each day that expires after the time specified in Paragraph 4.02 for Substantial Completion until the Work is substantially complete. After Substantial Completion, if Contractor shall neglect, refuse, or fail to complete the remaining Work within the Contract Time or any proper extension thereof granted by Owner, Contractor shall pay Owner $_____ for each day that expires after the time specified in Paragraph 4.02 for completion and readiness for final payment until the Work is completed and ready for final payment.

NOTES TO USER

   1.   *Where failure to reach a Milestone on time is of such consequence that the assessment of liquidated damages for failure to reach one or more Milestones on time is to be provided, appropriate amending or supplementing language should be inserted here.*

ARTICLE 5 - CONTRACT PRICE

5.01 Owner shall pay Contractor for completion of the Work in accordance with the Contract Documents an amount in current funds equal to the sum of the amounts determined pursuant to Paragraphs 5.01.A, 5.01.B, and 5.01.C below:

A. For all Work other than Unit Price Work, the Cost of the Work plus a Contractor's fee for overhead and profit, both of which shall be determined as provided in Articles 6 and 7 below, subject to additions and deletions as provided in the Contract Documents and subject to the limitations set forth in Article 8 below.

B. For all Unit Price Work, an amount equal to the sum of the established unit price for each separately identified item of Unit Price Work times the estimated quantity of that item as indicated in this Paragraph 5.01.B:

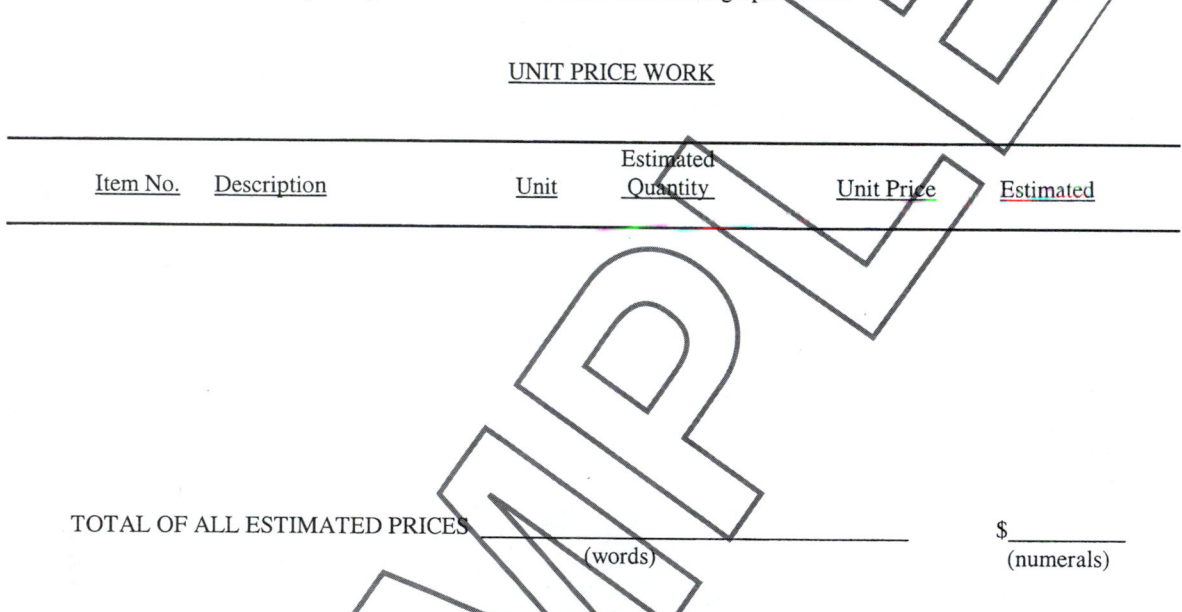

UNIT PRICE WORK

Item No.	Description	Unit	Estimated Quantity	Unit Price	Estimated

TOTAL OF ALL ESTIMATED PRICES _____ $_____
(words)   (numerals)

As provided in Paragraph 11.03 of the General Conditions, estimated quantities are not guaranteed, and determinations of actual quantities and classifications are to be made by Engineer as provided in Paragraph 9.07 of the General Conditions. Unit prices have been computed as provided in Paragraph 11.03 of the General Conditions.

C. For all Work, at the prices stated in Contractor's Bid, attached hereto as an exhibit.

NOTES TO USER

1. Depending upon the particular project bid form used, use A-5.01.A alone, A-5.01.A and A-5.01.B together, A-5.01.B alone, or A-5.01.C alone, deleting those not used and renumbering accordingly. If A-5.01.C is used, Contractor's Bid is attached as an exhibit and listed in A-14.

ARTICLE 6 - COST OF THE WORK

6.01 Cost of the Work shall be determined as provided in Paragraph 11.01 of the General Conditions.

ARTICLE 7 – CONTRACTOR'S FEE

7.01   Contractor's fee shall be determined as follows:

A.   A fee based on the following percentages of the various portions of the Cost of the Work:

1.   Payroll costs (see Paragraph 11.01.A.1 of General Conditions) _____.

2.   Material and equipment costs (see Paragraph 11.01.A.2 of General Conditions) _____.

3.   Amounts paid to subcontractors (see Paragraph 11.01.A.3 of General Conditions) _____.

4.   Amounts paid to special consultants (see Paragraph 11.01.A.4 of General Conditions) _____.

5.   Supplemental costs (see Paragraph 11.01.A.5 of General Conditions) _____.

6.   No fee will be payable on the basis of costs itemized in Paragraph 11.01.B of the General Conditions.

7.   The provisions in Paragraph 11.01.C of the General Conditions will apply only to changes in the Work.

B.   Contractor guarantees that the maximum amount payable by Owner in accordance with Paragraph 7.01.A as a percentage fee will not exceed $_____, subject to increases or decreases for changes in the Work as provided in Paragraph 9.01.B below.

[or]

C.   A fixed fee of $_____, which shall be subject to increases or decreases for changes in the Work as provided in Paragraph 9.01.A below.

NOTES TO USER

Depending on the fee agreement selected or negotiated, select 7.01.A (percentage fee), or 7.01.A and 7.01.B (percentage fee up to guaranteed maximum), or 7.01.C (fixed fee).

ARTICLE 8 - GUARANTEED MAXIMUM PRICE

8.01   Contractor guarantees that the maximum amount payable (Guaranteed Maximum Price) by Owner for the sum of the Cost of the Work plus Contractor's fee under Article 7 will not exceed $_____, subject to increases or decreases for changes in the Work. The Guaranteed Maximum Price will not apply to Unit Price Work.

ARTICLE 9 - CHANGES IN THE CONTRACT PRICE

9.01   The amount of any increases or decreases in Contractor's fee, in any Guaranteed Maximum Price, or in any guaranteed maximum fee which results from a Change Order shall be set forth in the applicable Change Order subject to the following:

A.   If Contractor's fee is a fixed fee, any increase or decrease in the Contractor's fee resulting from net additions or decreases in the Cost of the Work shall be determined in accordance with Paragraph 12.01.C of the General Conditions.

[or]

A.   If Contractor's fee is a percentage fee not subject to any guaranteed maximum limitation, Contractor's fee will adjust automatically as the Cost of the Work changes.

NOTES TO USER

Select one of the two provisions above for Paragraph 9.01.A.

B.   Wherever there is a Guaranteed Maximum Price or Fee:

1.   In the case of net additions in the Work, the amounts of any increase in either guaranteed maximum shall be determined in accordance with Paragraphs 11.01 through 11.02, inclusive, of the General Conditions.

2.   In the case of net deletions in the Work, the amount of any such decrease shall be determined in accordance with Paragraph 11.02.C of the General Conditions, and any Guaranteed Maximum (Price or Fee) shall be reduced by mutual agreement.

ARTICLE 10 - PAYMENT PROCEDURES

10.01 Submittal and Processing of Payments

A.   Contractor shall submit Applications for Payment in accordance with Article 14 of the General Conditions. Applications for Payment will indicate the amount of Contractor's fee then payable. Applications for Payment will be processed by Engineer as provided in the General Conditions.

10.02 Progress Payments; Retainage

A.   Owner shall make progress payments on account of the Contract Price on the basis of Contractor's Applications for Payment as recommended by Engineer on or about the _____ day of each month during construction as provided in Paragraphs 10.02.A.1 and 10.02.A.2 below. All such payments will be measured by the schedule of values established as provided in Paragraph 2.07.A of the General Conditions (and in the case of Unit Price Work based on the number of units completed) or, in the event there is no schedule of values, as provided in the General Requirements:

1.   For Cost of Work: Progress payments on account of the Cost of the Work will be made:

a.   Prior to Substantial Completion, progress payments will be made in an amount equal to the percentage indicated below but, in each case, less the aggregate of payments previously made and less such amounts as Engineer may determine or Owner may withhold, including but not limited to liquidated damages, in accordance with Paragraph 14.02 of the General Conditions:

(1)   _____ percent Cost of Work completed (with the balance being retainage). If the Work has been 50 percent completed as determined by Engineer, and if the character and progress of the Work have been satisfactory to Owner and Engineer, Owner, on recommendation of Engineer, may determine that as long as the character and progress of the Work remain satisfactory to them, there will be no retainage; and

(2)   _____ percent of cost of materials and equipment not incorporated in the Work (with the balance being retainage).

b.   Upon Substantial Completion, Owner shall pay an amount sufficient to increase total payments to Contractor to _____ percent of the Work completed, less such amounts as Engineer shall determine in accordance with Paragraph 14.02.B.5 of the General Conditions and less _____ percent of Engineer's estimate of the value of Work to be completed or corrected as shown on the tentative list of items to be completed or corrected attached to the certificate of Substantial Completion.

2. For Contractor's fee: Progress payments on account of the Contractor's fee will be made:

a. If Contractor's fee is a fixed fee, payments prior to Substantial Completion will be in an amount equal to _____ percent of such fee earned to the date of the approved Application for Payment (less in each case payments previously made on account of such fee) based on the progress of the Work measured by the schedule of values established as provided in Paragraph 2.07.B of the General Conditions (and in the case of Unit Price Work on the number of units completed), and upon Substantial Completion in an amount sufficient to increase total payments to Contractor on account of his fee to _____ percent of Contractor's fee. In the event there is no schedule of values the progress of the Work will be measured as provided in the General Requirements.

b. If Contractor's fee is a percentage fee, payments prior to Substantial Completion will be in an amount equal to _____ percent of such fee (less in each case payments previously made on account of such fee) based on the Cost of the Work completed, and upon Substantial Completion in an amount sufficient to increase total payments to Contractor on account of that fee to _____ percent of Contractor's fee.

10.03 Final Payment

A. Upon final completion and acceptance of the Work in accordance with Paragraph 14.07 of the General Conditions, Owner shall pay the remainder of the Contract Price as recommended by Engineer as provided in said Paragraph 14.07.

ARTICLE 11 - INTEREST

11.01 All moneys not paid when due as provided in Article 14 of the General Conditions shall bear interest at the rate of _____ percent per annum.

ARTICLE 12 – CONTRACTOR'S REPRESENTATIONS

12.01 In order to induce Owner to enter into this Agreement Contractor makes the following representations:

A. Contractor has examined and carefully studied the Contract Documents and the other related data identified in the Bidding Documents.

B. Contractor has visited the Site and become familiar with and is satisfied as to the general, local, and Site conditions that may affect cost, progress, and performance of the Work.

C. Contractor is familiar with and is satisfied as to all federal, state, and local Laws and Regulations that may affect cost, progress, and performance of the Work.

D. Contractor has carefully studied all: (1) reports of explorations and tests of subsurface conditions at or contiguous to the Site and all drawings of physical conditions in or relating to existing surface or subsurface structures at or contiguous to the Site (except Underground Facilities) which have been identified in the Supplementary Conditions as provided in Paragraph 4.02 of the General Conditions and (2) reports and drawings of a Hazardous Environmental Condition, if any, at the Site which has been identified in the Supplementary Conditions as provided in Paragraph 4.06 of the General Conditions.

NOTE TO USER
1. If the reports and/or drawings referred to in A-12.01.D do not exist, either modify A-12.01.D or delete A-12.01.D and renumber accordingly.

E. Contractor has obtained and carefully studied (or assumes responsibility for having done so) all additional or supplementary examinations, investigations, explorations, tests, studies, and data concerning conditions (surface, subsurface, and Underground Facilities) at or contiguous to the Site which may affect cost, progress, or performance of the Work or which relate to any aspect of the means, methods, techniques, sequences, and procedures of construction to be employed by Contractor, including applying the specific means, methods, techniques, sequences, and procedures of construction, if any, expressly required by the Contract Documents to be employed by Contractor, and safety precautions and programs incident thereto.

1.   *If the reports and/or drawings referred to in A-12.01.D do not exist, delete the phrase "additional or supplementary" in the first sentence of Paragraph A-12.01.E.*

F. Contractor does not consider that any further examinations, investigations, explorations, tests, studies, or data are necessary for the performance of the Work at the Contract Price, within the Contract Times, and in accordance with the other terms and conditions of the Contract Documents.

G. Contractor is aware of the general nature of work to be performed by Owner and others at the Site that relates to the Work as indicated in the Contract Documents.

H. Contractor has correlated the information known to Contractor, information and observations obtained from visits to the Site, reports and drawings identified in the Contract Documents, and all additional examinations, investigations, explorations, tests, studies, and data with the Contract Documents.

I. Contractor has given Engineer written notice of all conflicts, errors, ambiguities, or discrepancies that Contractor has discovered in the Contract Documents, and the written resolution thereof by Engineer is acceptable to Contractor.

J. The Contract Documents are generally sufficient to indicate and convey understanding of all terms and conditions for performance and furnishing of the Work.

ARTICLE 13 - ACCOUNTING RECORDS

13.01 Contractor shall check all materials, equipment, and labor entering into the Work and shall keep such full and detailed accounts as may be necessary for proper financial management under this Agreement, and the accounting methods shall be satisfactory to Owner. Owner shall be afforded access to all Contractor's records, books, correspondence, instructions, drawings, receipts, vouchers, memoranda, and similar data relating to the Cost of the Work and Contractor's fee. Contractor shall preserve all such documents for a period of three years after the final payment by Owner.

ARTICLE 14 - CONTRACT DOCUMENTS

14.01 Contents

A. The Contract Documents consist of the following:

1. This Agreement (pages 1 to _____, inclusive).

2. Performance bond (pages _____ to _____, inclusive).

3. Payment bond (pages _____ to _____, inclusive).

4. Other bonds (pages _____ to _____, inclusive).

   a. _____ (pages _____ to _____, inclusive).

   b. _____ (pages _____ to _____, inclusive).

   c. _____ (pages _____ to _____, inclusive).

344 Appendix D

5. General Conditions (pages _____ to _____, inclusive).

6. Supplementary Conditions (pages _____ to _____, inclusive).

7. Specifications as listed in the table of contents of the Project Manual.

8. Drawings consisting of a cover sheet and _____ sheets with each sheet bearing the following general title: _____ [or] the Drawings listed on attached sheet index.

9. Addenda (numbers _____ to _____, inclusive).

10. Exhibits to this Agreement (enumerated as follows):

    a. Contractor's Bid (pages _____ to _____, inclusive).

    b. Documentation submitted by Contractor prior to Notice of Award (pages _____ to _____, inclusive).

    c. _____

11. The following which may be delivered or issued on or after the Effective Date of the Agreement and are not attached hereto:

    a. Notice to Proceed (pages _____ to _____, inclusive).

    b. Work Change Directives.

    c. Change Order(s).

B. The documents listed in Paragraph 14.01.A are attached to this Agreement (except as expressly noted otherwise above).

C. There are no Contract Documents other than those listed above in this Article 14.

D. The Contract Documents may only be amended, modified, or supplemented as provided in Paragraph 3.04 of the General Conditions.

ARTICLE 15 - MISCELLANEOUS

15.01 Terms

A. Terms used in this Agreement will have the meanings stated in the General Conditions and the Supplementary Conditions.

15.02 Assignment of Contract

A. No assignment by a party hereto of any rights under or interests in the Contract will be binding on another party hereto without the written consent of the party sought to be bound; and, specifically but without limitation, moneys that may become due and moneys that are due may not be assigned without such consent (except to the extent that the effect of this restriction may be limited by law), and unless specifically stated to the contrary in any written consent to an assignment, no assignment will release or discharge the assignor from any duty or responsibility under the Contract Documents.

15.03 Successors and Assigns

    A.   Owner and Contractor each binds itself, its partners, successors, assigns, and legal representatives to the other party hereto, its partners, successors, assigns, and legal representatives in respect to all covenants, agreements, and obligations contained in the Contract Documents.

15.04 Severability

    A.   Any provision or part of the Contract Documents held to be void or unenforceable under any Law or Regulation shall be deemed stricken, and all remaining provisions shall continue to be valid and binding upon Owner and Contractor, who agree that the Contract Documents shall be reformed to replace such stricken provision or part thereof with a valid and enforceable provision that comes as close as possible to expressing the intention of the stricken provision.

15.05 Other Provisions

NOTES TO USER
1.   *If Owner intends to assign a procurement contract (for goods and services) to Contractor, see Note to User at Article 23 of Suggested Instructions to Bidders for Procurement Contracts (EJCDC No. P-200, 2000 Edition) for provisions to be inserted in this Article.*
2.   *Insert other provisions here if applicable.*

IN WITNESS WHEREOF, Owner and Contractor have signed this Agreement in duplicate. One counterpart each has been delivered to Owner and Contractor. All portions of the Contract Documents have been signed or identified by Owner and Contractor or on their behalf.

NOTES TO USER

1. See I-21 and correlate procedures for format and signing between the two documents.

This Agreement will be effective on _____, _____ (which is the Effective Date of the Agreement).

OWNER: CONTRACTOR:

                                                _____

By:_____   By: _____

Title: _____   Title: _____

Address for giving notices: Address for giving notices:

_____   _____

_____   _____

_____   _____

(If Owner is a corporation, attach evidence of authority to sign. If Owner is a public body, attach evidence of authority to sign and resolution or other documents authorizing execution of Owner-Contractor Agreement.)

License No.: _____

(Where applicable)

Agent for service of process: _____

_____

(If Contractor is a corporation or a partnership, attach evidence of authority to sign)

Appendix E

Arrow Notation Scheduling Calculations

E.1.   CPM CALCULATIONS (ARROW NOTATION)

In making calculations with arrow notation, the arrow and its two associated nodes have attributes that are formally defined as symbols for mathematical purposes. This formal notation associated with the arrow is shown in Figure E.1.

The left-hand node on the arrow represents the event time at which the activity begins. It is referred to as the i node. The right-hand node represents the end time of the activity. It is referred to as the j node. Associated with each node is an earliest time, which is shown as T^E_i for the i node and T^E_j for the j node. Similarly, each node can have a latest event time, which is shown in the figure as T^L_i for the i node and T^L_j for the j node. This establishes four events, two associated with starting and two with ending nodes, which are of interest in calculating the critical path of the network. The duration of the activity (as shown in the figure) is given as t_{ij}. Because the starting and ending nodes in arrow notation are referred to as i and j, arrow notation is sometimes referred to as i-j notation.

A schematic diagram representing the application of the forward-pass algorithm using arrow notation is shown in Figure E.2. The objective of the forward-pass algorithm is to calculate the earliest point in time at which a given event can occur. That is, the algorithm calculates the earliest event time of a given node. The earliest event time for a given node is controlled by the earliest event times of each of the set of events that precede it. The algorithm is given as follows:

$$T^E_j = \max_{\substack{\text{All } i \\ i \in M}} [T^E_i + t_{ij}]$$

where M is the set of all i events that immediately precede j.

The earliest event time for a given node j is controlled by the earliest event times of each of the i nodes that precede it. Each i node plus the duration of the associated activity, t_{ij} which links it to the j node, must be investigated. The maximum of the preceding i node early event times plus the durations of the appropriate activity ij controls the earliest time at which a given event j, can occur.

To demonstrate this, consider Figure E.2. Node 30 is preceded by nodes 22, 25, and 26. The durations of the activities emanating from each of these nodes are as follows:

$$
\begin{array}{ll}
\text{Act } 22,30 & t_{22,30} = 7 \text{ days} \\
\text{Act } 25,30 & t_{25,30} = 2 \text{ days} \\
\text{Act } 26,30 & t_{26,30} = 6 \text{ days}
\end{array}
$$

The earliest event times for each of the preceding nodes are as follows:

$$
\begin{array}{l}
T^E_{22} = 10 \\
T^E_{25} = 13 \\
T^E_{26} = 15
\end{array}
$$

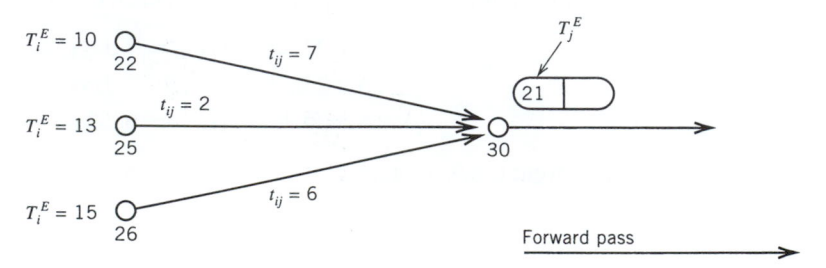

Figure E.1 Arrow notation symbols.

The algorithm for the early event time for node 30 is then

$$T^E{}_{30} = \max(T^E{}_{22} + t_{22,30}, T^E{}_{25} + t_{25,30}, T^E{}_{26} + t_{26,30})$$

or

$$T^E{}_{30} = \max(10 + 7, 13 + 2, 15 + 6) = 21$$

E.2. CALCULATING THE EARLY EVENT TIMES (ARROW NOTATION)

In order to understand how the forward-pass algorithm is applied, consider the arrow notation network model of the small gas station in Figure E.3. In order to record the calculated values of the early event time ($T^E{}_i$), a partitioned oval is located above each node. The calculated early event time is recorded in the left side of the oval. During the backward pass, the late event times for each node will be recorded in the right side.

The forward-pass algorithm is applied repetitively starting with the source node (node A) and moving from left to right in a "bootstrapping" fashion. The starting node A is given an early event time of zero (0). Moving to node B, the set of preceding events consists of only one event. Therefore, $T^E{}_B$ is max ($T^E{}_A + t_{AB}$) = max ($0 + 10$) = 10. Calculations for all of the nodes are shown in Table E.1. The values for each node are shown in Figure E.3.

The earliest time at which each activity can begin is given by the $T^E{}_i$ value for the i node associated with the activity of interest. In addition to this information, it is now clear that the minimum duration of the project is 96 days since the earliest time at which node S can be realized has been calculated as 96 time units.

E.3. BACKWARD-PASS ALGORITHM (ARROW NOTATION)

A schematic diagram representing the application of the backward-pass algorithm is shown in Figure E.4. The backward-pass algorithm calculates the latest time at which each event can occur. The latest event time for a node i is controlled by the latest event times of the set of events that follow it. The late event time of each j node minus the duration of the associated activity, ij, must be investigated. The minimum of the following j node late event times minus the duration of activity ij controls the latest time at which the i event can occur. To demonstrate this, consider Figure E.4.

Figure E.2 Schematic of forward-pass calculation.

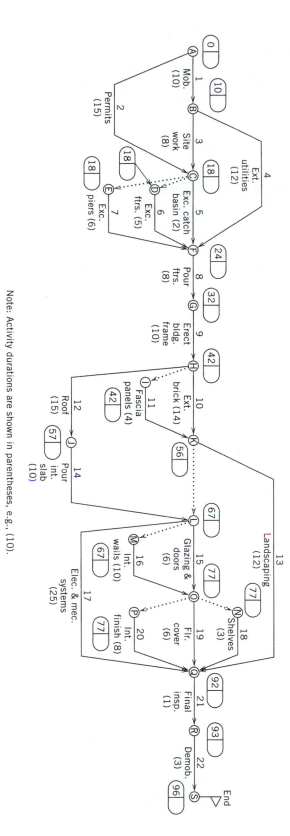

Figure E.3 Expanded Project Model with Early Event Times.

Note: Activity durations are shown in parentheses, e.g., (10).

Table E.1 Calculation of Early Event Times

Node	Formula	Numerical Value	T_i^E
A	N/A	N/A	0
B	$T_B^E = \max(T_A^E + t_{AB})$	$\max(0 + 10)$	10
C	$T_C^E = \max(T_B^E + t_{BC}, T_A^E + t_{AC})$	$\max(10 + 8, 0 + 15)$	18
D	$T_D^E = \max(T_C^E + t_{CD})$	$\max(18 + 0)$	18
E	$T_E^E = \max(T_C^E + t_{CE})$	$\max(18 + 0)$	18
F	$T_F^E = \max(T_B^E + t_{BF}, T_C^E + t_{CF}, T_D^E + t_{DF},$ $T_E^E + t_{EF})$	$\max(10 + 12, 18 + 2, 18 + 5, 18 + 6)$	24
G	$T_G^E = \max(T_F^E + t_{FG})$	$\max(24 + 8)$	32
H	$T_H^E = \max(T_G^E + t_{GH})$	$\max(32 + 10)$	42
I	$T_I^E = \max(T_H^E + t_{HI})$	$\max(42 + 0)$	42
J	$T_J^E = \max(T_H^E + t_{HJ})$	$\max(42 + 15)$	57
K	$T_K^E = \max(T_H^E + t_{HK}, T_I^E + t_{IK})$	$\max(42 + 14, 42 + 4)$	56
L	$T_L^E = \max(T_J^E + t_{JL}, T_K^E + t_{KL})$	$\max(57 + 10, 56 + 0)$	67
M	$T_M^E = \max(T_L^E + t_{LM})$	$\max(67 + 0)$	67
N	$T_N^E = \max(T_O^E + t_{ON})$	$\max(77 + 0)$	77
O	$T_O^E = \max(T_L^E + t_{LO}, T_M^E + t_{MO})$	$\max(67 + 6, 67 + 10)$	77
P	$T_P^E = \max(T_O^E + t_{OP})$	$\max(77 + 0)$	77
Q	$T_Q^E = \max(T_K^E + t_{KQ}, T_N^E + t_{NQ}, T_O^E + t_{OQ},$ $T_P^E + t_{PQ}, T_L^E + t_{LO})$	$\max(56 + 12, 77 + 3, 77 + 6, 77 + 8, 67 + 25)$	92
R	$T_R^E = \max(T_Q^E + t_{QR})$	$\max(92 + 1)$	93
S	$T_S^E = \max(T_R^E + t_S)$	$\max(93 + 3)$	96

Node i, labeled 18, is followed by nodes 21, 23, and 25. The durations of the associated ij activities are:

$$
\begin{aligned}
\text{Act } 18,21 \quad & t_{18,21} = 12 \\
\text{Act } 18,23 \quad & t_{18,23} = 3 \\
\text{Act } 18,25 \quad & t_{18,25} = 10
\end{aligned}
$$

The latest event times for each of the following nodes are as follows:

$$
\begin{aligned}
T^L_{21} &= 52 \\
T^L_{23} &= 46 \\
T^L_{25} &= 53
\end{aligned}
$$

The expression for the late event time of node 18 is:

$$
T^L_{18} = \min(52 - 12, 46 - 3, 53 - 10) = 40
$$

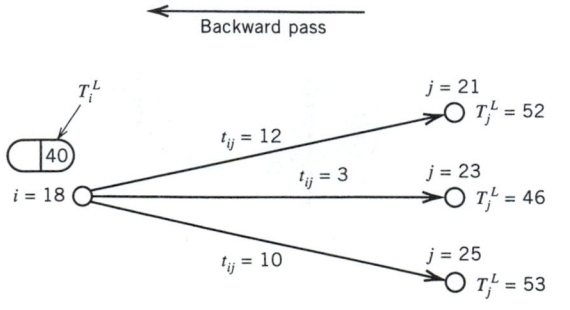

Figure E.4 Schematic of backward-pass algorithm.

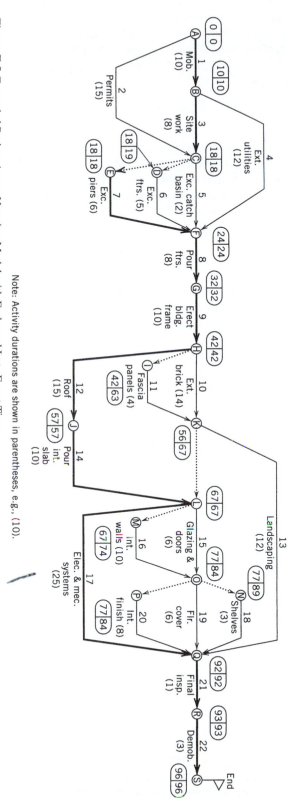

Figure E.5 Expanded Project Arrow Notation Model with Early and Late Event Times.

Note: Activity durations are shown in parentheses, e.g., (10).

Again considering the small gas station arrow notation network, the same bootstrapping approach used to work stepwise through the network is used to determine late event times (see Figure E.5). In this case we start at the last node (right side) of the network and step from right to left. In order to commence calculations, a late event time is needed for node S. The late event time of S will be set to the early event time (i.e., 96). Since we wish to complete the project in the minimum period of time, it is logical to set the early and late event times of S to 96. If we were to use a greater number, the finish date for the project would be extended. Since 96 time units (e.g., days) is the minimum duration of the project, a lesser value is not feasible.

Walking the network from right to left, we would start with node R. Node R is followed only by node S. The backward-pass algorithm reduces to:

$$T^L_R = \min(96 - 3) = 93$$

Similarly, the late event time of Node Q can be calculated as 92. Again, the late event times for nodes N and P can be calculated as:

$$T^L_N = \min(92 - 3) = 89$$
$$T^L_P = \min(92 - 8) = 84$$

Node O is followed by nodes N, and Q. Therefore, its late event time is calculated as:

$$T^L_O = \min(89 - 0, 84 - 0, 92 - 6) = 84$$

All of the late event times are shown in the right side of the ovals above each node in Figure E.5. It should be noted that the late event time for the beginning node A is 0. The early and late event times of the source or beginning node must be equal (i.e., zero). Otherwise, a mistake in calculating the backward-pass values has occurred.

E.4. IDENTIFYING THE CRITICAL PATH

As previously stated, the set of critical activities that form the critical path(s) cannot be delayed without causing an extension of the project duration. Therefore, they can be identified as the activities that have early and late event times (associated with their i and j nodes) that are equal. Activities with i and j node early and late times that are not equal can be delayed a certain amount without extending the duration of the project. By looking at Figure E.5 it can be determined that the following activities are critical.

AB, BC, CE, EF, FG, GH, HJ, JL, LQ, QR, RS

This is both the critical and longest path through the network. The duration of this path must be equal to the minimum project duration calculated using the forward-pass algorithm. All other paths will have durations that are less than the minimum project duration. Check to see that the critical path has a duration of 96 and that all other paths have a duration less than 96 time units.

Appendix F

AGC Builders Association of Chicago: Typical Agent Job Descriptions

PROJECT MANAGER

A. General Functions

The project manager in the construction industry is usually an "inside" and "outside" man. The position may vary considerably from company to company. The project manager in some companies may be an estimator, and expeditor, and even handle some duties normally done by the job superintendent, while with other companies he may merely supervise superintendents.

B. Detailed Functions

1. May procure the invitation to bid on jobs.
2. May, when working as an estimator, prepare bids.
3. May handle the legal requirements for a contract.
4. May negotiate the specialty contractor's arrangements and agreements.
5. Set up completion schedules by bar graph or critical path method.
6. Supervise subcontractors and coordinate their material deliveries.
7. Arrange for sufficient manpower for the project.
8. Supervise superintendents on the job—"walk the job" each day to see progress being made and, during this time, review the work with a superintendent.
9. Control the movement of workers from one job to another.
10. Arrange for permits from the city, county, and so forth.
11. Hire and fire superintendents, foremen, engineers, and other personnel under his supervision.
12. Set up occupancy dates for buildings.
13. Act as public relations representative.
14. Coordinate with architect and owner requested revisions or errors found in drawings.

ESTIMATOR

A. General Functions

An estimator makes as close an estimate as possible of what the costs will be. In order to do so, he must itemize all of the building materials and calculate labor costs for the entire project—the cost estimate may also include a percentage for profit, though this may be done with or by top management.

B. Detailed Functions

1. Mail or telephone bid proposals to subcontractors.
2. Follow up with subcontractors on submission of their bids.
3. Review bid with subcontractors.
4. Prior to bid, inspect job site to determine access and that the land is the same as on the plans. Look for water conditions and other problems that might arise.
5. Analyze plans and specifications, that is, "learn the job."
6. Make a takeoff for each type of work to be done by general contractor forces.
7. Does takeoff for subcontractors when necessary.
8. May sit in on owner, architect, and contractor conferences.
9. May check on other estimator's work or have his work checked by another estimator.
10. Price the quantity takeoffs.
11. Read prints, noting discrepancies.
12. Make itemized lists of prices for materials.
13. Review and preview subcontractors' bids.
14. In some companies, purchase steel, lumber, and all other materials necessary for the job.
15. Compute a percentage for overhead and profit, which would be added to estimated cost.
16. Arrive at final bid price or cost price.
17. Prepare change order (estimates cost of changes) as needed or required—if major item and not handled by field personnel.
18. Expedite distribution of plans, including general and mechanical.
19. Serve in quality control capacity, because of position in purchasing, and plan review.
20. Make bar graphs, network, or CPM for scheduling.
21. Make cost breakdown of work performed by company forces for cost control purposes.

EXPEDITOR

A. General Functions

An expeditor may schedule or coordinate job material requirements. He serves as a troubleshooter when there is a breakdown in delivery schedule. He foresees problems by reviewing plans and specifications of the subcontractors and coordinating these with the plans and specifications of the architect.

B. Detailed Functions

1. In some companies receives the plans and specifications and breaks the specifications down by trade.

2. In some companies writes to all subcontractors advising them what is necessary to do on their plans.

3. Follows up on drawings, that is, shop drawings or the detailed drawings of project.

4. Submits drawings to the architect after having checked them to see if they match, that the job is correct, that the materials used are those specified, and analyzes the drawings.

5. Maintains constant follow-up on plans and drawings to ensure documents reach the proper place at the proper time.

6. Distributes approved plans to subs or to anyone else who should get them. Has to order enough plans from subcontractors so that entrusted parties will have sufficient documentation for project schedule.

7. In some companies establishes delivery time for materials, equipment, or labor, based on when they will be required and when they can be acquired, and determines the lead time required for acquisition.

8. May follow a CPM printout, make out delivery schedules, use a bar graph method or the critical path method. Makes sure the shop items are on the critical path method or the bar graph.

9. In some companies does small buying such as purchasing mailboxes, signs, and finish items.

10. Maintains constant follow-up to ensure that schedule is accomplished.

11. Checks all incoming tests to ensure they meet specifications.

12. In some companies checks the concrete design, that is, the mix or fixed formula of the concrete used.

13. Plans material delivery and schedules with job superintendents.

14. Keeps in contact each day with subcontractors.

15. Writes memos as needed to architects, superintendents, subcontractors, and so forth.

16. Follows up daily on trouble areas, that is, those places where delivery of materials may be lagging.

17. In some companies accumulates change order information.

18. Generally troubleshoots, especially for delivery problems.

EQUIPMENT SUPERINTENDENT

A. General Functions

Maintains and repairs equipment owned by company. To do this, he supervises garage and yard personnel and coordinates delivery of equipment to the specific job sites and expedites repairs and deliveries.

B. Detailed Functions

1. Supervises, maintains, and repairs.

2. Purchases parts for maintenance and repairs.

3. Expedites repairs of equipment that cannot be done in the company garage.

4. Keeps detailed records of equipment, including maintenance costs for each piece of equipment.

5. Keeps track of equipment, that is, which job is using it at any given time.

6. Makes recommendations concerning purchase of new equipment.

7. Provides delivery of equipment to job sites, helps to plan the time, provides means of delivery to the site, and provides equipment setup at site.

8. Keeps weekly repair costs on his crew.

9. Prepares an annual budget for operation.

FIELD SUPERINTENDENT

A. General Functions

Builds the building. Manages men and materials on the job site so that the project is built for profit. Coordinates schedules so that men and materials are available to promote efficient erection of the building at a profit level.

B. Detailed Functions

1. "Learns the building." Studies plans and specifications so that he can plan the work to be accomplished.

2. Tries to anticipate problems.

3. Studies the costs.

4. Arranges scheduling and manufacture of building parts or components.

5. Coordinates building when the manufactured items become available for the building.

6. Does survey and layout work or supervises technical or field engineer who does this.

7. Keeps constant check on all trades, overseeing workmanship and materials.

8. Hires and fires workmen.

9. Supplies information to accounting department so that records of costs can be maintained.

10. Supervises men directly or indirectly (i.e., through the foreman).

11. May be responsible for deliveries.

12. Is responsible for drawings and seeing that drawings are made of changes or incomplete items.

13. Arranges for plan changes as needed.

14. May be responsible for written schedules or physical schedules.

15. Does on-the-spot estimating (material or labor).

16. May price out extra items or charges.

17. Does limited buying (supplies and items missed by the purchasing department).

18. Makes daily safety inspections.

19. May record daily field activities in a log.

MECHANICAL SUPERINTENDENT

A. General Functions

The mechanical superintendent coordinates subcontractor's work with that of the general contractor to ensure that project remains on schedule and quality is maintained.

B. Detailed Functions

1. Compiles listing of major mechanical electrical equipment required.
2. Expedites shop drawings and equipment deliveries.
3. Assists in preparation of project schedules.
4. Prepares weekly progress reports on electrical and mechanical work.
5. Coordinates subcontractors' work with general contractor.
6. Checks schedule to ensure project is on schedule.
7. Supervises general contractor's work done for subcontractors (equipment production, excavations, etc.).
8. Processes and distributes shop drawings.
9. Supervises, inspects, and evaluates work performed by subcontractors—ensures there is compliance with plans and specifications.
10. Supervises project closely to ensure that the owner is getting his money's worth on subcontractor work.

SCHEDULING ENGINEER (FIELD ENGINEER)

A. General Functions

Scheduling engineer schedules and coordinates. He serves as a troubleshooter when there is a breakdown in delivery schedule. He maintains a constant follow-up on the schedule to ensure progress as previously planned.

B. Detailed Functions

1. Receives plans and specifications and breaks them down by trade.
2. Writes to all subcontractors telling them when their work is necessary on the schedule.
3. Expedites follow-up for drawings, that is, shop drawings or detailed drawings. Checks with own staff for follow-up.
4. Keeps a close follow-up to ensure that plans and drawings reach the right people at the right time.
5. Establishes delivery times for materials, equipment, or labor, based on when they can be acquired, and determines the lead time required for acquisitions.
6. Makes out delivery schedules, using a bar graph method or critical path method. Makes sure that the shop items are on the bar graph or CPM.
7. Discusses material delivery and scheduling with job superintendent.
8. Keeps in touch with subcontractors as needed.
9. Writes memos as needed to superintendents, subcontractors, and so forth.

10. Follows up daily on trouble areas, where delivery of materials may be lagging.

11. Generally troubleshoots.

TIMEKEEPER

A. General Functions

A timekeeper is primarily concerned with maintaining cost control of labor force on a project. He maintains payroll records and may also maintain records on material deliveries.

B. Detailed Functions

1. Ensures that the men are on the job, checks what specific tasks they are performing, and checks this against job sheets given to him daily by the foreman.

2. Checks with the foreman to determine exact job and classification of work each man is doing so that the work can be coded and entered against the correct amount.

3. Walks the job a few times each day.

4. Computes previous day's work sheets to obtain costs.

5. Projects daily costs to determine if work was completed within the allocated budget.

6. Talks over costs with superintendent.

7. Posts workers' hours to the payroll on a daily basis.

8. Types a cost report each week. In some companies this may be done by central office staff.

9. Types payroll each week. In some companies this may be done by central office staff.

10. Types paychecks each week. In some companies this may be done by central office staff.

11. Types all back charges and time tickets.

12. Estimates costs of requests from subcontractors for sheds, shanties, carpenters, and concrete work performed for them.

13. May compile subcontractors' invoices for payment and discuss these with architect to determine accuracy.

14. Codes all delivery tickets to maintain costs on all building parts.

15. Keeps records of all reinforcing steel deliveries.

16. Records all concrete pours.

17. May assist superintendent by ordering labor, lumber, and other materials.

18. On certain big load days may call the union halls for extra men; will sign these men for the day and pay them by check at night.

19. Signs up all new workmen (W-4 forms, applications, etc.) and submits originals to central office.

20. Enters new employees' names and proper wage rate for the particular trade on payroll.

21. Types monthly report on welfare and pension. In some companies this may be done at the central office.

22. Balances the payroll and types it each Monday. Submits it to the main office so that checks can be made out and returned to the job site by Wednesday. In some companies this may be done at central office.

23. May travel to various job sites and perform same duties for each of the projects.

24. On projects involving federal funds, he collects payroll data from subcontractors for submission to the government in compliance with their regulations.

25. Maintains time record on company truck drivers when material deliveries are made.

26. May supervise "time checkers" on larger project.

Appendix G

AGC Standard Form Construction Subcontract

THE ASSOCIATED GENERAL CONTRACTORS OF AMERICA

INSTRUCTIONS FOR COMPLETION OF
AGC DOCUMENT NO. 650
STANDARD FORM OF AGREEMENT
BETWEEN CONTRACTOR AND SUBCONTRACTOR
(Where the Contractor Assumes the Risk of Owner Payment)
This document is endorsed by The Associated Specialty Contractors, Inc. (ASC)

1998 EDITION (FINAL)

The Standard Form of Agreement Between Contractor and Subcontractor (Where the Contractor Assumes the Risk of Owner Payment), AGC Document No. 650 (AGC 650), has been written to be generally compatible with both The American Institute of Architects General Conditions of the Contract for Construction, A201-1997, and AGC Document No. 200, Standard Form of Agreement and General Conditions Between Owner and Contractor (Where the Contract Price is a Lump Sum), 1997 Edition. In this document, AGC 650, payment to the Subcontractor is not conditioned on the Contractor having received, from the Owner, payment for Subcontract Work satisfactorily performed. See the discussion in these Instructions of Article 8, below. AGC Document No. 655, Standard Form of Agreement Between Contractor and Subcontractor (Where the Contractor and Subcontractor Share the Risk of Owner Payment), can be used when conditioned payment is valid in the jurisdiction and elected by the parties.

This document replaces AGC Document No. 650i, which was published on an interim basis for use until publication of this final version. AGC 650 benefited from an inclusive development process in which contractors, subcontractors and others offered comments and constructive feedback on its language.

Among the participants was The Associated Specialty Contractors, Inc. (ASC), an umbrella organization composed of the following eight specialty contractor groups: Mason Contractors Association of America, Mechanical Contractors Association of America, National Electrical Contractors Association, National Insulation Association, National Roofing Contractors Association, Painting and Decorating Contractors of America, Plumbing-Heating-Cooling Contractors—National Association, and Sheet Metal and Air Conditioning Contractors' National Association.
ASC has approved and endorsed this document, AGC 650.

SPECIAL NOTE: Care should be taken to ensure that the terms of this Subcontract Agreement coordinate with the general conditions of the prime contract. Particular attention should be paid to coordinating indemnity and insurance provisions.

GENERAL INSTRUCTIONS

Standard Form

These instructions are for the information and convenience of the users of AGC 650, 1998 Edition. They are not part of the Agreement nor a commentary on or interpretation of the contract form. It is the intent of the parties to a par-

ticular agreement that controls its meaning and not that of the writers and publishers of the standard form. As a standard form, this Agreement has been designed to establish the relationship of the parties in the standard situation. Recognizing that every project is unique, modifications may be required. See the following recommendations for modifications.

Legal and Insurance Counsel

This Agreement has important legal and insurance consequences. Consultation with an attorney and an insurance adviser is encouraged with respect to its completion or modification.

COMPLETING THE AGREEMENT

Completing Blanks

Diamonds in the margins indicate provisions requiring the parties to fill in blanks with information.

Modifications

Supplemental conditions, provisions added to the printed agreement, may be adopted by reference. It is always best for supplements to be attached to the agreement. Provisions in the printed document that are not to be included in the agreement may be deleted by striking through the word, sentence or paragraph to be omitted. It is recommended that unwanted provisions not be blocked out so that the deleted materials are illegible. The parties should be clearly aware of the material deleted from the standard form.

It is a good practice for both parties to sign and date all modifications and supplements.

Photocopying the Completed Agreement

The purchaser of this copyrighted document may make up to nine (9) photocopies of a completed document, whether signed or unsigned, for distribution to appropriate parties in connection with a specific project. Any other reproduction of this document in any form is strictly prohibited, unless the purchaser has obtained the prior written permission of The Associated General Contractors of America.

OBTAINING ADDITIONAL INFORMATION

To obtain additional information about AGC standard form contract documents and the AGC Contract Documents Program, contact AGC at 333 John Carlyle Street, Suite 200, Alexandria, VA 22314; phone (703) 548-3118; fax (703) 548-3119, or visit AGC's web site at www.agc.org.

AGC 650

The provisions of this standard form Agreement are organized by article closely following the order in the AGC 200 and approximately in the same order as in the AIA A201-1997 to facilitate determining the coordination of these documents. For example, the responsibilities of the parties are in the early articles; time, price, changes and payment are discussed in that order, and dispute resolution and miscellaneous provisions are in the last few articles.

Article 1 AGREEMENT

The date of the Agreement and identification of the parties and the Project are essential information to be accurately inserted in this article.

Article 2 SCOPE OF WORK

The relationship of the parties, the extent of the Agreement, and the definitions of some terms are described in this Article.

2.1 This Paragraph obligates the Subcontractor to perform the Subcontract Work in accordance with and reasonably inferable from that "indicated" in the Subcontract Documents, consistent with the Progress Schedule, and under the general direction of the Contractor.

Article 3 SUBCONTRACTOR'S RESPONSIBILITIES

3.1 This Paragraph addresses mutuality in the flow-down of rights and responsibilities.

3.3 This Paragraph addresses the Subcontractor's responsibilities for review and analysis of the Subcontract Documents. The Subcontractor's comparison of drawings, specifications, and other documents is "solely for the purpose of facilitating the Subcontract Work" and not for the purpose of discovery of errors, inconsistencies or omissions."

3.4 This Paragraph addresses site visitation. Prior to performing the Work, the Subcontractor is responsible for visiting the Project site and "shall conduct a visual inspection of the Project site to become generally familiar with local conditions and to correlate site observations with the Subcontract Documents."

3.8 Design delegation is an important new concept in the AIA A201-1997. This Paragraph addresses design delegation consistent with the A201-1997. Whether or not design has been delegated will be determined in the relevant drawings and specifications. The Designer is retained by the Subcontractor as permitted by the law of the jurisdiction. The Designer is named in Subparagraph 3.8.2. See Subparagraph 9.2.3 for the Professional Liability Insurance issues related to the delegated design.

3.11 The Subcontractor designates a representative.

3.13 This Paragraph states the Subcontractor's responsibility for cleanup relative to the Contractor's and others' responsibilities.

3.14 The importance of the safe performance of the Sub-contractor's Work is emphasized in this section. The Sub-contractor is required to designate an individual at the site as the Subcontractor's safety representative (3.14.6). Estab-lishment of a safety program by the Contractor does not relieve the Subcontractor of its safety responsibilities (3.14.9). The Contractor and Subcontractor assume a mutual indemnification obligation for fines or penalties aris-ing from safety violations (3.14.9).

3.22.2 This Subparagraph addresses the Subcontractor's responsibility to correct defective work.

3.27 Subcontract Bond requirements, if required, are inserted in blanks at this Paragraph.

3.30 Confidentiality provisions in the Owner-Contractor agreement flow down to the Subcontractor.

3.32 Provisions relative to labor relations must be inserted or incorporated by reference. Legal counsel is rec-ommended. If there are no labor provisions, the parties should so indicate in order that it is clear their omission was not an oversight.

Article 4 CONTRACTOR'S RESPONSIBILITIES

4.1 The Contractor designates a representative.

4.2 The Contractor indicates whether a payment bond has been provided the Owner. If so, the bond will be made available for the Subcontractor to review and copy.

4.3 The Owner's ability to pay is, obviously, crucial to the successful completion of the Project. It also can be the source of problems between the Contractor and Subcon-tractor. These provisions give the Subcontractor the right to request the information the Contractor has about the Project financing. If the Subcontractor does not receive information about the Owner's ability to pay as required in the Contract Documents, the Subcontractor may request the information from the Owner and/or the Owner's lender. The Subcon-tractor also has a right to request and receive any changes in the information relating to the Owner's financial capability.

AGC Document No. 690, *Guidelines for Obtaining Owner Financial Information, AGC Document No. 690.1, Owner Financial Questionnaire,* and AGC 3574, Private Work: Man-aging the Risks, a joint publication with the National Associ-ation of Surety Bond Producers, offer help on assessing and managing the financial risks inherent in contracting with pri-vate owners.

Article 5 PROGRESS SCHEDULE

5.2 Although the Contractor prepares the schedule for performance of the work, the Progress Schedule provision requires communication between the Subcontractor and the Contractor with regard to the schedule.

5.3 There is a distinction between Claims Relating to the Owner (5.3.2) and Claims Relating to the Contractor (5.3.4).

5.4 The Contractor and Subcontractor waive claims for consequential damages. Similarly, the Subcontractor shall obtain from its sub-subcontractors mutual waivers of conse-quential damages that correspond to the Subcontractor's waiver of consequential damages herein. Also, this Para-graph provides, "to the extent the Owner-Contractor agree-ment provides for a mutual waiver of consequential damages by the Owner and the Contractor, damages for which the Contractor is liable to the Owner, including those related to Subparagraph 9.1.1 are not consequential damages for the purpose of this waiver. Similarly, to the extent the Subcon-tractor-sub-subcontractor agreement provides for a mutual waiver of consequential damages by the Owner and the Contractor, damages for which the Subcontractor is liable to lower-tiered parties due to the fault of the Owner or Con-tractor are not consequential damages for the purpose of this waiver."

Article 6 SUBCONTRACT AMOUNT

The Subcontract Amount provisions offer specific options: fixed price, unit prices, or time and material rates.

Article 7 CHANGES IN THE SUBCONTRACT WORK

7.2 There is provision for the Construction Change Directive concept, to the extent it is provided for in the Sub-contract Documents.

7.3 This Paragraph addresses the Subcontractor's responsibility for unknown conditions encountered at the site.

Article 8 PAYMENT

8.1 The Subcontractor is required to submit a schedule of values satisfactory to the Contractor.

There are provisions for Retainage (8.2.2), and Stored Materials (8.2.4).

8.2.5 This Subparagraph does not condition the Subcon-tractor's payment on receipt of payment by the Contractor from the Owner. Progress payments to the Subcontractor are to be made within seven days after the Contractor receives payment from the Owner for the Subcontractor's Work. If, through no fault of the Subcontractor, the Owner does not pay the Contractor for the Subcontractor's Work, the Contractor assumes that liability "within a reasonable time." A "reasonable time" enables the Contractor to attempt to secure the payment from the Owner; what is "reasonable" for a particular project will depend on a variety of project-specific factors.

8.2.7 The Subcontractor's application for payment may be rejected for any of the seven reasons indicated. The Contractor must give the Subcontractor written notice of the specific reasons for disapproving an application.

8.3 The Final Payment provisions at Subparagraphs 8.3.3 and 8.3.4 are consistent with the Progress Payment provisions.

8.6 This Paragraph addresses payment use restrictions placed on the Contractor and Subcontractor.

8.8 This Paragraph addresses partial lien waivers and affidavits and states that the Subcontractor shall not be compelled to provide an unconditional waiver of lien or claim prior to receiving payment.

Article 9 INDEMNITY, INSURANCE AND WAIVER OF SUBROGATION

9.1 This Paragraph addresses claims for bodily injury and property damage on a comparative fault basis. Contractual indemnification is governed by state law. The states differ as to the types of indemnification agreements they will enforce. Consultation with insurance and legal counsel with knowledge of the jurisdiction is recommended.

9.2.2 The minimum limits of the Subcontractor's liability insurance are to be provided in an exhibit to this Agreement.

9.2.3 If design has been delegated to the Subcontractor as described in Paragraph 3.3, professional liability insurance coverage is provided as defined in this Paragraph.

9.2.6 Completed Operations coverage requirements are described in this Paragraph.

Article 10 CONTRACTOR'S RIGHT TO PERFORM SUBCONTRACTOR'S RESPONSIBILITIES AND TERMINATION OF AGREEMENT

10.1 This Paragraph governs the Contractor's recourse when the Subcontractor fails to perform, including notice to cure and termination of the Subcontractor by the Contractor for cause. If the Contractor performs work under these provisions or subcontracts its performance, it has the right to use the Subcontractor's materials and equipment at the project site to complete the Subcontractor's Work (10.1.3).

10.7 This Paragraph includes reasonable overhead and profit on the Subcontract Work not executed, and other costs incurred by reason of such action, in the Subcontractor's recoverable costs for the Contractor's wrongful exercise of rights.

Other provisions govern the event of Bankruptcy (10.2), Suspension by the Owner (10.3), Termination by the Owner (10.4), the Contingent Assignment of this Agreement (10.5), and Suspension by the Contractor (10.6).

Article 11 DISPUTE RESOLUTION

The parties are encouraged to settle their disputes through direct discussions. If these discussions are not successful, the subcontract provides for mediation as a condition precedent to any other form of binding dispute resolution. Any disputes not resolved by mediation are to be decided by the dispute resolution process selected in the agreement between the Owner and the Contractor.

11.5 This Paragraph addresses Disputes Between Contractor and Subcontractor.

Article 12 MISCELLANEOUS PROVISIONS

These provisions include the Governing Law (12.1), Severability (12.2), and Other Provisions and Documents (12.5).

Article 13 EXISTING SUBCONTRACT DOCUMENTS

The exhibits in this Article may include the Subcontract Work, Drawings, Specifications, General and other conditions, the Progress Schedule, alternates and unit prices, temporary services, insurance or others as appropriate.

AGC DOCUMENT NO. 650 • STANDARD FORM OF AGREEMENT BETWEEN CONTRACTOR AND SUBCONTRACTOR
(Where the Contractor Assumes the Risk of Owner Payment)
© 1998, The Associated General Contractors of America

HBP

THE ASSOCIATED GENERAL CONTRACTORS OF AMERICA

AGC DOCUMENT NO. 650
STANDARD FORM OF AGREEMENT
BETWEEN CONTRACTOR AND SUBCONTRACTOR
(Where the Contractor Assumes the Risk of Owner Payment)
This document is endorsed by The Associated Specialty Contractors, Inc. (ASC)

The Associated Specialty Contractors, Inc. (ASC) is an umbrella organization composed of the following eight specialty contractor groups: Mason Contractors Association of America, Mechanical Contractors Association of America, National Electrical Contractors Association, National Insulation Association, National Roofing Contractors Association, Painting and Decorating Contractors of America, Plumbing-Heating-Cooling Contractors-National Association, and Sheet Metal and Air Conditioning Contractors' National Association. ASC has approved and endorsed this document, AGC 650.

TABLE OF ARTICLES

This Agreement has important legal and insurance consequences. Consultation with an attorney and an insurance consultant is encouraged with respect to its completion or modification.

AGC DOCUMENT NO. 650
STANDARD FORM OF AGREEMENT
BETWEEN CONTRACTOR AND SUBCONTRACTOR
(Where the Contractor Assumes the Risk of Owner Payment)

ARTICLE 1

AGREEMENT

This Agreement is made this _____ day of _____, ◆

in the year _____, by and between the ◆

CONTRACTOR ◆
(Name and Address)

and the
SUBCONTRACTOR ◆
(Name and Address)

for services in connection with the
SUBCONTRACT WORK ◆

for the following
PROJECT ◆

whose
OWNER is ◆
(Name and Address)

The **ARCHITECT/ENGINEER** for the Project is ◆
(Name and Address)

Notice to the parties shall be given at the above addresses.

2

ARTICLE 2

SCOPE OF WORK

2.1   SUBCONTRACT WORK The Contractor contracts with the Subcontractor as an independent contractor to provide all labor, materials, equipment and services necessary or incidental to complete the work described in Article 1 for the Project in accordance with, and reasonably inferable from, that which is indicated in the Subcontract Documents, and consistent with the Progress Schedule, as may change from time to time. The Subcontractor shall perform the Subcontract Work under the general direction of the Contractor and in accordance with the Subcontract Documents.

2.2   CONTRACTOR'S WORK The Contractor's work is the construction and services required of the Contractor to fulfill its obligations pursuant to its agreement with the Owner (the Work). The Subcontract Work is a portion of the Work.

2.3   SUBCONTRACT DOCUMENTS The Subcontract Documents include this Agreement, the Owner-Contractor agreement, special conditions, general conditions, specifications, drawings, addenda, Subcontract Change Orders, amendments and any pending and exercised alternates. The Contractor shall make available to the Subcontractor, prior to the execution of the Subcontract Agreement, copies of the Subcontract Documents to which the Subcontractor will be bound. The Subcontractor similarly shall make copies of applicable portions of the Subcontract Documents available to its proposed subcontractors and suppliers. Nothing shall prohibit the Subcontractor from obtaining copies of the Subcontract Documents from the Contractor at any time after the Subcontract Agreement is executed. The Subcontract Documents existing at the time of the execution of this Agreement are set forth in Article 13.

2.4   CONFLICTS In the event of a conflict between this Agreement and the other Subcontract Documents, this Agreement shall govern.

2.5   EXTENT OF AGREEMENT Nothing in this Agreement shall be construed to create a contractual relationship between persons or entities other than the Contractor and Subcontractor. This Agreement is solely for the benefit of the parties, represents the entire and integrated agreement between the parties, and supersedes all prior negotiations, representations, or agreements, either written or oral.

2.6   DEFINITIONS

.1   Wherever the term *Progress Schedule* is used in this Agreement, it shall be read as Project Schedule when that term is used in the Subcontract Documents.

.2   Whenever the term *Change Order* is used in this Agreement, it shall be read as Change Document when that term is used in the Subcontract Documents.

.3   Unless otherwise indicated, the term *Day* shall mean calendar day.

ARTICLE 3

SUBCONTRACTOR'S RESPONSIBILITIES

3.1   OBLIGATIONS The Contractor and Subcontractor are hereby mutually bound by the terms of this Subcontract. To the extent the terms of the prime contract between the Owner and Contractor apply to the work of the Subcontractor, then the Contractor hereby assumes toward the Subcontractor all the obligations, rights, duties, and redress that the Owner under the prime contract assumes toward the Contractor. In an identical way, the Subcontractor hereby assumes toward the Contractor all the same obligations, rights, duties, and redress that the Contractor assumes toward the Owner and Architect under the prime contract. In the event of an inconsistency among the documents, the specific terms of this Subcontract shall govern.

3.2   RESPONSIBILITIES The Subcontractor agrees to furnish its best skill and judgment in the performance of the Subcontract Work and to cooperate with the Contractor so that the Contractor may fulfill its obligations to the Owner. The Subcontractor shall furnish all of the labor, materials, equipment, and services, including but not limited to, competent supervision, shop drawings, samples, tools, and scaffolding as are necessary for the proper performance of the Subcontract Work. The Subcontractor shall provide the Contractor a list of its proposed subcontractors and suppliers, and be responsible for taking field dimensions, providing tests, obtaining required permits related to the Subcontract Work and affidavits, ordering of materials and all other actions as required to meet the Progress Schedule.

3.3   INCONSISTENCIES AND OMISSIONS The Subcontractor shall make a careful analysis and comparison of the drawings, specifications, other Subcontract Documents and information furnished by the Owner relative to the Subcontract Work. Such analysis and comparison shall be solely for the purpose of facilitating the Subcontract Work and not for the discovery of errors, inconsistencies or omissions in the Subcontract Documents nor for ascertaining if the Subcontract Documents are in accordance with applicable laws, statutes, ordinances, building codes, rules or regulations. Should the Subcontractor discover any errors, inconsistencies or omissions in the Subcontract Documents, the Subcontractor shall report such discoveries to the Contractor in writing within three (3) days. Upon receipt of notice, the Con-

3

to be taken, and the Subcontractor shall comply with the Contractor's instructions. If the Subcontractor performs work knowing it to be contrary to any applicable laws, statutes, ordinances, building codes, rules or regulations without notice to the Contractor and advance approval by appropriate authorities, including the Contractor, the Subcontractor shall assume appropriate responsibility for such work and shall bear all associated costs, charges, fees and expenses necessarily incurred to remedy the violation. Nothing in this Paragraph 3.3 shall relieve the Subcontractor of responsibility for its own errors, inconsistencies and omissions.

3.4   SITE VISITATION Prior to performing any portion of the Subcontract Work, the Subcontractor shall conduct a visual inspection of the Project site to become generally familiar with local conditions and to correlate site observations with the Subcontract Documents. If the Subcontractor discovers any discrepancies between its site observations and the Subcontract Documents, such discrepancies shall be promptly reported to the Contractor.

3.5   INCREASED COSTS AND/OR TIME The Subcontractor may assert a Claim as provided in Article 7 if Contractor's clarifications or instructions in responses to requests for information are believed to require additional time or cost. If the Subcontractor fails to perform the reviews and comparisons required in Paragraph 3.3 and 3.4, above, to the extent the Contractor is held liable to the Owner because of the Subcontractor's failure, the Subcontractor shall pay the costs and damages to the Contractor that would have been avoided if the Subcontractor had performed those obligations.

3.6   COMMUNICATIONS Unless otherwise provided in the Subcontract Documents and except for emergencies, Subcontractor shall direct all communications related to the Project to the Contractor.

3.7   SUBMITTALS

3.7.1 The Subcontractor promptly shall submit for approval to the Contractor all shop drawings, samples, product data, manufacturers' literature and similar submittals required by the Subcontract Documents. The Subcontractor shall be responsible to the Contractor for the accuracy and conformity of its submittals to the Subcontract Documents. The Subcontractor shall prepare and deliver its submittals to the Contractor in a manner consistent with the Progress Schedule and in such time and sequence so as not to delay the Contractor or others in the performance of the Work. The approval of any Subcontractor submittal shall not be deemed to authorize deviations, substitutions or changes in the requirements of the Subcontract Documents unless express written approval is obtained from the Contractor and Owner authorizing such deviation, substitution or change. In the

event that the Subcontract Documents do not contain submittal requirements pertaining to the Subcontract Work, the Subcontractor agrees upon request to submit in a timely fashion to the Contractor for approval any shop drawings, samples, product data, manufacturers' literature or similar submittals as may reasonably be required by the Contractor, Owner or Architect.

3.7.2 The Contractor, Owner, and Architect are entitled to rely on the adequacy, accuracy and completeness of any professional certifications required by the Subcontract Documents concerning the performance criteria of systems, equipment or materials, including all relevant calculations and any governing performance requirements.

3.8   DESIGN DELEGATION

3.8.1 If the Subcontract Documents (1) specifically require the Subcontractor to provide design services and (2) specify all design and performance criteria, the Subcontractor shall provide those design services necessary to satisfactorily complete the Subcontract Work. Design services provided by the Subcontractor shall be procured from licensed design professionals retained by the Subcontractor as permitted by the law of the place where the Project is located (the Designer). The Designer's signature and seal shall appear on all drawings, calculations, specifications, certifications, Shop Drawings and other submittals prepared by the Designer. Shop Drawings and other submittals related to the Subcontract Work designed or certified by the Designer, if prepared by others, shall bear the Subcontractor's and the Designer's written approvals when submitted to the Contractor. The Contractor shall be entitled to rely upon the adequacy, accuracy and completeness of the services, certifications or approvals performed by the Designer.

3.8.2 If the Designer is an independent professional, the design services shall be procured pursuant to a separate agreement between the Subcontractor and the Designer. The Subcontractor-Designer agreement shall not provide for any limitation of liability, except to the extent that consequential damages are waived pursuant to Paragraph 5.4, or exclusion from participation in the multiparty proceedings requirement of Paragraph 11.4. The Designer(s) is (are) ___

_____

_____. The Subcontractor shall notify the ◆ Contractor in writing if it intends to change the Designer. The Subcontractor shall be responsible for conformance of its design with the information given and the design concept expressed in the Subcontract Documents. The Subcontractor shall not be responsible for the adequacy of the performance or design criteria required by the Subcontract Documents.

3.8.3 The Subcontractor shall not be required to provide design services in violation of any applicable law.

4

3.9   TEMPORARY SERVICES Subcontractor's responsibilities for temporary services are set forth in Exhibit
_____.

3.10   COORDINATION The Subcontractor shall:

.1   cooperate with the Contractor and all others whose work may interface with the Subcontract Work;

.2   specifically note and immediately advise the Contractor of any such interface with the Subcontract Work; and

.3   participate in the preparation of coordination drawings and work schedules in areas of congestion.

3.11   SUBCONTRACTOR'S REPRESENTATIVE The Subcontractor shall designate a person, subject to Contractor's approval, who shall be the Subcontractor's authorized representative. This representative shall be the only person to whom the Contractor shall issue instructions, orders or directions, except in an emergency. The Subcontractor's representative is _____
_____

who is agreed to by the Contractor.

3.12   TESTS AND INSPECTIONS The Subcontractor shall schedule all required tests, approvals and inspections of the Subcontract Work at appropriate times so as not to delay the progress of the work. The Subcontractor shall give proper written notice to all required parties of such tests, approvals and inspections. The Subcontractor shall bear all expenses associated with tests, inspections and approvals required of the Subcontractor by the Subcontract Documents which, unless otherwise agreed to, shall be conducted by an independent testing laboratory or entity approved by the Contractor and Owner. Required certificates of testing, approval or inspection shall, unless otherwise required by the Subcontract Documents, be secured by the Subcontractor and promptly delivered to the Contractor.

3.13   CLEANUP

3.13.1 The Subcontractor shall at all times during its performance of the Subcontract Work keep the Work site clean and free from debris resulting from the Subcontract Work. Prior to discontinuing the Subcontract Work in an area, the Subcontractor shall clean the area and remove all its rubbish and its construction equipment, tools, machinery, waste and surplus materials. Subcontractor shall make provisions to minimize and confine dust and debris resulting from its construction activities. The Subcontractor shall not be held responsible for unclean conditions caused by others.

3.13.2 If the Subcontractor fails to commence compliance with cleanup duties within forty-eight (48) hours after written notification from the Contractor of non-compliance, the Contractor may implement appropriate cleanup measures without further notice and the cost thereof shall be deducted from any amounts due or to become due the Subcontractor.

3.14   SAFETY

3.14.1 The Subcontractor is required to perform the Subcontract Work in a safe and reasonable manner. The Subcontractor shall seek to avoid injury, loss or damage to persons or property by taking reasonable steps to protect:

.1   employees and other persons at the site;

.2   materials and equipment stored at the site or at offsite locations for use in performance of the Work; and

.3   all property and structures located at the site and adjacent to work areas, whether or not said property or structures are part of the Project or involved in the Work.

3.14.2 The Subcontractor shall give all required notices and comply with all applicable rules, regulations, orders and other lawful requirements established to prevent injury, loss or damage to persons or property.

3.14.3 The Subcontractor shall implement appropriate safety measures pertaining to the Subcontract Work and the Project, including establishing safety rules, posting appropriate warnings and notices, erecting safety barriers, and establishing proper notice procedures to protect persons and property at the site and adjacent to the site from injury, loss or damage.

3.14.4 The Subcontractor shall exercise extreme care in carrying out any of the Subcontract Work which involves explosive or other dangerous methods of construction or hazardous procedures, materials or equipment. The Subcontractor shall use properly qualified individuals or entities to carry out the Subcontract Work in a safe and reasonable manner so as to reduce the risk of bodily injury or property damage.

3.14.5 Damage or loss not insured under property insurance which may arise from the performance of the Subcontract Work, to the extent of the negligence attributed to such acts or omissions of the Subcontractor, or anyone for whose acts the Subcontractor may be liable, shall be promptly remedied by the Subcontractor. Damage or loss attributable to the acts or omissions of the Contractor and not to the Subcontractor shall be promptly remedied by the Contractor.

AGC DOCUMENT NO. 650 • STANDARD FORM OF AGREEMENT BETWEEN CONTRACTOR AND SUBCONTRACTOR
(Where the Contractor Assumes the Risk of Owner Payment)

3.14.6 The Subcontractor is required to designate an individual at the site in the employ of the Subcontractor who shall act as the Subcontractor's designated safety representative with a duty to prevent accidents. Unless otherwise identified by the Subcontractor in writing to the Contractor, the designated safety representative shall be the Subcontractor's project superintendent.

3.14.7 The Subcontractor has an affirmative duty not to overload the structures or conditions at the site and shall take reasonable steps not to load any part of the structures or site so as to give rise to an unsafe condition or create an unreasonable risk of bodily injury or property damage. The Subcontractor shall have the right to request, in writing, from the Contractor loading information concerning the structures at the site.

3.14.8 The Subcontractor shall give prompt written notice to the Contractor of any accident involving bodily injury requiring a physician's care, any property damage exceeding Five Hundred Dollars ($500.00) in value, or any failure that could have resulted in serious bodily injury, whether or not such an injury was sustained.

3.14.9 Prevention of accidents at the site is the responsibility of the Contractor, Subcontractor, and all other subcontractors, persons and entities at the site. Establishment of a safety program by the Contractor shall not relieve the Subcontractor or other parties of their safety responsibilities. The Subcontractor shall establish its own safety program implementing safety measures, policies and standards conforming to those required or recommended by governmental and quasi-governmental authorities having jurisdiction and by the Contractor and Owner, including, but not limited to, requirements imposed by the Subcontract Documents. The Subcontractor shall comply with the reasonable recommendations of insurance companies having an interest in the Project, and shall stop any part of the Subcontract Work which the Contractor deems unsafe until corrective measures satisfactory to the Contractor shall have been taken. The Contractor's failure to stop the Subcontractor's unsafe practices shall not relieve the Subcontractor of the responsibility therefor. The Subcontractor shall notify the Contractor immediately following an accident and promptly confirm the notice in writing. A detailed written report shall be furnished if requested by the Contractor. Each party to this Agreement shall indemnify the other party from and against fines or penalties imposed as a result of safety violations, but only to the extent that such fines or penalties are caused by its failure to comply with applicable safety requirements.

3.15   PROTECTION OF THE WORK The Subcontractor shall take necessary precautions to properly protect the Subcontract Work and the work of others from damage caused by the Subcontractor's operations. Should the Subcontractor cause damage to the Work or property of the Owner, the Contractor or others, the Subcontractor shall promptly remedy such damage to the satisfaction of the Contractor, or the Contractor may remedy the damage and deduct its cost from any amounts due or to become due the Subcontractor, unless such costs are recovered under applicable property insurance.

3.16   PERMITS, FEES, LICENSES AND TAXES The Subcontractor shall give timely notices to authorities pertaining to the Subcontract Work, and shall be responsible for all permits, fees, licenses, assessments, inspections, testing and taxes necessary to complete the Subcontract Work in accordance with the Subcontract Documents. To the extent reimbursement is obtained by the Contractor from the Owner under the Owner-Contractor agreement, the Subcontractor shall be compensated for additional costs resulting from taxes enacted after the date of this Agreement.

3.17   ASSIGNMENT OF SUBCONTRACT WORK The Subcontractor shall not assign the whole nor any part of the Subcontract Work without prior written approval of the Contractor.

3.18   HAZARDOUS MATERIALS To the extent that the Contractor has rights or obligations under the Owner-Contractor agreement or by law regarding hazardous materials as defined by the Subcontract Document within the scope of the Subcontract Work, the Subcontractor shall have the same rights or obligations.

3.19   MATERIAL SAFETY DATA (MSD) SHEETS The Subcontractor shall submit to the Contractor all Material Safety Data Sheets required by law for materials or substances necessary for the performance of the Subcontract Work. MSD sheets obtained by the Contractor from other subcontractors or sources shall be made available to the Subcontractor by the Contractor.

3.20   LAYOUT RESPONSIBILITY AND LEVELS The Contractor shall establish principal axis lines of the building and site, and benchmarks. The Subcontractor shall lay out and be strictly responsible for the accuracy of the Subcontract Work and for any loss or damage to the Contractor or others by reason of the Subcontractor's failure to lay out or perform Subcontract Work correctly. The Subcontractor shall exercise prudence so that the actual final conditions and details shall result in alignment of finish surfaces.

3.21   WARRANTIES The Subcontractor warrants that all materials and equipment furnished under this Agreement shall be new, unless otherwise specified, of good quality, in conformance with the Subcontract Documents, and free from defective workmanship and materials. Warranties shall commence on the date of Substantial Completion of the Work or a designated portion.

6

3.22   UNCOVERING/CORRECTION OF SUBCONTRACT WORK

3.22.1   UNCOVERING OF SUBCONTRACT WORK

3.22.1.1   If required in writing by the Contractor, the Subcontractor must uncover any portion of the Subcontract Work which has been covered by the Subcontractor in violation of the Subcontract Documents or contrary to a directive issued to the Subcontractor by the Contractor. Upon receipt of a written directive from the Contractor, the Subcontractor shall uncover such work for the Contractor's or Owner's inspection and restore the uncovered Subcontract Work to its original condition at the Subcontractor's time and expense.

3.22.1.2   The Contractor may direct the Subcontractor to uncover portions of the Subcontract Work for inspection by the Owner or Contractor at any time. The Subcontractor is required to uncover such work whether or not the Contractor or Owner had requested to inspect the Subcontract Work prior to it being covered. Except as provided in Clause 3.22.1.1, this Agreement shall be adjusted by change order for the cost and time of uncovering and restoring any work which is uncovered for inspection and proves to be installed in accordance with the Subcontract Documents, provided the Contractor had not previously instructed the Subcontractor to leave the work uncovered. If the Subcontractor uncovers work pursuant to a directive issued by the Contractor, and such work upon inspection does not comply with the Subcontract Documents, the Subcontractor shall be responsible for all costs and time of uncovering, correcting and restoring the work so as to make it conform to the Subcontract Documents. If the Contractor or some other entity for which the Subcontractor is not responsible caused the nonconforming condition, the Contractor shall be required to adjust this Agreement by change order for all such costs and time.

3.22.2   CORRECTION OF WORK   If the Architect or Contractor rejects the Subcontract Work or the Subcontract Work is not in conformance with the Subcontract Documents, the Subcontractor shall promptly correct the Subcontract Work whether it had been fabricated, installed or completed. The Subcontractor shall be responsible for the costs of correcting such Subcontract Work, any additional testing, inspections, and compensation for services and expenses of the Architect and Contractor made necessary by the defective Subcontract Work.

3.22.2.2   In addition to the Subcontractor's obligations under Paragraph 3.21, the Subcontractor agrees to promptly correct, after receipt of a written notice from the Contractor, all Subcontract Work performed under this Agreement which proves to be defective in workmanship or materials within a period of one year from the date of Substantial Completion of the Subcontract Work or for a longer period of time as may

be required by specific warranties in the Subcontract Documents. Substantial Completion of the Subcontract Work, or of a designated portion, occurs on the date when construction is sufficiently complete in accordance with the Subcontract Documents so that the Owner can occupy or utilize the Project, or a designated portion, for the use for which it is intended. If, during the one-year period, the Contractor fails to provide the Subcontractor with prompt written notice of the discovery of defective or nonconforming Subcontract Work, the Contractor shall neither have the right to require the Subcontractor to correct such Subcontract Work nor the right to make claim for breach of warranty. If the Subcontractor fails to correct defective or nonconforming Subcontract Work within a reasonable time after receipt of notice from the Contractor, the Contractor may correct such Subcontract Work pursuant to Subparagraph 10.1.1.

3.22.3   The Subcontractor's correction of Subcontract Work pursuant to this Paragraph 3.22 shall not extend the one-year period for the correction of Subcontract Work, but if Subcontract Work is first performed after Substantial Completion, the one-year period for corrections shall be extended by the time period after Substantial Completion and the performance of that portion of Subcontract Work. The Subcontractor's obligation to correct Subcontract Work within one year as described in this Paragraph 3.22 does not limit the enforcement of Subcontractor's other obligations with regard to the Agreement and the Subcontract Documents.

3.22.4   If the Subcontractor's correction or removal of Subcontract Work destroys or damages completed or partially completed work of the Owner, the Contractor or any separate contractors, the Subcontractor shall be responsible for the cost of correcting such destroyed or damaged construction.

3.22.5   If portions of Subcontract Work which do not conform with the requirements of the Subcontract Documents are neither corrected by the Subcontractor nor accepted by the Contractor, the Subcontractor shall remove such Subcontract Work from the Project site if so directed by the Contractor.

3.23   MATERIALS OR EQUIPMENT FURNISHED BY OTHERS   In the event the scope of the Subcontract Work includes installation of materials or equipment furnished by others, it shall be the responsibility of the Subcontractor to exercise proper care in receiving, handling, storing and installing such items, unless otherwise provided in the Subcontract Documents. The Subcontractor shall examine the items provided and report to the Contractor in writing any items it may discover that do not conform to requirements of the Subcontract Documents. The Subcontractor shall not proceed to install non-conforming items without further instructions from the Contractor. Loss or damage due to acts or omissions of the Subcontractor shall be deducted from any amounts due or to become due the Subcontractor.

7

3.24   SUBSTITUTIONS No substitutions shall be made in the Subcontract Work unless permitted in the Subcontract Documents, and only upon the Subcontractor first receiving all approvals required under the Subcontract Documents for substitutions.

3.25   USE OF CONTRACTOR'S EQUIPMENT The Subcontractor, its agents, employees, subcontractors or suppliers shall use the Contractor's equipment only with the express written permission of the Contractor's designated representative and in accordance with the Contractor's terms and conditions for such use. If the Subcontractor or any of its agents, employees, subcontractors or suppliers utilize any of the Contractor's equipment, including machinery, tools, scaffolding, hoists, lifts or similar items owned, leased or under the control of the Contractor, the Subcontractor shall defend, indemnify and be liable to the Contractor as provided in Article 9 for any loss or damage (including bodily injury or death) which may arise from such use, except to the extent that such loss or damage is caused by the negligence of the Contractor's employees operating the Contractor's equipment.

3.26   WORK FOR OTHERS Until final completion of the Subcontract Work, the Subcontractor agrees not to perform any work directly for the Owner or any tenants, or deal directly with the Owner's representatives in connection with the Subcontract Work, unless otherwise approved in writing by the Contractor.

3.27   SUBCONTRACT BONDS

3.27.1 The Subcontractor ❑ shall ❑ shall not furnish to the Contractor, as the named Obligee, appropriate surety bonds to secure the faithful performance of the Subcontract Work and to satisfy all Subcontractor payment obligations related to Subcontract Work.

3.27.2 If a performance or payment bond, or both, are required of the Subcontractor under this Agreement, the bonds shall be in a form and by a surety mutually agreeable to the Contractor and Subcontractor, and in the full amount of the Subcontract Amount, unless otherwise specified.

3.27.3 The Subcontractor shall be reimbursed, without retainage, for the cost of any required performance or payment bonds simultaneously with the first progress payment. The reimbursement amount for the subcontractor bonds shall not exceed _____ percent (_____%) of the Subcontract Amount, which sum is included in the Subcontract Amount.

3.27.4 In the event the Subcontractor shall fail to promptly provide any required bonds, the Contractor may terminate this Agreement and enter into a subcontract for the balance of the Subcontract Work with another subcontractor. All Con-

tractor costs and expenses incurred by the Contractor as a result of said termination shall be paid by the Subcontractor.

3.28   SYSTEMS AND EQUIPMENT STARTUP With the assistance of the Owner's maintenance personnel and the Contractor, the Subcontractor shall direct the check-out and operation of systems and equipment for readiness, and assist in their initial startup and the testing of the Subcontract Work.

3.29   COMPLIANCE WITH LAWS The Subcontractor agrees to be bound by, and at its own cost comply with, all federal, state and local laws, ordinances and regulations (the Laws) applicable to the Subcontract Work, including but not limited to, equal employment opportunity, minority business enterprise, women's business enterprise, disadvantaged business enterprise, safety and all other Laws with which the Contractor must comply. The Subcontractor shall be liable to the Contractor and the Owner for all loss, cost and expense attributable to any acts of commission or omission by the Subcontractor, its employees and agents resulting from the failure to comply with Laws, including, but not limited to, any fines, penalties or corrective measures, except as provided in Subparagraph 3.14.9.

3.30   CONFIDENTIALITY To the extent the Owner-Contractor agreement provides for the confidentiality of any of the Owner's proprietary or otherwise confidential information disclosed in connection with the performance of this Agreement, the Subcontractor is equally bound by the Owner's confidentiality requirements.

3.31   ROYALTIES, PATENTS AND COPYRIGHTS The Subcontractor shall pay all royalties and license fees which may be due on the inclusion of any patented or copyrighted materials, methods or systems selected by the Subcontractor and incorporated in the Subcontract Work. The Subcontractor shall defend, indemnify and hold the Contractor and Owner harmless from all suits or claims for infringement of any patent rights or copyrights arising out of such selection. The Subcontractor shall be liable for all loss, including all costs, expenses, and attorneys' fees, but shall not be responsible for such defense or loss when a particular design, process or product of a particular manufacturer or manufacturers is required by the Subcontract Documents. However, if the Subcontractor has reason to believe that a particular design, process or product required by the Subcontract Documents is an infringement of a patent, the Subcontractor shall promptly furnish such information to the Contractor or be responsible to the Contractor and Owner for any loss sustained as a result.

3.32   LABOR RELATIONS (Insert here any conditions, obligations or requirements relative to labor relations and their effect on the project. Legal counsel is recommended.) ◆

8

ARTICLE 4

CONTRACTOR'S RESPONSIBILITIES

4.1 CONTRACTOR'S REPRESENTATIVE The Contractor shall designate a person who shall be the Contractor's authorized representative. The Contractor's representative shall be the only person the Subcontractor shall look to for instructions, orders and/or directions, except in an emergency. The Contractor's representative is _____

_____

_____ . ◆

4.2 PAYMENT BOND REVIEW The Contractor ❑ has ❑ has not provided the Owner a payment bond. The Contractor's payment bond for the Project, if any, shall be made available by the Contractor for review and copying by the Subcontractor. ◆ ◆

4.3 OWNER'S ABILITY TO PAY

4.3.1 The Subcontractor shall have the right upon request to receive from the Contractor such information as the Contractor has obtained relative to the Owner's financial ability to pay for the Work, including any subsequent material variation in such information. The Contractor, however, does not warrant the accuracy or completeness of the information provided by the Owner.

4.3.2 If the Subcontractor does not receive the information referenced in Subparagraph 4.3.1 with regard to the Owner's ability to pay for the Work as required by the Contract Documents, the Subcontractor may request the information from the Owner and/or the Owner's lender.

4.4 CONTRACTOR APPLICATION FOR PAYMENT Upon request, the Contractor shall give the Subcontractor a copy of the most current Contractor application for payment reflecting the amounts approved and/or paid by the Owner for the Subcontract Work performed to date.

4.5 INFORMATION OR SERVICES The Subcontractor is entitled to request through the Contractor any information or services relevant to the performance of the Subcontract Work which is under the Owner's control. To the extent the Contractor receives such information and services, the Contractor shall provide them to the Subcontractor. The Contractor, however, does not warrant the accuracy or completeness of the information provided by the Owner.

4.6 STORAGE AREAS The Contractor shall allocate adequate storage areas, if available, for the Subcontractor's materials and equipment during the course of the Subcontract Work. Unless otherwise agreed upon, the Contractor shall reimburse the Subcontractor for the additional costs of having to relocate such storage areas at the direction of the Contractor.

4.7 TIMELY COMMUNICATIONS The Contractor shall transmit to the Subcontractor, with reasonable promptness, all submittals, transmittals, and written approvals relative to the Subcontract Work. Unless otherwise specified in the Subcontract Documents, communications by and with the Subcontractor's subcontractors, materialmen and suppliers shall be through the Subcontractor.

4.8 USE OF SUBCONTRACTOR'S EQUIPMENT The Contractor, its agents, employees or suppliers shall use the Subcontractor's equipment only with the express written permission of the Subcontractor's designated representative and in accordance with the Subcontractor's terms and conditions for such use. If the Contractor or any of its agents, employees or suppliers utilize any of the Subcontractor's equipment, including machinery, tools, scaffolding, hoists, lifts or similar items owned, leased or under the control of the Subcontractor, the Contractor shall defend, indemnify and be liable to the Subcontractor as provided in Article 9 for any loss or damage (including bodily injury or death) which may arise from such use, except to the extent that such loss or damage is caused by the negligence of the Subcontractor's employees operating the Subcontractor's equipment.

ARTICLE 5

PROGRESS SCHEDULE

5.1 TIME IS OF THE ESSENCE Time is of the essence for both parties. They mutually agree to see to the performance of their respective obligations so that the entire Project may be completed in accordance with the Subcontract Documents and particularly the Progress Schedule as set forth in Exhibit _____ . ◆

5.2 SCHEDULE OBLIGATIONS The Subcontractor shall provide the Contractor with any scheduling information proposed by the Subcontractor for the Subcontract Work. In consultation with the Subcontractor, the Contractor shall prepare the schedule for performance of the Work (the Progress Schedule) and shall revise and update such schedule, as necessary, as the Work progresses. Both the Contractor and the Subcontractor shall be bound by the Progress Schedule. The Progress Schedule and all subsequent changes and additional details shall be submitted to the Subcontractor promptly and reasonably in advance of the required performance. The Contractor shall have the right to determine and, if necessary, change the time, order and priority in which the various portions of the Work shall be performed and all other matters relative to the Subcontract Work.

5.3 DELAYS AND EXTENSIONS OF TIME

5.3.1 OWNER CAUSED DELAY Subject to Subparagraph 5.3.2, if the commencement and/or progress of the Subcontract Work is delayed without the fault or responsibil-

9

ity of the Subcontractor, the time for the Subcontract Work shall be extended by Subcontract Change Order to the extent obtained by the Contractor under the Subcontract Documents, and the Progress Schedule shall be revised accordingly.

5.3.2 CLAIMS RELATING TO OWNER The Subcontractor agrees to initiate all claims for which the Owner is or may be liable in the manner and within the time limits provided in the Subcontract Documents for like claims by the Contractor upon the Owner and in sufficient time for the Contractor to initiate such claims against the Owner in accordance with the Subcontract Documents. At the Subcontractor's request and expense to the extent agreed upon in writing, the Contractor agrees to permit the Subcontractor to prosecute a claim in the name of the Contractor for the use and benefit of the Subcontractor in the manner provided in the Subcontract Documents for like claims by the Contractor upon the Owner.

5.3.3 CONTRACTOR CAUSED DELAY Nothing in this Article shall preclude the Subcontractor's recovery of delay damages caused by the Contractor.

5.3.4 CLAIMS RELATING TO CONTRACTOR The Subcontractor shall give the Contractor written notice of all claims not included in Subparagraph 5.3.2 within seven (7) days of the Subcontractor's knowledge of the facts giving rise to the event for which claim is made; otherwise, such claims shall be deemed waived. All unresolved claims, disputes and other matters in question between the Contractor and the Subcontractor not relating to claims included in Subparagraph 5.3.2 shall be resolved in the manner provided in Article 11.

5.3.5 DAMAGES If the Subcontract Documents provide for liquidated or other damages for delay beyond the completion date set forth in the Subcontract Documents, and such damages are assessed, the Contractor may assess a share of the damages against the Subcontractor in proportion to the Subcontractor's share of the responsibility for the delay. However, the amount of such assessment shall not exceed the amount assessed against the Contractor. This Paragraph 5.3 shall not limit the Subcontractor's liability to the Contractor for the Contractor's actual delay damages caused by the Subcontractor's delay.

5.4 MUTUAL WAIVER OF CONSEQUENTIAL DAMAGES

5.4.1 To the extent the Owner-Contractor agreement provides for a mutual waiver of consequential damages by the Owner and the Contractor, the Contractor and Subcontractor waive claims against each other for consequential damages arising out of or relating to this Agreement, including to the extent provided in the Owner-Contractor agreement,

damages for principal office expenses and the compensation of personnel stationed there; loss of financing, business and reputation; and for loss of profit. Similarly, the Subcontractor shall obtain from its sub-subcontractors mutual waivers of consequential damages that correspond to the Subcontractor's waiver of consequential damages herein. To the extent applicable, this mutual waiver applies to consequential damages due to termination by the Contractor or the Owner in accordance with this Agreement or the Owner-Contractor agreement. To the extent the Owner-Contractor agreement does not preclude the award of liquidated damages, nothing contained in this Paragraph 5.4 shall preclude the imposition of such damages, if applicable in accordance with the requirements of the Subcontract Documents.

5.4.2 To the extent the Owner-Contractor agreement provides for a mutual waiver of consequential damages by the Owner and the Contractor, damages for which the Contractor is liable to the Owner including those related to Subparagraph 9.1.1 are not consequential damages for the purpose of this waiver. Similarly, to the extent the Subcontractor-sub-subcontractor agreement provides for a mutual waiver of consequential damages by the Owner and the Contractor, damages for which the Subcontractor is liable to lower-tiered parties due to the fault of the Owner or Contractor are not consequential damages for the purpose of this waiver.

ARTICLE 6

SUBCONTRACT AMOUNT

As full compensation for performance of this Agreement, Contractor agrees to pay Subcontractor in current funds for the satisfactory performance of the Subcontract Work subject to all applicable provisions of the Subcontract:

(a) the fixed-price of _____ ◆
Dollars ($ _____) subject ◆
to additions and deductions as provided for in the Subcontract Documents; and/or

(b) unit prices in accordance with the attached schedule of Unit Prices and estimated quantities, which is incorporated by reference and identified as Exhibit _____ ; and/or ◆

(c) time and material rates and prices in accordance with the attached Schedule of Labor and Material Costs which is incorporated by reference and identified as Exhibit _____ . ◆

The fixed-price, unit prices and/or time and material rates and prices are referred to as the Subcontract Amount.

10

ARTICLE 7

CHANGES IN THE SUBCONTRACT WORK

7.1   SUBCONTRACT CHANGE ORDERS When the Contractor orders in writing, the Subcontractor, without nullifying this Agreement, shall make any and all changes in the Subcontract Work which are within the general scope of this Agreement. Any adjustment in the Subcontract Amount or Subcontract Time shall be authorized by a Subcontract Change Order. No adjustments shall be made for any changes performed by the Subcontractor that have not been ordered by the Contractor. A Subcontract Change Order is a written instrument prepared by the Contractor and signed by the Subcontractor stating their agreement upon the change in the Subcontract Work.

7.2   CONSTRUCTION CHANGE DIRECTIVES To the extent that the Subcontract Documents provide for Construction Change Directives in the absence of agreement on the terms of a Subcontract Change Order, the Subcontractor shall promptly comply with the Construction Change Directive and be entitled to apply for interim payment if the Subcontract Documents so provide.

7.3   UNKNOWN CONDITIONS If in the performance of the Subcontract Work the Subcontractor finds latent, concealed or subsurface physical conditions which differ materially from those indicated in the Subcontract Documents or unknown physical conditions of an unusual nature, which differ materially from those ordinarily found to exist, and not generally recognized as inherent in the kind of work provided for in this Agreement, the Subcontract Amount and/or the Progress Schedule shall be equitably adjusted by a Subcontract Change Order within a reasonable time after the conditions are first observed. The adjustment which the Subcontractor may receive shall be limited to the adjustment the Contractor receives from the Owner on behalf of the Subcontractor, or as otherwise provided under Subparagraph 5.3.2.

7.4   ADJUSTMENTS IN SUBCONTRACT AMOUNT If a Subcontract Change Order requires an adjustment in the Subcontract Amount, the adjustment shall be established by one of the following methods:

.1   mutual acceptance of an itemized lump sum;

.2   unit prices as indicated in the Subcontract Documents or as subsequently agreed to by the parties; or

.3   costs determined in a manner acceptable to the parties and a mutually acceptable fixed or percentage fee; or

.4   another method provided in the Subcontract Documents.

7.5   SUBSTANTIATION OF ADJUSTMENT If the Subcontractor does not respond promptly or disputes the method of adjustment, the method and the adjustment shall be determined by the Contractor on the basis of reasonable expenditures and savings of those performing the Work attributable to the change, including, in the case of an increase in the Subcontract Amount, an allowance for overhead and profit of the percentage provided in Paragraph 7.6. The Subcontractor may contest the reasonableness of any adjustment determined by the Contractor. The Subcontractor shall maintain for the Contractor's review and approval an appropriately itemized and substantiated accounting of the following items attributable to the Subcontract Change Order:

.1   labor costs, including Social Security, health, welfare, retirement and other fringe benefits as normally required, and state workers' compensation insurance;

.2   costs of materials, supplies and equipment, whether incorporated in the Subcontract Work or consumed, including transportation costs;

.3   costs of renting machinery and equipment other than hand tools;

.4   costs of bond and insurance premiums, permit fees and taxes attributable to the change; and

.5   costs of additional supervision and field office personnel services necessitated by the change.

7.6   Adjustments shall be based on net change in Subcontractor's reasonable cost of performing the changed Subcontract Work plus, in case of a net increase in cost, an agreed upon sum for overhead and profit not to exceed _____ percent (_____%). ♦

7.7   NO OBLIGATION TO PERFORM The Subcontractor shall not perform changes in the Subcontract Work until a Subcontract Change Order has been executed or written instructions have been issued in accordance with Paragraphs 7.2 and 7.9.

7.8   EMERGENCIES In an emergency affecting the safety of persons and/or property, the Subcontractor shall act, at its discretion, to prevent threatened damage, injury or loss. Any change in the Subcontract Amount and/or the Progress Schedule on account of emergency work shall be determined as provided in this Article.

11

7.9 INCIDENTAL CHANGES The Contractor may direct the Subcontractor to perform incidental changes in the Subcontract Work which do not involve adjustments in the Subcontract Amount or Subcontract Time. Incidental changes shall be consistent with the scope and intent of the Subcontract Documents. The Contractor shall initiate an incidental change in the Subcontract Work by issuing a written order to the Subcontractor. Such written notice shall be carried out promptly and are binding on the parties.

ARTICLE 8

PAYMENT

8.1 SCHEDULE OF VALUES As a condition to payment, the Subcontractor shall provide a schedule of values satisfactory to the Contractor not more than fifteen (15) days from the date of execution of this Agreement.

8.2 PROGRESS PAYMENTS

8.2.1 APPLICATIONS The Subcontractor's applications for payment shall be itemized and supported by substantiating data as required by the Subcontract Documents. If the Subcontractor is obligated to provide design services pursuant to Paragraph 3.8, Subcontractor's applications for payment shall show the Designer's fee and expenses as a separate cost item. The Subcontractor's application shall be notarized if required and if allowed under the Subcontract Documents may include properly authorized Subcontract Construction Change Directives. The Subcontractor's progress payment application for the Subcontract Work performed in the preceding payment period shall be submitted for approval of the Contractor in accordance with the schedule of values if required and Subparagraphs 8.2.2, 8.2.3, and 8.2.4. The Contractor shall incorporate the approved amount of the Subcontractor's progress payment application into the Contractor's payment application to the Owner for the same period and submit it to the Owner in a timely fashion. The Contractor shall immediately notify the Subcontractor of any changes in the amount requested on behalf of the Subcontractor.

8.2.2 RETAINAGE The rate of retainage shall be _____ percent (_____%). which is equal ◆ to the percentage retained from the Contractor's payment by the Owner for the Subcontract Work. If the Subcontract Work is satisfactory and the Subcontract Documents provide for reduction of retainage at a specified percentage of completion, the Subcontractor's retainage shall also be reduced when the Subcontract Work has attained the same percentage of completion and the Contractor's retainage for the Subcontract Work has been so reduced by the Owner.

8.2.3 TIME OF APPLICATION The Subcontractor shall submit progress payment applications to the Contractor no

later than the _____ day of each payment ◆ period for the Subcontract Work performed up to and including the _____ day of the payment ◆ period indicating work completed and, to the extent allowed under Subparagraph 8.2.4, materials suitably stored during the preceding payment period.

8.2.4 STORED MATERIALS Unless otherwise provided in the Subcontract Documents, and if approved in advance by the Owner, applications for payment may include materials and equipment not incorporated in the Subcontract Work but delivered to and suitably stored at the site or at some other location agreed upon in writing. Approval of payment applications for such stored items on or off the site shall be conditioned upon submission by the Subcontractor of bills of sale and applicable insurance or such other procedures satisfactory to the Owner and Contractor to establish the Owner's title to such materials and equipment, or otherwise to protect the Owner's and Contractor's interest including transportation to the site.

8.2.5 TIME OF PAYMENT Progress payments to the Subcontractor for satisfactory performance of the Subcontract Work shall be made no later than seven (7) days after receipt by the Contractor of payment from the Owner for the Subcontract Work. If payment from the Owner for such Subcontract Work is not received by the Contractor, through no fault of the Subcontractor, the Contractor will make payment to the Subcontractor within a reasonable time for the Subcontract Work satisfactorily performed.

8.2.6 PAYMENT DELAY If the Contractor has received payment from the Owner and if for any reason not the fault of the Subcontractor, the Subcontractor does not receive a progress payment from the Contractor within seven (7) days after the date such payment is due, as defined in Subparagraph 8.2.5, or, if the Contractor has failed to pay the Subcontractor within a reasonable time for the Subcontract Work satisfactorily performed, the Subcontractor, upon giving seven (7) days' written notice to the Contractor, and without prejudice to and in addition to any other legal remedies, may stop work until payment of the full amount owing to the Subcontractor has been received. The Subcontract Amount and Time shall be adjusted by the amount of the Subcontractor's reasonable and verified cost of shutdown, delay, and startup, which shall be effected by an appropriate Subcontractor Change Order.

8.2.7 PAYMENTS WITHHELD The Contractor may reject a Subcontractor payment application or nullify a previously approved Subcontractor payment application, in whole or in part, as may reasonably by necessary to protect the Contractor from loss or damage based upon:

.1 the Subcontractor's repeated failure to perform the Subcontract Work as required by this Agreement;

12

.2   loss or damage arising out of or relating to this Agreement and caused by the Subcontractor to the Owner, Contractor or others to whom the Contractor may be liable;

.3   the Subcontractor's failure to properly pay for labor, materials, equipment or supplies furnished in connection with the Subcontract Work;

.4   rejected, nonconforming or defective Subcontract Work which has not been corrected in a timely fashion;

.5   reasonable evidence of delay in performance of the Subcontract Work such that the Work will not be completed within the Subcontract Time, and that the unpaid balance of the Subcontract Amount is not sufficient to offset the liquidated damages or actual damages that may be sustained by the Contractor as a result of the anticipated delay caused by the Subcontractor.

.6   reasonable evidence demonstrating that the unpaid balance of the Subcontract Amount is insufficient to cover the cost to complete the Subcontract Work;

.7   third party claims involving the Subcontractor or reasonable evidence demonstrating that third party claims are likely to be filed unless and until the Subcontractor furnishes the Contractor with adequate security in the form of a surety bond, letter of credit or other collateral or commitment which are sufficient to discharge such claims if established.

The Contractor shall give written notice to the Subcontractor, at the time of disapproving or nullifying an application for payment stating its specific reasons for such disapproval or nullification. When the above reasons for disapproving or nullifying an application for payment are removed, payment will be made for amounts previously withheld.

8.3   FINAL PAYMENT

8.3.1   APPLICATION Upon acceptance of the Subcontract Work by the Owner and the Contractor and receipt from the Subcontractor of evidence of fulfillment of the Subcontractor's obligations in accordance with the Subcontract Documents and Subparagraph 8.3.2, the Contractor shall incorporate the Subcontractor's application for final payment into the Contractor's next application for payment to the Owner without delay, or notify the Subcontractor if there is a delay and the reasons therefor.

8.3.2   REQUIREMENTS Before the Contractor shall be required to incorporate the Subcontractor's application for final payment into the Contractor's next application for payment, the Subcontractor shall submit to the Contractor:

.1   an affidavit that all payrolls, bills for materials and equipment, and other indebtedness connected with the Subcontract Work for which the Owner or its property or the Contractor or the Contractor's surety might in any way be liable, have been paid or otherwise satisfied;

.2   consent of surety to final payment, if required;

.3   satisfaction of required closeout procedures;

.4   certification that insurance required by the Subcontract Documents to remain in effect beyond final payment pursuant to Clauses 9.2.3.1 and 9.2.6 is in effect and will not be cancelled or allowed to expire without at least thirty (30) days' written notice to the Contractor unless a longer period is stipulated in this Agreement;

.5   other data, if required by the Contractor or Owner, such as receipts, releases, and waivers of liens to the extent and in such form as may be designated by the Contractor or Owner;

.6   written warranties, equipment manuals, startup and testing required in Paragraph 3.28; and

.7   as-built drawings if required by the Subcontract Documents.

8.3.3   TIME OF PAYMENT Final payment of the balance due of the Subcontract Amount shall be made to the Subcontractor within seven (7) days after receipt by the Contractor of final payment from the Owner for such Subcontract Work.

8.3.4   FINAL PAYMENT DELAY If the Owner or its designated agent does not issue a certificate for final payment or the Contractor does not receive such payment for any cause which is not the fault of the Subcontractor, the Contractor shall promptly inform the Subcontractor in writing. The Contractor shall also diligently pursue, with the assistance of the Subcontractor, the prompt release by the Owner of the final payment due for the Subcontract Work. At the Subcontractor's request and expense, to the extent agreed upon in writing, the Contractor shall institute reasonable legal remedies to mitigate the damages and pursue payment of the Subcontractor's final payment including interest. If final payment from the Owner for such Subcontract Work is not received by the Contractor, through no fault of the Subcontractor, the Contractor will make payment to the Subcontractor within a reasonable time.

13

8.3.5 WAIVER OF CLAIMS Final payment shall constitute a waiver of all claims by the Subcontractor relating to the Subcontract Work, but shall in no way relieve the Subcontractor of liability for the obligations assumed under Paragraphs 3.21 and 3.22, or for faulty or defective work or services discovered after final payment.

8.4 LATE PAYMENT INTEREST Progress payments or final payment due and unpaid under this Agreement, as defined in Subparagraphs 8.2.5, 8.3.3 and 8.3.4, shall bear interest from the date payment is due at the prime rate prevailing at the place of the Project. However, if the Owner fails to timely pay the Contractor as required under the Owner-Contractor agreement through no fault or neglect of the Contractor, and the Contractor fails to timely pay the Subcontractor as a result of such nonpayment, the Contractor's obligation to pay the Subcontractor interest on corresponding payments due and unpaid under this Agreement shall be extinguished by the Contractor promptly paying to the Subcontractor the Subcontractor's proportionate share of the interest, if any, received by the Contractor from the Owner on such late payments.

8.5 CONTINUING OBLIGATIONS Provided the Contractor is making payments on or has made payments to the Subcontractor in accordance with the terms of this Agreement, the Subcontractor shall reimburse the Contractor for any costs and expenses for any claim, obligation or lien asserted before or after final payment is made that arises from the performance of the Subcontract Work. The Subcontractor shall reimburse the Contractor for costs and expenses including attorneys' fees and costs and expenses incurred by the Contractor in satisfying, discharging or defending against any such claims, obligation or lien including any action brought or judgment recovered. In the event that any applicable law, statute, regulation or bond requires the Subcontractor to take any action prior to the expiration of the reasonable time for payment referred in Subparagraph 8.2.5 in order to preserve or protect the Subcontractor's rights, if any, with respect to mechanic's lien or bond claims, then the Subcontractor may take that action prior to the expiration of the reasonable time for payment and such action will not create the reimbursement obligation recited above nor be in violation of this Agreement or considered premature for purposes of preserving and protecting the Subcontractor's rights.

8.6 PAYMENT USE RESTRICTION Payments received by the Subcontractor shall be used to satisfy the indebtedness owed by the Subcontractor to any person furnishing labor or materials, or both, for use in performing the Subcontract Work through the most current period applicable to progress payments received from the Contractor before it is used for any other purpose. In the same manner, payments received by the Contractor from the Owner for the Subcontract Work shall be dedicated to payment to the Subcontractor. This provision shall bear on this Agreement only,

and is not for the benefit of third parties. Moreover, it shall not be construed by the parties to this Agreement or third parties to require that dedicated sums of money or payments be deposited in separate accounts, or that there be other restrictions on commingling of funds. Neither shall these mutual covenants be construed to create any fiduciary duty on the Subcontractor or Contractor, nor create any tort cause of action or liability for breach of trust, punitive damages, or other equitable remedy or liability for alleged breach.

8.7 PAYMENT USE VERIFICATION If the Contractor has reason to believe that the Subcontractor is not complying with the payment terms of this Agreement, the Contractor shall have the right to contact the Subcontractor's subcontractors and suppliers to ascertain whether they are being paid by the Subcontractor in accordance with this Agreement.

8.8 PARTIAL LIEN WAIVERS AND AFFIDAVITS As a prerequisite for payments, the Subcontractor shall provide, in a form satisfactory to the Owner and Contractor, partial lien or claim waivers in the amount of the application for payment and affidavits covering its subcontractors and suppliers for completed Subcontract Work. Such waivers may be conditional upon payment. In no event shall Contractor require the Subcontractor to provide an unconditional waiver of lien or claim, either partial or final, prior to receiving payment or in an amount in excess of what it has been paid.

8.9 SUBCONTRACTOR PAYMENT FAILURE Upon payment by the Contractor, the Subcontractor shall promptly pay its subcontractors and suppliers the amounts to which they are entitled. In the event the Contractor has reason to believe that labor, material or other obligations incurred in the performance of the Subcontract Work are not being paid, the Contractor may give written notice of a potential claim or lien to the Subcontractor and may take any steps deemed necessary to assure that progress payments are utilized to pay such obligations, including but not limited to the issuance of joint checks. If upon receipt of notice, the Subcontractor does not (a) supply evidence to the satisfaction of the Contractor that the moneys owing have been paid; or (b) post a bond indemnifying the Owner, the Contractor, the Contractor's surety, if any, and the premises from a claim or lien, the Contractor shall have the right to withhold from any payments due or to become due to the Subcontractor a reasonable amount to protect the Contractor from any and all loss, damage or expense including attorneys' fees that may arise out of or relate to any such claim or lien.

8.10 SUBCONTRACTOR ASSIGNMENT OF PAYMENTS The Subcontractor shall not assign any moneys due or to become due under this Agreement, without the written consent of the Contractor, unless the assignment is intended to create a new security interest within the scope of Article 9 of the Uniform Commercial Code. Should the Subcontractor assign all or any part of any moneys due or to

14

become due under this Agreement to create a new security interest or for any other purpose, the instrument of assignment shall contain a clause to the effect that the assignee's right in and to any money due or to become due to the Subcontractor shall be subject to the claims of all persons, firms and corporations for services rendered or materials supplied for the performance of the Subcontract Work.

8.11   PAYMENT NOT ACCEPTANCE Payment to the Subcontractor does not consitute or imply acceptance of any portion of the Subcontract Work.

ARTICLE 9

INDEMNITY, INSURANCE AND WAIVER OF SUBROGATION

9.1   INDEMNITY

9.1.1   INDEMNITY To the fullest extent permitted by law, the Subcontractor shall defend, indemnify and hold harmless the Contractor, the Contractor's other subcontractors, the Architect/Engineer, the Owner and their agents, consultants and employees (the Indemnitees) from all claims for bodily injury and property damage that may arise from the performance of the Subcontract Work to the extent of the negligence attributed to such acts or omissions by the Subcontractor, the Subcontractor's subcontractors or anyone employed directly or indirectly by any of them or by anyone for whose acts any of them may be liable.

9.1.2   NO LIMITATION ON LIABILITY In any and all claims against the Indemnitees by any employee of the Subcontractor, anyone directly or indirectly employed by the Subcontractor or anyone for whose acts the Subcontractor may be liable, the indemnification obligation shall not be limited in any way by any limitation on the amount or type of damages, compensation or benefits payable by or for the Subcontractor under workers' compensation acts, disability benefit acts or other employee benefit acts.

9.2   INSURANCE

9.2.1   SUBCONTRACTOR'S INSURANCE Before commencing the Subcontract Work, and as a condition of payment, the Subcontractor shall purchase and maintain insurance that will protect it from the claims arising out of its operations under this Agreement, whether the operations are by the Subcontractor, or any of its consultants or subcontractors or anyone directly or indirectly employed by any of them, or by anyone for whose acts any of them may be liable.

9.2.2   MINIMUM LIMITS OF LIABILITY The Subcontractor shall maintain at least the limits of liability in a company satisfactory to the Contractor as set forth in Exhibit _____. ◆

9.2.3   PROFESSIONAL LIABILITY INSURANCE

9.2.3.1   PROFESSIONAL LIABILITY INSURANCE The Subcontractor shall require the Designer(s) to maintain Project Specific Professional Liability Insurance with a company satisfactory to the Contractor, including contractual liability insurance against the liability assumed in Paragraph 3.8, and including coverage for any professional liability caused by any of the Designer's(s) consultants. Said insurance shall have specific minimum limits as set forth below:

Limit of $ _____ per claim. ◆
General Aggregate of $ _____ for the ◆
subcontract services rendered.

The Professional Liability Insurance shall contain prior acts coverage sufficient to cover all subcontract services rendered by the Designer. Said insurance shall be continued in effect with an extended period of _____ years following final payment to the Designer. ◆

Such insurance shall have a maximum deductible amount of $ _____ per ◆
occurrence. The deductible shall be paid by the Subcontractor or Designer.

9.2.3.2 The Subcontractor shall require the Designer to furnish to the Subcontractor and Contractor, before the Designer commences its services, a copy of its professional liability policy evidencing the coverages required in this paragraph. No policy shall be cancelled or modified without thirty (30) days' prior written notice to the Subcontractor and Contractor.

9.2.4   NUMBER OF POLICIES Commercial General Liability Insurance and other liability insurance may be arranged under a single policy for the full limits required or by a combination of underlying policies with the balance provided by an Excess or Umbrella Liability Policy.

9.2.5   CANCELLATION, RENEWAL AND MODIFICATION The Subcontractor shall maintain in effect all insurance coverages required under this Agreement at the Subcontractor's sole expense and with insurance companies acceptable to the Contractor. The policies shall contain a provision that coverage will not be cancelled or not renewed until at least thirty (30) days' prior written notice has been given to the Contractor. Certificates of insurance showing required coverage to be in force pursuant to Subparagraph 9.2.2 shall be filed with the Contractor prior to commencement of the Subcontract Work. In the event the Subcontractor fails to obtain or maintain any insurance coverage required under this Agreement, the Contractor may purchase such coverage as desired for the Contractor's benefit and charge the expense to the Subcontractor, or terminate this Agreement.

15

9.2.6 CONTINUATION OF COVERAGE The Subcontractor shall continue to carry Completed Operations Liability Insurance for at least _____ years after either ninety (90) days following Substantial Completion of the Work or final payment to the Contractor, whichever is earlier. The Subcontractor shall furnish the Contractor evidence of such insurance at final payment and one year from final payment.

9.2.7 BUILDER'S RISK INSURANCE

9.2.7.1 Upon written request of the Subcontractor, the Contractor shall provide the Subcontractor with a copy of the Builder's Risk policy of insurance or any other property or equipment insurance in force for the Project and procured by the Owner or Contractor. The Contractor will advise the Subcontractor if a Builder's Risk policy of insurance is not in force.

9.2.7.2 If the Owner or Contractor has not purchased Builder's Risk insurance satisfactory to the Subcontractor, the Subcontractor may procure such insurance as will protect the interests of the Subcontractor, its subcontractors and their subcontractors in the Subcontract Work.

9.2.7.3 If not covered under the Builder's Risk policy of insurance or any other property or equipment insurance required by the Subcontract Documents, the Subcontractor shall procure and maintain at the Subcontractor's own expense property and equipment insurance for the Subcontract Work including portions of the Subcontract Work stored off the site or in transit, when such portions of the Subcontract Work are to be included in an application for payment under Article 8.

9.2.8 WAIVER OF SUBROGATION

9.2.8.1 The Contractor and Subcontractor waive all rights against each other, the Owner and the Architect/Engineer, and any of their respective consultants, subcontractors, and sub-subcontractors, agents and employees, for damages caused by perils to the extent covered by the proceeds of the insurance provided in Clause 9.2.7.1 except such rights as they may have to the insurance proceeds. The Subcontractor shall require similar waivers from its subcontractors.

9.2.9 ENDORSEMENT If the policies of insurance referred to in this Article require an endorsement to provide for continued coverage where there is a waiver of subrogation, the owners of such policies will cause them to be so endorsed.

ARTICLE 10

CONTRACTOR'S RIGHT TO PERFORM SUBCONTRACTOR'S RESPONSIBILITIES AND TERMINATION OF AGREEMENT

10.1 FAILURE OF PERFORMANCE

10.1.1 NOTICE TO CURE If the Subcontractor refuses or fails to supply enough properly skilled workers, proper materials, or maintain the Progress Schedule, or fails to make prompt payment to its workers, subcontractors or suppliers, or disregards laws, ordinances, rules, regulations or orders of any public authority having jurisdiction, or otherwise is guilty of a material breach of a provision of this Agreement, the Subcontractor shall be deemed in default of this Agreement. If the Subcontractor fails within three (3) days after written notification to commence and continue satisfactory correction of the default with diligence and promptness, then the Contractor without prejudice to any other rights or remedies, shall have the right to any or all of the following remedies:

.1 supply workers, materials, equipment and facilities as the Contractor deems necessary for the completion of the Subcontract Work or any part which the Subcontractor has failed to complete or perform after written notification, and charge the cost, including reasonable overhead, profit, attorneys' fees, costs and expenses to the Subcontractor;

.2 contract with one or more additional contractors to perform such part of the Subcontract Work as the Contractor determines will provide the most expeditious completion of the Work, and charge the cost to the Subcontractor as provided under Clause 10.1.1.1; and/or

.3 withhold any payments due or to become due the Subcontractor pending corrective action in amounts sufficient to cover losses and compel performance to the extent required by and to the satisfaction of the Contractor.

In the event of an emergency affecting the safety of persons or property, the Contractor may proceed as above without notice, but the Contractor shall give the Subcontractor notice promptly after the fact as a precondition of cost recovery.

10.1.2 TERMINATION BY CONTRACTOR If the Subcontractor fails to commence and satisfactorily continue correction of a default within three (3) days after written notification issued under Subparagraph 10.1.1, then the Contractor may, in lieu of or in addition to Subparagraph

16

10.1.1, issue a second written notification, to the Subcontractor and its surety, if any. Such notice shall state that if the Subcontractor fails to commence and continue correction of a default within seven (7) days of the written notification, the Agreement will be deemed terminated. A written notice of termination shall be issued by the Contractor to the Subcontractor at the time the Subcontractor is terminated. The Contractor may furnish those materials, equipment and/or employ such workers or subcontractors as the Contractor deems necessary to maintain the orderly progress of the Work. All costs incurred by the Contractor in performing the Subcontract Work, including reasonable overhead, profit and attorneys' fees, costs and expenses, shall be deducted from any moneys due or to become due the Subcontractor. The Subcontractor shall be liable for the payment of any amount by which such expense may exceed the unpaid balance of the Subcontract Amount. At the Subcontractor's request, the Contractor shall provide a detailed accounting of the costs to finish the Subcontract Work.

10.1.3   USE OF SUBCONTRACTOR'S EQUIPMENT If the Contractor performs work under this Article, either directly or through other subcontractors, the Contractor or other subcontractors shall have the right to take and use any materials, implements, equipment, appliances or tools furnished by, or belonging to the Subcontractor and located at the Project site for the purpose of completing any remaining Subcontract Work. Immediately upon completion of the Subcontract Work, any remaining materials, implements, equipment, appliances or tools not consumed or incorporated in performance of the Subcontract Work, and furnished by, belonging to, or delivered to the Project by or on behalf of the Subcontractor, shall be returned to the Subcontractor in substantially the same condition as when they were taken, normal wear and tear excepted.

10.2.   BANKRUPTCY

10.2.1   TERMINATION ABSENT CURE If the Subcontractor files a petition under the Bankruptcy Code, this Agreement shall terminate if the Subcontractor or the Subcontractor's trustee rejects the Agreement or, if there has been a default, the Subcontractor is unable to give adequate assurance that the Subcontractor will perform as required by this Agreement or otherwise is unable to comply with the requirements for assuming this Agreement under the applicable provisions of the Bankruptcy Code.

10.2.2   INTERIM REMEDIES If the Subcontractor is not performing in accordance with the Progress Schedule at the time a petition in bankruptcy is filed, or at any subsequent time, the Contractor, while awaiting the decision of the Subcontractor or its trustee to reject or to assume this Agreement and provide adequate assurance of its ability to perform, may avail itself of such remedies under this Article as are reasonably necessary to maintain the Progress

Schedule. The Contractor may offset against any sums due or to become due the Subcontractor all costs incurred in pursuing any of the remedies provided including, but not limited to, reasonable overhead, profit and attorneys' fees. The Subcontractor shall be liable for the payment of any amount by which costs incurred may exceed the unpaid balance of the Subcontract Price.

10.3   SUSPENSION BY OWNER Should the Owner suspend the Work or any part which includes the Subcontract Work and such suspension is not due to any act or omission of the Contractor, or any other person or entity for whose acts or omissions the Contractor may be liable, the Contractor shall notify the Subcontractor in writing and upon receiving notification the Subcontractor shall immediately suspend the Subcontract Work. In the event of Owner suspension, the Contractor's liability to the Subcontractor shall be limited to the extent of the Contractor's recovery on the Subcontractor's behalf under the Subcontract Documents. The Contractor agrees to cooperate with the Subcontractor, at the Subcontractor's expense, in the prosecution of any Subcontractor claim arising out of an Owner suspension and to permit the Subcontractor to prosecute the claim, in the name of the Contractor, for the use and benefit of the Subcontractor.

10.4   TERMINATION BY OWNER Should the Owner terminate its contract with the Contractor or any part which includes the Subcontract Work, the Contractor shall notify the Subcontractor in writing within three (3) days of the termination and upon written notification, this Agreement shall be terminated and the Subcontractor shall immediately stop the Subcontract Work, follow all of Contractor's instructions, and mitigate all costs. In the event of Owner termination, the Contractor's liability to the Subcontractor shall be limited to the extent of the Contractor's recovery on the Subcontractor's behalf under the Subcontract Documents. The Contractor agrees to cooperate with the Subcontractor, at the Subcontractor's expense, in the prosecution of any Subcontractor claim arising out of the Owner termination and to permit the Subcontractor to prosecute the claim, in the name of the Contractor, for the use and benefit of the Subcontractor, or assign the claim to the Subcontractor.

10.5   CONTINGENT ASSIGNMENT OF THIS AGREEMENT The Contractor's contingent assignment of this Agreement to the Owner, as provided in the Owner-Contractor agreement, is effective when the Owner has terminated the Owner-Contractor agreement for cause and has accepted the assignment by notifying the Subcontractor in writing. This contingent assignment is subject to the prior rights of a surety that may be obligated under the Contractor's bond, if any. Subcontractor consents to such assignment and agrees to be bound to the assignee by the terms of this Agreement, provided that the assignee fulfills the obligations of the Contractor.

AGC DOCUMENT NO. 650 • STANDARD FORM OF AGREEMENT BETWEEN CONTRACTOR AND SUBCONTRACTOR
(Where the Contractor Assumes the Risk of Owner Payment)
© 1998, The Associated General Contractors of America

10.6 SUSPENSION BY CONTRACTOR The Contractor may order the Subcontractor in writing to suspend all or any part of the Subcontract Work for such period of time as may be determined to be appropriate for the convenience of the Contractor. Phased Work or interruptions of the Subcontract Work for short periods of time shall not be considered a suspension. The Subcontractor, after receipt of the Contractor's order, shall notify the Contractor in writing in sufficient time to permit the Contractor to provide timely notice to the Owner in accordance with the Owner-Contractor agreement of the effect of such order upon the Subcontract Work. The Subcontract Amount or Progress Schedule shall be adjusted by Subcontract Change Order for any increase in the time or cost of performance of this Agreement caused by such suspension. No claim under this Paragraph shall be allowed for any costs incurred more than fourteen (14) days prior to the Subcontractor's notice to the Contractor. Neither the Subcontract Amount nor the Progress Schedule shall be adjusted for any suspension, to the extent that performance would have been suspended, due in whole or in part to the fault or negligence of the Subcontractor or by a cause for which Subcontractor would have been responsible. The Subcontract Amount shall not be adjusted for any suspension to the extent that performance would have been suspended by a cause for which the Subcontractor would have been entitled only to a time extension under this Agreement.

10.7 WRONGFUL EXERCISE If the Contractor wrongfully exercises any option under this Article, the Contractor shall be liable to the Subcontractor solely for the reasonable value of Subcontract Work performed by the Subcontractor prior to the Contractor's wrongful action, including reasonable overhead and profit on the Subcontract Work performed, less prior payments made, together with reasonable overhead and profit on the Subcontract Work not executed, and other costs incurred by reason of such action.

10.8 TERMINATION BY SUBCONTRACTOR If the Subcontract Work has been stopped for thirty (30) days because the Subcontractor has not received progress payments or has been abandoned or suspended for an unreasonable period of time not due to the fault or neglect of the Subcontractor, then the Subcontractor may terminate this Agreement upon giving the Contractor seven (7) days' written notice. Upon such termination, Subcontractor shall be entitled to recover from the Contractor payment for all Subcontract Work satisfactorily performed but not yet paid for, including reasonable overhead, profit and attorneys' fees, costs and expenses. However, if the Owner has not paid the Contractor for the satisfactory performance of the Subcontract Work through no fault or neglect of the Contractor, and the Subcontractor terminates this Agreement under this Article because it has not received corresponding progress payments, the Subcontractor shall be entitled to recover from the Contractor, within a reasonable period of time following termination, payment for all Subcontract Work satisfactorily

performed but not yet paid for, including reasonable overhead and profit. The Contractor's liability for any other damages claimed by the Subcontractor under such circumstances shall be extinguished by the Contractor pursuing said damages and claims against the Owner, on the Subcontractor's behalf, in the manner provided for in Subparagraphs 10.3 and 10.4 of this Agreement.

ARTICLE 11
DISPUTE RESOLUTION

11.1 INITIAL DISPUTE RESOLUTION If a dispute arises out of or relates to this Agreement or its breach, the parties shall endeavor to settle the dispute first through direct discussions. If the dispute cannot be resolved through direct discussions, the parties shall participate in mediation under the Construction Industry Mediation Rules of the American Arbitration Association before recourse to any other form of binding dispute resolution. The location of the mediation shall be the location of the Project. Once a party files a request for mediation with the other party and with the American Arbitration Association, the parties agree to commence such mediation within thirty (30) days of filing of the request. Either party may terminate the mediation at any time after the first session, but the decision to terminate must be delivered in person to the other party and the mediator. Engaging in mediation is a condition precedent to any other form of binding dispute resolution.

11.2 WORK CONTINUATION AND PAYMENT Unless otherwise agreed in writing, the Subcontractor shall continue the Subcontract Work and maintain the Progress Schedule during any dispute resolution proceedings. If the Subcontractor continues to perform, the Contractor shall continue to make payments in accordance with this Agreement.

11.3 NO LIMITATION OF RIGHTS OR REMEDIES Nothing in this Article shall limit any rights or remedies not expressly waived by the Subcontractor which the Subcontractor may have under lien laws or payment bonds.

11.4 MULTIPARTY PROCEEDING The parties agree that to the extent permitted by Subcontract Document all parties necessary to resolve a claim shall be parties to the same dispute resolution proceeding. To the extent disputes between the Contractor and Subcontractor involve in whole or in part disputes between the Contractor and the Owner, disputes between the Subcontractor and the Contractor shall be decided by the same tribunal and in the same forum as disputes between the Contractor and the Owner.

11.5 DISPUTES BETWEEN CONTRACTOR AND SUBCONTRACTOR In the event that the provisions for resolution of disputes between the Contractor and the Owner contained in the Subcontract Documents do not permit con-

18

solidation or joinder with disputes of third parties, such as the Subcontractor, resolution of disputes between the Subcontractor and the Contractor involving in whole or in part disputes between the Contractor and the Owner shall be stayed pending conclusion of any dispute resolution proceeding between the Contractor and the Owner. At the conclusion of those proceedings, disputes between the Subcontractor and the Contractor shall be submitted again to mediation pursuant to Paragraph 11.1. Any disputes not resolved by mediation shall be decided in the manner selected in the agreement between the Owner and the Contractor.

11.6   COST OF DISPUTE RESOLUTION The cost of any mediation proceeding shall be shared equally by the parties participating. The prevailing party in any dispute arising out of or relating to this Agreement or its breach that is resolved by a dispute resolution procedure designated in the Subcontract Documents shall be entitled to recover from the other party reasonable attorneys' fees, costs and expenses incurred by the prevailing party in connection with such dispute resolution process.

ARTICLE 12

MISCELLANEOUS PROVISIONS

12.1   GOVERNING LAW This Agreement shall be governed by the law in effect at the location of the Project.

12.2   SEVERABILITY The partial or complete invalidity of any one or more provisions of this Agreement shall not affect the validity or continuing force and effect of any other provision.

12.3   NO WAIVER OF PERFORMANCE The failure of either party to insist, in any one or more instances, upon the performance of any of the terms, covenants or conditions of this Agreement, or to exercise any of its rights, shall not be construed as a waiver or relinquishment of term, covenant, condition or right with respect to further performance.

12.4   TITLES The titles given to the Articles of this Agreement are for ease of reference only and shall not be relied upon or cited for any other purpose.

12.5   OTHER PROVISIONS AND DOCUMENTS Other provisions and documents applicable to the Subcontract Work are set forth in Exhibit _____.

12.6   JOINT DRAFTING The parties expressly agree that this Agreement was jointly drafted, and that they both had opportunity to negotiate its terms and to obtain the assistance of counsel in reviewing its terms prior to execution. Therefore, this Agreement shall be construed neither against nor in favor of either party, but shall be construed in a neutral manner.

19

ARTICLE 13

EXISTING SUBCONTRACT DOCUMENT

As defined in Paragraph 2.3, the following Exhibits are a part of this Agreement.

EXHIBIT _____ ◆   The Subcontract Work, _____ pages. ◆

EXHIBIT _____ ◆   The Drawings, Specifications, General and other conditions, addenda and other information. (Attach a complete listing by title, date and number of pages.)

EXHIBIT _____ ◆   Progress Schedule, _____ pages. ◆

EXHIBIT _____ ◆   Alternates and Unit Prices, include dates when alternates and unit prices no longer apply, _____ pages. ◆

EXHIBIT _____ ◆   Temporary Services, stating specific responsibilities of the Subcontractor, _____ pages. ◆

EXHIBIT _____ ◆   Insurance Provisions, _____ pages. ◆

EXHIBIT _____ ◆   Other Provisions and Documents, _____ pages. ◆

This Agreement is entered into as of the date entered in Article 1.

CONTRACTOR: _____ ◆

ATTEST: _____ ◆   BY: _____ ◆

PRINT NAME: _____ ◆

PRINT TITLE: _____ ◆

SUBCONTRACTOR: _____ ◆

ATTEST: _____ ◆   BY: _____ ◆

PRINT NAME: _____ ◆

PRINT TITLE: _____ ◆

11/01
20

Appendix H

Interest Tables

5% FACTORS INTEREST

	Single Payment		Uniform Series				
	Compound Amount Factor CAF	Present Worth Factor PWSP	Compound Amount Factor USCA	Sinking Fund Factor SFF	Present Worth Factor PWUS	Capital Recovery Factor CRF	
n	Given P to Find F $(1+i)^n$	Given F to Find P $\dfrac{1}{(1+i)^n}$	Given A to Find F $\dfrac{(1+i)^n-1}{i}$	Given F to Find A $\dfrac{i}{(1+i)^n-1}$	Given A to Find P $\dfrac{(1+i)^n-1}{i(1+i)^n}$	Given P to Find A $\dfrac{i(1+i)^n}{(1+i)-1}$	n
1	1.050	0.9524	1.000	1.00001	0.952	1.05001	1
2	1.102	0.9070	2.050	0.48781	1.859	0.53781	2
3	1.158	0.8638	3.152	0.31722	2.723	0.36722	3
4	1.216	0.8227	4.310	0.23202	3.546	0.28202	4
5	1.276	0.7835	5.526	0.18098	4.329	0.23098	5
6	1.340	0.7462	6.802	0.14702	5.076	0.19702	6
7	1.407	0.7107	8.142	0.12282	5.786	0.17282	7
8	1.477	0.6768	9.549	0.10472	6.463	0.15472	8
9	1.551	0.6446	11.026	0.09069	7.108	0.14069	9
10	1.629	0.6139	12.578	0.07951	7.722	0.12951	10
11	1.710	0.5847	14.206	0.07039	8.306	0.12039	11
12	1.796	0.5568	15.917	0.06283	8.863	0.11283	12
13	1.886	0.5303	17.712	0.05646	9.393	0.10646	13
14	1.980	0.5051	19.598	0.05103	9.899	0.10103	14
15	2.079	0.4810	21.578	0.04634	10.380	0.09634	15
16	2.183	0.4581	23.657	0.04227	10.838	0.09227	16
17	2.292	0.4363	25.840	0.03870	11.274	0.08870	17
18	2.407	0.4155	28.132	0.03555	11.689	0.08555	18
19	2.527	0.3957	30.538	0.03275	12.085	0.08275	19
20	2.653	0.3769	33.065	0.03024	12.462	0.08024	20
21	2.786	0.3589	35.718	0.02800	12.821	0.07800	21
22	2.925	0.3419	38.504	0.02597	13.163	0.07597	22
23	3.071	0.3256	41.429	0.02414	13.488	0.07414	23
24	3.225	0.3101	44.500	0.02247	13.798	0.07247	24
25	3.386	0.2953	47.725	0.02095	14.094	0.07095	25
26	3.556	0.2812	51.112	0.01957	14.375	0.06956	26
27	3.733	0.2679	54.667	0.01829	14.643	0.06829	27
28	3.920	0.2551	58.400	0.01712	14.898	0.06712	28
29	4.116	0.2430	62.320	0.01605	15.141	0.06605	29
30	4.322	0.2314	66.436	0.01505	15.372	0.06505	30
35	5.516	0.1813	90.316	0.01107	16.374	0.06107	35
40	7.040	0.1421	120.794	0.00828	17.159	0.05828	40
50	11.467	0.0872	209.336	0.00478	18.256	0.05478	50
75	38.830	0.0258	756.594	0.00132	19.485	0.05132	75
100	131.488	0.0076	2609.761	0.00038	19.848	0.05038	100

6% FACTORS INTEREST

n	Single Payment			Uniform Series			n
	Compound Amount Factor CAF	Present Worth Factor PWSP	Compound Amount Factor USCA	Sinking Fund Factor SFF	Present Worth Factor PWUS	Capital Recovery Factor CRF	
	Given P to Find F $(1+i)^n$	Given F to Find P $\dfrac{1}{(1+i)^n}$	Given A to Find F $\dfrac{(1+i)^n-1}{i}$	Given F to Find A $\dfrac{i}{(1+i)^n-1}$	Given A to Find P $\dfrac{(1+i)^n-1}{i(1+i)^n}$	Given P to Find A $\dfrac{i(1+i)^n}{(1+i)^n-1}$	
1	1.060	0.9434	1.000	1.00001	0.943	1.06001	1
2	1.124	0.8900	2.060	0.48544	1.833	0.54544	2
3	1.191	0.8396	3.184	0.31411	2.673	0.37411	3
4	1.262	0.7921	4.375	0.22860	3.465	0.28860	4
5	1.338	0.7473	5.637	0.17740	4.212	0.23740	5
6	1.419	0.7050	6.975	0.14337	4.917	0.20337	6
7	1.504	0.6651	8.394	0.11914	5.582	0.17914	7
8	1.594	0.6274	9.897	0.10104	6.210	0.16104	8
9	1.689	0.5919	11.491	0.08702	6.802	0.14702	9
10	1.791	0.5584	13.181	0.07587	7.360	0.13587	10
11	1.898	0.5268	14.971	0.06679	7.887	0.12679	11
12	2.012	0.4970	16.870	0.05928	8.384	0.11928	12
13	2.133	0.4688	18.882	0.05296	8.853	0.11296	13
14	2.261	0.4423	21.015	0.04759	9.295	0.10759	14
15	2.397	0.4173	23.275	0.04296	9.712	0.10296	15
16	2.540	0.3937	25.672	0.03895	10.106	0.09895	16
17	2.693	0.3714	28.212	0.03545	10.477	0.09545	17
18	2.854	0.3503	30.905	0.03236	10.828	0.09236	18
19	3.026	0.3305	33.759	0.02962	11.158	0.08962	19
20	3.207	0.3118	36.785	0.02719	11.470	048719	20
21	3.399	0.2942	39.992	0.02501	11.764	0.08501	21
22	3.603	0.2775	43.391	0.02305	12.041	0.08305	22
23	3.820	0.2618	46.994	0.02128	12.303	0.08128	23
24	4.049	0.2470	50.814	0.01968	12.550	0.07968	24
25	4.292	0.2330	54.863	0.01823	12.783	0.07823	25
26	4.549	0.2198	59.154	0.01690	13.003	0.07690	26
27	4.822	0.2074	63.704	0.01570	13.210	0.07570	27
28	5.112	0.1956	68.526	0.01459	13.406	0.07459	28
29	5.418	0.1846	73.637	0.01358	13.591	0.07358	29
30	5.743	0.1741	79.055	0.01265	13.765	0.07265	30
35	7.686	0.1301	111.430	0.00897	14.498	0.06897	35
40	10.285	0.0972	154.755	0.00646	15.046	0.06646	40
50	18.419	0.0543	290.321	0.00344	15.762	0.06344	50
75	79.051	0.0127	1300.852	0.00077	16.456	0.06077	75
100	339.269	0.0029	5637.809	0.00018	16.618	0.06018	100

7% FACTORS INTEREST

	Single Payment		Uniform Series				
	Compound Amount Factor CAF	Present Worth Factor PWSP	Compound Amount Factor USCA	Sinking Fund Factor SFF	Present Worth Factor PWUS	Capital Recovery Factor CRF	
n	Given P to Find F $(1+i)^n$	Given F to Find P $\dfrac{1}{(1+i)^n}$	Given A to Find F $\dfrac{(1+i)^n-1}{i}$	Given F to Find A $\dfrac{i}{(1+i)^n-1}$	Given A to Find P $\dfrac{(1+i)^n-1}{i(1+i)^n}$	Given P to Find A $\dfrac{i(1+i)^n}{(1+i)^n-1}$	n
1	1.070	0.9346	1.000	1.00000	0.935	1.07000	1
2	1.145	0.8734	2.070	0.48310	1.808	0.55310	2
3	1.225	0.8163	3.215	0.31106	2.624	0.38105	3
4	1.311	0.7629	4.440	0.22523	3.387	0.29523	4
5	1.403	0.7130	5.751	0.17389	4.100	0.24389	5
6	1.501	0.6663	7.153	0.13980	4.766	0.20980	6
7	1.606	0.6228	8.654	0.11555	5.389	0.18555	7
8	1.718	0.5820	10.260	0.09747	5.971	0.16747	8
9	1.838	0.5439	11.978	0.08349	6.515	0.15349	9
10	1.967	0.5084	13.816	0.07238	7.024	0.14238	10
11	2.105	0.4751	15.783	0.06336	7.499	0.13336	11
12	2.252	0.4440	17.888	0.05590	7.943	0.12590	12
13	2.410	0.4150	20.140	0.04965	8.358	0.11965	13
14	2.579	0.3878	22.550	0.04435	8.745	0.11435	14
15	2.759	0.3625	25.129	0.03980	9.108	0.10980	15
16	2.952	0.3387	27.887	0.03586	9.447	0.10586	16
17	3.159	0.3166	30.840	0.03243	9.763	0.10243	17
18	3.380	0.2959	33.998	0.02941	10.059	0.09941	18
19	3.616	0.2765	37.378	0.02675	10.336	0.09675	19
20	3.870	0.2584	40.995	0.02439	10.594	0.09439	20
21	4.140	0.2415	44.864	0.02229	10.835	0.09229	21
22	4.430	0.2257	49.005	0.02041	11.061	0.09041	22
23	4.740	0.2110	53.435	0.01871	11.272	0.08871	23
24	5.072	0.1972	58.175	0.01719	11.469	0.08719	24
25	5.427	0.1843	63.247	0.01581	11.654	0.08581	25
26	5.807	0.1722	68.675	0.01456	11.826	0.08456	26
27	6.214	0.1609	74.482	0.01343	11.987	0.08343	27
28	6.649	0.1504	80.695	0.01239	12.137	0.08239	28
29	7.114	0.1406	87.344	0.01145	12.278	0.08145	29
30	7.612	0.1314	94.458	0.01059	12.409	0.08059	30
35	10.676	0.0937	138.233	0.00723	12.948	0.07723	35
40	14.974	0.0668	199.628	0.00501	13.332	0.07501	40
50	29.456	0.0339	406.511	0.00246	13.801	0.07246	50
75	159.866	0.0063	2269.516	0.00044	14.196	0.07044	75
100	867.644	0.0012	12380.633	0.00008	14.269	0.07008	100

8% FACTORS INTEREST

	Single Payment		Uniform Series				
	Compound Amount Factor CAF	Present Worth Factor PWSP	Compound Amount Factor USCA	Sinking Fund Factor SFF	Present Worth Factor PWUS	Capital Recovery Factor CRF	
n							n
	Given P to Find F $(1+i)^n$	Given F to Find P $\dfrac{1}{(1+i)^n}$	Given A to Find F $\dfrac{(1+i)^n-1}{i}$	Given F to Find A $\dfrac{i}{(1+i)^n-1}$	Given A to Find P $\dfrac{(1+i)^n-1}{i(1+i)^n}$	Given P to Find A $\dfrac{i(1+i)^n}{(1+i)^n-1}$	
1	1.080	0.9259	1.000	1.00000	0.926	1.08000	1
2	1.166	0.8573	2.080	0.48077	1.783	0.56077	2
3	1.260	0.7938	3.246	0.30804	2.577	0.38804	3
4	1.360	0.7350	4.506	0.22192	3.312	0.30192	4
5	1.469	0.6806	5.867	0.17046	3.993	0.25046	5
6	1.587	0.6302	7.336	0.13632	4.623	0.21632	6
7	1.714	0.5835	8.923	0.11207	5.206	0.19207	7
8	1.851	0.5403	10.637	0.09402	5.747	0.17402	8
9	1.999	0.5003	12.487	0.08008	6.247	0.16008	9
10	2.159	0.4632	14.486	0.06903	6.710	0.14903	10
11	2.332	0.4289	16.645	0.06008	7.139	0.14008	11
12	2.518	0.3971	18.977	0.05270	7.536	0.13270	12
13	2.720	0.3677	21.495	0.04652	7.904	0.12652	13
14	2.937	0.3405	24.215	0.04130	8.244	0.12130	14
15	3.172	0.3152	27.152	0.03683	8.559	0.11683	15
16	3.426	0.2919	30.324	0.03298	8.851	0.11298	16
17	3.700	0.2703	33.750	0.02963	9.122	0.10963	17
18	3.996	0.2503	37.450	0.02670	9.372	0.10670	18
19	4.316	0.2317	41.446	0.02413	9.604	0.10413	19
20	4.661	0.2146	45.761	0.02185	9.818	0.10185	20
21	5.034	0.1987	50.422	0.01983	10.017	0.09983	21
22	5.436	0.1839	55.456	0.01803	10.201	0.09803	22
23	5.871	0.1703	60.892	0.01642	10.371	0.09642	23
24	6.341	0.1577	66.764	0.01498	10.529	0.09498	24
25	6.848	0.1460	73.105	0.01368	10.675	0.09368	25
26	7.396	0.1352	79.953	0.01251	10.810	0.09251	26
27	7.988	0.1252	87.349	0.01145	10.935	0.09145	27
28	8.627	0.1159	95.337	0.01049	11.051	0.09049	28
29	9.317	0.1073	103.964	0.00962	11.158	0.08962	29
30	10.062	0.0994	113.281	0.00883	11.258	0.08883	30
35	14.785	0.0676	172.313	0.00580	11.655	0.08580	35
40	21.724	0.0460	259.050	0.00386	11.925	0.08386	40
50	46.900	0.0213	573.753	0.00174	11.233	0.08174	50
75	321.190	0.0031	4002.378	0.00025	11.461	0.08025	75
100	2199.630	0.0005	27482.879	0.00004	11.494	0.08004	100

Appendix I

Plans for Small Gas Station

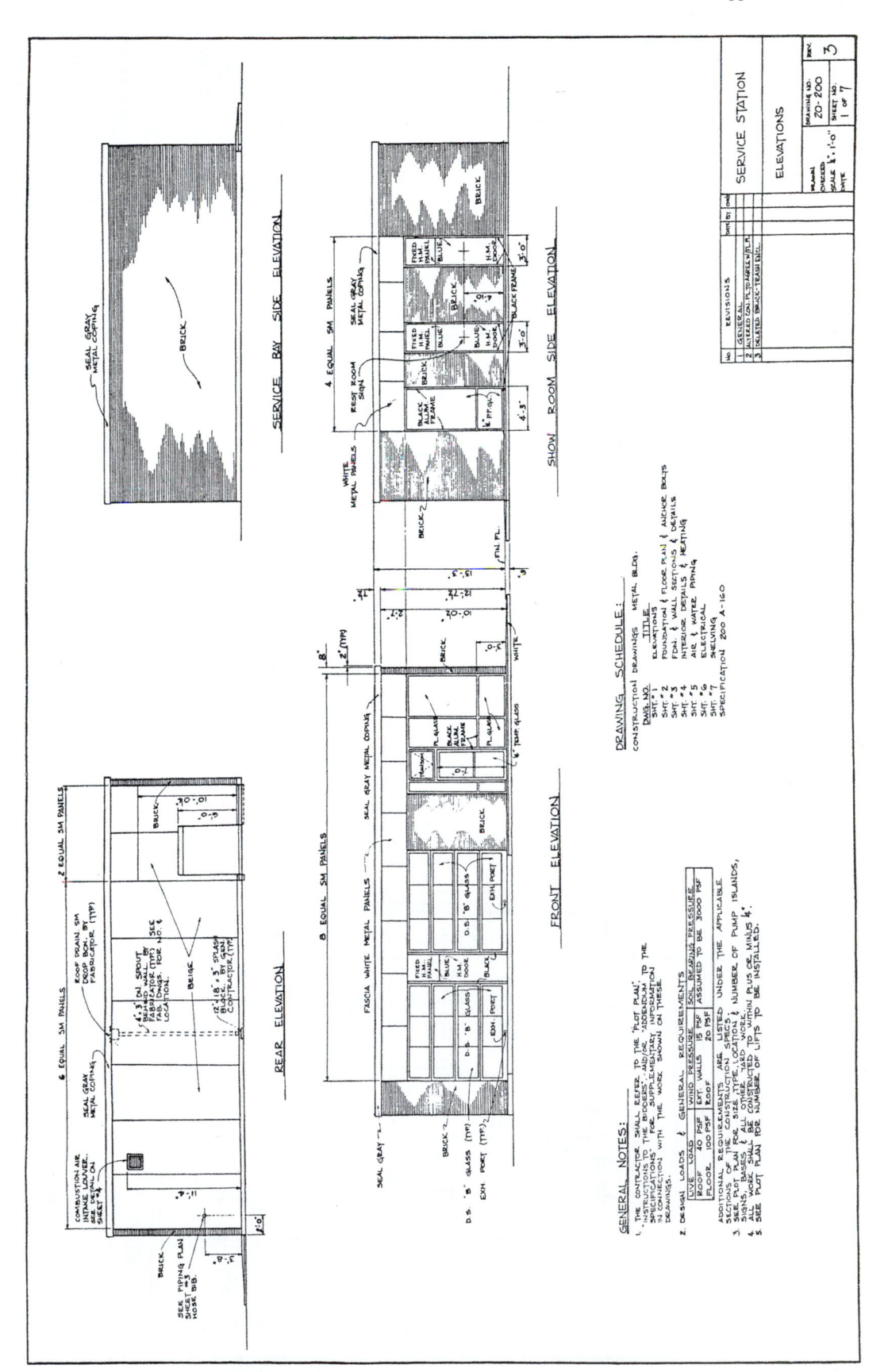

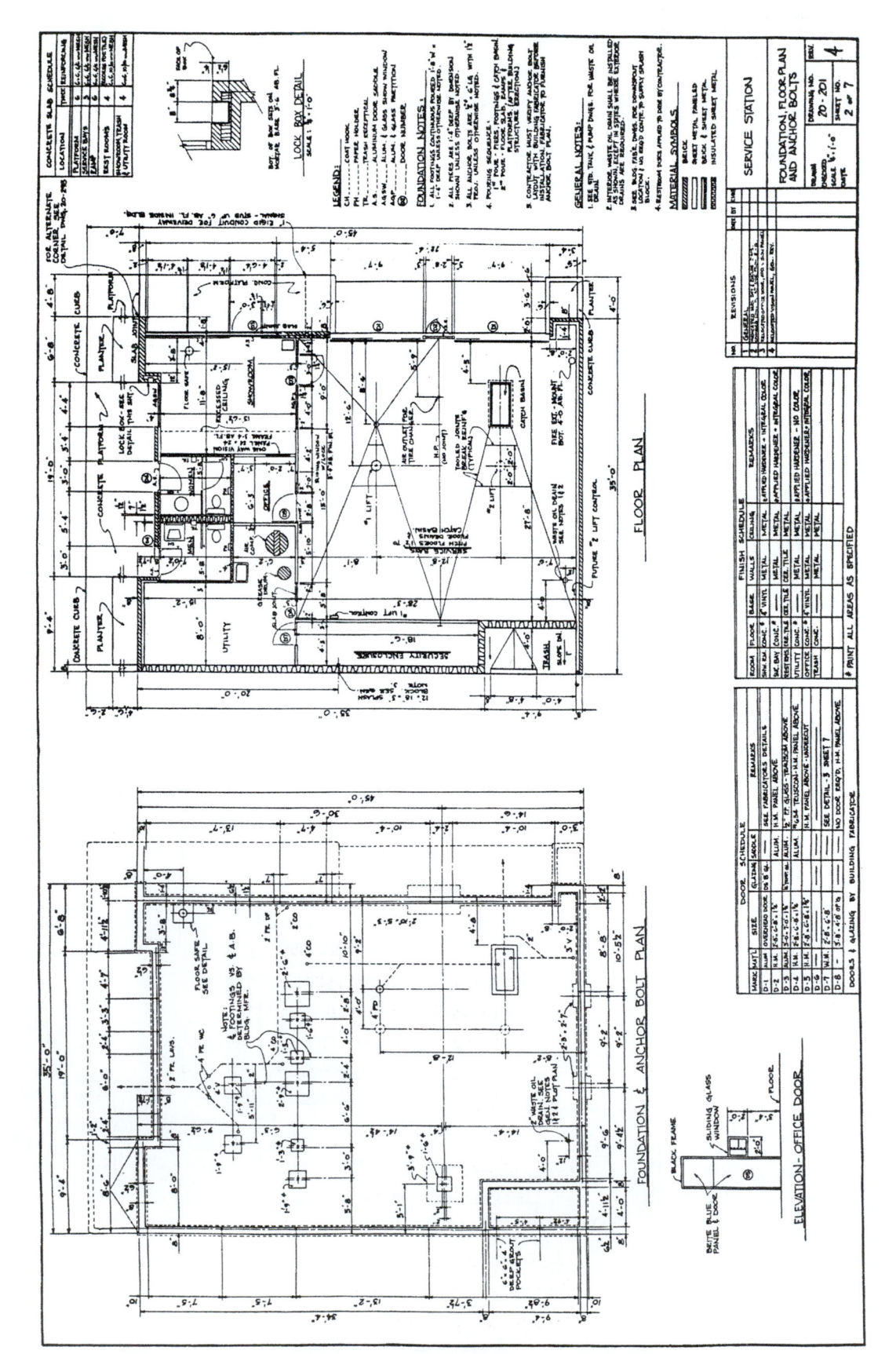

Appendix J

Site Reconnaissance Checklist

GENERAL CONSIDERATIONS

    A. What features are native to topography and climate?

    B. What is required for construction method selected?

    C. What features are needed to support construction force?

    D. What features might encroach on local society or environment?

    A. Features native to topography and climate

      1. Actual topography (excessive grades, etc.)

      2. Elevation

      3. Geology (soil characteristics, rock, etc.)

      4. Ground cover

      5. Excessive seasonal effects

      6. Wind direction

      7. Natural defenses

      8. Drainage

      9. Subsurface water conditions

      10. Seismic zones

    B. Features required that contribute to construction method

      1. Accessibility to site (rail, road, water)

      2. Labor availability (skill, cost, attitude)

      3. Material availability (salvage, cost, attitude)

      4. Locate borrow pits (gravel, sand, base, fill)

      5. Locate storage areas, plant sites

      6. Alternate building, campsites

      7. General working room about site

      8. Location of existing structures and utilities

      9. Conflicts with existing structures and utilities

      10. Overhead

      11. Disposal areas

      12. Land usage

      13. Local building practices

    C. Features to support construction force

      1. Billeting/shelter

      2. Food (also on-job meals)

      3. Special equipment

      4. Clothing

      5. Communications

      6. Local hazards

      7. Fire/security protection available

      8. Local customs/culture

      9. Potable H_2O

    10. Sanitary facilities (also for job)

    11. Entertainment

    12. Small stores

    13. Medical

    14. Banking, currency

    15. Transportation

    16. Local maintenance available

D. Features that might encroach on local society or environment

      1. Noise

      2. Dust

      3. Blasting

      4. Hauling over roads

      5. Use of water

      6. Burning (smoke)

      7. Drainage (create problems)

      8. Flight operations

      9. Disposal areas

    10. Utility disruption

    11. Relocation problems

    12. Work hours

    13. Economy impact

    14. Community attitude

    15. Security

    16. Political

Appendix K(1)

The Cumulative Normal Distribution Function[†]

$$\Phi(z) = \int_{-\infty}^{z} \phi(t)\,dt \qquad \text{for} \quad (-\infty < z \leq 0)$$

z	·00	·01	·02	·03	·04	·05	·06	·07	·08	·09
− ·0	·5000	·4960	·4920	·4880	·4840	·4801	·4761	·4721	·4681	·4641
− ·1	·4602	·4562	·4522	·4483	·4443	·4404	·4364	·4325	·4286	·4247
− ·2	·4207	·4168	·4129	·4090	·4052	·4013	·3974	·3936	·3897	·3859
− ·3	·3821	·3783	·3745	·3707	·3669	·3632	·3594	·3557	·3520	·3483
− ·4	·3446	·3409	·3372	·3336	·3300	·3264	·3228	·3192	·3156	·3121
− ·5	·3085	·3050	·3015	·2981	·2946	·2912	·2877	·2843	·2810	·2776
− ·6	·2743	·2709	·2676	·2643	·2611	·2578	·2546	·2514	·2483	·2451
− ·7	·2420	·2389	·2358	·2327	·2297	·2266	·2236	·2206	·2177	·2148
− ·8	·2119	·2090	·2061	·2033	·2005	·1977	·1949	·1922	·1894	·1867
− ·9	·1841	·1814	·1788	·1762	·1736	·1711	·1685	·1660	·1635	·1611
−1·0	·1587	·1562	·1539	·1515	·1492	·1469	·1446	·1423	·1401	·1379
−1·1	·1357	·1335	·1314	·1292	·1271	·1251	·1230	·1210	·1190	·1170
−1·2	·1151	·1131	·1112	·1093	·1075	·1056	·1038	·1020	·1003	·09853
−1·3	·09680	·09510	·09342	·09176	·09012	·08851	·08691	·08534	·08379	·08226
−1·4	·08076	·07927	·07780	·07636	·07493	·07353	·07215	·07078	·06944	·06811
−1·5	·06681	·06552	·06426	·06301	·06178	·06057	·05938	·05821	·05705	·05592
−1·6	·05480	·05370	·05262	·05155	·05050	·04947	·04846	·04746	·04648	·04551
−1·7	·04457	·04363	·04272	·04182	·04093	·04006	·03920	·03836	·03754	·03673
−1·8	·03593	·03515	·03438	·03362	·03288	·03216	·03144	·03074	·03005	·02938
−1·9	·02872	·02807	·02743	·02680	·02619	·02559	·02500	·02442	·02385	·02330
−2·0	·02275	·02222	·02169	·02118	·02068	·02018	·01970	·01923	·01876	·01831
−2·1	·01786	·01743	·01700	·01659	·01618	·01578	·01539	·01500	·01463	·01426
−2·2	·01390	·01355	·01321	·01287	·01255	·01222	·01191	·01160	·01130	·01101
−2·3	·01072	·01044	·01017	$\cdot0^2 9903$	$\cdot0^2 9642$	$\cdot0^2 9387$	$\cdot0^2 9137$	$\cdot0^2 8894$	$\cdot0^2 8656$	$\cdot0^2 8424$
−2·4	$\cdot0^2 8198$	$\cdot0^2 7976$	$\cdot0^2 7760$	$\cdot0^2 7549$	$\cdot0^2 7344$	$\cdot0^2 7143$	$\cdot0^2 6947$	$\cdot0^2 6756$	$\cdot0^2 6569$	$\cdot0^2 6387$
−2·5	$\cdot0^2 6210$	$\cdot0^2 6037$	$\cdot0^2 5868$	$\cdot0^2 5703$	$\cdot0^2 5543$	$\cdot0^2 5386$	$\cdot0^2 5234$	$\cdot0^2 5085$	$\cdot0^2 4940$	$\cdot0^2 4799$
−2·6	$\cdot0^2 4661$	$\cdot0^2 4527$	$\cdot0^2 4396$	$\cdot0^2 4269$	$\cdot0^2 4145$	$\cdot0^2 4025$	$\cdot0^2 3907$	$\cdot0^2 3793$	$\cdot0^2 3681$	$\cdot0^2 3573$
−2·7	$\cdot0^2 3467$	$\cdot0^2 3364$	$\cdot0^2 3264$	$\cdot0^2 3167$	$\cdot0^2 3072$	$\cdot0^2 2980$	$\cdot0^2 2890$	$\cdot0^2 2803$	$\cdot0^2 2718$	$\cdot0^2 2635$
−2·8	$\cdot0^2 2555$	$\cdot0^2 2477$	$\cdot0^2 2401$	$\cdot0^2 2327$	$\cdot0^2 2256$	$\cdot0^2 2186$	$\cdot0^2 2118$	$\cdot0^2 2052$	$\cdot0^2 1988$	$\cdot0^2 1926$
−2·9	$\cdot0^2 1866$	$\cdot0^2 1807$	$\cdot0^2 1750$	$\cdot0^2 1695$	$\cdot0^2 1641$	$\cdot0^2 1589$	$\cdot0^2 1538$	$\cdot0^2 1489$	$\cdot0^2 1441$	$\cdot0^2 1395$
−3·0	$\cdot0^2 1350$	$\cdot0^2 1306$	$\cdot0^2 1264$	$\cdot0^2 1223$	$\cdot0^2 1183$	$\cdot0^2 1144$	$\cdot0^2 1107$	$\cdot0^2 1070$	$\cdot0^2 1035$	$\cdot0^2 1001$
−3·1	$\cdot0^3 9676$	$\cdot0^3 9354$	$\cdot0^3 9043$	$\cdot0^3 8740$	$\cdot0^3 8447$	$\cdot0^3 8164$	$\cdot0^3 7888$	$\cdot0^3 7622$	$\cdot0^3 7364$	$\cdot0^3 7114$
−3·2	$\cdot0^3 6871$	$\cdot0^3 6637$	$\cdot0^3 6410$	$\cdot0^3 6190$	$\cdot0^3 5976$	$\cdot0^3 5770$	$\cdot0^3 5571$	$\cdot0^3 5377$	$\cdot0^3 5190$	$\cdot0^3 5009$
−3·3	$\cdot0^3 4834$	$\cdot0^3 4665$	$\cdot0^3 4501$	$\cdot0^3 4342$	$\cdot0^3 4189$	$\cdot0^3 4041$	$\cdot0^3 3897$	$\cdot0^3 3758$	$\cdot0^3 3624$	$\cdot0^3 3495$
−3·4	$\cdot0^3 3369$	$\cdot0^3 3248$	$\cdot0^3 3131$	$\cdot0^3 3018$	$\cdot0^3 2909$	$\cdot0^3 2803$	$\cdot0^3 2701$	$\cdot0^3 2602$	$\cdot0^3 2507$	$\cdot0^3 2415$
−3·5	$\cdot0^3 2326$	$\cdot0^3 2241$	$\cdot0^3 2158$	$\cdot0^3 2078$	$\cdot0^3 2001$	$\cdot0^3 1926$	$\cdot0^3 1854$	$\cdot0^3 1785$	$\cdot0^3 1718$	$\cdot0^3 1653$
−3·6	$\cdot0^3 1591$	$\cdot0^3 1531$	$\cdot0^3 1473$	$\cdot0^3 1417$	$\cdot0^3 1363$	$\cdot0^3 1311$	$\cdot0^3 1261$	$\cdot0^3 1213$	$\cdot0^3 1166$	$\cdot0^3 1121$
−3·7	$\cdot0^3 1078$	$\cdot0^3 1036$	$\cdot0^4 9961$	$\cdot0^4 9574$	$\cdot0^4 9201$	$\cdot0^4 8842$	$\cdot0^4 8496$	$\cdot0^4 8162$	$\cdot0^4 7841$	$\cdot0^4 7532$
−3·8	$\cdot0^4 7235$	$\cdot0^4 6948$	$\cdot0^4 6673$	$\cdot0^4 6407$	$\cdot0^4 6152$	$\cdot0^4 5906$	$\cdot0^4 5669$	$\cdot0^4 5442$	$\cdot0^4 5223$	$\cdot0^4 5012$
−3·9	$\cdot0^4 4810$	$\cdot0^4 4615$	$\cdot0^4 4427$	$\cdot0^4 4247$	$\cdot0^4 4074$	$\cdot0^4 3908$	$\cdot0^4 3747$	$\cdot0^4 3594$	$\cdot0^4 3446$	$\cdot0^4 3304$
−4·0	$\cdot0^4 3167$	$\cdot0^4 3036$	$\cdot0^4 2910$	$\cdot0^4 2789$	$\cdot0^4 2673$	$\cdot0^4 2561$	$\cdot0^4 2454$	$\cdot0^4 2351$	$\cdot0^4 2252$	$\cdot0^4 2157$
−4·1	$\cdot0^4 2066$	$\cdot0^4 1978$	$\cdot0^4 1894$	$\cdot0^4 1814$	$\cdot0^4 1737$	$\cdot0^4 1662$	$\cdot0^4 1591$	$\cdot0^4 1523$	$\cdot0^4 1458$	$\cdot0^4 1395$
−4·2	$\cdot0^4 1335$	$\cdot0^4 1277$	$\cdot0^4 1222$	$\cdot0^4 1168$	$\cdot0^4 1118$	$\cdot0^4 1069$	$\cdot0^4 1022$	$\cdot0^5 9774$	$\cdot0^5 9345$	$\cdot0^5 8934$
−4·3	$\cdot0^5 8540$	$\cdot0^5 8163$	$\cdot0^5 7801$	$\cdot0^5 7455$	$\cdot0^5 7124$	$\cdot0^5 6807$	$\cdot0^5 6503$	$\cdot0^5 6212$	$\cdot0^5 5934$	$\cdot0^5 5668$
−4·4	$\cdot0^5 5413$	$\cdot0^5 5169$	$\cdot0^5 4935$	$\cdot0^5 4712$	$\cdot0^5 4498$	$\cdot0^5 4294$	$\cdot0^5 4098$	$\cdot0^5 3911$	$\cdot0^5 3732$	$\cdot0^5 3561$
−4·5	$\cdot0^5 3398$	$\cdot0^5 3241$	$\cdot0^5 3092$	$\cdot0^5 2949$	$\cdot0^5 2813$	$\cdot0^5 2682$	$\cdot0^5 2558$	$\cdot0^5 2439$	$\cdot0^5 2325$	$\cdot0^5 2216$
−4·6	$\cdot0^5 2112$	$\cdot0^5 2013$	$\cdot0^5 1919$	$\cdot0^5 1828$	$\cdot0^5 1742$	$\cdot0^5 1660$	$\cdot0^5 1581$	$\cdot0^5 1506$	$\cdot0^5 1434$	$\cdot0^5 1366$
−4·7	$\cdot0^5 1301$	$\cdot0^5 1239$	$\cdot0^5 1179$	$\cdot0^5 1123$	$\cdot0^5 1069$	$\cdot0^5 1017$	$\cdot0^6 9680$	$\cdot0^6 9211$	$\cdot0^6 8765$	$\cdot0^6 8339$
−4·8	$\cdot0^6 7933$	$\cdot0^6 7547$	$\cdot0^6 7178$	$\cdot0^6 6827$	$\cdot0^6 6492$	$\cdot0^6 6173$	$\cdot0^6 5869$	$\cdot0^6 5580$	$\cdot0^6 5304$	$\cdot0^6 5042$
−4·9	$\cdot0^6 4792$	$\cdot0^6 4554$	$\cdot0^6 4327$	$\cdot0^6 4111$	$\cdot0^6 3906$	$\cdot0^6 3711$	$\cdot0^6 3525$	$\cdot0^6 3348$	$\cdot0^6 3179$	$\cdot0^6 3019$

Example: $\Phi(-3\cdot57) = \cdot0^3 1785 = 0\cdot0001785$.

[†] By permission from A. Hald, *Statistical Tables, and Formulas*, John Wiley & Sons, Inc., New York, 1952.

The Cumulative Normal Distribution Function[†]

$$\Phi(z)=\int_{-\infty}^{z}\phi(t)dt \qquad \text{for } (0 \leqq z < \infty)$$

z	·00	·01	·02	·03	·04	·05	·06	·07	·08	·09
·0	·5000	·5040	·5080	·5120	·5160	·5199	·5239	·5279	·5319	·5359
·1	·5398	·5438	·5478	·5517	·5557	·5596	·5636	·5675	·5714	·5753
·2	·5793	·5832	·5871	·5910	·5948	·5987	·6026	·6064	·6103	·6141
·3	·6179	·6217	·6255	·6293	·6331	·6368	·6406	·6443	·6480	·6517
·4	·6554	·6591	·6628	·6664	·6700	·6736	·6772	·6808	·6844	·6879
·5	·6915	·6950	·6985	·7019	·7054	·7088	·7123	·7157	·7190	·7224
·6	·7257	·7291	·7324	·7357	·7389	·7422	·7454	·7486	·7517	·7549
·7	·7580	·7611	·7642	·7673	·7703	·7734	·7764	·7794	·7823	·7852
·8	·7881	·7910	·7939	·7967	·7995	·8023	·8051	·8078	·8106	·8133
·9	·8159	·8186	·8212	·8238	·8264	·8289	·8315	·8340	·8365	·8389
1·0	·8413	·8438	·8461	·8485	·8508	·8531	·8554	·8577	·8599	·8621
1·1	·8643	·8665	·8686	·8708	·8729	·8749	·8770	·8790	·8810	·8830
1·2	·8849	·8869	·8888	·8907	·8925	·8944	·8962	·8980	·8997	·90147
1·3	·90320	·90490	·90658	·90824	·90988	·91149	·91309	·91466	·91621	·91774
1·4	·91924	·92073	·92220	·92364	·92507	·92647	·92785	·92922	·93056	·93189
1·5	·93319	·93448	·93574	·93699	·93822	·93943	·94062	·94179	·94295	·94408
1·6	·94520	·94630	·94738	·94845	·94950	·95053	·95154	·95254	·95352	·95449
1·7	·95543	·95637	·95728	·95818	·95907	·95994	·96080	·96164	·96246	·96327
1·8	·96407	·96485	·96562	·96638	·96712	·96784	·96856	·96926	·96995	·97062
1·9	·97128	·97193	·97257	·97320	·97381	·97441	·97500	·97558	·97615	·97670
2·0	·97725	·97778	·97831	·97882	·97932	·97982	·98030	·98077	·98124	·98169
2·1	·98214	·98257	·98300	·98341	·98382	·98422	·98461	·98500	·98537	·98574
2·2	·98610	·98645	·98679	·98713	·98745	·98778	·98809	·98840	·98870	·98899
2·3	·98928	·98956	·98983	·$9^2$0097	·$9^2$0358	·$9^2$0613	·$9^2$0863	·$9^2$1106	·$9^2$1344	·$9^2$1576
2·4	·$9^2$1802	·$9^2$2024	·$9^2$2240	·$9^2$2451	·$9^2$2656	·$9^2$2857	·$9^2$3053	·$9^2$3244	·$9^2$3431	·$9^2$3613
2·5	·$9^2$3790	·$9^2$3963	·$9^2$4132	·$9^2$4297	·$9^2$4457	·$9^2$4614	·$9^2$4766	·$9^2$4915	·$9^2$5060	·$9^2$5201
2·6	·$9^2$5339	·$9^2$5473	·$9^2$5604	·$9^2$5731	·$9^2$5855	·$9^2$5975	·$9^2$6093	·$9^2$6207	·$9^2$6319	·$9^2$6427
2·7	·$9^2$6533	·$9^2$6636	·$9^2$6736	·$9^2$6833	·$9^2$6928	·$9^2$7020	·$9^2$7110	·$9^2$7197	·$9^2$7282	·$9^2$7365
2·8	·$9^2$7445	·$9^2$7523	·$9^2$7599	·$9^2$7673	·$9^2$7744	·$9^2$7814	·$9^2$7882	·$9^2$7948	·$9^2$8012	·$9^2$8074
2·9	·$9^2$8134	·$9^2$8193	·$9^2$8250	·$9^2$8305	·$9^2$8359	·$9^2$8411	·$9^2$8462	·$9^2$8511	·$9^2$8559	·$9^2$8605
3·0	·$9^2$8650	·$9^2$8694	·$9^2$8736	·$9^2$8777	·$9^2$8817	·$9^2$8856	·$9^2$8893	·$9^2$8930	·$9^2$8965	·$9^2$8999
3·1	·$9^3$0324	·$9^3$0646	·$9^3$0957	·$9^3$1260	·$9^3$1553	·$9^3$1836	·$9^3$2112	·$9^3$2378	·$9^3$2636	·$9^3$2886
3·2	·$9^3$3129	·$9^3$3363	·$9^3$3590	·$9^3$3810	·$9^3$4024	·$9^3$4230	·$9^3$4429	·$9^3$4623	·$9^3$4810	·$9^3$4991
3·3	·$9^3$5166	·$9^3$5335	·$9^3$5499	·$9^3$5658	·$9^3$5811	·$9^3$5959	·$9^3$6103	·$9^3$6242	·$9^3$6376	·$9^3$6505
3·4	·$9^3$6631	·$9^3$6752	·$9^3$6869	·$9^3$6982	·$9^3$7091	·$9^3$7197	·$9^3$7299	·$9^3$7398	·$9^3$7493	·$9^3$7585
3·5	·$9^3$7674	·$9^3$7759	·$9^3$7842	·$9^3$7922	·$9^3$7999	·$9^3$8074	·$9^3$8146	·$9^3$8215	·$9^3$8282	·$9^3$8347
3·6	·$9^3$8409	·$9^3$8469	·$9^3$8527	·$9^3$8583	·$9^3$8637	·$9^3$8689	·$9^3$8739	·$9^3$8787	·$9^3$8834	·$9^3$8879
3·7	·$9^3$8922	·$9^3$8964	·$9^4$0039	·$9^4$0426	·$9^4$0799	·$9^4$1158	·$9^4$1504	·$9^4$1838	·$9^4$2159	·$9^4$2468
3·8	·$9^4$2765	·$9^4$3052	·$9^4$3327	·$9^4$3593	·$9^4$3848	·$9^4$4094	·$9^4$4331	·$9^4$4558	·$9^4$4777	·$9^4$4988
3·9	·$9^4$5190	·$9^4$5385	·$9^4$5573	·$9^4$5753	·$9^4$5926	·$9^4$6092	·$9^4$6253	·$9^4$6406	·$9^4$6554	·$9^4$6696
4·0	·$9^4$6833	·$9^4$6964	·$9^4$7090	·$9^4$7211	·$9^4$7327	·$9^4$7439	·$9^4$7546	·$9^4$7649	·$9^4$7748	·$9^4$7843
4·1	·$9^4$7934	·$9^4$8022	·$9^4$8106	·$9^4$8186	·$9^4$8263	·$9^4$8338	·$9^4$8409	·$9^4$8477	·$9^4$8542	·$9^4$8605
4·2	·$9^4$8665	·$9^4$8723	·$9^4$8778	·$9^4$8832	·$9^4$8882	·$9^4$8931	·$9^4$8978	·$9^5$0226	·$9^5$0655	·$9^5$1066
4·3	·$9^5$1460	·$9^5$1837	·$9^5$2199	·$9^5$2545	·$9^5$2876	·$9^5$3193	·$9^5$3497	·$9^5$3788	·$9^5$4066	·$9^5$4332
4·4	·$9^5$4587	·$9^5$4831	·$9^5$5065	·$9^5$5288	·$9^5$5502	·$9^5$5706	·$9^5$5902	·$9^5$6089	·$9^5$6268	·$9^5$6439
4·5	·$9^5$6602	·$9^5$6759	·$9^5$6908	·$9^5$7051	·$9^5$7187	·$9^5$7318	·$9^5$7442	·$9^5$7561	·$9^5$7675	·$9^5$7784
4·6	·$9^5$7888	·$9^5$7987	·$9^5$8081	·$9^5$8172	·$9^5$8258	·$9^5$8340	·$9^5$8419	·$9^5$8494	·$9^5$8566	·$9^5$8634
4·7	·$9^5$8699	·$9^5$8761	·$9^5$8821	·$9^5$8877	·$9^5$8931	·$9^5$8983	·$9^6$0320	·$9^6$0789	·$9^6$1235	·$9^6$1661
4·8	·$9^6$2067	·$9^6$2453	·$9^6$2822	·$9^6$3173	·$9^6$3508	·$9^6$3827	·$9^6$4131	·$9^6$4420	·$9^6$4696	·$9^6$4958
4·9	·$9^6$5208	·$9^6$5446	·$9^6$5673	·$9^6$5889	·$9^6$6094	·$9^6$6289	·$9^6$6475	·$9^6$6652	·$9^6$6821	·$9^6$6981

Example: $\Phi(3\cdot57) = \cdot9^3 8215 = 0\cdot9998215$.

[†] By permission from A. Hald, *Statistical Tables, and Formulas*, John Wiley & Sons, Inc., New York, 1952.

Bibliography

ADRIAN, J. J., *Construction Accounting: Financial, Managerial, Auditing and Tax*, Stipes Publishing, Champaign, IL, 1998.

AHUJA, H. N., S. P. DOZZI, and S. M. ABOURIZK, Project Management – *Techniques in Planning and Controlling Construction Projects*, 2nd edition, John Wiley and Sons, Inc., New York, 1994.

ANTILL, J. M. and R. W. WOODHEAD, *Critical Path Methods in Construction Practice*, 3rd edition, John Wiley and Sons, Inc., New York, 1982.

Associated General Contractors of America (AGC), *Project Delivery Systems for Construction*, AGC, Washington, D.C. 2004.

AU, T. and T. P. AU, *Engineering Economics for Capital Investment Analysis*, 2nd edition, Prentice Hall, Englewood Cliffs, NJ, 1992.

BARRIE, D. S. and B. S. PAULSON, *Professional Construction Management*, 3rd edition, McGraw-Hill, Inc., New York, 1992.

BONNY, JOHN B., and JOSEPH P. FREIN, *Handbook of Construction Management and Organization*, Von Nostrand Reinhold Co., New York, 1973.

Building (The) Estimator's Reference Book, 23rd edition, The Frank R. Walker Co., Chicago, IL, 1989.

Caterpillar Performance Handbook, 35th edition, Caterpillar Tractor Co., Peoria, IL, 2004.

CLOUGH, R. H. and G. A. SEARS, *Construction Contracting*, 6th edition, John Wiley and Sons, Inc., New York, 1994.

CLOUGH, R. H., SEARS, G. A. and SEARS, S. K., *Construction Project Management*, 4th Edition, John Wiley and Sons, Inc. New York, 2000.

COLLIER, C. A. and W. B. LEDBETTER, *Engineering Economic and Cost Analysis*, 2nd edition, Harper & Row, Publishers, New York, 1988.

COLLIER, K., *Construction Contracts*, 3rd edition, Prentice Hall, Englewood Cliffs, NJ, 2001.

COLLIER, N. S., COLLIER, C. A. and HALPERIN, D. A., *Construction Funding, The Process of Real Estate Development*, Third Edition, John Wiley & Sons, Inc., New York, 2002.

Credit Reports (individual subscription), Building Construction Division, Dunn and Bradstreet, Inc., New York, NY.

CULP, G. and R. ANNE SMITH, *Managing People (Including Yourself) for Project Success*, Van Nostrand Reinhold, New York, 1992.

DALLAVIA, L., *Estimating General Construction Costs*, 2nd edition, F. W. Dodge Co., New York, 1957.

DE KRUIF, P., Microbe Hunters, Harcourt, Brace and Co., New York, 1932.

Dodge Reports (daily publication), F. W. Dodge Corp. (Division of McGraw-Hill, Inc.), New York, NY.

DORSEY, ROBERT W., *Project Delivery Systems for Building Construction*, Associated General Contractors of America, Washington, D.C., 1997.

ESTEY, M., *The Unions*, Harcourt, Brace Jovanovich, New York, 1967.

FISK, E. R., *Construction Project Administration*, 6th Edition, John Wiley and Sons, Inc., New York, 2000.

FRIEDRICH, A. J., editor, *Sons of Martha*, American Society of Civil Engineers, New York, 1989.

GIBB, T. W., Jr., *Building Construction in the Southeastern United States*, Report presented to the School of Civil Engineering, Georgia Institute of Technology, Atlanta, 1975.

GOULD, FREDERICK E., *Managing the Construction Process*, 2nd Edition, Prentice-Hall, Upper Saddle River, N.J., 2000.

GOULD, FREDERICK E. and JOYCE, NANCY E., *Construction Project Management,* Professional Edition, Prentice-Hall, Upper Saddle River, New Jersey, 2002.

HALPIN, D. W. and R. D. NEATHAMMER, *"Construction Time Overruns," Technical Report P-16*, Construction Engineering Research Laboratory, Champaign, IL, 1973.

HALPIN, D. W., *"CYCLONE – Method for Modeling Job Site Processes,"* Journal of the Construction Division, American Society of Civil Engineers, Vol. 103, No. C03, Proc. Paper 13234, September 1977, pp. 489–499.

HALPIN, D. W., *Financial and Cost Concepts for Construction Management*, John Wiley and Sons, Inc., New York, 1985.

HALPIN, D. W. *MicroCYCLONE User's Manual*, Learning Systems, Inc., West Lafayette, IN, 1992.

HALPIN, D. W. and L. S. RIGGS, *Planning and Analysis of Construction Operations*, John Wiley and Sons, Inc., New York, 1992.

HARRIS, R. B. *Precedence and Arrow Networking Techniques for Construction*, John Wiley & Sons, Inc., New York, 1978.

HINZE, JIMMIE H., *Construction Planning and Scheduling*, Prentice-Hall, Upper Saddle River, N.J., 1998.

HINZE, JIMMIE, *Construction Contracts*, 2nd Edition, McGraw-Hill, New York, 2001.

HOOVER, HERBERT, *The Memoirs of Herbert Hoover – Years of Adventure 1874–1920*, The MacMillan Company, New York, 1951.

JACKSON, JOHN HOWARD, *Contract Law in Modern Society*, West Publishing Co., St. Paul, 1973.

KELLY, J. E. and WALKER, M. R., *Critical Path Planning and Scheduling,* Proceedings of the Eastern Joint Computer Conference, December, 1959, pp. 160–173.

KENIG, M., Coordinator, *Project Delivery Systems for Construction*, Associated General Contractors of America, Washington, D.C., 2004.

KERZNER, H., *Project Management – A System's Approach to Planning, Scheduling and Controlling*, 3rd edition, Van Nostrand Reinhold, New York, 1989.

LANG, H. J. and M. DE COURSEY, *Profitability Accounting and Bidding Strategy for Engineering and Construction Management*, Van Nostrand Reinhold Co., New York, 1983.

LEVITT, R. E. and N. M. SAMELSON, *Construction Safety Management*, 2nd edition, John Wiley and Sons, Inc., New York, 1993.

LEVY, S. M., *Project Management in Construction*, 2nd edition, McGraw-Hill, Inc., New York, 1994.

LEVY, S. M., *Build Operate Transfer*, John Wiley and Sons, Inc., New York, 1996.

Manual of Accident Prevention in Construction, Publication No. 100.2, The Associated General Contractors of America, Washington, DC, 1992.

McCULLOUGH, D., *The Great Bridge*, Touchstone Books, Simon and Schuster, New York, 1972.

McCULLOUGH, D., *The Path Between the Seas*, Touchstone Books, Simon and Schuster, New York, 1977.

MENHEERE, S. C. M. and POLLAIS, S. N., *Case Studies on Build Operate Transfer*, Delft University of Technology, Delft, The Netherlands, 1996.

MILLS, D. Q., *Labor Management Relations*, 5th edition, McGraw-Hill, Inc., New York, 1994.

MODER, J. J., C. R. PHILLIPS, and E. W. DAVIS, *Project Management with CPM, PERT and Precedence Diagramming*, 3rd edition, Van Nostrand Reinhold, New York, 1983.

NAVARETTE, PABLO F., *Planning, Estimating, and Control of Chemical Construction Projects*, Marcel Dekker, Inc., New York, Basel, Hong Kong, 1995.

NEIL, J. M., *Construction Cost Estimating for Project Control*, Prentice-Hall, Inc., Englewood Cliffs, NJ, 1983.

NEWHOUSE, ELIZABETH, L., editor, *The Builders – Marvels of Engineering*, National Geographic Society, Washington, DC, 1992.

NUNNALLY, S. W., *Construction Methods and Management*, 5th edition, Prentice Hall, Englewood Cliffs, NJ, 2001.

OBERLENDER, G. D., *Project Management for Engineering and Construction*, McGraw-Hill, Inc., New York, 1993.

O'BRIEN, J. J., *Preconstruction Estimating*, McGraw-Hill, Inc., New York, 1994.

OGLESBY, C. H. PARKER, and G. HOWELL, *Productivity Improvement in Construction*, McGraw-Hill, Inc., New York, 1988.

PEURIFOY, R. L. and G. D. OBERLENDER, *Estimating Construction Costs*, 4th edition, McGraw-Hill, Inc., New York, 1989.

PEURIFOY, R. L. and OBERLENDER, G. D., *Estimating Construction Costs*, Fifth Edition, McGraw-Hill, New York, 2002.

POAGE, W. S., *The Building Professional's Guide to Contract Documents*, R. S. Means Company, Inc., Kingston, MA, 1990.

Richardson General Construction Estimating Standards, Richardson Engineering Service, Inc., Mesa, AZ, published annually.

RITZ, G. J., *Total Construction Project Management*, McGraw-Hill, Inc., New York, 1994.

R. S. Means Company, *Building Construction Cost Data*, 54th annual edition, R. S. Means Co., Inc., Kingston, MA, 1995 (published annually).

SCHEXNAYDER, C. J. and MAYO, R. E., *Construction Management Fundamentals*, McGraw-Hill, New York, 2004.

SHUETTE, S. D. and R. W. LISKA, *Building Construction Estimating*, McGraw-Hill, Inc., New York, 1994.

SPINNER, M. P., *Project Management Principles and Practices*, Prentice-Hall, Inc., Upper Saddle River, New Jersey, 1997.

TWOMEY, T. R., *Understanding the Legal Aspects of Design/Build*, R. S. Means Co., Inc., Kingston, MA, 1989.

VAUGHN, R. C. and S. R. BORGMANN, *Legal Aspects of Engineering*, 5th edition, Kendall/Hunt Publishing Co., Dubuque, IA, 1993.

Index